U0124596

当代科学技术哲学论丛

卷4

主编　成素梅

当代科学哲学文献选读

（英汉对照）

成素梅　计海庆／主编

科学出版社

北京

图书在版编目(CIP)数据

当代科学哲学文献选读：英汉对照／成素梅，计海庆主编 . —北京：科学出版社，2013.7

（当代科学技术哲学论丛／成素梅主编）

ISBN 978-7-03-037730-2

Ⅰ.①当⋯ Ⅱ.①成⋯②计⋯ Ⅲ.①科学哲学–文集–英、汉 Ⅳ.①N02-53

中国版本图书馆 CIP 数据核字（2013）第 120851 号

丛书策划：胡升华

责任编辑：邹 聪 郭勇斌 王昌凤／责任校对：郑金红

责任印制：赵德静／封面设计：黄华斌

编辑部电话：010–64035853

E-mail：houjunlin@ mail. sciencep. com

科学出版社 出版

北京东黄城根北街 16 号

邮政编码：100717

http://www. sciencep. com

中国科学院印刷厂 印刷

科学出版社发行 各地新华书店经销

*

2013 年 7 月第 一 版 开本：B5（720×1000）

2013 年 7 月第一次印刷 印张：29 1/2

字数：558 000

定价：**128. 00 元**

（如有印装质量问题，我社负责调换）

总　序

　　梅森在他的《自然科学史》一书的导言中指出："科学有两个历史根源。首先是技术传统，它将实际经验与技能一代代传下来，使之不断发展。其次是精神传统，它把人类的理想与思想传下来并发扬光大……这两种传统在文明以前就存在了……在青铜时代的文明中，这两种传统大体上好像是各自分开的。一种传统由工匠保持下去，另一种传统由祭司、书吏集团保持下去，虽则后者也有他们自己一些重要的实用技术……在往后的文明中，这两种传统是分开的，不过这两种传统本身分化了，哲学家从祭司和书吏中分化出来，不同行业的工匠也各自分开……但总的说来，一直要到中古晚期和近代初期，这两种传统的各个成分才开始靠拢和汇合起来，从而产生一种新的传统，即科学传统。从此科学的发展比较独立了。科学的传统中由于包含有实践和理论的两个部分，它取得的成果也就具有技术和哲学两方面的意义。"[①]

　　显然，从梅森的观点看，科学在起源上是技术传统与哲学传统交汇的产物。然而，科学一旦产生并形成自己的独特传统之后，不仅反过来极大地影响了其根源，而且实质性地影响了远离这两个根源的其他领域。特别是，近几十年以来，当科学技术的发展由原初只是单纯地认识世界与改造世界，变成了当前的发展更需要考虑保护世界，同时日益接近于日常生活，越来越成为一项社会事业，乃至整个社会很有可能会变成一个巨大的社会实验室时，当以辩护科学为目标的英美哲学传统与以批判科学为宗旨的大陆哲学传统双双陷入困境时，当另辟蹊径、来势凶猛的关于科学技术的人文社会科学研究明显地给人留下反科学技术之嫌时，当整个哲学界对依靠科学技术发展推动社会进步的现代模式褒贬不一的讨论愈加激烈时……作为一门学科的"科学技术哲学"（philosophy of science and technology）也许会应运而生。

　　就当代哲学的发展而言，心灵哲学越来越与心理学的经验研究、神经科学、人工智能的发展内在地联系在一起；关于实在的本体论研究离不开以量子理论为

---

① 梅森. 自然科学史. 上海外国自然科学哲学著作编译组译. 上海：上海人民出版社，1977.6，7

基础的微观物理学的最新发展，也离不开对不可观察的心理结构和过程的假设与实验测试；与高新技术发展密切相关的网络伦理、环境伦理、干细胞伦理等已经成为伦理学关注的重要主题；关于社会心理、社会诚信等问题的哲学研究以及关于人性的哲学思考离不开围绕科学技术异化问题展开的一系列讨论。从这个意义上看，科学技术哲学恰好能提供架起抽象的哲学研究与前沿的科学技术研究之间的桥梁。

与传统的哲学研究相比，科学技术哲学研究不是通过先验的概念反思、日常语言的逻辑辨析以及提出概念真理的思想实验来获得知识并认知包括心灵在内的世界，也不是空洞地谈论规范人类行为的道德法则，而是通过综合考虑科学理论的基本假设、思想体系以及技术发展中的具体案例等复杂因素来研究哲学问题。在哲学框架内可能提出的关于科学技术的问题主要包括本体论、认识论和方法论问题（如实在论问题、证据对理论的非充分决定性问题、技术设计问题等），还有与科学技术的内容或方法直接相关的伦理问题或社会问题（如价值在科学技术中的作用问题、克隆技术和生化技术的合法应用问题等）。概念反思、语言分析和思想实验有助于提出假设，但不能用来评价假设。因此，必须把科学技术哲学与忽略科学技术发展的哲学明确地区分开来。

然而，强调科学技术哲学研究的经验性与实践性，并不意味着主张把哲学研究还原为经验研究，而是主张基于科学技术的当前发展，重新审视与回答传统的哲学问题。一方面，承认关于知识、实在、方法和伦理的哲学问题比经验科学与技术中的问题更具有普遍性和规范性；另一方面，主张对这些哲学问题的讨论要以科学技术的发展为基础。特别是，当科学的发展进入人类无法直接或间接观察的微观世界时，当人类的文明进入信息化时代时，技术已经不再只是单纯延伸人类感官的工具和充当人类认识世界、改造世界的手段，而是成为人类认识世界的一个必不可少的中介和人类生存、生活的基本条件，甚至正在成为人类超越自身感知阈限的有效手段（如在体内植入芯片）。在这种背景下，科学、技术、哲学事实上已经不可避免地在许多基本问题上相互纠缠在一起，很难彼此分离。如果说，科学的产生源于技术传统与哲学传统的交汇，那么，科学技术哲学的产生则是科学、技术、哲学三种传统汇集与衍生的结果，如关于量子测量解释的认识论争论、关于数字生命的实在性问题的争论、关于人类基因组序列带来的伦理问题的争论、关于体内植入芯片的工具平等问题的争论等。这些争论本身内在地蕴涵科学共同体在确立、维护与传播自己的学术见解时社会因素与修辞因素所起的作用。科学、技术、哲学三者之间的关系大致如下图所示。

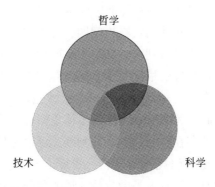

　　在上图中，哲学和科学、技术的两两相交之处，分别形成了科学哲学和技术哲学；科学与技术相交的区域表现了科学的技术化与技术的科学化，即技术趋向的科学研究（如量子计算）和科学趋向的技术研究（如生物技术、智能技术）；三者相交之处，形成了科学技术哲学。因此，在非常狭义的学理意义上，科学技术哲学不是科学哲学与技术哲学的简单综合，因为科学哲学主要是基于对科学理论的形成逻辑、与世界的关系、与证据的关系、与实验的关系、理论的变化等问题的剖析来讨论哲学问题，技术哲学主要是基于对技术设计、技术发明、技术评价、技术制品（即人工物）和技术应用等问题的研究来探讨相关的哲学问题。从上图可以看出，科学技术哲学是基于技术趋向的科学研究和科学趋向的技术研究来回答哲学问题，是科学、技术与哲学的问题重叠与互补研究。在科学技术哲学的研究中，哲学的认识论、本体论、方法论和伦理学问题是彼此关联的。

　　首先，在科学技术哲学中，两个重要的认识论问题是以技术为前提的科学研究是否能获得真理性知识的问题和如何合理评价理论的问题。从方法论的角度看，从经验到理论的归纳主义进路和从假说到证实的假设—演绎主义进路都过分简单。科学理论的形成是在基于假设的理论化、技术为主的实验和逻辑推理之间不断进行调整，最终达到反思平衡的一个动态负反馈过程。在这个过程中，理论与实验结果之间的关系不是单纯的归纳关系或演绎关系，而是一种说明关系。但是，说明关系预设了对说明本性的理解。例如，把说明理解为是语句之间的演绎关系、理论与数据之间的符合关系、机制与现象之间的本体论关系等。因此，关于说明的本性问题，既是一个认识论问题，也是一个本体论问题。

　　其次，在科学技术哲学中，最一般的本体论问题是，我们是否能够对不可能被直接观察的、只能通过技术手段间接地看到其效应的理论实体的存在性做出合理的辩护。例如，在量子理论中，我们是否应该相信量子物理学家用来解释物理现象的夸克、电子、光子等假定实体是真实存在的？或者，只是便于预言观察现象的谈话方式或工具？我们仅凭先验的推理根本无法解决围绕这个问题的实在论与反实在论之争。关于理论实体的实在论问题，必须与揭示量子力学的基本假设中的哲学基础联系起来，才能得到合理的解答。同样，心理学哲学中的实在论问题是，我们是否有或能够有好的根据相信，确实存在像规则和概念之类的心理表征。对于这个问题，只有与心理学、认知科学、神经科学的前沿研究结合起来，才能得到好的解答。因此，关于实在本性的本体论研究与关于知识的认识论研究之间存在着相互影响，即存在判断影响认知判断，反之亦然。

最后，伦理学虽然是一门规范的学科，表面上与经验性的科学技术相差甚远，但是当科学技术的研究触及人类的价值或道德判断问题时，伦理理论的研究就需要与人类的道德能力相一致。对人类道德能力的关注，不是以先验的概念构造为基础，而是以经验调查为基础。例如，如何解决当前心理学与神经科学实验中的知情同意问题。根据当前流行的人工智能研究进路，当把人的心理过程理解为受控于由生物物理机制建构的大脑过程，甚至把大脑过程理解为一种计算时，就很难把不道德的行为归属于意愿的失败，这显然对自由意志的概念提出了挑战。伦理学家在基于神经科学、人工智能等研究来讨论有没有自由意志的心灵本性和人们对自己的行动是否应该负有道德责任的问题时，伦理学就与本体论问题相互联系起来。

正是在这种意义上，我们可以说，"科学技术哲学"越来越成为当代哲学问题研究的核心。这是基于科学、技术、哲学发展的学理脉络对"科学技术哲学"存在的合法性与重要性的揭示。令人遗憾的是，到目前为止，这种意义上的科学技术哲学的形式体系还很不成熟，甚至没有引起学术界的关注。

我国的"科学技术哲学"这个概念最早是在 1987 年国务院学位委员会组织修改研究生学科目录时从素有"大口袋"之称的"自然辩证法"更名而来的。与自然辩证法的这种渊源关系，决定了我国的科学技术哲学，不同于前面描述的作为一门学科的科学技术哲学，而是具有学科群的特征。学术界通常把我国的科学技术哲学理解为对科学技术发展所提出的相关问题、基本要求和尖锐挑战的哲学回应，对整体的科学与技术及其各门分支学科所涉及的哲学问题进行批判式反思的一个学科群。

经过几十年的发展，我国的科学技术哲学研究在与国际接轨、关注我国现实问题的进程中，不断地发展与壮大，形成了以部门哲学、科学哲学、技术哲学、科学技术的人文社会科学研究以及社会科学哲学为基本方向的相对稳定的专业队伍，呈现出从抽象理论到生活实践，从单一到多元，从立足于部分到注重整体，从翻译到对话的发展特点。特别是自 21 世纪以来，我国的科学技术哲学在每个学科方向上都正在发生研究范式的转变、思维方式的转型和学术焦点与问题域的转移。那么，处于转变、转型和转移中的科学技术哲学将会"转"向何处？将会提出什么样的新问题？在不断地摒弃了小科学时代的科学观、技术观和哲学观之后，如何重建大科学时代的科学观、技术观和哲学观？作为学科群的科学技术哲学的不同分支领域，在深入研究的过程中，能否衍生出前面描述的作为一门学科的科学技术哲学？

陈昌曙先生在1995年发表的《科学技术哲学之我见》一文中，从学科名称的内涵与意义、学科分类及涵盖的学术交流活动三个方面，阐明了把"自然辩证法"更名为"科学技术哲学"所具有的必要性，然后指出："在我们的学科目录中，可以把科学技术哲学与自然辩证法作为同一的东西看待，但从学科的内容、层次看，似乎这两者又不是完全同一的；如果把当今出版和习用的《自然辩证法讲义》、《自然辩证法概论》原样不动地变换成为《科学技术哲学讲义》、《科学技术哲学概论》则未必相宜。科学技术哲学总应该有更深的哲学思考和更多的哲学色彩，而不全等于科学观与技术观。"[1]他主张"科学技术哲学"可能需要写出诸如"从哲学的观点看……"之类的内容，如"从哲学的观点看基础科学与技术科学"、"从哲学的观点看科学技术化、技术科学化与科学技术一体化"等。他认为："尽管科学与技术之间有着原则性的区别，尽管科学哲学与技术哲学有较多的差异，统一的科学技术哲学仍是可以设想的。"[2]

陈先生基于学科名称的内涵与意义提出探索统一的科学技术哲学的设想，与基于科学、技术、哲学发展的内在要求提出的探索科学技术哲学可能性的观点是相吻合的。如果"从哲学的观点看……"之类的内容是科学技术哲学研究的一条外在论的进路，那么从科学、技术、哲学研究的相交领域形成的科学技术哲学研究则是一条内在论的进路。在内在论者的进路中，哲学不再充当外在于科学技术研究的高高在上的指挥者，而是成为科学技术研究中离不开的参与者。这种哲学角色的转变，是当代大科学时代哲学研究的一个典型特征。例如，在认知科学的研究中，由科学家、工程师、医生、哲学家、企业家甚至政治家共同参与的会议并不少见。

在这里，哲学研究既不像逻辑经验主义者所说的那样，只是澄清科学命题的意义，更不像许多社会建构论者所追求的那样，已经被社会与文化研究取而代之，而是要求把思辨与先验的要素和实证与现实的问题结合起来，作为一种不同

---

① 陈昌曙.科学技术哲学之我见.科学技术与辩证法，1995，(3)：2
② 陈昌曙.科学技术哲学之我见.科学技术与辩证法，1995，(3)：3

的视角，参与科学技术研究。这是因为，当科学技术的发展离社会生活越来越近时，科学就不只是探索真理那么简单，技术也不只是作为改造世界的工具那么单纯。科学技术作为人类文明的成果，已经成为价值有涉的研究领域。在这种情况下，为了人类的和谐发展，凡是能够探索真理的科学研究都值得倡导吗？凡是能够用来按照人的意愿达到改造世界目标的技术都应该研制吗？专家提供的发展战略一定是完全合理的吗？人类究竟在为自己建构一个什么样的社会？作为社会的人在包括科学技术研究的一切社会活动中应该如何重建社会道德与社会信用？这些问题的提出就无疑为哲学家介入或参与科学技术的研究与发展提出了内在要求。

一言以蔽之，许多哲学问题需要深入科学技术的土壤，才能得到合理的解答。当代科学技术的发展在很大程度上需要嵌入哲学思考，才能达到更理性的发展。科学技术哲学既是从哲学视域把科学、技术、社会、政治、经济等因素整合起来思考问题的一门交叉的新型学科，也是把关于自然、社会与人的和谐发展作为研究核心的一门综合型学科。

《当代科学技术哲学论丛》的筹划与出版，正是试图为科学技术哲学的探索之路添砖加瓦，同时，也是上海社会科学院"科学技术哲学特色学科"多年来研究成果的展示。欢迎学界专家学者给予真诚的批评与指正。

成素梅

2011 年 8 月 10 日

前　言

　　本书由上海社会科学院哲学研究所为科技哲学专业的硕士研究生开设的科学哲学专业英语教学实践中精选出的十篇当代科学哲学方面的英文论文汇编而成。每篇论文都附有仅供参考的中文翻译。论文作者分别是美国、法国、挪威、瑞典等国的大学或科研院所的教授。相应的译文曾先后在《哲学分析》杂志上刊出过。通过对这些论文的阅读，读者一方面可以了解不同体裁的科学哲学论文的写作风格，另一方面能够在了解前沿学术动态的基础上系统地学习科学哲学英文写作与翻译的技巧，为进一步提高科学哲学英文文献的阅读能力和写作能力奠定基础。

　　更重要的是，这些论文反映了在当代科学哲学经历了自逻辑经验主义诞生以来以辩护科学为目标的传统科学哲学的发展，以及自科学知识社会学兴起以来以批判科学为宗旨的各种关于科学的人文社会科学研究（sciences studies）的发展之后，哲学界为了超越两者的局限，从而达到更合理地理解科学的目标所进行的努力，具有学术性与前沿性。这些论文包括综述型、分析论证型、应用研究型、历史考评型、评论型和演讲型六大类，在内容上大体可分为三大部分，前三篇文章强调了在实践中理解科学的重要性；接下来的三篇文章代表了一个新的科学哲学方向，即把科学哲学的研究从过去关注理论转向关注与专家的知识和技能相关的哲学研究；最后的四篇文章是技术哲学、认知科学哲学、医学哲学方面的应用类文章。

　　瑞典乌普萨拉大学科学史研究所所长奥托·斯巴姆教授的《19世纪的黄金数：一个科学事实的发展史》一文，基于对物理学家确立热功当量这一科学事实的整个过程的详尽考察，揭示了自牛顿力学诞生以来占据核心地位的自然哲学的研究方式在面对定量实验的发展时所发生的一些变化，以及定量实验如何改变了自然哲学家的注意力，并在19世纪的科学文化中占有一席之地的大致经过；基于对物理学家围绕焦耳于19世纪40年代在曼彻斯特完成的测量热功当量的经验知识之争的剖析，揭示了皇家学会确定的自然哲学家与试图成为绅士专家（gentlemen specialists）的仪器制造者、技工和实验者这一不断壮大的集体之间的权力之争，强调了以科学技能为基础的实验知识的社会地位与认识论地位。

美国科学哲学家保尔·费耶阿本德的《如何保护社会免受科学之害》是一篇公众演讲稿。其核心论点是批判现有教育体制中把科学当做一种宗教来对待的教育方式，提倡使科学不成为独裁者的教育改革。他认为，今天的科学显然完全不同于1650年的科学，当时的科学确实是解放和启蒙的工具，但今天的科学教育已经不再打算唤醒学生的批判能力，使其能够高瞻远瞩地看问题，而是成为一种教育灌输，完全缺乏批评，甚至今天的大多数科学家都缺乏创新观念，只是卷入了空洞的论文洪流之中。费耶阿本德强调实践的作用，在他看来，力学和光学在很大程度上归功于工匠，医学归功于助产士和巫婆。因此，教育的目标是保护学生的想象力，把他们领向生活，让他们学会如何与自然界和社会打交道。

美国斯坦福大学哲学系教授海伦·朗基诺的《〈科学革命的结构〉与科学中的女性主义革命》一文，系统地考察了库恩的重要著作《科学革命的结构》一书对女性主义研究和女性主义知识论的影响，描述了库恩的工作通向女性主义的前景，表明了库恩的思想如何为女性主义的科学哲学奠定了基础，最后从女性主义的视角讨论了库恩的不可通约性、范式和常规科学等观点的局限性。她认为，女性主义者求助于库恩是为了使她们能够合理地拒绝逻辑经验主义的主张。然而，认识论的多元主义虽然以库恩的见识为基础，但是，也用了某些不同的哲学原理。她自己提供的语境经验主义（contextual empiricism）放弃了用来说明不可通约性论点的意义和观察的理论负载的观点。语境经验主义没有把不同理论之间明显等价的经验适当性看成理论负载的问题，而是看成有助于数据和假说之间推理的背景假设方面的功能差异。

美国哲学家阿尔文·戈德曼的《专家：哪些是你应该信任的?》一文，是一篇很有分量的分析性说明文。该文从日常生活中常见的事例入手，通俗易懂地揭示了在当前社会中人们在评价专家所拥有的技能与知识时所带来的认识论问题，特别是，外行或新手如何在意见相左的两位专家的评价中做出选择时所遇到的现实问题，而这些问题恰好是传统的认识论和科学哲学所忽略的。在这篇文章中，作者剖析了五种情况：①支持一方观点和批评另一方观点的专家所提供的论证；②某一方得到了其他专家的认同；③对专家专长的"元专家"的评价（例如，专家的受教育情况及获奖证书等反映出的评价）；④专家的利益和偏见的证据；⑤专家过去"记录"的证据。他认为，关于这些问题的讨论，对"应用的"社会认识论提出了挑战。

美国科学哲学家埃文·赛林格和罗伯特·克里斯的《德雷福斯论专长：现象学分析的局限性》一文，是对德雷福斯兄弟基于现象学的研究和对大量日常生活中的操作说明书的分析总结出的技能获得模型的评价与讨论。作者认为，在当代生活中，许多经济、政治、科学和技术的决策都常规性地托付给了专家。然而，哲学家很少明确地讨论这一主题，尽管潜在的没有经过考察的专长概念通常蕴涵在像"权威性"、"殖民化"、"权力"和"理性的争论"之类的标题中。而德雷福斯是明确讨论这一概念的少数人之一。德雷福斯基于技能获得模型论证了专家的判断和行为是体知合一的人类行为表现的观点。但是，他的描述性模型和规范性要求缺乏解释学的敏感性，也是有缺陷的。

挪威哲学家达格芬·弗罗斯达尔的《胡塞尔和维特根斯坦论终极辩护》一文，对胡塞尔和维特根斯坦的终极辩护观进行了比较研究。他认为，胡塞尔的终极辩护观没有引起学界的注意。辩护有两个要素，知觉和证据的转移。胡塞尔早在 100 年前就阐述了这两个要素。胡塞尔与维特根斯坦虽然都深受詹姆斯的影响，但是，他们提供了不同的知觉理论，胡塞尔的知觉理论超越了维特根斯坦的知觉理论。他们的辩护观对反思平衡和生活世界的理解有许多相似之处，但他们在辩护的终极基础问题和生活世界有无确定性的基础问题上存在分歧。

美国技术哲学家安德鲁·芬伯格的《功能和意义：技术的双重面相》一文，认为马克思对市场合理性的批判、卢卡奇对科学知识合理性的批判和海德格尔的技术座架理论在马尔库塞那里发展成了一种技术的合理性批判，目的是在功能化占优势的技术理解中恢复意义的地位。当代技术哲学家伯格曼和辛普森，在后期海德格尔的影响下，也力图完成同样的任务，但这些方案都没有成功。通过分析当代信息技术发展中用户自发改进技术的案例可以发现，人们对技术的创新行动展现了从功能化中重新找回意义的可能。这种行动正是海德格尔眼中"此在"的原真性行动。

美国认知哲学家（也是浙江大学的客座教授）希拉里·科恩布利斯的《认识的能动性》一文，基于日常生活事例讨论了个人行为的批判性反思能力、反思信念与能动性之间的关系，以及与人的认识的能动性相关的真实欲望、心理异常、偏好等概念问题。他认为，人的批判反思能力与人的自由和责任相关。在哲学中，这种观点体现为对认识的能动性的主张。然而这种主张至少面临三种困难：①这种对知识反思的要求会导致无穷回归的严重后果；②有一种观点认为，由于反思为决断提供了可靠的保证，从而有助于区分异己的欲望和真正属于认知者的欲望，然而反思自身的可靠性却是有问题的；③认识的能动性主张不支持第一人称视角和谨慎思考。认识的能动性在信念的获得过程中与认识的责任感相关。这些观点不同程度地预设了某些关于反思的断言，而实际上，这些断言却与现今的最有用的证据相左。因此，目前对认识的能动性概念的诸多论述和辩护方式都是有问题的。

法国哲学家克劳德·德布鲁的《从哲学到医学的规范性概念的概述》是一篇有关"规范性"概念的产生与发展的综述性文章。作者认为，规范性观念是

20世纪哲学的一个发明，是指向有可能取得科技进步和改善人类生活的一个关键概念，规范性代表了一种特殊形式的意向性，它与对行动的判断、倾向和承诺相关。哲学家把规范性概念定义为创造与改变各种规范的力量。胡塞尔认为，规范性是在绝对的客观真理的规范控制下创造出来的一个观念世界。在现有的文献中，存在着实在论的、唯意志论的、反思认同的和潜在的普遍主义的关于规范性概念的四种哲学观。

挪威哲学家哈罗德·格里曼的《权力、信任和风险：关于权力问题缺失的一些反思》一文，讨论了医护人员与患者之间的权力、义务、责任等问题。作者认为，当前在医护人员中间进行的讨论中很少关注权力问题，这是一种很奇怪的现象。就关于适当的医患关系的讨论和关于信任的讨论而言，这种缺失带来了一些不良后果。因为在医疗保健体制中，权力问题的缺失妨碍了关于重要的制度形式的认真公开的道德讨论。此外，关于如何使医生与患者进行互动的一些提议，是相当不切实际的，这主要也是因为忽视了权力问题。

在本书即将出版之际，编者特别感谢每位作者与译者的辛勤劳动，感谢两位责任编辑郭勇斌先生和邹聪女士在编辑本书的过程中付出的心血，感谢上海社会科学院"科学技术哲学特色学科"在经费上的资助。此外，虽然，译者对原来刊发过的译文进行了进一步的修正，但仍难免存在不当之处，敬请读者批评指正。

编　者

2012 年 7 月 10 日

目　录

The Golden Number of the Century: The History of a Scientific Fact[①]

H. Otto Sibum

Das Wörtlein Thatsache ist noch jung. Ich weiß der Zeit ganz wohl zu erinnern, da es noch in niemands Munde war. Aber aus wessen Munde oder Feder es zuerst gekommen, daß weiß ich nicht... noch weniger weiß ich, wie es gekommen sein mag, daß dieses neue Wörtlein ganz wider das gewöhnliche Schicksal neuer Wörter in kurzer Zeit ein so gewaltiges Glück gemacht hat; noch, wodurch es eine so allgemeine Aufnahme verdient, da man in gewissen Schriften kein Blatt umschlagen kann, ohne auf eine Thatsache zu stoßen. (G. E. Lessing, 1778)

The ordinary crude mind has only two compartments, one for truth and one for error; ... the ideal scientific mind, however, has an infinite number. Each theory or law is in its proper compartment indicating the probability of its truth. As a new fact arrives the scientist changes it from one compartment to another so as, if possible, to always keep it in its proper relation to truth and error. (H. A. Rowland, 1899)

This paper aims to explore the history of a scientific fact commonly known as the "mechanical equivalent of heat". But what is a mechanical equivalent of heat? When does this "truth attested by authentic testimony" come to be known as a fact, when is it credited as scientific? The experience which is constitutive for the fact to investigate here was and is known since the beginning of human existence: If you rub your hands against

---

① This chapter summarises a long term research project undertaken by the author. Several arguments made here are substantiated in different self-contained articles published already in other journals but to which I refer to in this paper at the appropriate places. I would like to thank the organisers of this workshop for having provided me with the opportunity to present my current view of the development of this scientific fact. I am citing materials from archives at Glasgow University, the Royal Society, London, University of Manchester Institute of Technology, the Academy of Science in Vienna, Handschriftenabteilung of the Staats- und Universitätsbibliothek Göttingen, Manuskriptsammlung at the Deutsches Museum, Munich, the Eisenhower Library at Johns Hopkins University, and the Academy of Science at St Petersburg, Russia and am grateful for permission to do so.

each other you notice that they become warm. However it was only in the late 18th century that this relation between mechanical friction and heat become a matter of intensive natural philosophical investigation and controversy. In the early 19th century natural philosophers even expressed this relationship quantitatively. The experimentally established ratio between the mechanical work performed and the heat produced (measured as a difference in temperature) soon became a building block of the science of energy. The experiment quickly became regarded as a *proof of the principle of conservation of energy* and the ratio achieved the status of a *constant of nature*. Finally, around 1900, physicists began calling this exact fact the *golden number of that century* after it had become a conversion factor in the international system of units. This paper accordingly covers the period from the late 18th to the early 20th centuries, a time of considerable change within the development of the physical sciences. Furthermore, since the fact in question is a particular one, expressed numerically only, it is therefore not a coincidence that this narrative is simultaneously one about the establishment of precision measurement in the exact sciences of the 19th century. It concerns the establishment of quantitative experimentation as a specific form of experience.

Historians of the development of natural philosophy of the early modern period agree that the "fact" was an invention of the 17th century. Steven Shapin and Simon Schaffer, for example, argue that the laboratory at this time became a publicly accessible place in which authenticated experimental knowledge was produced—Matters of fact distinct from matters of opinion. [1] Lorraine Daston has shown that the contemporary epistemological tension between the conception that sense perception was concerned with particulars and that genuine knowledge depended on the recognition of universals, led natural philosophers to define facts as "nuggets of experience" detached from theory. Central features of this understanding of a fact, including its being "in principle immiscible with opinion, interpretation, and theory" endured over time and continued prevailing in the

---

[1] ST. Shapin, S. Schaffer, *Leviathan and the Air-Pump. Hobbes, Boyle, and the Experimental Life*. Princeton 1985, p. 39. On "factual knowledge" and truth see ST. Shapin, *A Social History of Truth. Civility and Science in Seventeenth-Century England*. Chicago, 1994. See also L. Fleck, *Genesis and Development of a Scientific Fact*, Chicago, London 1979; B. Latour, S. Woolgar, *Laboratory Life: The Construction of Scientific Facts*, Princeton, N. J. 1986.

period investigated here. [1]Numerical facts may be seen as excellent candidates for this purified form of experience. But as the paper will show this nugget of experience was always an integral part of the working knowledge of a rapidly changing scientific community which was in the process of establishing itself in the 19th century.

The following sections are ordered according to the changes in meaning this fact underwent. I will begin with a brief look at the second half of the 18th century, a time in which there was neither science nor a fact called the mechanical equivalent of heat. Instead I will situate experiences with heat and work, especially those practices of expressing them by means of numerical facts within the development of natural knowledge traditions. Despite their varying interpretations, historians agree that this period saw the rise of quantitative investigation. Furthermore, a change in manners in natural philosophy occurred around 1800: performing experiments became an increasingly private endeavour because the heightened sensitivity in measurements precluded the presence of witnesses. This rise of numerical facts went hand in hand with a distancing from sensuous technologies, emphasizing visual perception. Quantitative experimentation changed natural philosophers' attention and had to find it's place in this period of major cultural changes.

The second section gives a brief account of James Joule's experimental knowledge of the mechanical equivalent of heat performed during the 1840s in Manchester. His knowledge about the nature of heat was deeply entangled with the performance of experiments which could only have emerged at the intersection of apparently unconnected but locally available knowledge traditions. But the published account of this experiment in the *Philosophical Transactions* did not represent the author's knowledge. It only suggests that an exact fact distinct from any theoretical assumptions was established. The transformation of his experimental knowledge through the process of publication is an expression of a power relationship existing between established natural philosophers at the Royal Society and a rising collective of instrument makers, artisans and experimentalists seeking to become gentlemen specialists.

The third section follows the historical process by which this published fact acquired the status of the proof of conservation of energy and how it became the focus of a priority

---

① L. Daston, *Why are Facts Short?*, in this volume; L. Daston, *Baconian Facts, Academic Civility, and the Prehistory of Objectivity*, "Annals of Scholarship", VIII, nos. 3-4 (1991), pp. 337-364. For a discussion of experience and experiment in this period see P. Dear, *Discipline and Experience: The Mathematical Way in the Scientific Revolution*, Chicago 1995. For a discussion of the mentioned approaches towards historical facts see M. Poovey, *A History of the Modern Fact. Problems of Knowledge in the Sciences of Wealth and Society*, Chicago, London 1998, pp. 1-28.

dispute. Contrary of the conventional narrative of simultaneous discovery, I will argue that the still fragile social and epistemological status of experimental knowledge in the learned world sparked off claims about authorship and intellectual property rights and thereby helped construct the idea of simultaneous discovery. In this process, Joule's collaboration with William Thomson, later Lord Kelvin became significant.

The fourth section sketches a further transformation of the meaning of this fact, to the point when it was considered to be one of the most important constants of nature. Despite the development of the British science of energy and the related system of absolute units of measure, several attacks on the probable existence and the correct value of this number arose, leading the exact sciences into a crisis. One of the strategies to settle the disputes was to engineer evidence for this constant of nature. Henry A. Rowland, the American engineer-physicist, took this up as his first task at the physics laboratory of Johns Hopkins University. He settled the dispute for a while but his efforts in establishing the physical laboratory as a means of modern education drove a wedge between the exact sciences and the humanities. The effect of this was precisely to split matters of fact from matters of opinion.

In the concluding section I will briefly show how this exact fact finally became materialized as a conversion factor in the international system of units. On July 12th, 1894 the Senate and House of Representatives of the United States of America even legalized this constant as the unit of work. By that time the meaning of this scientific fact was no longer restricted to the laboratory, and it became the golden number of the century. Despite its importance even beyond the scientific community the meaning of the mechanical equivalent of heat changed again. In the expanding physics community with the new division of labour established theoretical physicists doubted that experiments to determine the mechanical equivalent of heat could have ever provided proof for this universal principle called conservation of energy.

1. Changing Attention: "On the Skill Necessary to Investigate Nature"

There are many Things capable of More and Less, which are perhaps not capable of Mensuration. Tastes, Smells, the Sensations of Heat and Cold...
(H. Miles, 1748)

La construction d'un mouvement perpétuel est absolument impossible: quand même le frottement, la résistance du milieu ne détruiroient point à longue l'effet de la force motrice... Ce genre de recherches a l'inconvénient d'être co ûteux, il a ruiné plus d'une famille, & souvent des Méchaniciens qui eussent pu rendre de grands services, y ont consumé leur fortune, leur temps & leur génie. (L'Academie Royale des Sciences, 1775)

On May 15th, 173 the French experimental natural philosopher Abbé Nollet gave his inaugural lecture, "On the skill necessary to investigate nature"[1], which nicely captured the transitional phase which quantitative experiment as a specific form of experiencing nature and attention was undergoing. I will take this lecture as a starting point for my investigation of the rising culture of precision and the establishment of numerical facts as a dominant form of experience in the late 18th and 19th centuries. Without going into detail here about this important document I would like to emphasize Nollet's reflections on the kind of attention required to perform experiments, the role of precision measurement in this endeavour, and the positioning of experimental natural philosophy within the Republic of Letters. These aspects were crucial for the change in practices of experimental natural philosophy and the related epistemological meaning of evidence. [2]On the experimentalist's attention he writes:

> However diligent or eager he is, he who seeks to discover nature's traces will never be able to contemplate what he is looking for in its entirety if even the slightest impression escapes his attention... The time, the place, the air's condition then, the size, duration, shape, colour, smell and indeed all the other properties accessible to the senses are so many circumstances which one should not only pay attention to, but also report on, except in the one case, that of apparent superfluity... One thus sees that a heightened degree of attention itself leads to a greater understanding [of things], and for this reason an observer should not let go of the object under his study until he has observed it most meticulously from all perspectives, and [considered] what it conceals

---

[1] J. A. Nollet, *Discours sur les dispositions et sur les qualités qu'il faut avoir pour faire du progrès dans l'étude de la physique expérimentale*, Paris 1753. German transl. Nollets *Rede von der nöthigen Geschicklichkeit zur Erforschung der Natur, welche er den 15. Mai 1753 bei dem Antritte seines öffentlichen Lehramtes, in dem Navarrischen Collegio gehalten*, Erfurt 1755.

[2] For a detailed account on the emergence of the experimentalist as a persona in the exact sciences of the late 18th and 19th century, the accompanying forms of disciplined attention and the epistemological problems raised by the establishment of this nonliterary knowledge tradition see my paper *Experimen talists in the Republic of Letters*, to appear in L. Daston and H. O. Sibum (eds.) *Scientific Personae*, "Science in Context", forthcoming.

within it and what surrounds it. [1]

For Nollet it was obvious that raising the degree of attention necessarily extended the scope of knowledge. But, as he insisted, this mode of investigation should be differentiated from the practices of natural history. The latter was a form of inquiry which completed the "inventory of our wealth" but did not investigate the "causes of what happens in the natural world". However, both were intimately linked with each other:

> For indeed, he who endeavours to investigate nature without understanding
> its history speaks at random and about things that he does not know in the least;
> but he who knows nothing else of nature than its history justly deserves a place
> among those natural philosophers who exercise their memory only. Accordingly, to
> practice experimental physics is nothing else than to investigate nature, not only
> with regards to its effects, but equally with the intention [of studying] the tools
> by which [nature's] effects are produced; in short it means to study what
> [nature] does, in order to be in a position to say how it does it. [2]

According to Nollet, naturalists therefore identified "bodies, of which the world is constituted, and each one of which has its devotee, to divide them in types and kinds, to notice the alteration taking place sometimes piece by piece, to recount the properties of all these things and their mutual relation". In contrast, the experimental physicist should draw his finetuned attention towards natural processes with "perspicacity". In many instances Nollet's lecture expressed the difficulties faced by "physique experimentale" when it positioned itself between the meditative and textoriented practices of traditional scholarship and those producing and collecting the goods of the world.

Measurement was another vexed issue addressed by experimental natural philosophers. Despite Nollet's scepticism towards the rise of precision measurement he nevertheless gave the "art of measurement" a small place in his experimental philosophy. But ultimately he saw a contradiction in being simultaneously an excellent measurer and a skilled natural philosopher. He remarked that "it would be dangerous for a natural philosopher to grow fond of the art of measuring". The tendency towards extreme precision and the representation of nature through mathematical signs were for him precarious developments:

> How many are there? In any case, those who do not want to concern
> themselves with common knowledge should not take circuitous paths, but all

---

① Nollet, *Rede von der nöthigen Geschicklichkeit...*, cit., pp. 37/38.

② Ivi, p. 12.

the same, they take great satisfaction from using algebraic signs to present things which could, without diminishing their value, be presented in a generally accessible manner. Such writings, which are bizarre when they are [in this way] cunningly inflated, show clearly enough that the little natural science they contain is only a pretext for [presenting] another science, in such a way as to make oneself more important?[1]

In his concluding remarks, Nollet defined the persona of the experimental natural philosopher, who should—following, to a certain extent, the model of the French engineer—apply his skills and insights to engage in useful projects, for example to investigate how to make water drinkable, to make the compass safer to use, to stop vegetables rotting, etc. This kind of natural philosopher was more valuable to him than those "arrogant scholars who imagine they can amaze us either by curiosities or rather the mere fancy of their selected material"[2].

Experimental natural philosophy of the Nollet type was already challenging these established knowledge traditions. But quantitative experimentation practised by the followup generation represented by Charles Augustin Coulomb, Antoine Lavoisier, Pierre-Simon Laplace *et al.* was an even greater challenge because it required an economy of the senses and a moral economy unfamiliar to members of the Republic of Letters. Around 1800 instruments of precision were employed in various cultural fields of production, but the key use was for the control and regulation of society and not—as one might expect—to fulfill the mathematicians' ambition to take command over subjects of natural philosophy.[3]Narratives about laboratory trials as well as travel experiences made at distant places increasingly showed numbers woven into their literary accounts, indicating degrees of heat, atmospheric pressures etc. But in order to make sense of these values an established regime of standardized instruments was required to share this new experience. Often such

---

① Ivi, p. 58.

② Ivi, p. 60.

③ See for example Thomas S. Kuhn's *The Function of Measurement in Modern Physical Sciences* in which he describes the period 1800 to 1850 as the Mathematization of Baconian Science in T. S. Kuhn, *The Essential Tension. Selected Studies in Scientific Tradition and Change.* Chicago 1977, pp. 178-224, 220. For a reverse interpretation, i. e. that natural philosophers made mixed mathematics useful for their investigations see J. Heilbron, *A Mathematicians' Mutiny, with Morals,* in P. Horwich (ed.) *World Changes: Thomas Kuhn and the Nature of Science.* Cambridge, Mass, 1993, pp. 81-129. For another twist of Kuhn's interpretation see I. Hacking, *The Taming of Chance,* Cambridge 1990, pp. 60-93. For a most recent account on the emergence of the culture of precision see M. N. Wise (ed.) *The Values of Precision.* Princeton 1995, and M. N. Bourguet, CH. Licoppe, H. O. Sibum (eds.) *Instruments, Travel and Science: Itineraries of Precision from 17th to 20th Century,* forthcoming.

a network did not exist. The case of thermometry is very telling in this respect. A thermometer in the mid-18th century could have up to eighteen different scales. Every investigator of nature seemed to have produced his own rational scale, and agreement about measurements was far from being achieved.[1] Furthermore, an increasing sensitivity in precision measurements may be observed, brought by new instruments, regarded as more reliable than the human senses. But this sensitivity required refined gestures of accuracy and more precautions to be taken while experimenting. Moreover, the usual gentlemanly practice of being present during trials then often thwarted the performance of delicate measurements.[2] The trial changed from being a public spectacle to a performance without any audience. In and outside the laboratory new forms of attention as well as of communication had to be developed which had not existed in the scholarly form of life.[3] The authority of books and correspondence and the usual face-to-face interaction between scholars were no longer practicable or sufficient to provide evidence for these numerical facts.[4] Numbers travelled easily from the laboratories and across cultural fields but the new practices of investigating numerical facts simultaneously shifted the conventional habit of trust in a person to trust in procedures. Quantifying experimentalists even began to distance themselves from sensuous experiences—while privately still practising them in their performances of experiment—demonstrating the

① Such a thermometer is held at University Museum, Utrecht; compare W. E. Knowles Middleton, *A History of the Thermometer and Its Use in Meteorology*, Baltimore 1966.

② See for example Charles Augustin Coulomb's torsian balance experiment. CH. Blondel, M. Dörries (eds), *Restaging Coulomb: usages, controverses et réplications autour de la balance de torsion*, Florence 1994.

③ For a persuasive account see M. N. Bourguet, *Landscape with Numbers—Natural History, Travel and Instruments (Mid-18th - Early 19th Century)*, in M. N. Bourguet, CH. Licoppe, H. O. Sibum (eds.) *Instruments, Travel and Science*, cit.

④ On the importance of witnessing and authority of books as means to establish evidence in this period see for example I. Hacking, *The Emergence of Probability: A Philosophical Study of Early Ideas about Probability, Induction and Statistical Inference*, Cambridge 1975, pp. 33/34 and S. Schaffer, *Self Evidence*, in "Critical Inquiry", XVIII (Winter 1992), pp. 327-362.

disembodiment of science, emphasizing visual perception and self registering devices. [1]Quantitative experimentation had to find its place in the academic form of life and both were about to assimilate one another. An expert culture was required which did not yet exist. [2]

Research on heat and work and their relation have to be seen within this context. It is neccessary to mention two aspects here: firstly, historians too often neglect the fact that most of what we know scientifically about heat we have learnt since the appearance of the thermometer. It might even be said that the history of the research on heat is to a large degree the history of thermometry, which became a prominent field of research in the second half of the 18th century. Secondly, measurement and the conceptualisation of work in natural philosophy and engineering in the 18th century equally deserves reworking. Enlightened philosophers developed various strategies to make sense of the world of work. The French military engineer Charles Augustin Coulomb as well as Antoine Lavoisier developed precision techniques to measure and evaluate human labour and their efficiency, implicitly assuming that the animal economy was a self-regulating system. Self-acting machinery—so visible in steam engine technology as well as in automata—became an ideal

---

[1] S. Schaffer argues that "we see a shift from the rituals of public performance towards the figure of the disembodied scientific genius; and we see the new insistence on the importance of self-registering material devices and instrumentation..." . S. Schaffer, *Self Evidence*, cit. , p. 330; On Lavoisier and his practices of distancing himself from sensuous technologies see L. Roberts, *The Death of the Sensuous Chemist: The "New" Chemistry and the Transformation of Sensuous Technology*, in "Studies in History and Philosophy of Science", XXVI, 4 (1995), pp. 503-529. For the distancing from sensuous experiences of the electrician see H. O. Sibum, *Charles Augustin Coulomb* (1736-1806), in K. Von Meyenn (ed.) *Die Grossen Physiker. Erster Band. Von Aristoteles bis Kelvin*, München 1997, pp. 243-262; for an excellent contribution to the formation of experimental natural philosophy see CH. Licoppe, *La Naissance de la pratique scientifique: Le discours de l'experience en France et en Angleterre* (1630-1800), Paris 1996.

[2] The values of precision measurement certainly differed in each European country. In 18th century England for example instrument makers who produced precision instrumentation like J. Dollond or I. Ramsden became Fellows of the Royal Society. But this common practice of crediting outstanding instrument makers investigations into precision measurement as valuable research in its own right changed in the first half of the 19th century. In that period hardly any instrument maker achieved this status. W. T. Ginn, *Philosophers and Artisans: The Relationship between Men of Science and Instrument Makers in London* 1820-1860, Ph. D. University of Kent at Canterbury 1991; for further comparative studies see in particular M. N. Wise, *The Values of Precision*, cit.

in enlightened reasoning about power and civilisation.①Inscription devices such as James Watts' indicator diagram, which graphically displayed the work done by a steam engine, as well as Morin and Poncelet's self-registering friction trolley mark the beginning of a new kind of objectivity practised in a rising expert culture and which enforced the process of disembodiment of science. In Morin's own words:

> The indications provided by the instrument must be obtained in a manner independent of the attention, of the will or of the prejudices of the observer, and thus consequently be furnished by the instrument itself by means of traces or of resulting materials, which subsist after the experiment. ②

Those dynamometrical measures of the performance of machines became the "mechanical currency" in this visual culture and thus may even have reinforced the emerging opinion that one cannot get any useful work for nothing. Hence the decision of the French Academy no longer to accept suggestions for *perpetua mobilia* which had—according to them—already ruined so many talented experimentalists and their families. But it is important to point out here that in these self-registering techniques of measurement, friction was nothing more than a measurable quantity whose nature was

---

① But neither of these investigations nor even Diderot's encyclopaedic project did fully succeed in extracting the knowledge embodied in artisanal work. On the role of automata see S. Schaffer, *Enlightened Automata*, in W. Clark, J. Golinski, S. Schaffer (eds.) *The Sciences in Enlightened Europe.* Chicago 1999, pp. 126-165; on Coulomb's project to study fatigue in human performance of work see C. A. Coulomb, *Résultats de plusieurs expériences destinées a déterminer la quantité d'action que les hommes peuvent fournir par leur travail journalier, suivante les differentes manières dont ils emploient leur forces*, in "Mémoire de l' Institut National des Sciences et Arts —Sciences mathematiques et physiques", Paris 1799, pp. 380- 428; H. O. Sibum, *Charles Augustin Coulomb* (1736-1806), cit. . On the encyclopaedic project see J. R. Pannabecker, *Representing Mechanical Arts in Diderot's Encyclopédie*, "Technology and Culture", XXXIX (1998), pp. 33-73; ST. L. Kaplan, C. J. Koepp (eds.) *Work in France. Representations, Meaning, Organization, and Practice.* Ithaca, London 1986; M. N. Wise, *Work and Waste I: Political Economy and Natural Philosophy in Nineteenth-Century Britain*, in "History of Science", 27 (1989), pp. 263-301. M. N. Wise, *Mediating Machines*, in "Science in Context", 2 (1988), pp. 77-113. Carlo Poni's article in this volume.

② A. Morin, *Notice sur divers appareils dynamométrique, propres à mesurer l' effort du travail développé par les moteurs animés ou inanimés, ou consommé par des machines de rotation, et sur un nouvel indicateur de la pression dans les cylindres des machines a vapeur*, Paris 1839, pp. 29-30; quoted after R. M. Brain, *The Graphic Method. Inscription, Visu-alisation, and Measurement in Nineteenth-Century Science and Culture*, Dissertation University of California, Los Angeles 1996. pp. 48-163. On the history of objectivity and the 19th century morality of self-restraint see also L. Daston, P. Galison, *The Image of Objectivity*, in "Representations", 40 (1992), pp. 81-128, 117ff.

not investigated. [1]And those investigators such as Count Rumford, who made use of experiments concerning the creation of heat by friction, turned this technique of heating water without fire into a marketable commodity. [2]This knowledge thereby remained in the realm of the mechanical arts. And even those who were interested from a natural philosophical point of view, such as Rumford in Munich or Henry Cavendish in England, did not challenge the dominant caloric theory of heat. In a manuscript Cavendish even suggested an experiment " to determine the *vis viva* necessary to give a given amount of sensible heat to a given body by alternately exposing a thermometer in the sun & shading", but he did not regard such a quantitative measure of the relation between heat and work to express a fact worthwhile to be put "into it's proper place in science"[3] .

2. Joule's Fact and Its Latent Meanings

Facts are things which must be felt; they cannot be learned from any description of them. (J. C. Maxwell, 1869)

In 1850 the Royal Society of London published a paper by James Prescott Joule, "On the mechanical equivalent of heat". For Joule this publication was a major

---

[1] On the measure of work by the French ingénieur-savant and the James Watts indicator diagram as the first self registering device to graphically display work see Brain, *The Graphic Method*, opus cit. ; M. N. Wise, *Visualising Work*, unpublished manuscript.

[2] Count Rumford most famously observed during his practice of canon boring heat developments effected through mechanical friction and concluded that heat was a kind of motion. But he was not suggesting a kinetic theory of heat because he had shown in other experiments that liquids did not possess the property to conduct these motions. ST. L. Wolff, *Benjamin Thomson, Sadi Carnot und Rudoph Clausisus*, in Meyenn, *Physiker* cit. , pp. 289-302, 291. B. Rumford, *An Inquiry concerning the Source of the Heat which is excited by Friction*, in "Philosophical Transactions of the Royal Society of London", LXXXVIII (1798), pp. 80-102. As Moritz Herrman Jacobi stated in 1856 in France MM. Beaumont constructed an apparatus on the Rumford experiment, patented the device in many countries by means of which steam is produced at a pressure of two and a half atmospheres, destined to accompany military expeditions "where there is always an abundance of muscular force but often a great want of means of existence, amongst which heat occupies one of the first places", M. H. Jacobi, *Sur la correlation des forces de la nature*, unpublished lecture manuscript from 1856, St Petersburg Academy of Sciences archive, Russia.

[3] Cavendish's draft on "Heat" from the 1780s including the suggestion for such an experiment is located under the reference number MG23, L6 in the Manuscript Division, Pre-Confederation Archives, Public Archives of Canada, Ottawa. For a detailed discussion of this draft see CH. Jungnickel, R. Mccormmack, *Cavendish. The Experimental Life*. Cranbury, NJ, 1999, pp. 400-423, 410. In the early 19th century in France it was Carnot who determined a mechanical equivalent of heat but didn't publish it either. U. Hoyer, *Über den Zusammenhang der Carnotschen Theorie mit der Thermodynamik*, "Archive for the History of the Exact Science", 1974, pp. 359-375.

achievement in getting credited as a "gentleman of science". Today this paper is seen as an important landmark in the development of the British science of energy because the fact stated in this publication became, in the second half of the 19th century, a building block of thermodynamics. ①In Joule's long and detailed paper the following conclusions appear:

I will therefore conclude by considering it as demonstrated by the experiments contained in this paper,

1st. That the quantity of heat produced by the friction of bodies, whether solid or liquid, is always proportional to the quantity of force expended. And

2nd. *That the quantity of heat capable of increasing the temperature of a pound of water (weighed in vacuo, and taken at between 55° and 60°) by 1° Fahr. requires for its evolution the expenditure of a mechanical force represented by the fall of 772 lb. through the space of one foot.* ②

However, a careful examination of the original manuscript, held at the Royal Society, London, shows that the published conclusion was not the one Joule had initially conceived. In the original version he stated that he had demonstrated:

by the experiments contained in this paper, —

1st. That the quantity of heat produced by the friction of bodies, whether solid or liquid, is always proportional and equivalent to the quantity of force expended. . .

2nd. That friction consists in the conversion of force into heat.

I consider that 779.692, the equivalent derived from the friction of water, is the most correct, both on account of the number of experiments tried, and the great capacity of the apparatus for heat. And since even in fluid friction it was impossible entirely to avoid vibrations and the production of a slight sound I prefer to state in round numbers as the result of the research,

That the quantity of heat capable of increasing the temperature of a lb of

---

① On Gentlemen of science see J. Morell and A. Thackray, *Gentlemen of Science. Early Years of the British Association for the Advancement of Science*, Oxford 1981. On Joule's role in the formation of the science of energy see D. S. L. Cardwell, *James Joule, A Biography*, Manchester 1989. C. Smith, *The Science of Energy. A Cultural History of Energy Physics in Victorian Britain*, London 1998; H. O. Sibum, *An Old Hand in a New System*, in J. P. Gaudielliere, I. Loewy, *Manufactures and the Production of Scientific Knowledge*, Houndmills 1998.

② J. P. Joule, *On the Mechanical Equivalent of Heat*, "Philosophical Transactions of the Royal Society", CXL (1850), pp. 61-82, printed also in Joule's *Scientific Papers*, I, London 1884-1887 [henceforth SPJ], pp. 298-328.

water (weighed in vacuo and taken at between 55° and 60°) by one degree
Fahr. , is equal to the mechanical force represented by the pressure of 772 lbs
through the space of one foot. [1]

The paper was considered for publication by the Royal Society only on condition
that the second conclusion, "friction consists in the conversion of force into heat" be
fully suppressed. In the end Joule accepted these drastic changes in order to get the
paper printed in the Philosophical Transactions but he wrote in the same year to the
Cambridge don George Gabriel Stokes:

> I beg your acceptance of the enclosed paper in which I have endeavoured to
> determine the mechanical equivalent of Heat with accuracy. The result at which I
> conceived I had arrived was that Friction consists in the conversion of Force into
> Heat; but the Committee of the R. S. having disapproved of such a deduction
> from the experiments I thought it best to withdraw it, although I think this view
> will ultimately be found to be the correct one. [2]

Joule's publication was crafted carefully by the scientific brewer in order to match
the ethos of science as understood by the distinguished readers of the Transactions. [3]But
for Joule the core meaning of this numerical fact was the interconvertibility of heat and
work and not just a proportionality of natural forces. This raises questions about the ways
in which the power relationship Joule experienced changed his knowledge claim and
about the nature of the latent meanings actually associated with Joule's fact. The referees
could not accept his claim of interconvertibility, which was based on a microscopic view
of the nature of heat. From their perspective, only a proportionality between mechanical
force and heat was demonstrated as measurable and therefore acceptable in science. By
making him revise his paper, the Royal Society made sure that Joule's experimental
research, which bore important theoretical consequences for the current understanding
of the caloric nature of heat, was freed from those knowledge claims violating the
current canon of physics and contaminating what they regarded as an exact fact. The

---

[1] Joule's original draft "On the Mechanical Equivalent of Heat" (1850) held at the Royal Society archive
PT. 37. 3. See also C. Smith, *Faraday as Referee of Joule' s Royal Society Paper* On the Mechanical Equivalent of
Heat, in "Isis", LXVII (1976), pp. 444-449.

[2] Joule to Stokes, July 3rd, 1850, CUL, Add. 7656 J75.

[3] For a detailed description of this process see H. O. Sibum, *Reworking the Mechanical Value of Heat:*
Instruments of Precision and Gestures of Accuracy in Early Victorian England, in "Studies in History and Philosophy of
Science", XXVI (1995), pp. 73-106.

revised paper aimed at demonstrating how in Victorian Science scientific facts were established as empirical evidence and, indeed, in this respect the determination of the mechanical equivalent of heat was publicly regarded as a masterpiece. But, for Joule, the knowledge deduced from his experiments was neither hypothetical nor speculative; this exact fact was the result of nearly ten years of experience in investigating the nature of heat. His local knowledge "that friction consists in the conversion of force into heat" emerged at the intersection of various Manchester knowledge traditions. I have elaborated elsewhere in some detail these hitherto neglected dimensions of Joule's research work of the late 1830s and 1840s.

In this paper I shall briefly refer to those results which help to substantiate the view that Joule's fact was an integral part of a still local but exact knowledge about the dynamic nature of heat. In a small collective of Manchester researchers which had succeeded in visualizing hidden working mechanisms of nature, Joule's factual knowledge was firmly established but its creator had yet to gain credibility as natural philosopher. In order to see the hidden dimensions of Joule's fact let us briefly recall the contemporary positions on heat and friction. Up to the 1840s most scholars treated heat as an immaterial substance called caloric. But a new regime of natural philosophers in England began to adopt French mathematics in which heat gained the status of a quantitative term but required no explanation as to its nature. However, several practitioners, often unaware of each others work, provided increasing evidence questioning this advanced theory of caloric. With regard to friction, engineers as well as users of steam engines used this phenomenon as a measure of the work lost in mechanical processes. Armchair philosophers on the other hand still treated friction in their books on mechanical philosophy as a mundane, negligible effect which should be kept at arms length from theory. And, as Crosbie Smith has shown, even Joule applied the engineers' concept of "economical duty", understood as "pounds raised to the height of one foot by the agency of one pound" of coal, as his standard to express numerically mechanical force and to calibrate the performance of machines.

As the son of the wealthiest brewer in Manchester Joule explored and participated in different Mancunian worlds of work. As such he could afford to choose research topics he thought of as most important to him and for his plan to become a gentlemen of science. Living in this industrial environment it is not astonishing to see this wealthy nineteen-year old, bourgeois brewer experimenting on the question of self-moving forces in nature and their imitation through self-acting machinery. In these days electricity and magnetism were regarded as most powerful natural forces and his research initially concentrated on electromagnetic engines. In the beginning it was a test of its economic feasibility—realizing it not to be a *perpetuum mobile*. Later, the machine gradually became a tool to investigate conversion processes of natural forces.

In the search for Joule's motif to investigate nature quantitatively one might be tempted to follow some historians' interpretations which link his work to the workshop of engineers. But the brewing culture of which he was an active member had its own particular demands for precision measurement and was the space for expressing quantitatively phenomena and their relations to each other. The Brewer's premises, controlled in all their operations by the fiscal state, especially the Excise system, were the places where brewers learnt to exchange accurate numerical data taken independently by the producer and state officials. This was the training ground for learning how to "trust in numbers". Furthermore, the brewing industry underwent major changes in the 1830s including an expansion of production to an industrial scale. As the scientific brewers argued, in the old system brewers produced "relative facts" by which they meant that brewing practices were tied to local conditions and were difficult to reproduce elsewhere. [1]As a result of this new economic situation scientific brewers made thermometers and decimal tables into indispensable practical and theoretical technologies. The Excise likewise regarded instrumental measures and the determination of absolute standards as the only means of establishing successfully their collecting system: Precision measurement was regarded as the means of establishing proof. Given the long history of the Excise providing the sinews of British power it is no wonder that already in 1842 the British government opened the "Excise

---

[1] Reproduction of brewing beer at different premises often presented great difficulties even when performed by the same brewer because every locale had its particulars, different brewing utensils, different materials etc. But even every brew at the same premise was a unique event because the grain, the malt, the temperature conditions were always different and therefore each brew required particular attention and adjustment of measures. In the process of turning brewing into a science through precision measurement several brewers argued that standard heats did not exist, that they only existed because of the Excise. R. Shannon, *A Practical Treatise on Brewing, Distilling and Rectifying*, London 1805, p. 57.

laboratory" which became an early Victorian institution of precision measurement. [1] Work performed there contributed to throw into disrepute forms of knowledge integral to the old brewer's sensuous order, judged to be unreliable in comparision. [2] The daily experience of brewing not only taught one to trust numbers, it also involved the observation of the continuous change and transfiguration of natural phenomena: the visible world of grain, hops, malt, wort and mash was one of constant generation, transformation and decay. Heat was "nature's main instrument" there but it was also an ephemeral, challenging one and, as many old brewers still argued, it was not quantifiable. [3] Joule's strong commitment to the new system of brewing in which thermometry, steam engine technology and mensuration were key elements, no doubt contributed to make him practice and believe in the importance of expressing experiences by numbers. In his exploration of the electro-motor, Joule finally realised that

we have therefore in magneto-electricity an agent capable by simple mechanical means of destroying or generating heat. In a subsequent part of this paper I shall make an attempt to connect heat with mechanical power in absolute numerical relations. [4]

His electro-magnetic experiments had shown that the concept of the materiality of

---

① J. Bateman, *The Excise Officer's Manual Being a Practical Introduction to the Business of Charging and Collecting the Duties under the Management of Her Majesty's Commissioners of Inland Revenue*, London: 1852; on the Military-Fiscal State and precision measurement see J. Brewer, *The Sinews of Power. War, Money and the English State, 1688-1783*, London 1994; S. Schaffer, *Golden Means: Assay Instruments and the Geography of Precision in the Guinea Trade*, in M. N. Bourguet, CH. Licoppe, H. O. Sibum, *Instruments, Travel and Science*, cit.; W. J. Ashworth, *Between the "Trader and the Public": defining production and measures in eighteenth century Britain*, unpublished manuscript. On the notion "Trust in Numbers" see T. Porter, *Trust in Numbers: the Pursuit of Objectivity in Science and Public Life*, Princeton, New Jersey 1995.

② On the excise laboratory see P. W. Hammond, H. Egan, *Weighed in the Balance. A History of the Laboratory of the Government Chemist*, London1992; H. O. Sibum, *Les Gestes de la Mesure. Joule, les pratiques de la brasserie et la science*, in "Annales Histoire, Science Sociales", IV-V (1998), pp. 745-774.

③ The term "nature's main instrument" Joule used in a lecture delivered in 1865 at Greenock in commemoration of the birth of James Watt, cited from CH. A. Parsons, *The Rise of Motive Power and the Work of Joule. Second Joule Memorial Lecture*, in "Memoirs and Proceeding of the Manchester Literary and Philosophical Society", LXVII (1922/23), pp. 17-29, 22. For the brewer's world of heat see G. A. Wigney, *An elementary Dictionary, or, Cyclopaediae for the use of Maltsters, Brewers, Distillers, Rectifiers, Vinegar Manufacturers and others*, Brighton 1838, H. O. Sibum, *Les Gestes*, cit.

④ J. P. Joule, *On the Caloric Effects of Magneto-Electricity, and on the Mechanical Value of Heat*, SPJ, pp. 123-159, p. 146, Italics by Joule.

heat no longer held because it could be destroyed or generated by mechanical means. ①It showed furthermore that there existed an intimate connection between heat and mechanical force. The machine now functioned as a displaying technology of effects when shifting between the micro and the macroworld, between electro-chemical processes and the world of mechanical work. Crucially this unintended drift in the course of his research not only provided Joule with an important empirical argument against the declining theory of caloric, but also tempted him to make an important generalisation, namely that there existed an *absolute* numerical relationship between heat and mechanical work, which in the end was the beginning of Joule's well-known experimental series to determine the mechanical equivalent of heat. At this point Joule realised that the current conflicting opinions about the nature of heat and friction could be resolved, and he had enough evidence to publish an attack on caloric theory. His conviction that an absolute measure expressing a hidden relationship controlling the changing phenomena of the visible world even was in tune with his religious belief. ②

But in order to establish this absolute measure, Joule had to demonstrate that this numerical fact existed independently of materials and procedures. The problematic subject in this endeavour was heat, or to be more precise, the practices of measuring heat. Joule was not simply applying metrology to the field of research on heat—as some have suggested. Joule was a bridge figure in the cultural process of the establishment of exact sciences in Britain. With him, this experimental culture of precision measurement developed into an art of its own kind. ③In order to provide further empirical evidence for his dynamical conception of heat, Joule had to advance thermometry itself because he

---

① It is important to mention here that caloric theory was already quite under attack by different researchers, see ST. G. Brush, *The Wave Theory of Heat. A Forgotten Stage in the Transition from Caloric Theory to Thermodynamics*, "The British Journal for the History of Science", V, no. 18 (1970), pp. 143-167, 147.

② "Everything may appear complicated and involved in the apparent confusion and intricacy of an almost endless variety of causes, effects, conversions and arrangements, yet is the most perfect regularity pre served— the whole being governed by the sovereign will of God"; J. P. Joule, *On Matter*, *Living Force and Heat. A Lecture at St Ann's Church Reading-Room*, SPJ, pp. 265-276, p. 273; for his religious back ground see D. S. L. Cardwell, *James Joule, A Biography*, Manchester 1989, p. 47 and pp. 10-13; Crosbie Smith suggested that Joule's perspective goes well with his self-description of a life-long Tory seeking for order and stability in society and therefore contrasted with Chalmer's evangelical vision and harmonized with John Playfairs's system of nature earlier in the century. C. Smith, *The Science of Energy. A Cultural History of Energy Physics in Victorian Britain*. London, 1998, p. 72.

③ On Joule as a bridge figure see D. S. L. Cardwell, *James Joule.*, cit. and H. O. Sibum, *Narrating by Numbers. Keeping an Account of Early Nineteenth Century Laboratory Experiences*, in L. Holmes, J. Renn, H. J. Rheinberger (eds.), *Reworking the Bench. Research Notebooks in the History of Science*, forthcoming.

was convinced that nearly every scientific claim about heat hinged on unreliable thermometer measurements.

Besides his long experience as a scientific brewer, another cultural resource available to him was the Manchester optician and instrument maker John Benjamin Dancer, a master of the emergent visual culture in early Victorian Britain. His first major breakthrough was the production of an achromatic microscope with unheard-of resolution. With the invention of photography, Dancer immediately succeeded in reducing regular photographic images of real objects to minute forms which could only be read by looking through a microscope. For Dancer this technique of shifting scales was part of a broader programme of producing a reliable means of orientation in the new visual culture which astronomers and microphysicist had opened up. [1]

From 1844 Joule spent several mornings in Dancer's workshop discussing problems of heat measurement and the design of a new sensitive mercury thermometer. This extremely sensitive device acted like Dancer's microscopes, which made latent images of the microcosm visible. Joule's thermometer displayed sensible heat which under normal circumstances would not leave any visual trace. Joule and Dancer thereby provided the means to shift scales within heat research which turned out to be crucial in the later experiments to determine the exact value of the mechanical equivalent of heat. Joule employed the sensitive thermometer for the first time in his research "On the changes of temperature produced by the rarefaction and condensation of air" and in all his subsequent experiments on the determination of the mechanical equivalent of heat. Joule further developed the skills in thermometry he had acquired in his fathers brewery to such an extent that his laboratory became a unique experiential space. [2] Furthermore, as Joule rightly stated, these advances in the sensitivity of precision measurements no longer allowed anymore the presence of witnesses. He had turned himself into a performer

---

[1] For a detailed discussion of the cultural technique of shifting scales and its various applications see H. O. Sibum, *Shifting Scales. Microstudies in Early Victorian Britain*; Max Planck Institute for the History of Science Preprint No. 171; on Dancer's microphotographs see B. Bracegirdle; J. B. Mccormick, *The Microscopic Photographs of J. B. Dancer. With illustrations from original Dancer negatives from the collection of A. L. E. Barron*, Chicago 1993; On the much broader investigation of this rising visual culture in this period see J. Crary, *Techniques of the Observer. On Vision and Modernity in the Nineteenth Century*, Cambridge, Mass., 1991.

[2] For further details on the cultural resources of this unique performance of measurement see Sibum, *Reworking*, cit.; the whole choreography of the performance of the temperature measurement was strongly moulded through the body techniques practiced in the brewing premise and even the experimental set-up of the paddle-wheel experiment was a small scale brewing model, H. O. Sibum, *Les Gestes*, cit.

without any audience. Joule faced severe problems because there was not outside the laboratory an expert culture in which people shared the practices, values and norms which Joule wished to establish. "I have always found a difficulty in making people believe that fractions of a degree could be measured with any great certainty."[1]

To conclude this section, let me reflect briefly on some mechanisms which explain why this numerical fact could have different meanings in these research collectives. Despite Joule's accurate determination of the mechanical equivalent of heat, members of the Royal Society could not accept his knowledge claim that friction consisted in the conversion of force into heat because this would have violated methodology and the canonical knowledge of heat as caloric then taught in the scholarly world. Furthermore, Joule's sensitive measurements precluded the presence of witnesses and because of the minimal temperature increases detected, it was already difficult enough to convey the extreme accuracy of Joule's experimental work. In order to avoid controversy and to provide further grounds for discussing this topic the referees prompted Joule to extract from his paper the most agreeable matter of fact, i. e. a measurable proportionality between heat and work. For Joule this was fatal. It was as if his knowledge deduced from years of experience was torn apart into an uncontroversial matter of fact and a matter of opinion. The former was established through precision measurement and the latter was probably regarded as an expression of Joule's lack of education in natural philosophy. For Joule, Dancer and other artisans, instrument makers and dilettanti who were actively taking part in promoting and defining experimental science as a part of the exact sciences, performing experiments was regarded as a complex form of deduction. Facts could not be split up into these two components. When he submitted his reworked paper to the Royal Society, despite having won the accreditation as a "gentlemen specialist", Joule must have felt as miserable as the artisans of the late 18th century whose working knowledge enlightened philosophers often translated into simple sets of practices.[2]

---

[1] Joule to Thomson, 7 November 1848, CUL Add 7342.

[2] On the problem of extracting factual knowledge from the artisanal culture see references made in foot note 17. The historical and epistemological problems Diderot et al. were facing when investigating artisanal cultures remain partly unresolved. From the materials discussed here it seems reasonable to suggest that experimentalists in the late 18th and 19th century who try to establish factual knowledge in the Republic of Letters experience comparable problems. This relation between knowledge and science deserves further historical investigation currently undertaken in the Research Group "Experimental History of Science" at the Max Planck Institute for the History of Science.

3. Energy Conservation: Establishing Proof and the Construction of Simultaneous Discovery

Liebig in his "Letters" has given the whole credit of the discovery of the mechanical equivalent to Mayer. This is not honest; for no one knows much better than Liebig what I have done on the subject. He is a warm partisan and is thus doubtless led astray. (Joule to Thomson, April 1859)

Joule's experience of becoming a gentleman specialist, with all the problems it entailed, is exemplary of the formation of the exact sciences in this period. We may even see this on a broader scale. The foundation of the British Association for the Advancement of Science (BAAS), modelled on the German "Naturforscher Gemeinschaft", marked the beginning of a social enterprise described by Morrell and Thackeray as the formation of a community of "Gentlemen of science". In Germany, the *Deutsche Physikalische Gesellschaft* was founded in 1845 to promote the physical sciences, its first generation consisting of a very heterogeneous group of artisans, craftsmen, merchants, and high school and university teachers. Simultaneously, universities slowly began to establish laboratories which changed conventional forms of teaching and in which new experimentalists could do research. [1]However, as the story of Joule and Dancer shows, it was important for learned societies with members of heterogeneous social backgrounds promoting the new ethos of experimental science that Dancer and Joule distinguished themselves from being "ingenious inventors". For example, when the brewer Joule published his work on electro-magnetic machines in Sturgeons *Annals* he soon realised that he was addressing practical men and artisans rather then gentlemanly natural philosophers. But in order not be seen as someone interested in commercial promotion and, instead, claiming scientific authority, he continuously reminded his readers that he was advancing experimental research on principles of nature, establishing quantitative laws, discussing sources of error, and

---

[1] On "devotee science" as a social movement see R. H. Kargon, *Science in Manchester. Enterprise and Expertise*. Manchester 1977; J. Morell, A. Thackray; *Gentlemen of Science*, cit.; D. Hoffmann (ed.), *Gustav Magnus und sein Haus*, Stuttgart 1995; A. Fiedler, *Die physikalische Gesellschaft zu Berlin. Vom lokalen naturwissenschaftlichen Verein zur nationalen Deutschen Physikalischen Gesellschaft*, Aachen 1998 (Dissertation Universität Halle).

measuring precise numerical details. [1]Dancer, who earned his income through his instrument shop, equally made sure that he followed the moral economy of the gentlemen of science, freely sharing their knowledge with the learned world. Thus, he never patented any of his inventions, although they would have provided him with a good financial income. He did not do so because that would have indicated that he was only interested in earning money rather than gaining natural knowledge. Within this context, Joule's decision in his capacity of President of *Manchester Literary and Philosophical Society* (1851) to start printing the date of submission of papers accepted for publication by the Society makes sense. By doing so he hoped to avoid priority disputes over discoveries, an annoying affair in which Joule himself was involved. Joule the experimentalist, whose knowledge was to a large degree of a nonliterary kind with its own modes of communication, clearly saw authorship and copyright as a pressing issue in the experimental sciences. As we will see it effected the further development of this numerical fact. According to Robert Kargon this indicated the dawn of a new age in which the devotee scientists demonstrated their expertise not only through their work but also by excluding the literati from the Manchester Society. [2]

Following Joule's publications on the mechanical value of heat, in particular his 1850 paper, the officially established matter of fact travelled quickly. But agreement about whether it existed, what it finally meant and what value it had, required a collective effort spanning the whole 19th century which demanded the firm establishment of a culture of laboratory sciences and precision measurement. [3]Immediately, there were a great variety of responses to "Joule's fact". At an Academy of science meeting in

[1] See C. Smith, *The Science of Energy*, cit. , pp. 57ff; on science as vocation not to live on, J. Morrell, A. Thackray, *Gentlemen of Science*, cit. , p. 33; for Joule's publication in Sturgeons Annals: see SPJ, 1884, cit. , pp. 1-53; J. P. Joule, *A Short Account of the Life and Writings of the Late Mr. William Sturgeon*, in "Memoirs of the Literary and Philosophical Society of Manchester", XIV (1857); I. R. Morus, *Frankenstein's Children: Electricity, Exhibition, and Experiment in Early-Nineteenth-Century London*, Princeton 1998.

[2] R. Kargon, *Science in Victorian Manchester*, cit. p. 79f. On Dancer see his autobiography, J. B. Dancer, *John Benjamin Dancer*, *F. R. A. S.* , 1812-1887, *An Autobiographical Sketch*, *with some Letters*, in "Memoirs and Proceedings of the Manchester Lit. And Phil. Society", CVII (1964/65), pp. 115-142.

[3] On the state of laboratory science in France see R. Fox, *Scientific Enterprise and the Patronage of Research in France*, 1800-1870, in R. Fox, *The Culture of Science in France*, 1700-1900, Aldershot 1992, pp. 442-437; for England see M. N. Wise, C. Smith, *Energy & Empire. A biographical study of Lord Kelvin*, Cambridge 1989; F. A. James (ed.), *The Development of Laboratory. Essays on the Place of Experiment in Industrial Civilisation.* Houndsmill 1989; for Germany see CH. Jungnickel, R. Mccormmick, *Cavendish*, cit. ; K. Olesko, *Physics as Calling: Discipline and Practice in the Königsberger Seminar for Physics*, Ithaca, London 1991.

Vienna the physiologist and member of the German Physical Society Brücke commented:

> Amongst all numbers which have been and will be examined next the numerical value to be investigated is of such high importance that no other can compete with it. In future there will be no part of the physical sciences in which it does not play an essential role. No less important is its meaning for practical applications by preparing the basis for the estimate of any work producing system. Simultaneously it will show us the limit above which we should not hold out any hopes of gaining more work. [1]

And indeed an invariable quantitative relation between heat and work would set the physio logical theory of combustion on solid grounds. The young Helmholtz was also enthusiastic although he did not yet trust the value, dismissing the circumstance that in his "literary replication" he had used the wrong conversion table of units. Rudolf Clausius likewise performed detailed literary replication and inserted the numerical fact into his equation representing the first law of thermodynamics. [2]Wilhelm Weber at Göttingen also saw this number to be important but he did not believe in Joule's methodological approach. He doubted that the measure of mechanical work could be the most precise standard against other natural forces could be calibrated. Of course, Weber preferred his own magnetometer to determine the absolute measure of magnetic declination as the most reliable standard to calibrate natural forces and to determine this equivalent. His colleague Franz Ernst Neumann in Königsberg suggested a workable

---

[1]   E. Brücke (1851), in J. J. Weyrauch (ed.), *Die Mechanik der Wärme in Gesammelten Schriften von Robert Mayer. Dritte ergänzte und mit historisch-literarischen Mitteilungen versehene Auflage*, Stuttgart 1893, p. 296.

[2]   The term "literary replication" was introduced by Jim Secord and designates reading and drawing practices in order to make sense of an experimental performance. J. Secord, *Extraordinary Experiment: Electricity and the Creation of Life in Victorian England*, in D. Gooding, T. Pinch, S. Schaffer (eds), *The Uses of Experiment. Studies in the Natural Sciences*, Cambridge, 1989; pp. 337-383, 347ff. On Helmholtz see F. Bevilacqua, *Helmholtz's "Ueber die Erhaltung der Kraft". The Emergence of a Theoretical Physicist*, in D. Cahan (Ed.) *Hermann von Helmholtz and the Foundations of Nineteenth-Century Science*, Berkeley, Los Angeles 1993; See especially R. Brain, M. N. Wise, *Muscles and Engines: Indicator Diagrams and Helmholtz's Graphical Method*, in L. Krüger (ed.), *Universalgenie Helmholtz. Rückblick nach 100 Jahren*, Berlin 1994, who argue that Helmholtz was preferring the graphical method as the more trustworthy means to establish the measure of work. On R. Clausius see his notebook entries, Deutsche Museum, Munich, and D. S. L. Cardwell, *From Watt to Clausisus. The Rise of Thermodynamics in the Early Industrial Age*, Ithaca 1971, p. 246.

experiment for it. [1]In Britain Joule became acquainted with William Thomson, later Lord Kelvin, at the *BAAS* meeting in Oxford in 1847 and from then on, a close collaboration started. Thomson had recognized that Joule's paddle-wheel experiment demonstrated an important fact; the quantitative relationship between heat and work. But in agreement with his brother James, the engineer, William concluded that James Joule "was wrong in many of his ideas, but he seems to have discovered some facts of extreme importance, as for instance that heat is developed by the fric [tion] of fluids in motion". This fact connected on the one hand to the brothers' concern with problems of losses of useful work in fluid friction and on the other hand to recent discussions with Stokes on hydrodynamic analogies for heat, magnetism and other physical phenomena. [2]Without going into detail here about this important exchange it suffices to mention that Thomson needed four years to finally accept Joule's "wrong ideas" about the general meaning of his established fact. [3]During those years Joule repeatedly wrote to Thomson that he was very surprised to still read that Thomson favoured the concept of caloric. Indeed, as lecture notes from 1849 show, Thomson initially praised Joule's thermometrical skills and interpreted his experiments in a rather peculiar way. But during their collaboration over the next decades he became a powerful spokesman for Joule's knowledge claim about the interconvertability of heat and work, and even made Joule's number the foundation of the British science of energy and the absolute system of units. In a lecture of 1863 he tells his students:

> Heat is motion. Heat is properly called a Dynamical Equivalent for Work spent; but the difficulty is that we cannot get back the work spent in producing heat. The reconciling of Joule's views with Carnot's was essential to the Dynamical Theory of Heat. [4]

However, in 1851, at the beginning of this founding period of thermodynamics

---

① W. Weber, *Zur Galvanometrie*, 1862, in W. Weber, R. Kohlrausch, *Fünf Abhandlungen über die absoulte elektrische Stromund Widerstandsmessungen*, Leipzig 1904, Ostwald's Klassiker No. 142, pp. 70-94, p. 88. On Neumann's lecture see F. E. Neumann, *Vorlesung: Mechanische Wärmetheorie. Vorlesungsmitschrift von Prof. Wild, D. Dorn from the year* 1867, Handschriftenabteilung Niedersäch-sische Staats-und Universitätsbibliothek Göttingen. MS F. E. Neumann 8 II, p. 237-418, 256; on Neumann see also K. Olesko, *Physics as Calling*, cit.

② William Thomson to Dr. James Thomson, 1 July 1847, T367, Kelvin Collection, Glasgow University Library (hereafter GUL), see also Smith, The Science of Energy, cit. , p. 79.

③ On the work to harmonise his theory with Joule's fact see in particular M. N. Wise, *William Thomson's Mathematical Route to Energy Conservation: A Case Study to the Role of Mathematics in Concept Formation*, "Historical Studies in the Physical and Biological Sciences", X (1979), pp. 49-83.

④ W. Thomson, Lecture notes by David Murray "Lecture LI, 14 April 1863", GUL, MS Murray 326.

and energy conservation the German physician Julius Robert Mayer claimed intellectual property rights on Joule's fact:

> The new subject [the mechanical equivalent of heat] began soon to excite the attention of learned men; but inasmuch as both at home and abroad the subject has been exclusively treated as a foreign discovery, I find myself compelled to make the claims to which priority entitles me; for although the few investigations which I have given to the public, and which have almost disappeared in the flood of communications which every day sends forth without leaving a trace behind, prove, by the very form of their publication, that I am not one who hankers after effect, it is not therefore to be assumed that I am willing to be deprived of intellectual property which documentary evidence proves to be mine. ①

Historians have discussed this episode as a case of simultaneous discovery of the conservation of energy in which several others were involved besides Mayer and Joule. ②I want to argue that the priority dispute discussed publicly and in private correspondence between Joule and Thomson was an expression of the difficulties experimentalists had in establishing scientific authority in a community which had yet to define its own new identity. Since the first half of the 19th century numerical facts gained a high reputation amongst these new scientists as "the foundations of an exact investigation of nature". However, Joule's and Mayer's determinations of the mechanical equivalent of heat were practiced in very different ways. The latter calculated the number from known experimental data whereas Joule measured the quantity in several conversion experiments. But neither of them described it as a proof of the conservation of energy. ③Rather, the combination of the exact fact with the clear idea of the conservation of energy was only gradually constructed

---

① R. Mayer, *Remarks on the Mechanical Equivalent of Heat*, "The London, Edinburgh and Dublin Philosophical Magazine and Journal of Science", Supplement to Vol. XXV. Fourth Series, No. 171, June 1863, p. 520. English translation of*Bemerkungen über das mechanische Äquivalent der Wärme*, Heilbronn und Leipzig, 1851.

② For the seminal paper see T. S. Kuhn, *Energy Conservation as an Example of Simultaneous Discovery*, in Kuhn, *The Essential Tension* cit. , pp. 66-104.

③ Several historians have by now questioned the concept of simultaneous discovery and provided local studies to show the particularities of these contributions. See for example Y. Elkana, *The Conservation of Energy: A Case of Simultaneous Discovery?*, "Archives Internationales d' Histoire des Sciences", 23-e an-née (1979), pp 31-60; K. Caneva, *Robert Mayer and the Conservation of Energy*, Princeton 1993; C. Smith, *The Science of Energy*, cit. ; H. O. Sibum, *Reworking* cit. ; On the notion of discovery see S. Schaffer, *Scientific Discoveries and the End of Natural Philosophy*, "Social Studies of Science" XVI (1986), pp. 387-420.

between 1850 and 1880. When Joule read about Mayer's priority claim in 1848 he replied

> that the simple fact of the heat produced by the compression of a gas [i. e. the sinking of a mercury column in a glass tube by which a gas is compressed, H. O. S.] did not warrant the conclusion that the heat was an exact equivalence to the force employed; because one may conceive of a gas formed of repulsive particles in which part of the force would be employed in forcing them partially nearer and only the reminder in generating heat. [1]

In the correspondence between Thomson, Tait and Joule, Mayer's claim for the existence of the mechanical equivalent of heat was described as "*mere speculation*" and only Joule's experimental work to provide empirical evidence being regarded as "*research*":

> There is also another point—It is openly admitted that Mayer's paper of 1842 was printed as a note, merely to secure an early date from which to count. I for my own part would never have published such a paper, because I have always held, after Herschel, that science has no greater bane than hasty generalisation. Mayer's paper was merely a speculation, mine was a research. The only influence Mayer could have was to retard the advance of the subject. This is my candid opinion, although I feel far too great respect and kinship for Mayer to say so in print. [2]

Now in order to understand why Joule and other British scientists could regard Mayer's work as being speculative and "pre scientific" and to judge the specificity of Joule's experimental practice with its peculiar moral values, it is necessary to focus a little more on Mayer's research practice and the way he derived the mechanical equivalent of heat from published experimental results. In his calculation he used as a basis the ratio of the capacity for heat of air under constant pressure to that under constant volume. By reflecting on an experiment in which a mercury column of a cubic centimetre and 76 cm height and 1033g weight was lifted 1/274 of its original volume he estimated that the warming of an equal weight of water from $0°C$ to $1°C$ corresponded to the fall of an equal weight from the height of about 365 metres. And he finally concluded, "the same result you will get if you take a simple or mixed gases for your calculation instead of atmospheric air". Yet in a footnote he was more precise and

---

[1] Joule to Thomson, March 18, 1851.
[2] Joule to Tait, July 25, 1863, Glasgow University Library, Kelvin papers J 173.

mentioned the precondition for this claim, *aconstant expansion coefficient* for all gases.

However, up to the 1840s the phenomenon of the expansion of gases was one of the major concerns amongst natural philosophers in France, England and Germany. By taking for granted that this expansion was constant for all gases Mayer completely neglected this research in the field of thermometry. Although for nearly forty years the gas expansion coefficient had been regarded as "eine der sichersten Zahlen der Physik" criticism of it arose amongst several researchers (Regnault, Rudberg, Magnus) in the 1840s. The most reliable value was then suggested by Regnault but he also concluded that the expansion coefficient was not exactly the same for all gases and therefore that Gay-Lussac's law was not generalizable. [1]Magnus rightly reminded his readers that this conflict in values would make it difficult to take the expansion of air as the measure of temperature as had been practised since the work of Dulong and Petit. This redetermination of the value of the expansion coefficient had several consequences: First, a practical one, the air thermometer used then as the French standard fell into disrepute, as it was no better than the mercury thermometer, whose working according to Regnault were "more or less a complicated function of heat increases" [2]. The theoretical implication was also expressed by Regnault: we still lack the means to measure absolute quantity of heat and with regard to this current state of our knowledge "we have little hope to discover simple laws of those phenomena which depend on these measures". From this perspective it seems obvious that Poggendorff, the editor of the German *Annalen der Physik*, was hesitant to publish Mayer's first account in which it was simply claimed that heat could be transformed into motion, and that thermal or material expansion was the phenomenological evidence for this. [3]It seems that Poggendorff decided to keep the paper until he knew more clearly what the outcome of the controversy about the reliable measure of thermal expansion would be. In the end, he refused to publish it all.

---

[1] G. Magnus, *Über die Ausdehnung der Gase duch die Wärme*, in "Poggendorff's Annalen der Physik und der Chemie", LV (1842), pp. 1-27; V. Regnault, *Untersuchung über die Ausdehnung der Gase*, in "Poggen-dorff's Annalen der Physik und der Chemie", LV (1842), pp. 391-414 u. 557-585, see also CH. Sichau, *Der Joule-Thomson Effekt. Der Versuch einer Replikation*, Physics Diploma Thesis, University of Oldenburg, 1995.

[2] V. Regnault, *Über den Vergleich des Luftthermometers mit dem Quecksilberthermometer*, in Poggendorff's Annalen der Physik und der Chemie', LVII (1842), p. 218.

[3] J. R. Mayer, *Über quantitative und qualitative Bestimmung der Kräfte*, send to editor of *Poggendorff's Annalen* on June 16, 1841, not published, found in Friedrich Zöllner archive published in J. R. Mayer, *Die Mechanik der Wärme. Sämtliche Schriften*, ed. by Hans Peter Münzenmayer, Heilbronn 1978.

In Manchester Joule performed his heat experiments with the new mercury thermometer he and Dancer had made which resolved the above-mentioned problems. Thomson, who was working at the time on a procedure to establish theoretically the absolute temperature scale, trusted Joule's thermometrical work fully, and his development of thermodynamics over the following years was strongly supported by Joule's laboratory work in Manchester. [1]In the rising debate over priority Thomson as well as Tait took sides with Joule and defended the empiricist position that Joule's more accurate experimental investigations of the value of the mechanical equivalent of heat demonstrated that "heat is the dynamical equivalent of work spent" and that "conservation of energy cannot be established but by experiment". [2]

4. Engineering Evidence: A Constant of Nature

It may be an object of curiosity to some of my readers to know how it happened that the measure of work came to be adopted... Professor Maxwell informs us that "if a body whose mass is one pound lifted one foot high in opposition to the force of gravity, a certain amount of work is done, and this quantity is known amongst engineers as a footpound". If this passage may be taken to afford the true clue, we can hardly fail to be struck with the originality of the suggestion that a great philosophical principle may be established by "engineering evidence". (Robert Moon, 1874)

British physicists who worked on the science of energy regarded the mechanical equivalent of heat as a proof of the conservation of energy and, as a true numerical representation of nature it was for them the key element in the absolute system of units. An exact value of the mechanical equivalent of heat would provide the quantitative unifying structure for the many phenomena displaying correlation of forces. Once this fact was es-

---

[1] For the intense collaboration of Joule and Dancer in producing the thermometer and the differences in tackling thermometry between Joule and W. Thomson see H. O. Sibum, *Shifting Scales* cit.. On the absolute thermometric scale see W. Thomson, *On a Absolute Thermometric Scale founded on Carnot's Theory of the Motive Power of Heat, and Calculated from Regnault's Observations*, in "Philosophical Magazine", XXXIII (1848), pp. 313-317.

[2] See especially the correspondence between Tait and Thomson, July 3, 1863 (T41) and J. Thomson and W. Thomson, Aug. 13, 1863 (T119) at GUL. For a detailed account of the correspondence concerning the Joule-Mayer controversy see J. T. Lloyd, *Background to the Joule-Mayer Controversy*, in "Notes and Records of the Royal Society", XXV, (1970), pp. 211-225.

tablished as a true representation of nature it made possible the calibration of all other natural forces such as electricity, magnetism and chemical forces, against mechanical work. It seemed that everything was at hand to set up this "coherent system", as the *BAAS* called their measures. [1]Their standards programme was intimately linked with the telegraphy industry of the British Empire and demanded the highest degree of accuracy and precision. Reports about the work from the 1860s to 1880s show clearly the challenge centering on precision measurement which lay in this highly competitive project. [2]For the development of an experimental thermodynamics and electrodynamics as an exact science, this extreme accuracy in measurement became an obligatory requirement. From 1851, precisely a year after Joule's publication, members of the Kew Observatory announced that they had taken steps towards the production of "instruments, under their own superintendence, for distribution to institutions and individuals who might require accurate standards of reference "[3] . Furthermore laboratories were set up in many Universities to learn and improve upon the appropriate use of instruments of precision. Not only in Britain but in the whole of Europe, the culture of precision reached its peak. By the 1870s, in science to know meant to measure. [4]

I often say that when you can measure what you are speaking about, and express it in numbers, you know something about it; but when you cannot measure it, when you cannot express it in numbers, your knowledge is of a meagre and unsatisfactory kind: it may be the beginning of knowledge, but you have scarcely, in your thoughts, advanced to the stage of science, whatever the matter may be. [5]

Yet despite the success in promoting the British science of energy, the universal principle of energy conservation was still called into question from several points of view:

---

① *Report of the Committee appointed by the British Association on Standards of Electrical Resistance.* in *British Association for the Advancement of Science Reports*, London 1864, pp. 111-177.

② M. N. Wise, C. Smith, *Energy & Empire. A biographical study of Lord Kelvin*, Cambridge 1989; S. Schaffer, *Late Victorian Metrology and its Instrumentation: A Manufactory of Ohm's*, in S. Cozzen, R. Budd, (eds.), *Invisible Connections*, Bellingham 1992, pp. 23- 65. S. Schaffer, *Accurate Measurement is an English Science*, in M. N. Wise, *The Values of Precision*, Princeton 1995, pp. 135-172. B. Hunt, *The Ohm is where Art is: British Telegraphic Engineers and the Development of Electrical Standards*, in "Osiris", IX (1994), pp. 48-63.

③ J. Welsh, *"On the Graduation of Standard Thermometers at the Kew Observatory"*, BAAS Report, (1853), pp. 34-36.

④ On first practical physics courses see F. Kohlrausch, *Leitfaden der praktischen Physik zunächst für das physikalische Praktikum in Göttingen*, Leipzig 1870.

⑤ W. Thomson, *Popular Lectures and Addresses*, 3 Vols. , London 1889, I, pp. 73-74.

The Oxford educated reverend Highton described the subject as

> of extreme importance both for the interpretation of physical phenomena
> and for determining what limits are assigned by the stern laws of Nature to the
> exercise of man's mechanical and scientific skill. [1]

Finally he and his colleague Gore argued that one could certainly state the equivalency of heat and force which Joule had demonstrated in several experiments. However, they were unwilling to accept an absolute limitation to man's and nature's productivity. They also suggested that this numerical fact was only valid in mechanics and not in the fields of electricity or magnetism. Therefore it would be wrong to deduce from it a universal principle. [2] In 1867 a major controversy started when Joule replicated his own experiments. [3] Without going into detail about this experiment it is important to point that Joule obtained a different value for the equivalent. He claimed that he had performed it with more accuracy then he had in the old ones. A fact which had thrown doubt upon the accuracy of the BAAS. [4] The British science of energy was running into a crisis. Even the first Cambridge Professor of Experimental Physics James Clerk Maxwell involved himself in redetermining the mechanical equivalent of heat with a new direct method designed by him.

> I think it a plan free from mechanical difficulties and in a lofty room with
> plenty of mercury and strong ironwork, and cherub aloft to read the level &
> the Thermometer and a monkey to carry up mercury to him (called Quicksilver
> Jack) the thing might go on for hours. [5]

This metaphorical description indicated Maxwell's doubts about the previous experiments: the heat radiation of the experimenter's body disqualified humans from doing the experiment to the extent that only disembodied cherubs could read temperatures with accuracy. But his underlying concern was the dynamical theory of heat and the experiment's implications about the nature of heat. Joule wrote in 1869 to Thomson:

[1] H. Highton, *On the Mechanical Equivalent of Heat*, 1871, in "Proceedings of Literary and Philosophical Society", X (1871), p. 147.

[2] H. Highton, *On the Relations between Chemical Change, Heat, and Force —With a Special View to the Economy of Electro-dynamic Engines*, in "The Quarterly Journal of Science and Annals of Mining, Metallurgy, Engineering, Industrial Arts, Manufactures, and Technology", I (1871), p. 77-94, 89.

[3] J. P. Joule, *Determination of the Dynamical Equivalent of Heat from the Thermal Effects of Electric Currents*, British Association for the Advancement of Science Reports, London 1867, pp. 512-522.

[4] W. Thomson, *Popular Lectures and Addresses*, 3 Vols. , London 1889, I, pp. 73-136, pp. 133/134.

[5] Maxwell to Tait, Dec. 23, 1867, Cambridge University Library, Add 7655 1b.

Our sudden post has frozen up the moral & intellectual virtue of our philosophers... This is the time for new proofs now the hermetical conclave are preparing to put down the dynamical theory as heretical. ①

Meanwhile, several strategies were developed to solve these problems, with Maxwell mediating between the groups involved. ②The American engineer and physicist Henry A. Rowland realised during a one-year stay in Europe that the different values experimentalists had obtained in their measurements of fundamental units shook the new energy physics to its fundaments. Rowland regarded a new determination of the mechanical equivalent of heat to be necessary to establish "one of the most important constants of nature"③ . Only an exact value in absolute measures would qualify as a true representation of nature and therefore provide the required sinews for the rising international community of modern physics. But to achieve this would require drastic changes in the practise of science, involving for example an intensive collaboration between industrialists, engineers and laboratory sciences. As the founding director of physics laboratory at John Hopkins University Rowland explained to the board of trustees, in modern times " it is useless to expect anything from extemporized apparatus"④ . Modern physics needed the highest standards from the workshop, its tools and machinery, and in particular excellent craftsmanship from the instrument maker, to provide the researcher with precision instruments of outstanding quality. Furthermore a course in physics well balanced between mathematical and experimental methods "will not only meet a long-felt need in this country and hoped that it will attract that class of students who would otherwise go to Germany to pursue their studies" but also train "physicists of precision" . Only such a regime would make "the modern order of things" work. Rowland coined this phrase on the occasion of the opening

---

① Joule to Thomson, Dec. 29, 1869, Cambridge University Library MSS Add 7342 J287.

② For studies of how these issues were resolved see H. O. Sibum, *An Old Hand in a New System*, cit. , pp. 23-57, H. O. Sibum, *Exploring the Margins of Precision*, Max Planck Institute for the History of Science Preprint No. 172, and C. Smith, *The Science of Energy*, cit.

③ H. A. Rowland, *On the Mechanical Equivalent of Heat, with Subsidary Researches on the Variation of the Mercurial from the Air Thermometer, and on the Variation of the Specific Heat of Water.* in "Proceedings of the American Academy of Arts and Sciences", 1879-1880, pp. 75-200.

④ H. A. Rowland, *Report to the Board of Trustees*, undated, JHU, Ms 6, p. 17-18. Although he was a great admirer of Faraday and other British scientists, Rowland wanted to make it clear that the "science of the future" had to go beyond those forms of experimental practice, based as that was on improvising performances. Compare H. O. Sibum, *An Old Hand in a New System*, cit.

of the new physics laboratory. The modern order of things began with Galileo, Rowland argued, the first researcher to trust his own reasoning only, because he had tested it experimentally. [1]In Rowland's laboratory the new generation of students would have to learn

> to test their knowledge constantly and thus see for themselves the sad results of vague speculation; they must learn by direct experiment that there is such a thing in the world as truth and that their own mind is most liable to error. They must try experiment after experiment and work problem after problem until they become men of action and not of theory. This, then, is the use of the laboratory in general education, to train the mind in right modes of thought by constantly bringing it in contact with absolute truth. [2]

In several addresses to the public he laid out how the physical laboratory as the means to modern education should look like and why it was so important for cultural development:

> Those who have studied the present state of education in the schools and colleges tell us that most subjects, including the sciences, are taught as an exercise to the memory. I myself have witnessed the melancholy sight in a fashionable school for young ladies of those who were born to be intellectual beings reciting page after page from memory, without any effort being made to discover whether they understood the subject or not... Words, mere words, are taught and a state of mind far distant from that above described is produced... The object of education in not only to produce a man who knows, but one who does; who makes his mark in the struggle of life and succeeds well in whatever he undertakes; who can solve problems of nature and of humanity as they arise, and who, when he knows he is right, can boldly convince the world of the fact. [3]

Rowland's remarks are characteristic of a historical process of the late 19th century in which modern experimental science became the model for cultural development. But as the case of Rowland shows, in this historical process a wedge was driven between the

---

①   H. A. Rowland, lecture notes (given at the opening of the physics laboratory), Rowland Manuscripts, John Hopkins University, Eisenhower Library, Ms. 6, Ser. 5.

②   H. A. Rowland, *The Physical Laboratory in Modern Education*, in *The Physical Papers of Henry Augustus Rowland*, Baltimore 1902, pp. 614-618, 617.

③   75Ivi. , p. 617.

humanities and the natural sciences, cutting off matters of fact from matters of opinion.

The facts and theories of our science are so much more certain than those
of history, of the testimony of ordinary people on which the facts of ordinary
history or of legal evidence rest. [1]

In establishing the constant of nature, the engineer-physicist Rowland identified
thermometical practices as the weak chain in all previous attempts to determine the value
of the mechanical equivalent of heat. In a letter to Gilman he wrote that his standard air
thermometer "is the embodiment of my whole work". His publication and the archive
materials show that he possessed an intimate knowledge of the quality of
thermometers. He even regarded this investigation to be of such a great importance that
he proposed to set up a "subdepartment of standards where comparisons were made in
absolute measure". Shortly after obtaining his first experimental results he concluded
that:

Of all the directions in which the department may finally expand, that of
thermometry remains very prominent... the air thermometer has been taken as
the standard and all comparisons will be reduced to the final absolute standard
of this perfect gas thermometer. [2]

For his research on the mechanical equivalent of heat, Rowland did extensive work
on the comparison of thermometers then available in Europe and America. He knew that
in order to achieve international recognition for scientific knowledge in matters of heat,
a standards laboratory was an important measure to take. When he wrote up his research
on the mechanical equivalent of heat for publication he already felt absolutely certain
about his leading position amongst physicists and attempted to define a new standard of
accuracy for the coming millennium.

I have the highest respect for the accuracy of Joule's work, and regard
him as a model whom in this respect of care and accuracy which younger
physicists would do well to consider. And about one thing we may be certain,
that when the scientific millenium is reached, when only *physicists of precision*
remain, we shall... not have to wait again thirty years to have two physicists
agree on what is the true value of such an important quantity as the mechanical

---

[1] H. A. Rowland, *The Highest Aim of the Physicist. Address delivered as the President of the American Physical Society, at its meeting in New York, October* 28, *1899*, in *The Physical Papers*, cit., pp. 668-678, 676.

[2] H. A. Rowland, manuscript *Physical Laboratory*, Johns Hopkins University. Comparison of Standards. Circular No. I, p. 6. JHU Ms. 6, Ser. 5.

equivalent of heat. [1]

He went on to establish this modern standard, impressing the scientific world with the fact that a great philosophical principle could be demonstrated by engineering evidence. [2]

5. The Golden Number of that Century

The mechanical equivalent of heat, the number of units of work necessary to raise 1 pound of water 1° in temperature, has, with much reason, been called the golden number of that century. (Mendenhall, October 1901)

Rowland's publication in 1880 did not mark the end of the process of establishing the mechanical equivalent of heat as a distinct matter of fact. Two lines of development are worth noting here. Attempts to stabilise the value of the mechanical equivalent of heat continued as long as physicists developed new methods which seemed promising for the improvement of the accuracy of the established value. This is a characteristic of modern science: to constantly reevaluate the physical standards which are the tools of every working scientist. [3]This kind of research was first institutionalized with the establishment of national laboratories like the Physikalisch Technische Reichsanstalt. [4]In the last decades of the 19th century this exact fact finally became materialized as a conversion factor in the international system of units. On July 12th, 1894 the Senate and House of Representatives of the United States of America made use of this constant to establish the legal unit of work in the system of units of electrical measures:

The unit of work shall be the Joule, which is equal to ten million units of work in the centimeter-gram-second system, and which is practically

---

[1]   Rowland's unfinished concluding sentences and parts of sentences are to be found in his draft on *Appendix to Paper on the Mechanical Equivalent of Heat*, *Containing theComparison with Dr. Joule's Thermometer*, emphasis by H. O. Sibum, JHU, Ms. 6, Box 39, Series 5. For the published version see H. A. Rowland, *Appendix... cit.*, "Proceedings of the American Academy of Arts and Sciences", XVI (1880), pp. 38-45.

[2]   For a detailed exploration of Henry Rowland's contribution to establish this fact see H. O. Sibum, *Exploring the Margins* cit. and H. O. Sibum, *An Old Hand in a new System*, cit.

[3]   See for example E. H. Griffiths, *The Value of the Mechanical Equivalent of Heat*, *deduced from some Experiments performed with the view of establishing the relation between the Electrical and Mechanical Units*; *together with an Investigation into the Capacity for Heat of Water at different Temperatures*, in "Philosophical Transactions of the Royal Society of London", vol. 184 (1894), pp. 361-504; O. Reynolds and W. H. Moorby, *On the Mechanical Equivalent of Heat*, "Philosophical Transactions of the Royal Society of London (A)" vol. 190 (1898), pp. 301-422.

[4]   D. Cahan, *An Institute for an Empire: the Physikalische Technische Reichsanstalt* 1871-1918, Cambridge 1989.

equivalent to the energy expended in one second by an international ampere in an international ohm. [1]

By that time the meaning of this scientific fact was no longer restricted to the laboratory. It had become an integral part of the rapidly expanding "networks of power" and labelled "the golden number of that century" —as a student of Rowland once called it. [2]Indeed, 19th century investigators involved in determining the value of this number would have agreed with this description. In a private correspondence Robert Mayer, for example, wrote:

> Die Physik, beziehungsweise die Mechanik, kann mit dem ihr ebenfalls wohl bekannten Kilogramm allein nicht weit springen; in der Dynamik, und dies ist offenbar das weitaus wichtigste Gebiet der Physik, findet das blosse Kilogramm, das blosse statische Druckmass, keine Verwendung mehr, hier ist das Kilogrammeter das neuerdings allenthalben eingeführte Einheitsmass des Untersuchungsobjektes, der sogenannten lebendigen Kraft. Den unproduktiven Druck haben wir umsonst, die Kraft aber oder das sogenannte Kilogrammeter kostet immer Geld. . . Der große Newton hat eben das Kilogrammeter noch nicht gekannt, und die Wissenschaft schreitet am Ende über alle Autoritäten fort. . . Ein statisches Aequivalent, d. h. eine konstante numerische Bezeichnung zwischen Wärme und Druck, zwischen Kalorie und Kilogramm, existiert aber

---

① *Committee Report to the President of the National Academy of Sciences*, *New York City*, *February 9*, *1895*. Rowland Papers, Eisenhower Library archive, Johns Hopkins University. The "Joule" is the unit of work or energy in the International system of units (SI). The International System is called the SI, using the first two initials of its French name Système International d'Unités. The key agreement is the Treaty of the Meter (Convention du Mètre), signed in Paris on May 20, 1875. 48 nations have now signed this treaty, including all the major industrialized countries. The United States is a charter member of this metric club, having signed the original document back in 1875: Today the Joule is defined to be the work done by a force of one Newton acting to move an object through a distance of one meter in the direction in which the force is applied. Equivalently, since kinetic energy is one half the mass times the square of the velocity, one joule is the kinetic energy of a mass of two kilograms moving at a velocity of 1 m/sec. It equals 10^7 ergs, or approximately 0, 7377 foot-pounds. In electrical terms, the joule equals one wattsecond—i. e., the energy released in one second by a current of one ampere through a resistance of one ohm.

② TH. C. Mendenhall, *Commemoration of Prof. Henry A. Rowland*, in "Johns Hopkins University Circulars", XXI, No. 154, December 1901, reprinted in *Annual Report of the Board of Regents of the Smithsonian Institution*, *Showing the Operations*, *Expenditures*, *and Conditions of the Institution for the Year Ending June* 30, 1901; Washington 1902, pp. 739-753, 743. For the legal act see Rowland's *Report to the President of the National Academy of Sciences*, Johns Hopkins University Library archive, Rowlands manuscript collection. On the notion of "networks of power" see T. P. Hughes, *Networks of Power. Electrification in Western Society 1880-1930*, Baltimore 1983.

nicht, wohl aber die Aequivalentenzahl zwischen Kilogrammeter und Kalorie. ①

This golden number fundamentally shaped humans perception. As the president of the *Physikalische Technische Reichsanstalt*, Professor Friedrich Kohlrausch stated in 1900, all things (Sache) will be perceived as to their particular effect inherent in these bodies. Wood, coal, food amongst others belong to this class of things and its value is determined through the available energy or work it provided. In future we will become consumers of energy instead of bodies. Therefore German jurisdiction should act quickly in order to legalise the act of declaring energy as a "Sache". ②

The second line of development was equally important. At the end of the 19th century, experimentalists in Europe and North America had successfully established the laboratory sciences as a defined space within the academic landscape. ③Practical physics courses with the aim of reading and communicating factual knowledge through the performance of experiments had become an integral part of the University curriculum. Teaching manuals of practical physics were published and apparatuses to demonstrate the mechanical equivalent of heat became standard equipment in laboratories of

---

① Mayer to Mohr, April 28, 1868, in J. J. Weyrauch, *Kleinere Schriften und Briefe von Robert Mayer. Nebst Mittheilungen aus seinem Leben*, Stuttgart 1893, pp. 419-420. Today the numerical fact finally links the calorie (the c. g. s. unit of heat) to the joule (the unit of mechanical energy), equal to 4. 1868 joules per calorie. In the SI system of units this fact which measures heat and all forms of energy in joules (so that the mechanical equivalent of heat is 1) is even redundant.

② "Die neuere Entwicklung des Kleinverkehrs in Werten liegt zum großen Teil in dem Übergang von dem Verkehr in Körpern zu dem in Energien. Und diese Umgestaltung ist noch lange nicht am Ende angelangt nicht das abgelaufene, sondern das kommende Jahrhundert wird als der Zeitraum zu bezeichnen sein, in welchem diese Frucht naturwissenschaftlicher Erkenntnis reift"; F. Kohlrausch, *Die Energie oder Arbeit und die Anwendungen des elektrischen Stromes*, Leipzig 1900, pp. 68-69. One should know that Kohlrausch wrote this book on the occasion of several court trials concerning criminal acts of illegally consuming electrical energy occurring in Germany. Kohlrausch thought it was an effect of bad public education in matters of science of energy.

③ With regard to the historical development of laboratory science, certainly in the 17th and 18th century Universities began making space for physical cabinets. But with regard to the absence of budgets and training as well as status it is reasonable to argue that only in late 19th century did experimentalists fully succeed in materially and ideally establishing experimental science as a scholarly discipline in its own right. For a survey of the 18th century physical cabinets see J. L. Heilbron, *Elements of Early Modern Physics*. Berkeley 1982, pp. 139ff; W. Clark, *The Hero of Knowledge*, to appear with University of Califronia Press; for the 19th century survey see for example F. A. J. L. James (ed.) *The Development of the Laboratory*. cit.

the physical sciences. ①At the time of this material embodiment of the numerical fact into the working environment of the experimental physicists, the meaning of this scientific fact changed again. This change in meaning occurred within a new emerging subculture of science, called theoretical physics, of which Max Planck was a member of the first generation and an important spokesman. For him everything humans saw, heard and felt were facts— "Tatsachen an die kein Skeptiker rütteln kann". Even socalled sensory illusions were facts and the word meant that we are yet lacking the appropriate power of reasoning to deduce the correct conclusion from that sense impression. ②But

the development of theoretical physics, up to the present, is the unification of its systems which has been obtained by a certain elimination of the anthropomorphous elements, particularly the specific senseperceptions. Seeing, however, that the sensations are acknowledged to be the startingpoint of all physical research, this deliberate departure from the fundamental premises must appear astonishing, if not paradoxical. Yet there is hardly a fact in the history of physics so obvious now as this, and, in truth, there must be undoubted advantages in such self-alienations. ③

With regard to the principle of energy conservation he was very anxious not to perpetuate the bold claim that the determination of the mechanical equivalent of heat was a proof of that principle. He argued that physicists cannot provide empirical evidence for this universal principle but only deduce its universality logically from the fact that a perpetuum mobile *does not* exist. And this is still in accordance with all human experience.

Thus perpetual motion became of as farreaching importance in physics as the philosopher's stone in chemistry, though the advantages to science were derived from the negative rather than from the positive results of the experi-

---

① On practical physics text books see for example F. Kohlrausch, *Leitfaden der praktischen Physik zunächst für das physikalische Praktikum in Göttingen*, Leipzig 1870; on teaching apparatus see J. Puluj, *Über einen Schulapparat zur Bestimmung des mechanischen Wärmeäquivalents*, in "Annalen der Physik und Chemie", XVII (1876), pp. 437-446, W. E. Ayrton, W. C. Haycraft, *Students simple Apparatus for determining the mechanical equivalent of heat*, in "Proceedings of the Physical Society", VIII (1894-1895), pp. 295-309, compare also G. J. N. Gooday, *The Morals of Energy Metering: Constructing and deconstructing the precision of the Victorian electrical engineer's ammeter and voltmeter*, in M. N. Wise, *The Values of Precision*, cit, 239-282.

② M. Planck, *Sinn und Grenzen der exakten Wissenschaft*, Leipzig 1967, p. 6.

③ M. Planck, *A Survey of Physical Theory*, New York 1960, p. 4.

ments. Today we talk of the principle of energy without any reference to the human or the technical point of view. We say that the total energy of an enclosed system of bodies is a quantity whose value cannot be increased or decreased by any reactions within the system. We no longer think of making the validity of this theorem dependent upon the refinements of methods which we possess at present to get experimental proof on the question of perpetual motion. In this generalisation, which cannot be rigidly proved, but which forces itself upon our attention, lies the above-mentioned emancipation from the anthropomorphous elements. ①

---

① Ivi, p. 6. For his (non)-treatment of the mechanical equivalent of heat see M. Planck, *Das Prinzip der Erhaltung der Energie*, Leipzig, Berlin 1924, (fifth ed.) which was written on the occasion of a prize competition set out by the Philosophical Faculty of the University of Göttingen in 1887. He received the second prize (no first prize winner at all) and the committee remarked "insbesondere hat der Verfasser eine kritische Besprechung derjenigen Experimentaluntersuchungen unterlassen, auf welchen unsere Kenntnis von dem numerischen Werte des mechanischen Wärmeäquivalents beruht" (p. XI). See also p. 148ff. where he discusses the potentials of inductive and deductice methods to establish proof. In his Treatise on Thermodynamics he certainly refers to the 19th century experiments on the determination of the mechanical equivalent of heat emphasising the variety of apparatuses used which provide a crucial condition for the principle of energy conservation. M. Planck, *Treatise on Thermodynamics*, Berlin 1897, reprint New York 1945, p. 43.

19 世纪的黄金数：一个科学事实的发展史[①]

奥托·斯巴姆[*]

"事实"（tatsache）这个词依然很新。我非常清楚地记得还没有人谈起它的时候。但是，我不知道，是谁说出了这个词或是谁第一次写下这个词……更不知道这个新词为何有悖于常理，在如此短的时间内，取得巨大的成功。我也不知道，它通过什么方式得到了如此普遍的接受，以至于你在某些书的每一页都会碰到一个事实。（Lessing，1778）

普通的原始思维（crude mind）只有两个象限（compartment），一个是真理，一个是错误；……然而，理想的科学思维（scientific mind）却有无数个象限。每一个理论和定律都在其适当的象限表明它为真的可能性。当一个新事实出现时，科学家会把它从一个象限变到另一个象限，如果可能的话，以使它总是与真理和错误保持适当的联系。（Rowland，1899）

本文旨在研究通常称之为"热功当量"的科学事实史。但什么是热功当量呢？这个"由权威的证言所证明的真理"什么时候开始被称为一个事实呢？它什么时候被誉为是科学的呢？自从人类诞生以来，构成这里研究的这个事实的经验过去和现在都广为人知：如果你摩擦双手，就会发现双手会发热。然而直到18世纪末，机械摩擦与热之间的关系，才成为强化自然哲学研究和争论的一个问题。19世纪初，自然哲学家甚至定量地表述了这种关系。由实验确立的所做的机械功与所产生（作为一个温度差来测量）的热量之比，不久就成为能量科学的基石。这个实验很快被视为是对能量守恒定理的一个证明，而且这一比值取得了自然常数的地位。最后，大约在1900年，当这一比值成为国际单位制中的

---

① 本文是对作者长期研究的一个项目的总结。这里给出的一些论证，在其他杂志刊发的几篇文章中得到了证实，我在本文适当的地方提到了这些文章。我引证的资料来自格拉斯哥大学、英国皇家学会、曼彻斯特理工大学、维也纳科学院、哥廷根国家图书馆和大学图书馆手抄本部、慕尼黑德国博物馆的手稿藏室、约翰·霍普金斯大学的艾森豪威尔图书馆和俄罗斯圣彼得堡科学院，我对他们的许可深表谢意。

* 奥托·斯巴姆（H. Otto Sibum），瑞典乌普萨拉大学科学史研究所所长；汉斯·劳辛（Hans Rausing），科学史教授。本文为首发稿。——译者注

换算因子后，物理学家开始把这个精确的事实称为那个世纪的黄金数（golden number）。本文据此叙述了 18 世纪末到 20 世纪初这段时期，即物理学的发展变化很大的一个时期。此外，既然所讨论的事实很特殊，只用数字表示，那么，这种叙述同时也是关于 19 世纪精密科学中确立精确测量的一种叙述，就并非巧合了。它关系到把定量实验确立为一种特殊的实验形式。

研究近代早期自然哲学发展史的历史学家认定，"事实"是 17 世纪的一个发明。例如，史蒂文·夏平（ST. Shapin）和西蒙·谢弗（S. Schaffer）认为，这一时期的实验室成为产生权威实验知识（事实不同于意见）的易接近的公共场所。① 罗瑞安·达斯顿（L. Daston）已经表明，感官知觉涉及特殊性，而真知依赖于对共性的认识，这两个概念之间存在的当代认识论的张力，致使自然哲学家将事实定义为脱离理论的"经验块"（nuggets of experience）。对事实的这种理解（包括"原则上不能把事实与意见、解释和理论相混淆"）的核心特征，一直持续地存在着，而且，在这里所考察的这段时期内仍然很流行。② 数值事实（numerical fact）可以被看成是这种纯化实验形式的极佳候选者。但本文将表明，科学共同体在 19 世纪建立的过程中是变化莫测的，这个经验块总是这个变化的科学共同体的动作性知识的一个组成部分。

下面几部分根据这一事实经历的意义变化来编排。第一部分简要考虑 18 世纪下半叶，当时，既没有科学也没有被称为热功当量的事实。相反，我将把关于热和功的经验，特别是根据数值事实来表示这些经验的那些实践，置于自然知识的发展传统中。历史学家同意，这一时期见证了定量研究的兴起，尽管他们的解释是可变的。此外，自然哲学的研究方式大约在 1800 年发生了变化：做实验越来越成为一种私人努力，因为高度敏感的测量杜绝目击者在场。这种数值事实的兴起与对感官技术（强调视知觉）的疏远相伴而行。定量实验不仅改变了自然哲学家的注意力，而且必须在这一时期的大变革中找到其位置。

第二部分简要描述了詹姆斯·焦耳（J. Joule）于 19 世纪 40 年代在曼彻斯特

---

① ST. Shapin, S. Schaffer, *Leviathan and the Air-Pump. Hobbes, Boyle, and the Experimental Life.* Princeton, 1985, p. 39. 关于"事实性知识"和真理，参见 ST. Shapin, *A Social History of Truth. Civility and Science in SeventeenthCentury England.* Chicago, 1994。另参见 L. Fleck, *Genesis and Development of a Scientific Fact.* Chicago, London, 1979; B. Latour, S. Woolgar, *Laboratory Life: The Construction of Scientific Facts.* Princeton, N. J., 1986.

② L. Daston, *Why are Facts Short?*, in this volume; L. Daston, "Baconian Facts, Academic Civility, and the Prehistory of Objectivity", *Annals of Scholarship*, VIII, nos. 3-4 (1991), pp. 337-364. 关于这一时期对经验与实验的讨论，参见 P. Dear, *Discipline and Experience: The Mathematical Way in the Scientific Revolution.* Chicago, 1995。关于涉及的将事实历史化的进路，参见 M. Poovey, *A History of the Modern Fact. Problems of Knowledge in the Sciences of Wealth and Society.* Chicago, London, 1998, pp. 1-28.

完成的测量热功当量的经验知识。他关于热的本性的知识与所做的实验深深地纠缠在一起，这些实验只可能出现在貌似无关，实则局部有用的那些知识传统的交汇处。但在《哲学学报》（*Philosophical Transactions*）发表的那个实验报告并不代表作者的知识，它只表明，确立了与任何理论假设都截然不同的一个精确事实。他的实验知识在整个发表过程中的这种转变，体现了皇家学会确定的自然哲学家与试图成为绅士专家（gentlemen specialists）的仪器制造者、技工和实验者这一不断壮大的集体之间的权力关系。

第三部分追述了公布的这个事实据此取得了证明能量守恒的地位，以及它如何成为优先权之争的焦点的历史经过。与同时发现的传统叙述相反，我将主张，在当时的学术界，尽管实验知识的社会地位和认识论地位非常脆弱，但它仍旧引发关于著作权和知识产权的争议，并有助于建构同时发现的观点。在这个过程中，焦耳与汤姆孙（即后来的开尔文勋爵）的合作变得十分重要。

第四部分略述了这一事实的意义的进一步转化，在这个关键时刻，它被认为是最重要的自然常数之一。尽管英国的能量科学和相关的测量的绝对单位制取得了进一步发展，但是，还是产生了对这个数字的可能的存在性和正确值的一些抨击，致使精密科学陷入危机。工程师为这个自然常数提供的证据，成为解决这些争论的策略之一。美国工程师兼物理学家亨利·罗兰（H. A. Rowland）把这作为他在约翰·霍普金斯大学物理实验室的首要任务。他在很短的时间内就解决了争论，但是，他把建立物理实验室作为近代教育的一种手段的努力，造成了精密科学与人文科学之间的分裂。这么做的效果恰好是把事实与意见分离开来。

在结尾部分，我将简要表明，这一精确事实最后如何具体化为国际单位制中的一个换算因子。1894 年 7 月 24 日，美国参众两院的议员甚至把这个常数合法化为功的单位。此后，这一科学事实的意义不再局限于实验室，而变成了 19 世纪的黄金数。尽管它的重要性完全超越了科学共同体，但是，热功当量的意义再次发生了改变。不断壮大的物理学界已经确立了新的劳动分工，理论物理学家对热功当量的测定实验可能为能量守恒的普遍原理提供证明充满怀疑。

一、改变注意力："论研究自然界必需的技能"

有许多东西，或许不能精确测量，只能是差不多。味觉、嗅觉、冷热的感觉……（Miles, 1748）

永恒运动的建构是绝对不可能的：即使摩擦和介质的阻力从长远来看也不会摧毁驱动力……这样的研究具有花钱多的缺点，它已经毁掉了不止一个派系（famille），而且，贡献很大的技工经常消耗他们的财富、时间和天赋。（L'Academie Royale des Sciences，1775）

1753 年 5 月 15 日，法国实验哲学家阿贝·诺莱特（A. Nollet）发表了他的就职演讲"论研究自然界必需的技能"①，这个演讲很好地抓住了作为体验自然和注意力的一种特殊方式的定量实验所经历的过渡期。我将把这个演讲作为我研究 18 世纪末和 19 世纪兴起的精密文化和把数值事实确立为经验的主导形式的出发点。这里不详述这个重要文献，我喜欢强调，诺莱特对做实验所要求的那种注意力、在这种努力过程中精确测量所起的作用，以及实验的自然哲学在"文人共和国"（Republic of Letters）② 中的地位进行的反思。这些方面对于实验自然哲学家实践的改变和相关的证据的认识论意义是至关重要的③。关于实验者的注意力，他写道：

> 试图发现自然踪迹的人，不论他多么勤奋或努力，即使忽略最微不足道的印象，也不能对所探索的东西作出全面沉思……时间、地点、当时的空气状况、大小、持续时间、形状、颜色、气味，甚至感官可接近的所有其他特性，全都是人们不仅应该注意而且应该报告的情况，明显多余的情况除外……人们因此看到，高度关注本身致使对（事物）作出更好的理解，而且由于这个原因，观察者只有从所有的视角都最细致地观察了对象，并（考虑了）它的内部隐藏着什么及它的环境是什么时，才应该在他的研究中放开这个对象。④

对于诺莱特来说，提高关注度必然会扩大知识范围是显而易见的。但正如他所坚持认为的那样，这种观察模式应该不同于自然史的实践，后者是完善"我们的财富清单"的一种探究形式，但并未研究"自然事件发生的原因"。然而，两者彼此密切地联系在一起：

> 因为努力研究自然界又不了解其历史的人，的确是在随意谈论他一无所知的事情；但一个人如果只知道自然史，他就只配在那些只用记忆

---

① J. A. Nollet, *Discours sur les dispositions et sur les qualités qu'il faut avoir pour faire du progrès dans l'étude de la physique expérimentale*. Paris 1753. German transl. Nollets *Rede von der nothigen Geschicklichkeit zur Erforschung der Natur, welche er den 15. Mai 1753 bei dem Antritte seines öffentlichen Lehramtes, in dem Navarrischen Collegio gehalten*. Erfurt 1755.

② "文人共和国"不是一个国家，是启蒙时代欧洲和北美的文人学者通过书信与印刷品联合而成的一个松散的具有共同兴趣的研究群体。——校者注

③ 关于 18 世纪末和 19 世纪精密科学中实验者作为某种角色出现、伴随的训练注意力的形式和建立这种非书面的知识传统引发的认识论问题的详细的描述，参见我的论文 "Experimentalists in the Republic of Letters, to appear", in L. Daston and H. O. Sibum (eds.) "Scientific Personae", *Science in Context*. forthcoming.

④ Nollet. *Rede von der nothigen Geschicklichkeit. . .*, cit., pp. 37/38.

力的自然哲学中占有一席之地。因此，从事实验物理学完全是在研究自然界，它不仅涉及自然界的效应，而且同样涉及（研究）用来产生（自然界的）效应之工具的意图；简言之，这意味着研究（自然界）做什么，以便能够说出自然界是如何做的。[①]

根据诺莱特的观点，博物学家（naturalists）识别"构成世界的那些物体及拥有爱好者的每一个物体，对它们进行分类，注意有时逐渐发生的改变，详述所有这些物体的属性及其相互关系"。相比之下，实验物理学家应该是根据"洞察力"激发他对自然过程的适当注意。诺莱特的演讲举了许多例子，表达了当"物理实验"置身于思辨和重视文本活动的传统学术与产生并收集世界的材料的活动之间时所面临的困难。

测量是由实验自然哲学家提出的又一个烦恼。尽管对兴起的精确测量持怀疑态度，诺莱特仍然在他的实验哲学中为"测量的艺术"留了一点点空间。但是，最终他看到了在同时成为一名优秀的测量者和一名有技能的自然哲学家时的矛盾。他评论说："逐渐喜欢测量的艺术，对自然哲学家来说，是很危险的。"对于他来说，极度精确的趋势和用数学符号表征自然界是危险的发展：

> 有多少呢？在任何情况下，不想关心常识的那些人，不应该采取迂回路线，但尽管如此，他们在用代数符号表示事物时感到很满足，而这些事物本来能在不降低其价值的前提下，用一般可接近的方式来表示。这些作品，当（以这种方式）被狡猾地夸大了时，是怪诞的，它们足以清楚地表明，它们包含的少许自然科学只是（提出）另一种科学的一个借口，是以这样一种方式来使自身变得更加重要吗？[②]

在他的总结评论中，诺莱特定义了实验自然哲学家的角色，自然哲学家应该——在某种程度上遵循法国工程师的模式——应用他的技能与见识从事有用的项目，例如，研究如何使水能被饮用，如何使指南针能被安全使用，如何防止蔬菜腐烂，等等。对他而言，这类自然哲学家比那些"设想他们通过好奇，更确切地说，完全异想天开地挑选材料，使我们感到惊奇的傲慢学者"更有价值。[③]

诺莱特型的实验自然哲学家已经对这些已确立的知识传统提出了挑战，但是以库仑、拉瓦锡、拉普拉斯等人为代表的下一代科学家所进行的定量实验，恰好是更大的挑战，因为它需要"文人共和国"的成员所陌生的感官经济（economy of the senses）和道德经济（moral economy）。大约在 1800 年，精密仪器被用在各

---

① Ivi，p. 12.

② Ivi，p. 58.

③ Ivi，p. 60.

种各样的文化生产领域，但关键用途是为了控制和监管社会，而不是——像人们预期的那样——实现数学家统管自然哲学各门学科的抱负。[①] 关于实验室的实验和长途旅行体验的叙述越来越表明，数字已经融入到他们的书面描述之中，标出热度、大气压等。但为了理解这些数值，需要确立标准仪器的某种体制才能共享这种新的经验。通常，这样的交流网络并不存在，温度计就是一个有力的例子。18 世纪中叶的温度计可能多达 18 种不同的刻度。研究自然界的每一个人似乎都制作了他自己的合理刻度，而且，远没有达成测量共识。[②] 此外，可以看到，新的仪器使得精确测量时的灵敏度不断提高，这被认为比人的感官更可靠。但是，在做实验时，这种灵敏度需要提炼出准确的姿势和采取更多的防范措施。还有，通常试验时绅士般的在场行为经常会妨碍精细测量。[③] 实验从向公众的公开展示变成了没有任何观众的行为。在实验室内外，不得不发展学术生活形式中不存在的新的注意方式和交流方式。[④] 著作的权威性和学者之间的通信与通常面对面的互动不再适用，或者，不足以为这些数值事实提供证据。[⑤] 这些数值很容易从实验室传播开来，影响到不同的文化领域，然而，研究数值事实的新实践使信任人的传统习惯同时转向了信任程序。从事定量实验的人甚至开始使自己远离感官经验——尽管在他们做实验时仍然私下践行感官经验——以此来证实科学的

---

① 例如，参见 Thomas S. Kuhn's "The Function of Measurement in Modern Physical Sciences", in T. S. Kuhn, *The Essential Tension. Selected Studies in Scientific Tradition and Change*. Chicago, 1977, pp. 178-224, 220。在这篇文章中，他把 1800 年到 1850 年这个时期描述为培根科学的数学化。关于相反的解释，即使混合数学（mixed mathematics）对他们的研究很有用的自然哲学家的解释，参见 J. Heilbron, "A Mathematicians' Mutiny, with Morals", in P. Horwich（ed.）*World Changes：Thomas Kuhn and the Nature of Science*. Cambridge, Mass. 1993, pp. 81-129。关于对库恩解释的另一种歪曲，参见 I. Hacking, *The Taming of Chance*. Cambridge 1990, pp. 60-93。关于新兴的精密文化的解释，参见 M. N. Wise（ed.）*The Values of Precision*. Princeton, 1995；M. N. Bourguet, C. H. Licoppe, H. O. Sibum（eds.）*Instruments, Travel and Science：Itineraries of Precision from 17th to 20th Century*. forthcoming.

② 这种温度计仍保存在乌特勒支的大学博物馆；比较 W. E. Knowles Middleton, *A History of the Thermometer and its Use in Meteorology*, Baltimore, 1966.

③ 例如，参见库仑的转矩平衡实验。C. H. Blondel, M. Dorries（eds），*Restaging Coulomb：usages, controverses et réplications autour de la balance de torsion*. Florence, 1994.

④ 关于一个有说服力的解释，参见 M. N. Bourguet, "Landscape with Numbers—Natural History, Travel and Instruments（Mid-18th‐Early 19th Century）", in M. N. Bourguet, C. H. Licoppe, H. O. Sibum（eds.）*Instruments, Travel and Science*, cit.

⑤ 这一时期，作为确立证据手段的目击的重要性和图书的权威性，参见 I. Hacking, *The Emergence of Probability：A Philosophical Study of Early Ideas about Probability, Induction and Statistical Inference*. Cambridge, 1975, pp. 33/34；S. Schaffer, "Self Evidence" in *Critical Inaffquiry*, XVIII（Winter 1992）, pp. 327-362.

去身体化（*disembodiment*），并强调视知觉和自动记录仪。① 定量实验不得不在学术生活形式中找到自身的位置，而且，两者会相互同化。这需要一种前所未有的专家文化。②

对热与功及其相互关系的研究必须在这种语境中加以审视。这里有必要提及两个方面。第一，历史学家经常忽略这样的事实：我们关于热的大部分科学知识是自温度计出现以后才学到的。甚至可以说，对热的研究史在很大程度上就是测温技术史，测温技术在18世纪下半叶成为一个突出的研究领域。第二，18世纪的自然哲学和工程领域内的功的测量和概念化同样值得重新考虑。启蒙哲学家提出各种各样的策略理解功的领域。法国军事工程师库仑和拉瓦锡开发了衡量和评价人力及其效率的精确技术，潜在地假定，动物机体是一个自动调节系统。自动机器——在蒸汽机技术和自动机中是如此常见——在关于力量与文明的启蒙推理中变成了一种理想。③ 记录装置（inscription devices），比如，詹姆斯·瓦特的示功图（即用图表显示出蒸汽机所做的功），还有莫兰（Morin）和彭赛利（Poncelet）的自动记录

---

① 谢弗认为："我们看到了从公开表演仪式向象征去身体的（disembodied）科学天赋的一种转变；我们也看到了对自动记录的材料装置和仪表的新的坚持……" S. Schaffer, Self Evidence, cit. , p. 330；拉瓦锡及其远离感官技术的实践，参见 L. Roberts, "The Death of the Sensuous Chemist: The 'New' Chemistry and the Transformation of Sensuous Technology", in *Studies in History and Philosophy of Science*, XXVI, 4 (1995), pp. 503-529。关于电工对感官经验的远离，参见 H. O. Sibum, "Charles-Augustin Coulomb (1736-1806)", in K. Von Meyenn (ed.) *Die Grossen Physiker. Erster Band. Von Aristoteles bis Kelvin. München*, 1997, pp. 243-262；关于形成实验自然哲学的杰出贡献，参见 C. H. Licoppe, *La Naissance de la pratique scientifique: Le discours de l'experience en France et en Angleterre* (1630-1800). Paris, 1996.

② 在欧洲各国，精确测量得到的值无疑是不同的。例如，在18世纪的英国，像多朗德（J. Dollond）和拉姆斯登（I. Ramsden）那样的精密测量仪器制造者成为皇家学会的成员。但是，相信杰出的仪器制造者对精确测量的研究有其自身价值的风气，在19世纪前半叶发生了改变。在那个时期，几乎没有仪器制造者能达到这个地位。W. T. Ginn, *Philosophers and Artisans: The Relationship between Men of Science and Instrument Makers in London* 1820-1860. Ph. D. University of Kent at Canterbury, 1991；关于进一步的比较研究特别参见 M. N. Wise, *The Values of Precision*, cit.

③ 但这些研究，甚至狄德罗的百科全书计划，在从工匠的工作中提炼体型（embodied）知识时，都不完全成功。有关自动机的作用，参见 S. Schaffer, "Enlightened Automata", in W. Clark, J. Golinskei, S. Schaffer (eds.) *The Sciences in Enlightened Europe*. Chicago, 1999, pp. 126-165；有关库仑研究人在劳动时感到疲劳的方案，参见 C. A. Coulomb, "Résultats de plusieurs expériences destinées a déterminer la quantité d'action que les hommes peuvent fournir par leur travail journalier, suivante les differentes manières dont ils emploient leur forces", in *Mémoire de l'Institut National des Sciences et Arts —Sciences mathematiques et physiques*, Paris, 1799, pp. 380-428；H. Otto Sibum, *Charles Augustin Coulomb* (1736-1806), cit. 百科全书计划参见 J. R. Pannabecker, "Representing Mechanical Arts in Diderot's Encyclopédie", *Technology and Culture*, XXXIX (1998), pp. 33-73；ST. L. Kaplan, C. J. Koepp (eds.) *Work in France. Representations*, *Meaning*, *Organization*, *and Practice*. Ithaca, London, 1986；M. N. Wise, "Work and Waste I: Political Economy and Natural Philosophy in Nineteenth-Century Britain", in *History of Science*, 27 (1989), pp. 263-301。M. N. Wise, "Mediating Machines", in *Science in Context*, 2 (1988), pp. 77-113，波尼（Carlo Poni）的文章在这本文集中。

摩擦的小车，标志着在正在兴起的专家文化中所实践的一种新型客观性的开始，而且，用莫兰自己的话来说：这种客观性强行实施了科学的去身体化的过程。

　　仪器的指示一定是以与观察者的注意力、意志和偏见无关的方式得到的，因而是由仪器本身凭借实验后存在的轨迹或最终材料提供的。①

在这种视觉文化中，那些用机器进行的动力计测量成为"机械潮流"，因此这甚至强化了人们不能无故得到有用功的新兴看法。此后，法国科学院决定不再接受"永动机"的建议，依据这些建议，永动机已经毁掉了很多天才的实验者及其群体。但这里重要的是指出，在这些自动记录的测量技术中，摩擦只不过是其本性还没有被研究过的可测量的量。② 而且，那些研究者，比如，利用摩擦生热实验的朗福德（C. Rumford），把无火加热水的技术转变成一种有销路的商品。③ 这种知识因此保留在机械工艺领域。甚至从自然哲学的观点来看，对此有兴趣的那些人，比如，慕尼黑的朗福德和英国的哈利·卡文迪什（H. Cavendish），也没有对占统治地位的热质说提出挑战。卡文迪什在一篇手稿中甚至提出用一个实验"来决定，已知一个物体，通过在阳光下和阴暗处交替地用温度计测量，给出已知的可感热量所需要的活力"，但他并没有把热功关系的这样一种定量测量看成是表达了值得"在科学中占有适当位置"④ 这样一个事实。

---

　　① A. Morin, *Notice sur divers appareils dynamométrique, propres à mesurer l'effort du travail développé par les moteurs animés ou inanimés, ou consommé par des machines de rotation, et sur un nouvel indicateur de la pression dans les cylindres des machines a vapeur*, Paris 1839, pp. 29-30; quoted after R. M. BRAIN, *The Graphic Method. Inscription, Visualisation, and Measurement in Nineteenth-Century Science and Culture*, Dissertation University of California. Los Angeles, 1996. pp. 48-163. 关于客观性的历史和 19 世纪自我克制的道德规范，参见 L. Daston, P. Galison, "The Image of Objectivity", in *Representations*, 40（1992），pp. 81-128, 117ff.

　　② 这位法国工程师—学问家所做的对功的测量，以及詹姆斯·瓦特的第一台用图表示功的自动记录仪的图示可参见 Brain, The Graphic Method, opus cit.; M. N. Wise, *Visualising Work*, 未发表的手稿。

　　③ 朗德福伯爵在他通过机械摩擦产生令人厌烦的热的标准实践中，最著名的评论和结论是，热是一种运动。但他并未提出热的分子运动论，因为他在其他实验中表明，液体没有传导这些运动的特性。S. T. L. Wolff, "Benjamin Thomson, Sadi Carnot und Rudoph Clausius", in Meyenn, *Physiker cit.*, pp. 289-302, 291. B. Rumford, "An Inquiry concerning the Source of the Heat which is excited by Friction", in *Philosophical Transactions of the Royal Society of London*, LXXXVIII（1798），pp. 80-102. 正如 1856 年莫里茨·赫尔曼·雅可比在法国主张的那样，博蒙特（MM. Beaumont）依据朗福德的实验建造了一台仪器，这一装置在许多国家获得专利，借助于这个仪器，在两个半大气压下产生蒸汽，注定同时会用于军事考察，"在那里，总是有大量强健的部队，但时常极度缺乏生存手段，其中取暖是第一位的"，M. H. Jacobi, *Sur la correlation des forces de la nature*，自 1856 年以来未发表的讲稿，St Petersburg Academy of Sciences archive, Russia.

　　④ 卡文迪什从 18 世纪 80 年代开始绘制有关"热"的草图，包含这样的实验的建议位于手稿部的参考编号 MG23, L6 处，Pre-Confederation Archives, Public Archives of Canada, Ottawa. 关于这一草图的详细讨论，参见 C. H. Jungnickel, R. McCormmack, *Cavendish. The Experimental Life*. Cranbury, NJ, 1999, pp. 400-423, 410。在 19 世纪初，正是法国卡诺测定了热功当量，但也没有发表。U. Hoyer, *Über den Zusammenhang der Carnotschen Theorie mit der Thermodynamik*, "*Archive for the History of the Exact Science*". 1974, pp. 359-375.

二、焦耳的事实及其潜在含义

事实是必须被感觉到的东西，无法从对它们的任何描述中获得。
（Maxwell，1869）

1850 年，英国皇家学会发表了焦耳的一篇文章《论热功当量》。这篇文章是使焦耳获得"科学绅士"称号的一项主要成就。今天，这篇文章被看成是英国能量科学发展的一个重要里程碑，因为该文中陈述的事实，在 19 世纪下半叶成为热力学的基石。[①] 在焦耳详尽的长篇文章中，给出了下面的结论：

我认为，这篇文章中的实验证实了这一事实，因此断定：

第一，物体（不论固体或液体）摩擦产生的热量总是与所消耗的力的大小成正比。

第二，能使 1 磅水（在真空中称量，其温度在 55～60℃）的温度升高 1 华氏度需要的热量，就其发展来说，要求消耗 772 磅下落 1 英尺[②]所表示的机械功。[③]

然而，对保存在皇家学会的原手稿的仔细检查却表明，所发表的结论并非焦耳最初的设想。在原版本中他声明他证实了：

就这篇文章包含的实验而论，

第一，物体（不论固体或液体）摩擦产生的热量总是与所消耗的力的大小成正比……

第二，摩擦在于把力转化为热。

我认为，既根据实验次数，又因为仪器的热容量很大，所以，779.692（即水摩擦产生的当量）是最正确的。而且，既然流体摩擦时不可能完全避免振动和轻微的声响，所以，我宁愿取整数作为研究结果。

能使 1 磅水（真空中的重量，温度确定在 55～60℃）的温度升高 1 华氏度所需要的热量，相当于由 722 磅的压力所表示的机械力移动

---

① 关于科学的绅士，参见 J. Morell and A. Thackray, *Gentlemen of Science. Early Years of the British Association for the Advancement of Science.* Oxford 1981。有关焦耳在形成能量科学时所起的作用，参见 D. S. L. Cardwell, *James Joule, A Biography.* Manchester, 1989；C. Smith, *The Science of Energy. A Cultural History of Energy Physics in Victorian Britain.* London, 1998；H. O. Sibum, "An Old Hand in a New System", in J. P. Gaudielliere, I. Loewy, *Manufactures and the Production of Scientific Knowledge.* Houndmills, 1998.

② 1 磅＝0.4536 千克；1 英尺＝0.3048 米。——译者注

③ J. P. Joule, "On the Mechanical Equivalent of Heat", *Philosophical Transactions of the Royal Society*, CXL (1850), pp. 61-82, printed also in Joule's Scientific Papers, I, London 1884-1887 [henceforth SPJ], pp. 298-328.

1 英尺的距离。①

皇家学会考虑发表这篇文章的唯一条件是，完全取消第二个结论："摩擦在于把力转化为热。"最终，焦耳为了在《哲学学报》上发表这篇文章，接受了这些彻底的更改，但是，他在同年给剑桥的唐·乔治·加布里埃尔·斯托克斯（don George Gabriel Stokes）的信中写道：

> 我请求你们接受随信附上的这篇论文，在这篇文章中，我努力精确地确定了热功当量。我设想，我所得到的结果是，摩擦在于把力转化为热；但皇家学会的成员不赞成这样一个实验推论，因此我考虑，最好取消这个推论，尽管我认为，人们最终将发现这种观点是正确的。②

为了与学报的有名望的读者们所理解的科学的精神气质相匹配，科学的酿酒师（scientific brewer）对焦耳的论文作了精心的处理。③ 但对于焦耳来说，这个用数字表示的事实的核心意义在于热与功之间的相互转化性，而不仅仅是自然力之比。这就引发了关于下列诸方面的一些问题：焦耳经历的权力关系改变了他的知识主张（knowledge claim），以及与焦耳的事实实际相关的潜在意义的本性。论文审阅人无法接受他基于热本性的微观观点提出的相互转化性的主张。从他们的视角来看，在科学中，只有机械力与热之间的比值被证明是可测量的，因而是可接受的。皇家学会确信，让焦耳修改了他的文章，就使得他的实验研究摆脱了违反流行的物理学准则和损坏他们所认为的确切事实的那些知识主张，成为支持热的热质性的流行理解的重要的理论结果。所修改的论文的目标证实了，在维多利亚时代的科学中，在如何把科学事实确定为经验证据方面，热功当量的确定被公认为是一项杰作。但对焦耳来说，从他的实验中推断出的知识既不是假设的，也不是猜测的；这一确切事实是近十年来探究热本性的实验结果。他关于"摩擦在于把力转化为热"的地方性知识，在曼彻斯特的不同知识传统的交集中显示出来。我在其他地方较详细地阐述了对焦耳在 18 世纪 30 年代末和 40 年代的研究工作迄今仍忽略的那些维度。

我在本文中将简要提到有助于证实下列观点的那些结果：焦耳的事实尽管是地方性的，但却是关于热的动态本质这一确切知识的内在部分。曼彻斯特的研究者们已成功揭示了隐藏的自然界的运行机制，在这个小集体中，焦耳的事实知识已经得到了确认，但其创造者，作为自然哲学家，还没有获到可信性。为了看到

---

① 焦耳的手稿 "On the Mechanical Equivalent of Heat"（1850）保存于英国皇家学会档案馆 PT. 37. 3。另参见 C. Smith, "Faraday as Referee of Joule's Royal Society Paper On the Mechanical Equivalent of Heat", in *Isis*, LXVII (1976), pp. 444-449.

② Joule to Stokes, July 3rd, 1850, CUL, Add. 7656 J75.

③ 关于这个过程的详细描述，参见 H. O. Sibum, "Reworking the Mechanical Value of Heat: Instruments of Precision and Gestures of Accuracy in Early Victorian England", in *Studies in History and Philosophy of Science*, XXVI (1995), pp. 73-106.

焦耳事实那些隐藏的维度，让我们简要回顾当时人们关于热和摩擦的立场。直到19 世纪 40 年代，大多数学者仍把热看成是可称之为热质的非物质实体（immaterial substance）。但在英格兰，提倡新政（new regime）的自然哲学家开始采纳法国数学，在法国数学中，热取得了量化术语的地位，但并不要求说明其本性。然而，通常并不熟悉彼此工作的实践者，提供越来越多的证据质疑这种高深的热质说。一方面，就摩擦而言，工程师和蒸汽机的用户用这种现象作为对机械过程中功损失的一种量度。另一方面，一些思辨哲学家仍然把他们的机械哲学书中的摩擦视为应该远离理论的一种平凡的、可忽略的效应。而且，正如克罗斯比·史密斯（Crosbie Smith）已经表明的那样，焦耳恰好应用了工程师的"经济责任"的概念，这被理解为"1 磅煤的力量（agency）能使几磅重的物体升高 1 英尺"，就像他用数字表示机械力和校准机械性能的标准一样。

作为曼彻斯特最富裕的酿酒师的儿子，焦耳在曼彻斯特探索和参与了不同工作。就此而言，他有条件选择他认为对自己最重要的研究主题和使他成为一名科学绅士的研究计划。生活在这种工业环境中，看到这位 19 岁有钱的资产阶级酿酒师对自然界中的各种自动力（self-moving forces）的实验，以及通过自动机器来模仿这些自然力，并不令人惊讶。当时，人们认为电力和磁力是最强大的两种自然力，而且，最初焦耳的研究集中于各种电磁发动机。一开始只是检验其经济上的可行性——逐渐认识到电磁发动机不能成为永动机。后来，这种机器逐渐成为研究各种自然力的转化过程的一种工具。

为了寻求焦耳对自然界进行定量研究的动机，人们也许很想采纳一些历史学家的解释，这些解释把焦耳的工作与工程师的工作场所联系起来。不过，焦耳是酿酒文化的积极分子，酿酒文化本身对精确测量有自己的独特要求，并且，为定量地表达各种现象及其相互关系留有余地。酿酒场所，其所有运作都受国家财政部门，尤其是税收制的控制，正是在酒厂里，酿酒师们学会了交流生产者和政府官员各自采纳的精确数据。这是学会如何"相信数字"的训练场。此外，酿酒业在 18 世纪 30 年代经历了重大变革，生产扩大到产业化规模。正如科学的酿酒师们所认为的那样，旧体制下的酿酒师提供的是"相对事实"，他们的意思是说，酿酒实践与当地条件有关，因而这一经验很难被复制到其他地方。① 作为这种新的经济形势的结果，科学的酿酒师们使温度计和十进制表成为必不可少的实

---

① 在不同条件下酿酒的再生产时常出现很大困难，即使是同一个酿酒师酿造，也是如此，因为每个场所都有其特殊性，不同的酿造器具、不同的材料等。但即使条件相同，每次酿造也都是独一无二的事件，因为谷物、麦芽和温度条件总是不同的，因此，每次酿造需要特别注意和调整措施。在通过精确测量将酿酒转变为一门科学的过程中，一些酿酒师认为，不存在标准的热量，它们只是由于国产税务局而存在。R. Shannon, *A Practical Treatise on Brewing, Distilling and Rectifying*, London, 1805, p. 57.

践工具和理论工具。税务局同样把仪器测量和绝对标准的确立看成是成功建立其税收体制的唯一途径：把精确测量看成是确立证据的手段。已知支持英国实力的税务局的历史很悠久，难怪 1842 年英国政府就设立了"税务研究室"，这成为维多利亚初期的一个精确测量机构。① 那里做的工作有助于使不可信的知识形式成为老酿酒师的感知秩序的一部分，相比之下，这种感知秩序被判断为是不可靠的。② 日常的酿酒经历不仅教人相信数字，还包括观察自然现象的持续变化与变形：可见的谷物、啤酒花、麦芽、麦芽汁和麦芽浆不断地生成、变形和腐烂。在那里，热是"自然界的主要工具"，但它也是一种转瞬即逝的、富有挑战的工具，并且，正如许多老酿酒师仍然认为的那样，热是无法计量的。③ 焦耳对新的酿酒体系（其中的关键元素是温度计、蒸汽机技术和测量）的极力支持，无疑促使他实践和相信用数字表示经验的重要性。焦耳在探索电动马达时，最终认识到：

> 因此在电磁方面，我们拥有一种能通过简单机构手段产生或毁灭热的动力。我在本文的后面部分，将试图用绝对的数值关系把热量与机械力联系起来。④

焦耳的电磁实验已表明，热的物质性概念不再成立，因为热能通过机械手段得以产生或消灭。⑤ 它还进一步表明，热与机械力之间存在着一种密切的联系。

---

① J. Bateman, *The Excise Officer's Manual Being a Practical Introduction to the Business of Charging and Collecting the Duties under the Management of Her Majesty's Commissioners of Inland Revenue*. London, 1852；有关军队财政状况与精确测量，参见 J. Brewer, *The Sinews of Power. War, Money and the English State*, 1688—1783. London, 1994；S. Schaffer, "Golden Means: Assay Instruments and the Geography of Precision in the Guinea Trade", in M. N. Bourguet, C. H. Locoppe, H. O. Sibum, *Instruments, Travel and Science*, cit.；W. J. Ashworth, *Between the "Trader and the Public"*: *defining productionand measures in eighteenth-century Britain*, 未发表的手稿。关于"信任数字"的看法，参见 T. Porter, *Trust in Numbers: the Pursuit of Objectivity in Science and Public Life*. Princeton, New Jersey, 1995.

② 关于税务研究室，参见 P. W. Hammond, H. Egan, *Weighed in the Balance. A History of the Laboratory of the Government Chemist*. London, 1992；H. O. Sobum, *Les Gestes de la Mesure. Joule, les pratiques de la brasserie et la science*, in 'Annales Histoire, Science Sociales', IV-V (1998), pp. 745-774.

③ 1865 年，焦耳在格陵诺克纪念詹姆斯·瓦特诞生时发表的演讲中使用了"自然界的主要工具"的术语，引自 C. H. A. Parsons, "The Rise of Motive Power and the Work of Joule. Second Joule Memorial Lecture", in *Memoirs and Proceeding of the Manchester Literary and Philosophical Society*, LXVII (1922/23), pp. 17-29, 22. 关于酿酒师的热的领域，参见 G. A. Wigney, *An elementary Dictionary, or, Cyclopaediae for the use of Maltsters, Brewers, Distillers, Rectifiers, Vinegar Manufacturers and others*. Brighton, 1838, H. O. Sibum, Les Gestes, cit.

④ J. P. Joule, *On the Caloric Effects of Magneto-Electricity, and on the Mechanical Value of Heat*, SPJ, pp. 123-159, p. 146, Italics by Joule.

⑤ 这里重要的是提到，热质说很快就受到不同研究者的攻击，参见 S. T. G. Brush, "The Wave Theory of Heat. A Forgotten Stage in the Transition from Caloric Theory to Thermodynamics", *The British Journal for the History of Science*, V, no. 18 (1970), pp. 143-167, 147.

现在，当电磁机在微观世界与宏观世界之间、各种电－化学过程与机械力的世界之间转换时，它就充当了呈现各种效应的技术。关键是，在他的实验过程中，这种无意的转换，不仅为焦耳反对日趋衰落的热质说提供了一个经验证据，而且诱使他进行了一种重要的概括，也就是，热与机械功之间存在着一种绝对的数值关系，这最终成为焦耳测定热功当量的一系列著名实验的开端。在这一点上，焦耳意识到，当前关于热的本质与摩擦的冲突观点能够得以解决，而且，他有足够的证据，公开抨击热质说。焦耳坚信，一种绝对的测量展现出，有一种隐藏着的关系支配着可见世界的可变现象，他的这种信念正好与他的宗教信仰合拍。①

但为了确立这种绝对测量，焦耳不得不证明，这一数值事实的存在与实验材料和程序无关。在这种努力中，成问题的主题是热，或更准确地说，是测量热的那些实践。正如有人建议的那样，焦耳并不是简单地把计量应用于关于热的研究领域。在英国建立精密科学的文化进程中，焦耳是一位桥梁式的重要人物（bridge figure）。由于他的工作，这种精密测量的实验文化发展成为有自身特性的一种艺术。② 为了进一步为他的动态的热概念提供经验证据，焦耳不得不改进测温技术本身，因为他坚信，几乎每一个关于热的科学主张，都与不可靠的温度计的各种测量相关。

焦耳除了有长期作为一名科学的酿酒师的经历之外，还可利用的另一种文化资源是曼彻斯特的光学仪器制造者约翰·本杰明·丹瑟（J. B. Dancer），丹瑟是英国维多利亚初期新兴的视觉文化大师。他的第一个重大突破是生产出具有前所未有分辨率的消色差显微镜。随着摄影技术的发明，丹瑟很快成功地将实物的正常摄影图像缩小到只能通过显微镜观察才可被读取的微缩形式。对于丹瑟来说，这种比例转换技术是在由天文学家和微视物理学家已经开创的新的视觉文化

① "一切可能看起来很复杂并且陷入了原因、结果、转化和安排的几乎是无休止变化的明显的混淆和混乱之中，然而，最完美的规律性被保存下来——整个存在受到至高无上的上帝旨意的支配。" J. P. Joule, *On Matter, Living Force and Heat. A Lecture at St Ann's Church Reading-Room*, *SPJ*, pp. 265-276, p. 273；关于焦耳的宗教背景，参见 D. S. L. Cardwell, *James Joule*, *A Biography*. Manchester, 1989, p. 47 and pp. 10-13；史密斯建议说，焦耳的观点与他对终身追求社会秩序和稳定的保守分子的自我描述相一致，因此，与查尔姆（Chalmer）的新教派的幻想形成对比，与约翰·普莱费尔（John Playfairs）在那个世纪初的自然系统相协调。C. Smith, *The Science of Energy. A Cultural History of Energy Physics in Victorian Britain*. London, 1998, p. 72.

② 有关焦耳作为一个桥梁式的重要人物的文献，参见 D. S. L. Cardwell, James Joule. , cit. and H. O. Sibum, "Narrating by Numbers. Keeping an Account of Early Nineteenth Century Laboratory Experiences", in L. Holmes, J. Renn, H. J. Rheinberger (eds.), *Reworking the Bench. Research Notebooks in the History of Science*, forthcoming.

中提供可靠的定向手段这一更大计划的一部分。[①]

从 1844 年开始，焦耳在丹瑟的工作室花了几个上午的时间来讨论测量热的问题和灵敏的新的水银温度计的设计。这种极其灵敏的器件起到了像丹瑟的显微镜一样的作用，能够显示出潜在的缩影。焦耳的温度计能显示在通常情况下留不下视觉痕迹的可感热（sensible heat）。因此，焦耳和丹瑟提供了热研究中比例转换的方法，这一点在后来测定热功当量精确值的实验中是至关重要的。焦耳在他的"论通过空气的压缩与稀少产生的温度改变"的研究中，第一次使用了这种灵敏温度计，后来他在所有测定热功当量的实验中都使用这种温度计。焦耳进一步开发他在父亲的酿酒厂里学到的测温技术，竟然使他的实验室变成了独一无二的经验所。[②] 此外，正如焦耳所正确陈述的那样，在精确测量的灵敏度方面所取得的这些进步，今后不再允许目击者在场。他使自己变成了一位无观众的执行者。焦耳面临着一些严峻的问题，因为在实验室之外并不存在人们共享焦耳希望确立的这些实践、价值和标准的一种专家文化。"我总是难以使人相信我能很确定地测量小于1℃的热量。"[③]

为了结束这一部分，让我扼要地考虑说明为什么这一数值事实在这些研究集体中会有不同意义的某些机制。尽管焦耳精确地测定了热功当量，但皇家学会的成员无法接受他关于"摩擦在于把力转化为热"的断言，因为这违背了当时学术界教导的方法论和热质说的权威知识。此外，焦耳的灵敏测量排除了目击者的在场，而且，由于检测到最低温度持续上升，已经很难声称焦耳的实验工作是极度准确的。为了避免争论，并为讨论这一问题提供进一步的基础，审阅人促使焦耳从他的论文中提取出最能达成一致的事实，即功和热之间的可测量的比值。对于焦耳来说，这是极其不幸的，仿佛他根据几年的经验得到的知识，被分割成毫无争议的事实和意见，前者被通过精确测量来确立，而后者很可能被视为焦耳缺乏自然哲学教育的表现。对于焦耳及丹瑟等技师、仪器制造者和那些积极参与促进实验科学并把实验科学定义为精密科学的一部分的业余爱好者来说，他们把做实验看成一种复杂的推论形式，不能把事实分裂成这两部分。当焦耳把修改后的

① 关于转化比例的文化技巧及其各种应用的详细讨论，参见 H. O. Sibum, *Shifting Scales. Microstudies in Early Victorian Britain*；Max Planck Institute for the History of Science Preprint No. 171；有关丹瑟的微缩微影照片，参见 B. Bracegirdle, J. B. McCormick, *The Microsopic Photographs of J. B. Dancer. With illustrations from original Dancer negatives from the collection of A. L. E. Barron*. Chicago, 1993；对这一时期兴起的视觉文化的更广泛的研究，参见 J. Crary, *Techniques of the Observer. On Vision and Modernity in the Nineteenth Century*. Cambridge, Mass, 1991.

② 关于这种进行独特测量的文化资源的进一步详情，参见 Sibum, Reworking, cit.；进行温度测量的整个动作设计是非常明显地通过酿酒过程中实践的身体技巧来塑造的，甚至浆式叶轮实验的实验设置也是一个小规模的酿酒模型。Sibum, Les Gestes, cit.

③ Joule to Thomson, 7 November 1848, CUL Add 7342.

论文呈交给皇家学会时，尽管赢得了"绅士专家"的称号，可他一定感到与18世纪的那些技师一样痛苦，技师的应用知识启发了哲学家，却常常被解释成一些朴素的实践。①

三、能量守恒：确立证据和同时发现的建构

李比希在他的"书信"中已将发现热功当量的整个荣誉赋予迈尔（R. Mayer）。这是不诚实的；因为没有人比李比希更清楚我对这个问题的研究。他是一个热心的袒护者，因而无疑误入歧途。（1859年4月焦耳写给汤姆孙的信）

焦耳成为"绅士专家"的经历，包括所承担的一切问题，是这一时期形成精密科学的范例。我们甚至可以在一个更大的范围内看到这一点。英国科学促进会（BAAS）的创建，效仿了德国的"自然探索者协会"（Naturforscher Gemeinschaft），标志着某种社会事业的开始，这被莫雷尔（Morrell）和萨克雷（Thackray）描述为形成了"科学绅士"的共同体。在德国，为了促进物理学的发展，1845年成立了"德国物理学会"。第一代学会成员是由技师、技工、商人、高中和大学教师组成的。同时，各个大学慢慢地开始建立实验室，实验室改变了传统的教学方式，新的实验者可以在实验室里做研究。② 然而，正如焦耳和丹瑟的故事所表明的那样，对于学术团体和推动形成新的实验科学风气的不同社会背景的成员来说，重要的是，焦耳和丹瑟使他们自己与"富有独创性的发明者"区分开来。例如，当酿酒师焦耳把他关于电磁机的工作发表在斯图金（W. Sturgeon）的《电学年鉴》上时，他很快意识到，他是讲给实业家（practical men）和技师（artisans）听的，而不是讲给有绅士风度的自然哲学家听的。为了不被看成是对商业促销感兴趣，而相反是在追求科学的权威，他不断地提醒读者，他是在推进关于自然原理的实验研究、确立定量的定律、讨论误差的来源，

---

① 有关从手工艺的文化中抽取事实性知识的问题，参见前面脚注中的文献。狄德罗等人在研究手工艺文化时面临的历史问题和认识论问题仍然没有完全解决。从这里讨论的内容来看，似乎合理的建议是，在公开的信件中试图确立事实性知识的18世纪末和19世纪的实验者感受到类似的问题。知识与科学的关系值得进一步进行历史研究，当前，马克斯·普朗克科学史研究所的研究小组正在从事"科学实验史"研究。

② 有关作为一种社会运动的"爱好者科学"（devotee science），参见 R. H. Kargon, *Science in Manchester. Enterprise and Expertise.* Manchester, 1977；J. Morell, A. Thackray；*Gentlemen of Science*, cit.；D. Hoffmann（ed.），*Gustav Magnus und sein Haus.* Stuttgart, 1995；A. Fiedler, *Die physikalische Gesellschaft zu Berlin. Vom lokalen naturwissenschafilichen Verein zur nationalen Deutschen Physikalischen Gesellschaft.* Aachen, 1998（Dissertation Universitat Halle）.

以及测量精确的数值细节。① 从其仪器店里获得收入的丹瑟同样确保他遵守科学同仁的道德经济、与知识界免费分享他们的知识。因此，他从来没有转让他的那些发明，尽管那些发明本来可以为他提供一笔可观的经济收入。他之所以不这样做，是因为那会表明他只有兴趣赚钱，而不是获得自然知识。在这种语境中，焦耳以他的曼彻斯特文学与哲学学会主席（1851）的身份在递交给学会同意发表的论文上注明日期的决定是有意义的。他希望由此能避免关于发现的优先权的纠纷，这是焦耳卷入其中的令他厌烦的一件事。焦耳是实验者，他的知识很大程度上是非书面的，有其自身的交流方式，他显然把著作权和版权看成是实验科学中的迫切问题。正如我们将看到的那样，这影响了这一数值事实的进一步发展。根据罗伯特·卡尔根（R. Kargon）的观点，这预示着一个新时代的来临，在这个时代，作为"绅士科学家"，他们不仅通过自己的工作，而且采取在曼彻斯特学会将文人排除在外的方式，来证明他们的专长。②

在焦耳的关于热的机械值的出版物，特别是 1850 年的文章发表之后，由官方确认的事实很快传播开来。然而，这一事实是否存在，它最终意味着什么，它有什么样的价值，要想就这些问题达成一致，就需要一种集体的努力。这一努力贯穿了整个 19 世纪，并要求实验室科学和精密测量的文化的尽早确定。③ 很快，对"焦耳的事实"的不同回应出现了。在维也纳的一次科学学术会议上，生理学家兼德国物理学会成员布鲁克（Brücke）评论道：

> 在所有已被检验和接下来将被检验的数值中，我们所研究的数值是如此重要以至于没有其他数值能与它竞争。未来，它会在物理学的所有分支中发挥基本作用。同样重要的是，它由于为评估做功系统提供了基

---

① 参见 C. Smith, *The Science of Energy*, cit., pp. 57ff；作为一种职业而不是生活的科学，参见 J. Morrell, A. Thackray, *Gentlemen of Science*, cit., p. 33；关于焦耳在 *Sturgeons Annals* 上发表的文章，参见 SPJ, 1884, cit., pp. 1-53；J. P. Joule, "A Short Account of the Life and Writings of the Late Mr. William Sturgeon", in *Memoirs of the Literary and Philosophical Society of Manchester*, XIV（1857）；I. R. Morus, *Frankenstein's Children：Electricity, Exhibition, and Experiment in Early-Nineteenth-Century London*. Princeton, 1998.

② R. Kargon, *Science in Victorian Manchester*, cit. p. 79f. 关于丹瑟的观点，参见他的传记，J. B. Dancer, *John Benjamin Dancer*, F. R. A. S., 1812-1887, "An Autobiographical Sketch, with some Letters", in *Memoirs and Proceedings of the Manchester Lit. And Phil. Society*, CVII（1964/65）, pp. 115-142.

③ 有关法国实验室科学的状况，参见 R. Fox, "Scientific Enterprise and the Patronage of Research in France, 1800-1870", in R. Fox, *The Culture of Science in France*, 1700-1900. Aldershot, 1992, pp. 442-437；关于英国实验室科学的状况，参见 M. N. Wise, C. Smith, *Energy & Empire. A biographical study of Lord Kelvin*, Cambridge1989；F. A. James（ed.）, *The Development of Laboratory*, *Essays on the Place of Experiment in Industrial Civilisation*. Houndsmill, 1989；关于德国实验室科学的状况，参见 C. H. Jungnickel, R. McCormmick, *Cavendish*, cit.；K. Olesko, *Physics as Calling：Discipline and Practice in the Konigsberger Seminar for Physics*. Ithaca, London, 1991.

础，而具有实际应用价值。同时它将会向我们表明做功的极限，超出这个范围，我们就不应该坚持有望获得更多的功。[1]

的确，热功之间的一种不变的数量关系，使燃烧的生理理论建立在牢固的基础之上。由于年轻的亥姆霍兹在他的"书面答复"中使用了错误的单位转换表，所以他尽管非常有兴趣，却并不相信这一数值。鲁道夫·克劳修斯同样做了详细的书面摹写，并将这一事实数据插入到他表示热力学第一定律的方程中。[2] 哥廷根的威廉·韦伯也认为这个数字是很重要的，只不过，他不相信焦耳的方法论进路。他怀疑，机械功的测量能够成为用来校准其他自然力的最精确的标准。当然，韦伯更喜欢用自己的磁力计把对磁偏角的绝对测量确定为校准其他自然力和测定热功当量的最可靠的标准。他在格尼斯堡的同事纽曼（F. E. Neumann）为此设计了一个可使用的实验。[3] 在英国，焦耳于 1847 年在牛津召开的一次 BAAS 会议上结识了威廉·汤姆孙，即后来的开尔文勋爵，并且，从那以后，两人开始了密切的合作。汤姆孙承认，焦耳的浆–轮实验证明了一个重要的事实：热与功之间的定量关系。尽管与其工程师兄弟詹姆斯达成了一致，但汤姆孙断定，詹姆斯·焦耳"的思想中有许多错误，不过，他似乎发现了一些极为重要的事实，比如，热是由流体的摩擦产生的"。这一事实一方面与兄弟俩都关心流体摩擦中消耗有用功的问题相关，另一方面与他们不久前和斯托克斯讨论对热、磁等物理现

---

① E. Brücke（1851），in J. J. Weyrauch（ed.），*Die Mechanik der Warme in Gesammelten Schriften von Robert Mayer. Dritte erganzte und mit historisch-literarischen Mitteilungen versehene Auflage.* Stuttgart，1893，p. 296.

② "书面摹写"（literary replication）这个术语是由吉姆·西科德（Jim Secord）提出的，意指为了使实验行为有意义，读取和描绘实践。J. Secord，"Extraordinary Experiment：Electricity and the Creation of Life in Victorian England"，in D. Gooding，T. Pinch，S. Schaffer（eds），*The Uses of Experiment. Studies in the Natural Sciences.* Cambridge，1989；pp. 337-383，347ff. 有关亥姆霍兹的文献参见 F. Bevilacqua，"Helmholtz's 'Ueber die Erhaltung der Kraft'. The Emergence of a Theoretical Physicist"，in D. Cahan（Ed.）*Hermann von Helmholtz and the Foundations of Nineteenth-Century Science.* Berkeley，Los Angeles，1993；特别是参见，R. Brain，M. N. "Wise，Muscles and Engines：Indicator Diagrams and Helmholtz's Graphical Method"，in L. Krüger（ed.），*Universalgenie Helmholtz. Rückblick nach 100 Jahren.* Berlin，1994，他认为，亥姆霍兹喜欢把图示法作为更值得信赖的手段来确立功的测量。有关克劳修斯的文献，参见他的笔记记录，Deutsche Museum，Munich，and D. S. L. Cardwell，*From Watt to Clausisus. The Rise of Thermodynamics in the Early Industrial Age.* Ithaca，1971，pp. 246.

③ W. Weber，"Zur Galvanometrie"，1862，in W. Weber，R. Kohlrausch，*Fünf Abhandlungen über die absoulte elektrische Stromund Widerstandsmessungen*，Leipzig 1904，Ostwald's Klassiker No. 142，pp. 70-94，p. 88. 有关纽曼的讲稿参见 Franz Ernst Neumann，*Vorlesung：Mechanische Warmetheorie. Vorlesungsmitschrift von Prof. Wild，D. Dorn from the year 1867*，Handschriftenabteilung Niedersachsische Staatsund Universitatsbibliothek Gottingen. MS F. E. Neumann 8 II，p. 237-418，256；有关纽曼的文献也参见 K. Olesko，*Physics as Calling*，cit.

象进行流体力学的类推相关。① 这里无须详述这个重要的交流，只需提及汤姆孙用四年的时间才最终接受焦耳关于其既定事实的普遍意义的"错误思想"②。在这些年中，焦耳不断给汤姆孙写信说，当他看到汤姆孙仍然喜欢热质说时感到非常惊讶。确实，正如 1849 年的讲稿所表明的那样，汤姆孙起初称赞焦耳的温度计的技艺，却以一种相当特殊的方式解释焦耳的实验。然而在他们后来几十年的合作中，他成了焦耳关于热与功的互换性的知识主张的强有力的代言人，甚至使焦耳数成为英国能量科学和绝对单位制的基础。他在 1863 年的讲课中对学生说：

> 热是运动。热被恰当地称为消耗功的动态当量。但困难在于，我们无法再获得产生热所消耗的功。焦耳的观点与卡诺的观点相符合，对于热的动力学理论来说，是非常重要的。③

然而在 1851 年，即在创立热力学和能量守恒的初期，德国物理学家朱利叶斯·罗伯特·迈尔就焦耳的事实要求知识产权：

> 这个新问题（热功当量）不久就引起了有识之士的注意，但就国内外都把这个问题排他性地看成是外国人的发现而言，我感到自己不得不要求把优先权赋予我；尽管我公之于众的只是很少的一些研究，并且在每日海量的通信洪流中，这些研究也几乎无迹可寻。但正是这种发表的形式证明，我并不是一个贪图结果的人，然而并不能因此假设我愿意被剥夺文件证据证明是属于我的知识产权。④

历史学家以能量守恒同时被发现为例讨论了这段插曲，其中，除了迈尔和焦耳之外，还涉及其他人。⑤ 我想论证的是，公开的信件和焦耳与汤姆孙的私人通信中讨论的优先权之争，反映了实验者在尚未明确其身份的共同体中确立科学权威时所遇见的那些困难。自 19 世纪上半叶以来，"数值事实"作为"精确研究自然界的基础"在这些新的科学家当中赢得了很高的声誉。然而，焦耳和迈尔以完全不同的方式测定了热功当量。迈尔根据已知的实验数据计算出这个数字，而

---

① William Thomson to Dr. James Thomson, 1 July 1847, T367, Kelvin Collection, Glasgow University Library (hereafter GUL), 也参见 Smith, *The Science of Energy*, cit., p. 79.

② 有关他的理论与焦耳的事实之间的调和工作，特别参见 M. N. Wise, "William Thomson's Mathematical Route to Energy Conservation: A Case Study to the Role of Mathematics in Concept Formation", *Historical Studies in the Physical and Biological Sciences*, X (1979), pp. 49-83.

③ W. Thomson, Lecture notes by David Murray "Lecture LI, 14 April 1863", GUL, MS Murray 326.

④ R. Mayer, "Remarks on the Mechanical Equivalent of Heat", *The London, Edinburgh and Dublin Philosophical Magazine and Journal of Science*, Supplement to Vol. XXV. Fourth Series, No. 171, June 1863, p. 520. English translation of *Bemerkungen über das mechanische Aquivalent der Warme*, Heilbronn und Leipzig, 1851.

⑤ 关于这一开创性论文参见 T. S. Kuhn, "Energy Conservation as an Example of Simultaneous Discovery", in Kuhn, *The Essential Tension*, cit., pp. 66-104.

焦耳则是在几个转化实验中测量出大小。但两人都没有把这描述为能量守恒的证据。[①] 相反，这个精确的事实与能量守恒思想的结合，只是在 1850 年到 1880 年才逐步建立起来的。焦耳在 1848 年看到迈尔对优先权的要求时回应说：

> 压缩气体（也就是，通过气体中的水银柱的下降压缩气体，即 H.O.S.）产生热量这一简单事实并不能保证这样的结论：热量精确地等同于所用的力，因为人们可以把气体设想为是由相互排斥的粒子构成的，其中，一部分力使它们部分地相互靠近，而只有其余的力才产生热。[②]

在汤姆孙、泰特（Tait）和焦耳之间的通信中，迈尔关于存在热功当量的主张被描述为"只是推测"而已，只有焦耳提供经验证据的实验工作才能被视为"研究"。

> 还有一点——人们公认，迈尔在 1842 年的文章是作为笔记印刷的，只是为了获得一个较早的日期以显示其价值。就我个人而言，我从来没有发表过这样的文章，因为我总是追随赫舍尔（Herschel），坚持认为科学最大的祸根是草率地概括。迈尔的文章只是一种推测，我的文章是一种研究。迈尔可能具有的唯一影响是，阻碍了这一问题的进展。这是我坦率的观点，尽管我感到，以书面形式这么说，对迈尔很不尊敬与亲近。[③]

现在，为了理解焦耳等英国科学家为什么把迈尔的工作看成是推测的和"前科学的"，也为了评价焦耳具有特殊道德价值的实验工作的特殊性，有必要稍微关注一下迈尔的研究实践和他从公布的实验结果中推出热功当量的方式。在他的计算中，他把空气的定压热容与定容热容的比率作为基础。通过反思这样的实验，1 立方厘米的水银柱，高 76 厘米，重 1033 克，升高原容量的 1/274，他估计，把等重的水从 0℃加热到 1℃所需的热量，相当于等重的物体从 365 米的高度落下所做的功。他最后得出的结论是，"如果在你的计算中用单一气体或混合气体取代空气，你也会得到同样的结果"。但是，他在一个脚注中更精确地提到这种断言的条件是：所有气体的膨胀系数都不变。

---

① 一些历史学家至今都一直质疑同时发现的观念，并提供了局域性研究，来表明这些贡献的特殊性。例如，参见 Y. Elkana, *The Conservation of Energy: A Case of Simultaneous Discovery?*, Archives Internationales d'Histoire des Sciences, 23-e année（1979），pp 31-60；K. Caneva, *Robert Mayer and the Conservation of Energy*. Princeton, 1993；C. Smith, *The Science of Energy*, cit.；H. O Sibum, Reworking cit.；有关发现概念参见 S. Schaffer, "Scientific Discoveries and the End of Natural Philosophy", *Social Studies of Science* XVI（1986），pp. 387-420.

② Joule to Thomson, March 18, 1851.

③ Joule to Tait, July 25, 1863, Glasgow University Library, Kelvin papers J 173.

然而，直到 19 世纪 40 年代，气体膨胀系数才成为法国、英国和德国的自然哲学家关注的重大问题之一。迈尔由于把所有气体的膨胀系数都不变看成是理所当然的，所以，在测温技术领域完全忽视了这种研究。尽管近 40 年来气体膨胀系数一直被视为"最安全的物理学数字之一"，但在 18 世纪 40 年代，一些研究者［雷诺（Regnault）、里德伯（Rudberg）、马格努斯（Magnus）］仍对它提出了批评。当时，最可靠的值是由里德伯建议的，但他也得出结论说，对于所有的气体来说，膨胀系数并不是完全相同的，因此，盖·吕萨克定律没有普遍性。① 马格努斯正确地提醒他的读者，在数值方面的这种冲突使得很难用气体膨胀来测量温度，自从杜隆（Dulong）和珀蒂（Petit）的工作之后一直是这么做的。气体膨胀系数的重新测定有几种结果：首先，一种实际结果是，当时作为法国标准的空气温度计失去了信誉，因为它不如水银温度计好，根据雷诺的观点，水银温度计的运行"几乎是热量增加的复杂函数"②。雷诺也表达了理论意义：我们仍然没有手段来测量热的绝对量，而且，就我们知识的这种现状而言，"我们几乎没有希望发现依赖于这些测量的那些现象的各个简单定律"。从这个视角来看，德国《物理年鉴》的编辑波根多夫（Poggendorff）不愿发表迈尔的第一篇报道，似乎是显而易见的，这篇报道只断言，热能被转化为运动及热膨胀或材料膨胀是其现象学的证据。③ 波根多夫似乎决定，直到当他更清楚地知道关于热膨胀的可靠测量值的争议结果时，才发表这篇文章。最终，他完全拒绝发表该文。

在曼彻斯特，焦耳用他和丹瑟制造的水银温度计做实验，解决了上面提到的问题。当时正在研究从理论上建立绝对温标方法的汤姆孙，完全信任焦耳的温度计所起的作用，接下来的几年间，焦耳在曼彻斯特实验室的工作有力地支持了他对热力学的发展。④ 随着优先权争论的不断激烈，汤姆孙和泰特站在了焦耳一边，并为这样的经验立场辩护：焦耳对热功当量值的更精确的实验研究证明，

---

① G. Magnus, "über die Ausdehnung der Gase duch die Warme", in *Poggendorff's Annalen der Physik under Chemie*, LV（1842）, pp. 1-27；V. Regnault, "Untersuchung über die Ausdehnung der Gase", in *Poggendorff's Annalen der Physik und der Chemie*, LV（1842）, pp. 391-414 u. 557-585，也参见 C. H. Sichau *Der Joule-Thomson Effekt. Der Versuch einer Replikation*, Physics Diploma Thesis, University of Oldenburg, 1995.

② V. Regnault, über den Vergleich des Luftthermometers mit de Quecksilberthermometer, in *Poggendorff's Annalen der Physik und der Chemie*, LVII（1842）, p. 218.

③ J. R. Mayer, *über quantitative und qualitative Bestimmung der Krafte*, send to editor of *Poggendorff's Annalen* on June 16, 1841, not published, found in Friedrich Zollner archive published in J. R. MAYER, *Die Mechanik der Warme. Samtliche Schriften*, ed. by Hans Peter Münzenmayer, Heilbronn 1978.

④ 关于焦耳与丹瑟在制造温度计方面的紧密合作，以及焦耳与汤姆孙之间在处理温度测量方面的差异，参见 H. O. Sibum, *Shifting Scales* cit.. 绝对温标参见 W. Thomson, "On a Absolute Thermometric Scale founded on Carnot'sTheory of the Motive Power of Heat, and Calculated from Regnault's Observations", in *Philosophical Magazine*, XXXIII（1848）, pp. 313-317.

"热量是所耗功的动力学当量"，而且，"能量守恒只能通过实验来确立"①。

四、工程证据：一个自然常数

对我的一些读者来说，一个令人好奇的目标是，知道人们如何开始接受功的测量……麦克斯韦教授告诉我们："如果1磅的物体克服引力升高1英尺，就做了一定量的功，而且，工程师把这个量称为1尺磅。"如果可以采纳这段引文来提供真正的线索，我们很难不被下列建议的原创性所吸引：重要的哲学原理可能是通过"工程证据"确立的。（Moon，1874）

从事能量科学研究的英国物理学家把热功当量看成能量守恒的一个证据，并且，对他们而言，作为对自然界的真正的数字表征，它在绝对单位制中是关键因素。热功当量的精确值将为呈现各种力的相互关联的许多现象提供定量的统一结构。一旦这一事实被确立为对自然界的真表征，就有可能依据机械功校准所有其他的自然力，比如，电力、磁力和化学力。似乎眼前的一切将要创立这个"一致系统"，正如英国科学促进会对他们的测量所要求的那样。② 他们的标准化程序与大英帝国的电报产业密切相关，要求最高程度的精准性。1860~1880年的工作报告清楚地表明，在这个高度竞争的项目中提出的以精确测量为核心，已经受到了挑战。③ 对于作为精密科学的实验热力学和电动力学的发展来说，测量的极端精确成为一种强制性的必要条件。1851年，刚好在焦耳的论文发表一年之后，丘天文台（Kew Observatory）的人宣布，他们已经采取措施，"在他们自己的监督之下，走向仪器的生产，以便分配给需要精确参考标准的机构和个人"④。此外，许多大学建立了实验室，学习和改进精密仪器的适当用法。不仅在英国，而

① 泰特和汤姆孙之间的通信特别参见 July 3, 1863 (T41) and J. Thomson and W. Thomson, Aug. 13, 1863 (T119) at GUL. 关于涉及焦耳与迈尔争论的通信的详细说明，参见 J. T. Lloyd, "Background to the Joule-Mayer Controversy", in *Notes and Records of the Royal Society*, XXV, (1970), pp. 211-225.

② "Report of the Committee appointed by the British Association on Standards of Electrical Resistance", in *British Association for the Advancement of Science Reports*. London 1864, pp. 111-177.

③ M. N. Wise, C. Smith, *Energy & Empire. A biographical study of Lord Kelvin*. Cambridge, 1989; S. Schaffer, "Late Victorian Metrology and its Instrumentation: A Manufactory of Ohm's", in S. Cozzen, R. BUDD, (eds.), *Invisible Connections*. Bellingham, 1992, pp. 23-65; S. Schaffer, "Accurate Measurement is an English Science", in M. N. Wise, *The Values of Precision*. Princeton, 1995, pp. 135-172; B. Hunt, "The Ohm is where Art is: British Telegraphic Engineers and the Development of Electrical Standards", in *Osiris*, IX (1994), pp. 48-63.

④ J. Welsh, "On the Graduation of Standard Thermometers at the Kew Observatory", *BAAS Report*, (1853), pp. 34-36.

且在整个欧洲，精密文化达到了顶峰。到19世纪70年代，在科学中，想知道意味着去测量。①

> 我常说，当你能测量你所谈论的东西并用数字来表示它时，你才会对它有所了解；但当你不能测量它时，当你不能用数字表示它时，你的知识就属于贫乏的不能令人满意的类型：可能处于知识的初级阶段，但无论问题是什么，在你的思想中，你几乎不可能进步到科学的阶段。②

然而，尽管在促进英国的能量科学方面取得了成功，但能量守恒的普遍原理仍然能根据几种观点加以质疑：牛津德高望重的哈顿（Highton）把这个问题描述为

> 对于解释物理现象及确定由严格的自然法则给予人类机械技能和科学技能的应用的限制来说，都是极其重要的。③

最后，他和他的同事戈尔（Gore）认为，人们一定能阐明焦耳在几个实验中证明的热与力的等价性。然而，他们不情愿接受对人类生产力和自然生产力的某种绝对限制。他们也建议说，这个数值事实上只在力学中有效，在电学和磁学领域是无效的。因而，由此得出一个普遍原理是错误的。④ 在1867年，当焦耳重复自己的实验时，爆发了一场重要的争论。⑤这里不详述这个实验，重要的是指出，焦耳得到了不同的当量值。他声称，他此时做的实验比过去做的实验更精确。这一事实的抛出，引发了人们对皇家学会的准确性的怀疑。⑥ 英国的能量科学陷入了危机。甚至第一位剑桥大学的实验物理学教授麦克斯韦，也卷入了用他设计的新的直接方法对热功当量的重新测定中。

> 我认为，这是一个没有机械困难的方案，在有充足水银和坚固铁器的一间高耸的房屋内，一个小天使在头顶上读取同位的温度计，一只猴子把水银携带给他（叫做银色快手杰克），事情可能进行几个小时。⑦

这种隐喻描述表明了麦克斯韦对以前实验的怀疑：在只有无实体的天使才能

---

① 有关第一门实用物理学课程参见 F. Kohlrausch, *Leitfaden der praktischen Physik zunächst für das physikalische Praktikum in Göttingen. Leipzig*, 1870.

② W. Thomson, *Popular Lectures and Addresses*, 3 *Vols.* London, 1889, I, pp. 73-74.

③ W. Thomson, *Popular Lectures and Addresses*, 3 *Vols.* London, 1889, I, pp. 73-74.

④ 参见 H. Highton, "On the Relations between Chemical Change, Heat, and Force —With a Special View to the Economy of Electro-dynamic Engines", in *The Quarterly Journal of Science and Annals of Mining, Metallurgy, Engineering, Industrial Arts, Manufactures, and Technology*, I (1871), p. 77-94, 89.

⑤ J. P. Joule, "Determination of the Dynamical Equivalent of Heat from the Thermal Effects of Electric Currents", *British Association for the Advancement of Science Reports*. London, 1867, pp. 512-522.

⑥ W. Thomson, *Popular Lectures and Addresses*, 3 *Vols.* London, 1889, I, pp. 73-136, pp. 133/134.

⑦ Maxwell to Tait, Dec. 23, 1867, Cambridge University Library, Add 7655 1b.

读取准确温度的程度上，实验者的身体产生的热辐射，使人类失去了做实验的资格。但他潜在关心的是热的动力学理论和关于热的本性的实验意义。焦耳在1869年写给汤姆孙的信中说：

> 我们突如其来的理论发布已经使得哲学家们的道德和智力优势荡然无存……是提出新证据的时候了，现在，异端的秘密会议正准备将动力学理论贬为异教。[1]

同时，由于麦克斯韦在所卷入的不同群体之间进行调和，物理学家提供了解释这些问题的几个策略。[2] 美国工程师兼物理学家罗兰在欧洲逗留的一年中意识到，实验者在他们测量基本单位时得到的不同数值，从根本上动摇了新的能量物理学。罗兰把对热功当量的新的测定看成是确立"自然界的最重要的常数之一"所必要的。[3] 在绝对测量中只有一个精确值能界定为对自然界的真表征，因此为近代物理学国际共同体的兴起提供了必要的支撑。但为了实现这一目标，需要科学实践的剧变，例如，涉及实业家、工程师与实验室科学之间的密切合作。罗兰作为约翰·霍普金斯大学物理实验室的创办主任，向理事会成员解释说，在近代，"指望临时制作仪器是无用的"[4]。为了向研究者提供高质量的精密仪器，近代物理学需要最高标准的实验室、实验工具和器械，特别是，要求仪器制造者技艺精湛。此外，一门在数学方法与实验方法之间非常协调的物理学课程"将不仅满足这个国家的长期需要，并且有望对班里原本打算去德国追求学业的那些学生有吸引力"，而且可以培养"严谨的物理学家"。只有这样的体制，才能使"近代的事物秩序"起作用。罗兰在新的物理实验室的启用典礼会上新创了这个短语。罗兰认为，"近代的事物秩序"开始于伽利略，即第一位只相信自己推理的研究者，因为他能用实验检验推理。[5] 在罗兰的实验室里，新一代学生不得不学会

> 不断地检验他们的知识，从而亲自认清模糊推测的糟糕后果；他们

① Joule to Thomson, Dec 29, 1869, Cambridge University Library MSS Add 7342 J287.

② 关于这些问题如何解决的研究，参见 H. O. Sibum, "An Old Hand in a New System, cit., pp. 23-57, H. O. Sibum, Exploring the Margins of Precision", *Max Planck Institute for the History of Science Preprint*, No. 172, and C. Smith, The Science of Energy, cit.

③ H. A. Rowland, "On the Mechanical Equivalent of Heat, with Subsidary Researches on the Variation of the Mercurial from the Air Thermometer, and on the Variation of the Specific Heat of Water", in *Proceedings ofthe American Academy of Arts and Sciences*, 1879-1880, pp. 75-200.

④ H. A. Rowland, *Report to the Board of Trustees*, undated, JHU, Ms 6, p. 17-18. 尽管罗兰钦佩法拉第等英国科学家，但他希望明确，"未来的科学"必须超越那些以改进性能为基础的实验操作形式。比较 H. O. Sibum, *An Old Hand in a New System*, cit.

⑤ H. A. Rowland, *Lecture notes (given at the opening of the physics laboratory)*, Rowland Manuscripts, John Hopkins University, Eisenhower Library, Ms. 6, Ser. 5.

必须通过直接的实验了解到，世界上有真理，他们自己的心智很有可能出错。他们必须反复实验，不断解决问题，一直到他们成为行动者，而不是理论家。这就是实验室在通识教育中的作用，即不断地使心智与绝对真理联系起来，以正确的思维方式培育心智。[①]

在几次公开演讲中，他揭示出，作为近代教育手段的物理实验室应该看起来是怎样的，为什么对文化发展如此重要：

> 对中学和大学教育现状有所研究的人告诉我们，包括科学在内的许多学科的教学是在训练记忆力。我在一所时尚学校里，亲眼目睹了天生聪颖的小姐们凭记忆一页一页背诵的沮丧情景，她们没有付出任何努力发现自己是否理解了这门课……完全只教词语，并且产生了远离上面描述的一种心态……教育的目标是不仅要培养懂得理论的人，而且要培养付诸行动的人；培养在生活的斗争中留下印记，并且在所经历的一切中取得成功的人；培养能够解决所出现的自然界的问题和人类问题的人，以及当他知道自己正确时能勇敢地使世界相信事实的人。[②]

19世纪末，近代实验科学成为文化发展的典范，罗兰的评论反映了这一历史进程的特征，但正如罗兰的案例所表明的那样，在这个历史过程中，人文科学与自然科学分离开来，事实与意见分离开来。

> 我们的科学的事实与理论，比历史的事实与理论、普通的历史事实或法律证据所依赖的普通人的证言，更加确定。[③]

在确立这一自然常数时，工程师兼物理学家罗兰认定，测温技术实践是先前测定热功当量的所有努力中的薄弱环节。他在给吉尔曼（Gilmam）的信中写道，他的空气温度计"是我全部工作的体现"。他的出版物和档案材料表明，他精通温度计的特性。他甚至认为，这项研究是如此重要，以至于他建议，下设"标准认证分部，在这里，对绝对测量作出比较"。在获得了首批实验结果之后不久，他断定：

> 这个部门可能最终会扩展，在扩展的所有方向中，测温学依然很重要……以空气温度计为标准，所有的比较被简化为这种理想气体温度计

---

① H. A. Rowland, "The Physical Laboratory in Modern Education", in *The Physical Papers of Henry Augustus Rowland*. Baltimore, 1902, pp. 614-618, 617.

② Ivi., p. 617.

③ H. A. Rowland, "The Highest Aim of the Physicist. Address delivered as the President of the American Physical Society, at its meeting in New York, October 28, 1899", in *The Physical Papers*, cit., pp. 668-678, 676.

的最终的绝对标准。①

就他对热功当量的研究而言，罗兰对当时欧美所用的温度计进行了大量的比较研究。他知道，为了使关于热的科学知识得到国际认可，建立标准实验室是可被采纳的一项重要措施。当他把自己关于热功当量的研究写成文章发表时，他已经感到，他绝对有把握在物理学家当中处于领先地位，并试图为即将到来的新千年确定新的精确度标准。

> 我对焦耳工作的精确性抱有最崇高的敬意，我认为在谨慎和精确性方面，他是年轻物理学家最值得考虑的典范。我们可以肯定的一个问题是，当科学的千年来临之际，在只留下精确的物理学家之时，我们就……不必再等30年才有两位物理学家对热功当量之类重要的量的真值是什么达成一致。②

他继续确立这一近代标准，使科学界铭记这样一个事实：一个重要的哲学原理能够由工程证据来证明。③

五、19世纪的黄金数

> 热功当量，即使1磅水温度升高1度所必需的单位功的总数，非常有理由被称为那个世纪的黄金数。（Mendenhall，October 1901）

罗兰于1880年发表的文章并不标志着把热功当量确立为一个明确事实这个过程的结束。这里有两条发展线索值得注意。只要物理学家提出新方法，有望提高既定值的精确度，使热功当量值稳定下来的努力就会继续。近代科学的特征是：不断重新评价物理学的标准，这些标准是每一位在职科学家的工具。④ 随着

---

① H. A. Rowland, *Manuscript Physical Laboratory*, Johns Hopkins University. Comparison of Standards. Circular No. I, p. 6. JHU Ms. 6, Ser. 5.

② 罗兰未完成的结论性判断和部分判断在他的下列草稿中能找到：on *Appendix to Paper on the Mechanical Equivalent of Heat*, *Containing the Comparison with Dr. Joule's Thermometer*, emphasis by H. O. Sibum, JHU, Ms. 6, Box 39, Series 5；出版的版本参见 H. A. Rowland, Appendix..., cit., *Proceedings of the American Academy of Arts and Sciences*, XVI (1880), pp. 38-45.

③ 关于亨利·罗兰对确立这一事实的贡献的详细研究，参见 H. O. Sibum, *Exploring the Margins* cit. and H. O. Sibum, *An Old Hand in a new System*, cit.

④ 例如，参见 E. H. Griffiths, "The Value of the Mechanical Equivalent of Heat, deduced from some Experiments performed with the view of establishing the relation between the Electrical and Mechanical Units; together with an Investigation into the Capacity for Heat of Water at different Temperatures", in *Philosophical Transactions of the Royal Society of London*, vol. 184 (1894), pp. 361-504；O. Reynolds and W. H. Moorby, "On the Mechanical Equivalent of Heat", *Philosophical Transactions of the Royal Society of London* (A) vol. 190 (1898), pp. 301-422.

像德国帝国技术物理研究所那样的国家实验室的建立，这类研究首先被制度化。[1] 在 19 世纪的最后几十年里，这一确切事实最终落实成为国际单位制中的换算因子。1894 年 7 月 12 日，美国参众两院的议员在电气测量的单位制中使用这一常量确立了功的法定单位。

> 功的单位是焦耳，等于厘米–克–秒制中的 1000 万个单位功，在实践中，相当于 1 国际安培的电流通过 1 国际欧姆的电阻时 1 秒内消耗的电能。[2]

至此，这一科学事实的意义不再限于实验室，它已成为飞速发展"力量网络"（networks of power）的一个组成部分，并被贴上了"那个世纪的黄金数"的标签——正如罗兰的一位学生所称呼的那样。[3] 的确，19 世纪测定这一数值的研究者赞成这种描述，如罗伯特·迈尔在私人通信中所写的：

> 物理学或力学用知名的千克不可能跨越很远；在动力学中（而且这显然是到目前为止最重要的物理学领域），纯粹的千克，即纯粹的静压，不再有用，这里是千克–米（kilogrammeter），新近引入的研究对象，即所谓活力（lebendige kraft）的测量单位。我们无偿拥有无效的压力，但力或所谓千克–米总是有代价的……伟大的牛顿恰好不知道千克–米，最终，科学的进展超越了所有的权威……不存在一个静态的当量，即在热量与压力之间、卡路里与千克之间没有一个不变的数值关系，但无

---

[1] D. Cahan, *An Institute for an Empire：the Physikalische Technische Reichsanstalt* 1871-1918. Cambridge, 1989.

[2] *Committee Report to the President of the National Academy of Sciences*, New York City, February 9, 1895. Rowland Papers, Eisenhower Library archive, Johns Hopkins University. "焦耳"是国际单位制（SI）中功或能量的单位。国际单位制被称为 SI，采用了其法语名称 Système International d'Unités 前两个单词的首字母。关键的协议是 1875 年 5 月 20 日在巴黎签署的《公尺条约》（Convention du Mètre）。迄今已有 48 个国家签署了这一条约，包括所有主要的工业化国家在内。美国是这种公米制俱乐部的创办成员之一，1875 年已经签署了从前的原始文件：今天把焦耳定义为，1 牛顿的作用力，使一个物体在这个力的方向移动 1 米所做的功。相当于，既然动能等于质量的一半乘以速率的平方，所以，1 焦耳就是 2 千克的物体以 1 米/秒的速率运动时的动能，等于 10^7 尔格，或大约 0.7377 尺磅。用电学术语来说，焦耳等于 1 瓦/秒，也即，1 安培电流通过 1 欧姆电阻在 1 秒内释放的能量。

[3] T. H. C. Mendenhall, "Commemoration of Prof. Henry A. Rowland", in *Johns Hopkins University Circulars*, XXI, No. 154, December 1901, reprinted in *Annual Report of the Board of Regents of the Smithso-nian Institution, Showing the Operations, Expenditures, and Conditions of the Institution for the Year Ending June 30, 1901; Washington 1902*, pp. 739-753, 743. 关于这个法案，参见 Rowland's *Report to the President of the National Academy of Sciences*, Johns Hopkins University Library archive, Rowlands manuscript collection. 有关"力量网络"的观念，参见 T. P. Hughes, *Networks of Power. Electrification in Western Society* 1880-1930. Baltimore, 1983.

疑，在千克–米与卡路里之间存在当量。①

这一黄金数从根本上塑造了人的感知。德国帝国技术物理研究所所长弗里德里希·柯尔劳施（Friedrich Kohlrausch）教授在 1900 年声称，所有的东西（事物）都被作为它们体内固有的特殊效果来感知。木材、煤和其他食物都属于这类东西，其价值取决于其所提供的可用的能或功。未来，我们将成为能源的消费者，而非物体的消费者。因此，在德国管辖的范围内应该很快行动起来，以使声明能源为一种"事物"（sache）的法案合法化。②

第二条发展线索同样重要。在 19 世纪末，欧洲和北美的实验者成功地使实验科学在学术界占有一席之地。③以通过做实验来理解和交流事实性知识为目的的物理实验课，成为大学课程的一个组成部分。实验物理的教学手册出版，而且，证明热功当量的仪器成为物理科学实验室里的标准装备。④ 当这一数值事实具体体现为实验物理学家的工作环境时，这一科学事实的含义再次发生了改变。这种含义的变化发生在被称为理论物理学的新兴的科学亚文化当中，其中，马克斯·普朗克是第一代成员和重要的发言人。对他来说，人们看到、听到和感觉到的一切都是事实——"事实，没有一位怀疑论者能推翻它们"（tatsachen an die kein skeptiker rütteln kann）。甚至所谓的感官幻觉也是事实，这句话意味着，我

---

① Mayer to Mohr, April 28, 1868, in J. J. Weyrauch, *Kleinere Schriften und Briefe von Robert Mayer. Nebst Mittheilungen aus seinem Leben.* Stuttgart, 1893, p. 419-420. 今天，这一数值事实最终将卡（热的厘米克秒制单位）与焦耳（机械能的单位）联系起来，1 卡等于 4.1868 焦耳。在国际单位制中，用焦耳（为使热功当量是 1）衡量热量和所有形式的能量这一事实，甚至成了多余的。

② "价值的微观交易的新发展，在很大程度上，归于从物体交易转向能量交易。这个革新过程并没有达到其目标，不是本世纪，而是下一个世纪，被认定为是这一自然探索的果实成熟的时期。F. Kohlrausch, *Die Energie oder Arbeit und die Anwendungen des elektrischen Stromes.* Leipzig, 1900, p. 68-69. 人们应该知道，正值德国发生几起非法消耗电能的犯罪行为的法庭审判之际，柯尔劳施撰写了这本书。柯尔劳施认为，这是能量科学问题上糟糕的公共教育的结果。

③ 就实验室科学的历史发展而言，当然在 17 ~ 18 世纪的大学里开始为物理操作台（physical cabinets）腾出地方。但是就没有预算和训练及地位而言，有理由认为，直到 19 世纪晚期，实验者们才从内容上和理想上完全成功地把实验科学确立为一门独立的精深的学科。对 18 世纪物理操作台的概述，参见 J. L. Heilbron, *Elements of Early Modern Physics.* Berkeley, 1982, pp. 139ff; W. Clark, *The Hero of Knowledge*, to appear with University of Califronia Press; 19 世纪物理操作台的概述，参见 F. A. J. L. James (ed.) *The Development of the Laboratory.* cit.

④ 有关实用物理学教科书，参见，例如，F. Kohlrausch, *Leitfaden der praktischen Physik zunachst für das physikalische Praktikum in Gottingen.* Leipzig, 1870; 有关教学仪器，参见 J. Puluj, "Über einen Schulapparat zur Bestimmung des mechanischen Warmeaquivalents", in *Annalen der Physik und Chemie*, XVII (1876), pp. 437-446, W. E. Ayrton, W. C. Haycraft, "Students simple Apparatus for determining the mechanical equivalent of heat", in *Proceedings of the Physical Society*, VIII (1894-1895), pp. 295-309, 也比较 G. J. N. Gooday, "The Morals of Energy Metering: Constructing and deconstructing the precision of the Victorian electrical engineer's ammeter and voltmeter", in M. N. Wise, *The Values of Precision*, cit, 239-282.

们尚缺乏适当的推理能力从这种感官印象中推出正确结论。[①] 但是

> 迄今为止，理论物理学的发展是统一它的各个体系，这些体系是通过确定地排除拟人化的因素特别是特殊的感官感知而得到的。然而，鉴于感觉是整个物理学研究的起点，对根本前提的这种故意背离，看上去一定是令人惊讶的，如果不是悖论的话。然而，在物理学史上，几乎没有一个事实像现在这样明显，实际上，在这些自我疏离中，一定有毋庸置疑的优势。[②]

至于能量守恒原理，他很不愿意坚持这一鲁莽的断言，即热功当量的测定是这一原理的一个证据。他认为，物理学家无法为这一普遍原理提供经验证据，但只能从逻辑上根据不存在永动机的事实推导出它的普遍性。这仍然与全人类的经验相一致。

> 因此，永恒运动在物理学中的深远意义，如同点金石 (philosopher's stone)[③] 在化学中的深远意义一样，尽管科学的优势源于实验的否定结果，而不是肯定结果。今天，我们不参照任何人类的观点或技术的观点来谈论能量守恒原理。我们说，一个封闭的物体系统的总能量是一个有大小的量，在这个系统内发生的任何反应，都不会使它的值增加或减少。我们不再认为，这一定律的有效性依赖于对我们目前拥有的方法的改进，以便从实验上证明永恒运动问题。在这种我们无法严格证明而只能被迫予以注意的概括中，展现了上述从拟人化因素中的解放。[④]

（马小东 译，成素梅 校）

---

① M. Planck, *Sinn und Grenzen der exakten Wissenschaft*. Leipzig, 1967, p. 6.

② M. Planck, *A Survey of Physical Theory*. New York, 1960, p. 4.

③ 炼金术是中世纪的一种化学哲学思想与实践，是现代化学的雏形。其目标是将一些一般金属转化为黄金，发现灵丹妙药和长生不老之药。西方的炼金术士称丹药为点金石（philosopher's stone）。——校者注。

④ Ivi, p. 6. 他（没有）处理的热功当量，参见 M. Planck, *Das Prinzip der Erhaltung der Energie*. Leipzig, Berlin, 1924, (fifth ed.), 写于 1887 年哥廷根大学哲学系举办的一次有奖竞赛之际，他获得二等奖（一等奖空缺），委员会评论说"作者特别回避了对成为我们关于热功当量数值的知识基础的那些实验研究的批评讨论"（p. XI），也参见 p. 148ff，在这里，他讨论了用归纳法和演绎法确立证据的可能性；在这篇关于热力学 7 的论文中，他当然提及 19 世纪测定热功当量的实验，强调了所用的为能量守恒原理提供关键条件的各种设备。M. Planck, *Treatise on Thermodynamics*. Berlin, 1897, reprint New York 1945, p. 43.

How to Defend Society Against Science

Paul Feyerabend

Practitioners of a strange trails, friends, enemies, ladies and gentlemen: Before starting with my talk, let me explain to you how it came into existence.

About a year ago I was short of funds. So I accepted an invitation to contribute to a book dealing with the relation between science and religion. To make the book sell I thought l should make my contribution a provocative one and the most provocative statement one can make about the relation between science and religion is that science is a religion. Having made the statement the core of my article I discovered that lots of reasons, lots of excellent reasons, could be found for it. I enumerated the reasons, finished my article, and got paid. That was stage one.

Next I was invited to a Conference for the Defence of Culture. I accepted the invitation because it paid for my flight to Europe. I also must admit that I was rather curious. When I arrived in Nice I had no idea what I would say. Then while the conference was taking its course I discovered that everyone thought very highly of science and that everyone was very serious. So I decided to explain how one could defend culture from science. All the reasons collected in my article would apply here as well and there was no need to invent new things. I gave my talk, was rewarded with an outcry about my "dangerous and ill considered ideas", collected by ticket and went on to Vienna. That was stage number two.

Now I am supposed to address you. I have a hunch that in some respect you are very different from my audience in Nice. For one, you look much younger. My audience in Nice was full of professors, businessmen, and television executives, and the average age was about 58 1/2. Then I am quite sure that most of you are considerably to the left of some of the people in Nice. As a matter of fact, speaking somewhat superficially I might say that you are a leftist audience while my audience in Nice was a rightist audience. Yet despite all these differences you have some things in common. Both of you, I assume, respect science and knowledge. Science, of course, must be reformed and must be made less authoritarian. But once the reforms are carried out, it is a valuable source of knowledge that must not be contaminated by ideologies of a different kind. Secondly, both of you are serious people. Knowledge is a serious matter, for the Right as well as for the Left, and it must be pursued in a serious spirit. Frivolity is out, dedication and earnest application to the task at hand is in. These similarities are all I need for repeating my Nice talk to you with hardly any change. So, here it is.

1. Fairytales

I want to defend society and its inhabitants from all ideologies, science included. All ideologies must be seen in perspective. One must not take them too seriously. One must read them like fairytales which have lots of interesting things to say but which also contain wicked lies, or like ethical prescriptions which may be useful rules of thumb but which are deadly when followed to the letter.

Now, is this not a strange and ridiculous attitude? Science, surely, was always in the forefront of the fight against authoritarianism and superstition. It is to science that we owe our increased intellectual freedom vis-a-vis religious beliefs; it is to science that we owe the liberation of mankind from ancient and rigid forms of thought. Today these forms of thought are nothing but bad dreams—and this we learned from science. Science and enlightenment are one and the same thing even the most radical critics of society believe this. Kropotkin wants to overthrow all traditional institutions and forms of belief, with the exception of science. Ibsen criticises the most intimate ramifications of nineteenth-century bourgeois ideology, but he leaves science untouched. Levi-Strauss has made us realise that Western Thought is not the lonely peak of human achievement it was once believed to be, but he excludes science from his relativization of ideologies. Marx and Engels were convinced that science would aid the workers in their quest for mental and social liberation. Are all these people deceived? Are they all mistaken about the role of science? Are they all the victims of a chimaera?

To these questions my answer is a firm *Yes and No.*

Now, let me explain my answer.

My explanation consists of two parts, one more general, one more specific.

The general explanation is simple. Any ideology that breaks the hold a comprehensive system of thought has on the minds of men contributes to the liberation of man. Any ideology that makes man question inherited beliefs is an aid to enlightenment. A truth that reigns without checks and balances is a tyrant who must be overthrown and any falsehood that can aid us in the overthrow of this tyrant is to be welcomed. It follows that seventeenth-and eighteenth-century science indeed *was* an instrument of liberation and enlightenment. It does not follow that science is bound to *remain* such an instrument. There is nothing inherent in science or in any other ideology that makes it *essentially* liberating. Ideologies can deteriorate and become stupid religions. Look at Marxism. And that the science of today is very different from the science of 1650 is evident at the most superficial glance.

For example, consider the role science now plays in education. Scientific "facts" are taught at a very early age and in the very same manner in which religious "facts" were taught only a century ago. There is no attempt to waken the critical abilities of the pupil so that he may be able to see things in perspective. At the universities the situation is even worse, for indoctrination is here carried out in a much more systematic manner. Criticism is not entirely absent. Society, for example, and its institutions, are criticised most severely and often most unfairly and this already at the elementary school level. But science is excepted from the criticism. In society at large the judgement of the scientist is received with the same reverence as the judgement of bishops and cardinals was accepted not too long ago. The move towards "demythologization", for example, is largely motivated by the wish to avoid any clash between Christianity and scientific ideas. If such a clash occurs, then science is certainly right and Christianity wrong. Pursue this investigation further and you will see that science has now become as oppressive as the ideologies it had once to fight. Do not be misled by the fact that today hardly anyone gets killed for joining a scientific heresy. This has nothing to do with science. It has something to do with the general quality of our civilization. Heretics in science are still made to suffer from the *most severe* sanctions this relatively tolerant civilization has to offer.

But—is this description not utterly unfair? Have I not presented the matter in a very distorted light by using tendentious and distorting terminology? Must we not describe the situation in a very different way? I have said that science has become *rigid*, that it has ceased to be an instrument of *change* and *liberation*, without adding that it has found the *truth*, or a large part thereof. Considering this additional fact we realise, so the objection goes, that the rigidity of science is not due to human wilfulness. It lies in the nature of things. For once we have discovered the truth—what else can we do but follow it?

This trite reply is anything but original. It is used whenever an ideology wants to reinforce the faith of its followers. "Truth" is such a nicely neutral word. Nobody would deny that it is commendable to speak the truth and wicked to tell lies. Nobody would deny that—and yet nobody knows what such an attitude amounts to. So it is easy to twist matters and to change allegiance to truth in one's everyday affairs into allegiance to the Truth of an ideology which is nothing but the dogmatic defense of that ideology. And it is of course *not* true that we *have* to follow the truth. Human life is guided by many ideas. Truth is one of them. Freedom and mental independence are others. If Truth, as conceived by some ideologists, conflicts with freedom, then we have a choice. We may abandon freedom. But we may also abandon Truth. (Alternatively, we may adopt a more sophisticated idea of truth that no longer contradicts freedom; that was Hegel's solution.) My criticism of modern science is that it inhibits freedom of thought. If the reason is that it has found the truth and now follows it then I would say that there are better things than first finding, and then following such a monster.

This finishes the general part of my explanation.

There exists a more specific argument to defend the exceptional position science has in society today. Put in a nutshell the argument says (1) that science has finally found the correct method for achieving results and (2) that there are many results to prove the excellence of the method. The argument is mistaken—but most attempts to show this lead into a dead end. Methodology has by now become so crowded with empty sophistication that it is extremely difficult to perceive the simple errors at the basis. It is like fighting the hydra—cut off one ugly head, and eight formalizations take its place. In this situation the only answer is superficiality: when sophistication loses content then the only way of keeping in touch with reality is to be crude and superficial. This is what I intend to be.

2. Against Method

There is a method, says part (1) of the argument. What is it? How does it work? One answer which is no longer as popular as it used to be is that science works by collecting facts and inferring theories from them. The answer is unsatisfactory as theories never follow from facts in the strict logical sense. To say that they may yet be supported by facts assumes a notion of support that (a) does not show this defect and is (b) sufficiently sophisticated to permit us to say to what extent, say, the theory of relativity is supported by the facts. No such notion exists today, nor is it likely that it will ever be found (one of the problems is that we need a notion of support in which grey ravens can be said to support "All ravens are black"). This was realised by conventionalists and transcendental idealists who pointed out that theories shape and order facts and can therefore be retained come what may. They can be retained because the human mind either consciously or unconsciously carries out its ordering function. The trouble with these views is that they assume for the mind what they want to explain for the world, viz., that it works in a regular fashion. There is only one view which overcomes all these difficulties. It was invented twice in the nineteenth century, by Mill, in his immortal essay *On Liberty*, and by some Darwinists who extended Darwinism to the battle of ideas. This view takes the bull by the horns: theories cannot be justified and their excellence cannot be shown without reference to other theories. We may explain the *success* of a theory by reference to a more comprehensive theory (we may explain the success of Newton's theory by using the general theory of relativity); and we may explain our preference for it by comparing it with other theories.

Such a comparison does not establish the intrinsic excellence of the theory we have chosen. As a matter of fact, the theory we have chosen may be pretty lousy. It may contain contradictions, it may conflict with well-known facts, it may be cumbersome, unclear, *ad hoc* in decisive places, and so on. But it may still be better than any other theory that is available at the time. It may in fact be the best lousy theory there is. Nor are the standards of judgement chosen in an absolute manner. Our sophistication increases with every choice we make, and so do our standards. Standards compete just as theories compete and we choose the standards most appropriate to the historical situation in which the choice occurs. The rejected alternatives (theories; standards; "facts") are not eliminated. They serve as correctives (after all, we may have made the wrong choice) and they also explain the content of the preferred views (we understand relativity better when we understand the structure of its competitors; we know the full meaning of freedom only when we have an idea of life in a totalitarian state, of its advantages—and there are many advantages—as well as of its disadvantages). Knowledge so conceived is an ocean of alternatives channelled and subdivided by an ocean of standards. It forces our mind to make imaginative choices and thus makes it grow. It makes our mind capable of choosing, imagining, criticising.

Today this view is often connected with the name of Karl Popper. But there are some very decisive differences between Popper and Mill. To start with, Popper developed his view to solve a special problem of epistemology—he wanted to solve "Hume's problem. " Mill, on the other hand, is interested in conditions favourable to human growth. His epistemology is the result of a certain theory of man, and not the other way around. Also Popper, being influenced by the Vienna Circle, improves on the logical form of a theory before discussing it while Mill uses every theory in the form in which it occurs in science. Thirdly, Popper's standards of comparison are rigid and fixed, while Mill's standards are permitted to change with the historical situation. Finally, Popper's standards eliminate competitors once and for all: theories that are either not falsifiable or falsifiable and falsified have no place in science. Popper's criteria are clear, unambiguous, precisely formulated; Mill's criteria are not. This would be an advantage if science itself were clear, unambiguous, and precisely formulated. Fortunately, it is not.

To start with, no new and revolutionary scientific theory is ever formulated in a manner that permits us to say under what circumstances we must regard it as endangered: many revolutionary theories are unfalsifiable. Falsifiable versions do exist, but they are hardly ever in agreement with accepted basic statements: every moderately interesting theory is falsified. Moreover, theories have formal flaws, many of them contain contradictions, *ad hoc* adjustments, and so on and so forth. Applied resolutely, Popperian criteria would eliminate science without replacing it by anything comparable. They are useless as an aid to science. In the past decade this has been realised by various thinkers, Kuhn and Lakatos among them. Kuhn's ideas are interesting but, alas, they are much too vague to give rise to anything but lots of hot air. If you don't believe me, look at the literature. Never before has the literature on the philosophy of science been invaded by so many creeps and incompetents. Kuhn encourages people who have no idea why a stone falls to the ground to talk with assurance about scientific method. Now I have no objection to incompetence but I do object when incompetence is accompanied by boredom and self-righteousness. And this is exactly what happens. We do not get interesting false ideas, we get boring ideas or words connected with no ideas at all. Secondly, wherever one tries to make Kuhn's ideas more definite, one finds that they are *false*. Was there ever a period of normal science in the history of thought? No—and I challenge anyone to prove the contrary.

Lakatos is immeasurably more sophisticated than Kuhn. Instead of theories he considers research programmes which are sequences of theories connected by methods of modification, so-called heuristics. Each theory in the sequence may be full of faults. It may be beset by anomalies, contradictions, ambiguities. What counts is not the shape of the single theories, but the tendency exhibited by the sequence. We judge historical developments and achievements over a period of time, rather than the situation at a particular time. History and methodology are combined into a single enterprise. A research programme is said to progress if the sequence of theories leads to novel predictions. It is said to degenerate if it is reduced to absorbing facts that have been discovered without its help. A decisive feature of Lakatos' methodology is that such evaluations are no longer tied to methodological rules which tell the scientist either to retain or to abandon a research programme. Scientists may stick to a degenerating programme; they may even succeed in making the programme overtake its rivals and they therefore proceed rationally whatever they are doing (provided they continue calling degenerating programmes degenerating and progressive programmes progressive). This means that Lakatos offers *words* which *sound* like the elements of a methodology; he does not offer a methodology. There is no method according to the most advanced and sophisticated methodology in existence today. This finishes my reply to part (1) of the specific argument.

3. Against Results

According to part (2), science deserves a special position because it has produced *results*. This is an argument only if it can be taken for granted that nothing else has ever produced results. Now it may be admitted that almost everyone who discusses the matter makes such an assumption. It may also be admitted that it is not easy to show that the assumption is false. Forms of life different from science either have disappeared or have degenerated to an extent that makes a fair comparison impossible. Still, the situation is not as hopeless as it was only a decade ago. We have become acquainted with methods of medical diagnosis and therapy which are effective (and perhaps even more effective than the corresponding parts of Western medicine) and which are yet based on an ideology that is radically different from the ideology of Western science. We have learned that there are phenomena such as telepathy and telekinesis which are obliterated by a scientific approach and which could be used to do research in an entirely novel way (earlier thinkers such as Agrippa of Nettesheim, John Dee, and even Bacon were aware of these phenomena). And then—is it not the case that the Church saved souls while science often does the very opposite? Of course, nobody now believes in the ontology that underlies this judgement. Why? Because of ideological pressures identical with those which today make us listen to science to the exclusion of everything else. It is also true that phenomena such as telekinesis and acupuncture may eventually be absorbed into the body of science and may therefore be called "scientific" . But note that this happens only after a long period of resistance during which a science *not yet* containing the phenomena wants to get the upper hand over forms of life that contain them. And this leads to a further objection against part (2) of the specific argument. The fact that science has results counts in its favour only if these results were achieved by science alone, and without any outside help. A look at history shows that science hardly ever gets its results in this way. When Copernicus introduced a new view of the universe, he did not consult *scientific* predecessors, he consulted a crazy Pythagorean such as Philolaos. He adopted his ideas and he maintained them in the face of all sound rules of scientific method. Mechanics and optics owe a lot to artisans, medicine to midwives and witches. And in *our* own day we have seen how the interference of the state can advance science: when the Chinese communists refused to be intimidated by the judgement of experts and ordered traditional medicine back into universities and hospitals there was an outcry all over the world that science would now be ruined in China. The very opposite occurred: Chinese science advanced and Western science learned from it. Wherever we look we see that great scientific advances are due to outside interference which is made to prevail in the face of the most basic and most "rational" methodological rules. The lesson is plain: there does not exist a single argument that could be used to support the exceptional role which science today plays in society. Science has done many things, but so have other ideologies. Science often proceeds systematically, but so do other ideologies (just consult the records of the many doctrinal debates that took place in the Church) and, besides, there are no overriding rules which are adhered to under any circumstances; there is no " scientific methodology" that can be used to separate science from the rest. *Science is just one of the many ideologies that propel society and it should be treated as such* (this statement applies even to the most progressive and most dialectical sections of science) . What consequences can we draw from this result?

The most important consequence is that there must be a *formal separation between state and science* just as there is now a formal separation between state and church. Science may influence society but only to the extent to which any political or other pressure group is permitted to influence society. Scientists may be consulted on important projects but the final judgement must be left to the democratically elected consulting bodies. These bodies will consist mainly of laymen. Will the laymen be able to come to a correct judgement? Most certainly, for the competence, the complications and the successes of science are vastly exaggerated. One of the most exhilarating experiences is to see how a lawyer, who is a layman, can find holes in the testimony, the technical testimony, of the most advanced expert and thus prepare the jury for its verdict. Science is not a closed book that is understood only after years of training. It is an intellectual discipline that can be examined and criticised by anyone who is interested and that looks difficult and profound only because of a systematic campaign of obfuscation carried out by many scientists (though, I am happy to say, not by all). Organs of the state should never hesitate to reject the judgement of scientists when they have reason for doing so. Such rejection will educate the general public, will make it more confident, and it may even lead to improvement. Considering the sizeable chauvinism of the scientific establishment we can say: the more Lysenko affairs, the better (it is not the *interference* of the state that is objectionable in the case of Lysenko, but the *totalitarian* interference which kills the opponent rather than just neglecting his advice). Three cheers to the fundamentalists in California who succeeded in having a dogmatic formulation of the theory of evolution removed from the textbooks and an account of Genesis included. (But I know that they would become as chauvinistic and totalitarian as scientists are today when given the chance to run society all by themselves. Ideologies are marvelous when used in the companies of other ideologies. They become boring and doctrinaire as soon as their merits lead to the removal of their opponents.) The most important change, however, will have to occur in the field of education.

4. Education and Myth

The purpose of education, so one would think, is to introduce the young into life, and that means: into the *society* where they are born and into the *physical universe* that surrounds the society. The method of education often consists in the teaching of some *basic myth*. The myth is available in various versions. More advanced versions may be taught by initiation rites which firmly implant them into the mind. Knowing the myth, the grown-up can explain almost everything (or else he can turn to experts for more detailed information). He is the master of Nature and of Society. He understands them both and he knows how to interact with them. However, *he is not the master of the myth that guides his understanding.*

Such further mastery was aimed at, and was partly achieved, by the Presocratics. The Presocratics not only tried to understand the *world.* They also tried to understand, and thus to become the masters of, the *means of understanding the world.* Instead of being content with a single myth they developed many and so diminished the power which a well-told story has over the minds of men. The sophists introduced still further methods for reducing the debilitating effect of interesting, coherent, "empirically adequate" etc. etc. tales. The achievements of these thinkers were not appreciated and they certainly are not understood today. When teaching a myth we want to increase the chance that it will be understood (i. e. no puzzlement about any feature of the myth), believed, *and accepted.* This does not do any harm when the myth is counterbalanced by other myths: even the most dedicated (i. e. totalitarian) instructor in a certain version of Christianity cannot prevent his pupils from getting in touch with Buddhists, Jews and other disreputable people. It is very different in the case of science, or of rationalism where the field is almost completely dominated by the believers. In this case it is of paramount importance to strengthen the minds of the young, and " strengthening the minds of the young" means strengthening them *against* an easy acceptance of comprehensive views. What we need here is an education that makes people *contrary*, *counter-suggestive*, without making them incapable of devoting themselves to the elaboration of any single view. How can this aim be achieved?

It can be achieved by protecting the tremendous imagination which children possess and by developing to the full the spirit of contradiction that exists in them. On the whole children are much more intelligent than their teachers. They succumb, and give up their intelligence because they are bullied, or because their teachers get the better of them by emotional means. Children can learn, understand, and keep separate two to three different languages ("children" and by this I mean three to five-year olds, *not* eight year olds who were experimented upon quite recently and did not come out too well; why? because they were already loused up by incompetent teaching at an earlier age) . Of course, the languages must be introduced in a more interesting way than is usually done. There are marvellous writers in all languages who have told marvellous stories—let us begin our language teaching with *them* and not with " der Hund hat einen Schwanz" and similar inanities. Using stories we may of course also introduce " scientific " accounts, say, of the origin of the world and thus make the children acquainted with science as well. But science must not be given any special position except for pointing out that there are lots of people who believe in it. Later on the stories which have been told will be supplemented with "reasons", where by reasons I mean further accounts of the kind found in the tradition to which the story belongs. And, of course, there will also be contrary reasons. Both reasons and contrary reasons will be told by the experts in the fields and so the young generation becomes acquainted with all kinds of sermons and all types of wayfarers. It becomes acquainted with them, it becomes acquainted with their stories, and every individual can make up his mind which way to go. By now everyone knows that you can earn a lot of money and respect and perhaps even a Nobel Prize by becoming a scientist, so many will become scientists. They will *become* scientists *without having been taken in by the ideology of science*, they will *be* scientists *because they have made a free choice.* But has not much time been wasted on unscientific subjects and will this not detract from their competence once they have become scientists? Not at all! The progress of science, of good science depends on novel ideas and on intellectual freedom: science has very often been advanced by outsiders (remember that Bohr and Einstein regarded themselves as outsiders). Will not many people make the wrong choice and end up in a dead end? Well, that depends on what you mean by a "dead end". Most scientists today are devoid of ideas, full of fear, intent on producing some paltry result so that they can add to the flood of inane papers that now constitutes "scientific progress" in many areas. And, besides, what is more important? To lead a life which one has chosen with open eyes, or to spend one's time in the nervous attempt of avoiding what some not so intelligent people call "dead ends"? Will not the number of scientists decrease so that in the end there is nobody to run our precious laboratories? I do not think so. Given a choice many people may choose science, for a science that is run by free agents looks much more attractive than the science of today which is run by slaves, slaves of institutions and slaves of "reason". And if there is a temporary shortage of scientists the situation may always be remedied by various kinds of incentives. Of course, scientists will not play any predominant role in the society I envisage. They will be more than balanced by magicians, or priests, or astrologers. Such a situation is unbearable for many people, old and young, right and left. Almost all of you have the firm belief that at least *some* kind of truth has been found, that it must be preserved, and that the method of teaching I advocate and the form of society I defend will dilute it and make it finally disappear. You have this firm belief; many of you may even have reasons. *But what you have to consider is that the absence of good contrary reasons is due to a historical accident*; it does *not* lie in the nature of things. Build up the kind of society I recommend and the views you now despise (without knowing them, to be sure) will return in such splendour that you will have to work hard to maintain your own position and will perhaps be entirely unable to do so. You do not believe me? Then look at history. Scientific astronomy was firmly founded on Ptolemy and Aristotle, two of the greatest minds in the history of Western Thought. Who upset their well-argued, empirically adequate and precisely formulated system? Philolaos the mad and antediluvian Pythagorean. How was it that Philolaos could stage such a comeback? Because he found an able defender: Copernicus. Of course, you may follow your intuitions as I am following mine. But remember that your intuitions are the result of your "scientific" training where by science I also mean the science of Karl Marx. My training, or, rather, my non-training, is that of a journalist who is interested in strange and bizarre events. Finally, is it not utterly irresponsible, in the present world situation, with millions of people starving, others enslaved, downtrodden, in abject misery of body and mind, to think luxurious thoughts such as these? Is not freedom of choice a luxury under such circumstances? Is not the flippancy and the humour I Want to see combined with the freedom of choice a luxury under such circumstances? Must we not give up all self indulgence and *act*? Join together, and *act*? This is the most important objection which today is raised against an approach such as the one recommended by me. It has tremendous appeal, it has the appeal of unselfish dedication. Unselfish dedication—to what? Let us see!

We are supposed to give up our selfish inclinations and dedicate ourselves to the liberation of the oppressed. And selfish inclinations are what? They are our wish for maximum liberty of thought in the society in which we live *now*, maximum liberty not only of an abstract kind, but expressed in appropriate institutions and methods of teaching. This wish for concrete intellectual and physical liberty in our own surroundings is to be put aside for the time being. This assumes, first, that we do not need this liberty for our task. It assumes that we can carry out our task with a mind that is firmly closed to some alternatives. It assumes that the correct way of liberating others *has always been found* and that all that is needed is to carry it out. I am sorry, I cannot accept such doctrinaire self-assurance in such extremely important matters. Does this mean that we cannot act at all? It does not. But it means that *while acting we have to try to realise as much of the freedom I have recommended so that our actions may be corrected in the light of the ideas we get while increasing our freedom.* This will slow us down, no doubt, but are we supposed to charge ahead simply because some people tell us that they have found an explanation for all the misery and an excellent way out of it? Also we want to liberate people not to make them succumb to a new kind of slavery, *but to make them realise their own wishes*, however different these wishes may be from our own. Self-righteous and narrow-minded liberators cannot do this. As a rule they soon impose a slavery that is worse, because more systematic, than the very sloppy slavery they have removed. And as regards humour and flippancy the answer should be obvious. Why would anyone want to liberate anyone else? Surely not because of some *abstract* advantage of liberty but because liberty is the best way to free development *and thus to happiness.* We want to liberate people so that *they can smile.* Shall we be able to do this if we ourselves have forgotten how to smile and are frowning on those who still remember? Shall we then not spread another disease, comparable to the one we want to remove, the disease of puritanical self-righteousness? Do not object that dedication and humour do not go together—Socrates is an excellent example to the contrary. *The hardest task needs the lightest hand or else its completion will not lead to freedom but to a tyranny much worse than the one it replaces.*

如何保护社会免受科学之害

保尔·费耶阿本德[*]

各位创业者、朋友们、反对者们、女士们和先生们:

在开始讲演之前,请允许我向你们说明这次讲演的初衷。

大约在一年前,我由于缺钱,接受了一份邀请,来撰写一本研究科学与宗教关系的书。为了能让这本书成为畅销书,我想到,我应该使我的书稿具有挑衅性,人们对科学与宗教之间的关系作出的最有挑衅的陈述是,科学就是一种宗教。当使这种说法成为我文章的核心时,我发现,有许多理由,许多出色的理由,支持这个论点。我列举了这些理由来结束我的论题,并拿到了稿酬。这是阶段一。

接下来,我应邀出席一个为文化辩护的会议。我之所以接受邀请,是因为会议资助了我到欧洲的旅费。也必须承认,我相当好奇。当我到达尼斯时,我根本不知道要说些什么。然后,我发现,在会议期间,人人都很看重科学,人人都很认真。因此,我决定说明,人们应该如何保护文化免受科学之害。我的文章中收集到的所有理由,在这里也适用,不需要发明新东西。我的讲演换来了我前往维也纳的机票,其回报是,大家对我的"危险而草率的想法"提出了强烈的抗议。这是阶段二。

现在,我该向你们讲演了。我有一种预感,在某个方面,你们与我在尼斯的听众很不同。首先,你们看起来更加年轻。我在尼斯的听众全都是教授、商人、电视台台长,平均年龄大约是58.5岁。因此,我非常相信,你们中的大多数人比尼斯的大多数听众左很多。事实上,完全从表面上讲,我会说,你们是左派听众,而我在尼斯的听众是右派听众。然而,尽管有这些差别,你们还是有某些共同之处的。我假定,你们双方都尊重科学和知识。当然,我们必须变革科学,必须使科学不成为独裁者。但是,一旦进行变革,有价值的知识源一定不能被不同类型的意识形态所污染。其次,你们双方都是认真之人。对于左派和右派来说,

---

* 保尔·费耶阿本德(Paul Feyerabend),20世纪著名的奥地利裔美国籍科学哲学家。本文原文发表于《激进哲学》(*Radical Philosophy*)1975年夏季第11期,是作者于1974年11月在英国苏塞克斯大学举行的哲学学会上的一个演讲的修改稿。本译文的刊发得到了作者家人的允许与授权。——译者注

知识是一个严肃的问题，必须以认真的精神来追求。不能轻率，要有奉献精神，并以最真诚的态度对待当前的任务。这些相似之处决定了我有必要毫无改变地为你们重复我在尼斯的讲演。下面就是我的讲演。

一、童 话 故 事

我希望保护社会及其居民免受包括科学在内的所有意识形态之害。我们必须正确地看待所有的意识形态。人们一定不会很认真地接受它们。人们一定像童话故事或伦理规定一样解读它们，童话故事讲了许多有趣的事情，但也含有邪恶的谎言；伦理规定可能是有用的经验法则，但当照着做时，却是致命的。

难道现在这不是一种古怪而荒谬的态度吗？科学无疑总是冲在反对独裁主义和迷信的前沿。与宗教信仰相比，科学能使我们增进思想自由；科学使我们从陈旧而僵化的思想形式中解放出来。今天，这些思想形式都是噩梦——而且这是科学告诉我的。科学和启蒙运动是一回事——即使是最激进的社会批评家也相信这一点。克鲁泡特金（Kropotkin）希望推翻所有的传统制度和信仰形式，但科学除外。易卜生（H. J. Ibsen）批判了19世纪资产阶级意识形态的最直接的分歧，但他没有去碰科学。列维－斯特劳斯（Levi-Strauss）使我们意识到，西方思想（western thought）不是人们曾经相信的人类成就的孤峰（lovely peak），但他认为，意识形态都是相对的，科学除外。马克思和恩格斯确信，科学有助于工人们探索他们的心理解放和社会解放。所有这些人都受骗了吗？他们都误解了科学所起的作用了吗？他们都是幻想的受害者吗？

对这些问题而言，我的回答是坚定的，是和不是。

现在，请允许我说明我的答案。

我的说明由两部分组成，一部分较为一般，一部分较为特殊。

一般说明是简单的。任何一种意识形态，只要能打破人们持有的综合的思想体系，就都有助于人的解放；任何一种意识形态，只要能使人们对承袭的信念产生质疑，就都有助于启蒙。没有相互制衡就占有主导地位的真理，是必须被推翻的暴君，在推翻这个暴君的过程中，能对我们有帮助的任何错误，都是深受欢迎的。因而，17世纪和18世纪的科学确实是解放和启蒙的工具。但不能因此断定，科学一定依然还是这样一种工具。科学或任何其他的意识形态不可能生来就成为本质上的解放工具。意识形态会退化，会变成愚蠢的宗教。今天的科学显然完全不同于1650年的科学，这是最明显不过的事情。

例如，考虑当前科学在教育中所起的作用。孩子很小就接受科学"事实"的教育，其教育方式与19世纪教授宗教"事实"的方式完全一样，根本不打算唤醒学生的批判能力，使他们能够高瞻远瞩地看问题。在大学里，情况甚至更糟，因为大学里以非常系统化的方式进行教育灌输。不是完全缺乏批评。例如，对社会及其制度的批评最严重，通常最不公平，而且这已经波及小学层。但是，科学却免于批评。总的说来，在社会上，人们尊敬地接受科学家的判断，就像不久前尊敬地接受教皇和主教的判断一样。例如，"去神秘化"（demythologization）的发展趋势，在很大程度上，是希望避免基督教与科学观念之间的冲突。如果发生这样的冲突，那么，科学肯定是正确的，基督教是错误的。进一步追求这种研究，你会看到，科学现在已经变得与曾经斗争过的意识形态一样是强迫性的。今天，没有人因为加入科学的异端而惨遭杀害，千万不要被这种事实所误导。这与科学无关，与我们文明的一般品质相关。科学中的异端者仍然遭到了这种相对宽容的文明不得不给予的最严厉的制裁。

但是——这种描述是绝对公平的吗？难道我不是以很失真的眼光，用有偏见的和曲解的术语，来提出这个问题吗？难道我们不准以完全不同的方式来描述这种情况吗？我说科学已经成为僵化的，不再是变化与解放的工具，并没有增加真理的内容或大部分真理的内容。考虑到这个附加的事实，我们意识到，科学的僵化并不是由于人类的任性（反对意见认为如此），而在于问题的本质。因为一旦我们发现了真理——接下来我们能干什么呢？

这种老生常谈的回答没有原创性。只要一种意识形态希望强化其追随者的信仰，它就会被利用。"真理"是一个很漂亮的中性词。人人都赞成说实话，厌恶说假话。没有人否认这一点——可是，也没有人知道，这样一种态度意味着什么。因此，问题很容易被扭曲，人们在日常事务中拥护的真理变成了拥护一种意识形态的真理，这种真理只对那种意识形态进行教条式的辩护。于是，我们不得不遵循真理，这当然是不正确的。人类的生活受许多观念的引导。真理只是其中之一。其他还有自由和精神独立。正如某些空想家所设想的那样，如果真理与自由发生冲突，那么，我们只能有一种选择。我们可以放弃自由，但我们也可以放弃真理（在可选择的意义上，我们可以采纳不再与自由相冲突的更精致的真理观；那是黑格尔的解决方案）。我对现代科学的批评是，它抑制了思想自由。如果理由是，人们发现了真理，现在是遵循真理，那么，我将会说，有些东西比首先发现然后遵循这样一个怪物更好。

到此我的一般说明部分结束了。

为科学在当今社会具有的独特地位辩护，有一种更特殊的论证。一言以蔽之，这种论证认为：①科学最终发现了得到结果的正确方法；②有许多结果证明了这种方法是卓越的。这种论证被误解了——但最大的企图是表明，这会陷入僵局。方法论到如今已经变成了如此空洞的诡辩，使人很难在基础上觉察到简单的错误。这就像与九头蛇搏斗——砍掉一个凶恶的头，还有八种形式取而代之。在这种情况下，唯一的回答是浅薄的：当诡辩失去内容时，与实在保持联系的唯一方式将会是粗俗的和浮浅的。这正是我的意图所在。

二、反对方法

存在着一种方法，比如说，这种论证的部分①。它是什么呢？它是如何起作用的呢？不再像过去那样流行的一种回答是：科学通过收集事实并从事实推出理论来进行。当理论在严格的逻辑意义上不是由事实推论出来时，这种回答是不能令人满意的。理论仍然可以得到事实的支持，这一说法假设了一个支持概念：（a）这个概念没有表明这种缺点，（b）这个概念在多大程度上允许我们说，比如，相对论得到了事实的支持，是极其复杂的。这样的概念如今根本不存在，甚至以后也不可能找到（问题之一是，我们需要一个支持概念，在这里，灰色的乌鸦能被说成是支持了"所有的乌鸦都是黑的"）。约定主义者和超验的唯心主义者早已意识到这一点，他们指出，理论形成事实和整理事实，因此，无论如何，理论是能被提炼出来的。理论之所以能被提炼，是因为人类的思维有意识或无意识地贯彻它的有序功能。这些观点的困惑是，它们假设了希望说明世界的思维，即思维是以一种有规则的方式进行的。只有一种观点能克服所有这些困难。这种观点在19世纪被发明过两次，一次是穆勒在他的传世之作《论自由》一书中发明的；另一次是把达尔文主义扩展到观念战的达尔文主义者发明的。这种观点很有冒险精神：理论不可能得到辩护，而且，如果不参照其他理论，就无法表明这些理论的卓越。我们通过参照一个更全面的理论，能说明一个理论的成功（我们可以用广义相对论来说明牛顿理论的成功）；而且，我们通过把此理论与其他理论相比较，来说明我们对它的偏爱。

这样一种比较并没有确立我们所选择的理论的内在卓越性。事实上，我们所选择的理论可能是极其糟糕的。它可能包含着矛盾，它可能与众所周知的事实相矛盾，它在某些关键之处可能是繁杂的、不明确的、特设性的等。但它可能仍然比当时可利用的其他理论更好。它事实上可能是已有的最好的差理论。也不能以绝对的方式来选择判断标准。我们的辩解随着我们每次作出的选择的增加而增加，我们的标准也是如此。标准之间的竞争恰好像理论之间的竞争一样，我们选择最适合于我们作出选择的历史情况的标准。并没有排除那些被拒绝的替代选择（理论、标准、"事实"）。它们起到了正确的作用（毕竟，我们已经作出了错误的选择），它们也说明了所偏爱的观点的内容（当我们理解相对论的竞争者的结构时，我们更好地理解了相对论；只有当我们拥有极权主义国家的生活观时，我们才能知道自由的全部意义：它的优势——有许多优势——还有它的劣势）。如此构想的知识是受海量标准引导与细化的海量选择。它迫使我们的思维作出富有想象的选择，并因此而不断扩展。它使我们的思维具有选择、想象、批评的能力。

今天，这种观点通常与卡尔·波普尔（Karl Popper）的名字联系在一起。但波普尔与穆勒有着很关键的不同。首先，波普尔提出他的观点，是为了解决认识论的特殊问题——他想解决"休谟问题"。而穆勒对有利于人性成长的条件感兴趣。他的认识论是人的某种理论的结果，而不是反过来。其次，波普尔深受维也纳学派的影响，他在讨论理论之前，改进了理论的逻辑形式，而穆勒以科学中出现的形式运用每个理论。再次，波普尔的比较标准是僵化的和固定的，而穆勒的标准允许随历史情境的变化而变化。最后，波普尔的标准彻底地排除了竞争者：理论要么是不可证伪的，要么是可证伪的，被证伪的理论没有科学地位。波普尔的标准是清楚的、明确的，得到了精确的阐述；穆勒的标准则不是。如果科学本身是清楚的、明确的和被精确地阐述的，这本该是一种优势。幸好不是这种情况。

首先，我们从来没有以下列方式表述一个新的革命性的科学理论，即允许我们说，在什么情况下，我们必须把这个理论看成是危险的：许多革命性的理论是不可证伪的。可证伪的版本确实是存在的，但它们总是很难与公认的基本陈述相一致：每一个适当地令人感兴趣的理论都是可被证伪的。此外，理论有形式上的缺陷，其中，许多理论还含有矛盾、专门的调整，如此等等。坚决地应用的话，波普尔的标准就能排除科学，而不是通过任何比较来取代科学。这些标准不能用来促进科学的发展。在过去的10年间，不同的思想家，比如，库恩和拉卡托斯，已经意识到了这一点。库恩的观念是有趣的，但是，它们太模糊，甚至只是吹牛。如果你不相信我，可以查看一下文献。科学哲学的文献以前从来没有受到如此之多的奉承之人和不胜任者的干扰。库恩鼓励连为什么说一块石头会落向地面的原因都不知道的那些人有自信谈论科学方法。现在，我不是反对不胜任，而是反对伴有厌倦与自以为是的不胜任。而恰好发生了这种情况。我们没有获得令人感兴趣的错误观念，我们获得了令人厌倦的观念或毫无观念的词语。其次，凡是在人们试图使库恩的观念更明确的地方，人们就会发现，库恩的观念是错误的。在思想史上总是有常规科学时期吗？没有——并且，我向证明相反结论的任何人提出挑战。

拉卡托斯远比库恩老练。他考虑研究纲领，替代了考虑理论，研究纲领是通过修正方法（所谓的启发法）联系起来的一系列理论。这个序列中的每个理论都可能充满错误。它可能受到反常、矛盾、含糊的困扰。重要的不是单个理论的形成，而是这个序列所表现出的倾向。我们判断一段时期的历史发展和成就，而不是某一特定时间的情况。历史与方法论结合成为一项事业。如果理论系列导致了新颖的预言，一个研究纲领就被说成是进步的。如果一个研究纲领降低为引人入胜的事实，而这些事实不是在它的帮助下发现的，就被说成是退化的。拉卡托斯的方法论的一个关键特征是，这样的进化不再受制于告诉科学家保留或放弃一个研究纲领的方法论规则。科学家可以坚持一个退化的纲领；他们甚至可以成功地使这个纲领超过它的竞争者，因此，他们无论做什么，都会在理性意义上继续下去（倘若他们继续把退化的纲领称为是退化的，把进步的纲领称为是进步的）。这意味着，拉卡托斯提供了听起来像一种方法论要素的词语；他没有提供一种方法论。根据今天最先进和最精致的方法论的观点，根本就没有方法。这结束了我对特殊论证的部分①的答复。

三、反 对 结 果

根据部分②，科学应有特殊的地位，因为它产生了结果。只有在理所当然地认为其他理论总是不会产生结果时，这才是一种论证。现在大家承认，几乎讨论这个问题的每一个人都相信这样一个假设。大家也承认，很难表明这个假设是错误的。不同于科学的生活形式要么消失了，要么退化到不可能进行公正比较的程度。尽管如此，这种情况不像 10 年前那样没有希望。我们已经熟悉了下面的医学诊断和治疗的方法：这些方法是有效的（甚至也许比相应的西医更有效）并且以一种完全不同于西方科学的意识形态为基础。我们已经了解到，有些现象，比如，心灵感应和心灵致动，被科学的进路排除了，而且，我们能以全新的方式用这些现象来做研究［像内阿格里帕·冯·特斯海姆（Agrippa von Nettesheim）、约翰·迪伊（John Dee）甚至培根（Bacon）之类的早期思想家觉察到了这些现象］。于是——教会拯救灵魂，而科学通常反其道而行之，难道不是这种情况吗？当然，现在无人相信构成这种判断基础的本体论。为什么呢？因为意识形态的强制与今天使我们听从科学排除其他一切的那些意识形态相吻合。同样正确的是，像心灵致动和针灸之类的现象完全可能被并入到科学中，因而可以被称为"科学的"。但注意，在尚未包括这些现象的科学希望在包括这些现象的生活形式中占有上风期间，只有经过长期的反抗，这种情况才会发生。于是，这导致了进一步反对特殊论证的部分②。仅当没有任何外在帮助，并且只通过科学获得那些结果时，科学有结果这个事实，才能算作是科学的优势。审视历史可知，科学很少以这种方式获得它的结果。哥白尼在提出一种新的宇宙观时，并没有顾及科学的前辈，他着迷地参照毕达哥拉斯主义者，比如，菲罗劳斯（Philolaos）的观点。他采纳菲罗劳斯的观点，并且不顾所有科学方法的可靠规则而坚持这些观念。力学和光学在很大程度上归功于工匠，医学归功于助产士和巫婆。而且，在我们自己的时代，我们已经看到，国家的干预如何能促进科学：当中国共产党不受专家判断的威胁，命令中医回到大学和医院时，全世界的人都强烈抗议，说现在中国破坏了科学。但完全相反的情况出现了：中国的科学前进了，西方的科学向中国的科学学习。只要我们随便看看，我们就会明白，伟大的科学进步，归因于占优势的外界干预，而不顾最基本的和最"合理的"方法论规则。教训是明显的：没有一种单独的论证能用来支持科学在当今社会中所起的异常作用。科学做了许多事情，但其他意识形态也是如此。科学通常是系统地前进的，但其他意识形态也是如此（在教会里发生的许多教条的争论的记录，仅供参考），除此之外，根本没有在任何情况下都能坚持的压倒一切的规则；根本没有能用来把科学与其他一切分离开来的"科学方法论"。科学只是推动社会发展的许多意识形态之一，而且，就应该这样看待科学（这种陈述甚至适用于科学的最进步和最辩证的部分）。我们能从这种结果中得出什么结论呢？

最重要的结论是，国家与科学之间必须在形式上是分离的，就像国家与教会之间在形式上是分离的一样。只有在允许任何政治团体或其他施压组织影响社会的程度上，科学才能影响社会。关于重要的项目，可以咨询科学家，但最终的判断必须留给被民主地选出的顾问团。这些团体主要由外行组成。外行能够得出正确的判断吗？非常肯定地说，就胜任能力而言，科学的复杂化和科学的成功被极大地夸大了。最令人兴奋的经验之一是看一下一位外行律师如何能找出由最高级专家提供的技术性证词中的漏洞，为陪审团作出裁定做准备。科学不是只有通过几年训练之后才能被理解的谜。科学是一种智力训练（intellectual discipline），它能受到感兴趣之人的考察和批评，它看起来困难和深奥，只是因为许多科学家（尽管我高兴地说，不是所有的科学家）打了一场混淆的系统战役。当科学家有理由这么做时，国家机构应该毫不犹豫地拒绝科学家的判断。这样的拒绝将会教育公众，使公众更有信心，甚至可能会促进改革。考虑到科研机构的大沙文主义，我们会说："李森科事件"越多越好（这不是"李森科事件"中令人反感的国家干预，是灭掉对手而不是忽视其劝告的极权主义者的干预）。在加利福尼亚，原教旨主义者成功地对进化论作出了教条的阐述，他们的三种喝彩被从教科书中删除了，包括创世纪在内（但我知道，今天，当他们有机会操纵整个社会时，他们成为与科学家一样的沙文主义者和极权主义者。有些意识形态，当与其他意识形态一起运用时，是不可思议的。一旦它们的优势致使除掉其对手，它们就成为令人讨厌的和教条的）。然而，最重要的变化还是发生在教育领域。

四、教育与神话

人们应该认为，教育的目标是把年轻人领向生活，那就意味着：领向他们出生的社会和社会周围的物质世界。教育方法通常存在于某个基本神话的教义里。神话有各种不同的版本。更高级的版本可以通过入会仪式来教授，入会仪式稳固地使教义植入人心。知道了神话，长大后就几乎能说明一切（要不然，他会向专家求教更详细的信息）。他是自然界与社会的主人。他理解自然界与社会，他知道如何与它们打交道。然而，神话指导了他的理解，他却不是神话的主人。

苏格拉底之前的哲学家致力于并且部分地获得这种进一步的控制。苏格拉底之前的哲学家不仅试图理解世界，他们也试图领会理解世界的手段，因而成为理解世界的手段的主人。他们提出了许多神话，而不是满足于一种神话，因此，削弱了一个广泛传说的故事能深入人心的力量。这些诡辩学家还提出了进一步的方法，削弱有趣的、连贯的、"经验上适当的"等传说的微弱作用。这些思想家的成就没有得到重视，无疑今天也没有被理解。当我们教神话时，我们希望增加理解它（即对该神话的任何特征都不感到困惑）、相信它和接受它的机会。当该神话与其他神话相抗衡时，这没有任何坏处：在某个版本的基督教教义中，甚至最有奉献精神（即极权主义）的牧师也不能阻止他的学生接触佛教徒、犹太人等名誉不好的人。科学和理性主义的情况完全不同，在这里，整个领域几乎完全被信徒们所控制。在这种情况下，对于强化年轻人的心灵来说，这是至关重要的，"强化年轻人的心灵"意味着强化他们反对轻易地接受综合的观点。我们在这里所需要的教育是使人们唱反调、提出相反的建议，使他们有能力投身于阐述一种观点。这个目的如何才能达到呢？

通过保护孩子拥有的极大的想象力和通过开发他们身上充满矛盾的精神能够达到这个目的。总的来看，孩子们比他们的老师更聪明。他们放弃自己的智慧，是因为他们受到了威胁，或者，是因为他们的老师比他们更能用好情感的手段。孩子们能学习、理解和分别持有两三种语言（我说的"孩子"是指 3~5 岁的孩子，不是 8 岁的孩子，对 8 岁的孩子做实验是最近的事，效果不太好；为什么呢？因为他们已经在更小的时候被不合格的老师毁了）。当然，语言一定是以比通常更有趣的方式引入的。在所有的语言中，都有讲杰出故事的杰出作家——让我们用这些故事来教我们的语言，不是用"这条狗有尾巴"（der Hund hat einen Schwanz）等类似空洞的故事来教我们的语言。我们当然也可以用故事引入"科学的"解释，比如说，关于世界起源的科学解释，因此使得孩子很好地熟悉科学。但科学除了指出有许多相信它的人之外，一定给不出任何特殊的立场。后来，用"各种理由"补充这些传说的故事，我这里所说的理由是指在故事传统中进一步发现的那种解释。当然，也有相反的理由。在这个领域内，专家告诉学生各种正反理由，因此，年轻一代就会熟悉所有类型的说教和各种人。年轻一代熟悉了各种人，熟悉了他们的故事，每个人都能确定走哪条路。到如今，人人都知道，你可能很会赚钱且受人尊敬，甚至也许能成为一名科学家并荣获诺贝尔奖，这样，许多人将会成为科学家。他们只有不被科学的意识形态所欺骗，才能成为科学家，即他们之所以将是科学家是因为他们作出了自由的选择。但是，一旦他们成为科学家，就不会在非科学问题上花费很多时间吗？这不会有损于他们的胜任能力吗？根本不会！科学的进步，好科学的进步，依赖于新颖的观念和思想自由：科学经常是靠局外人推动的（记得玻尔和爱因斯坦把他们自己看成是局外人）。许多人都不会作出错误的选择，最终陷入僵局吗？大概这取决于你的"僵局"是什么意思。今天的大多数科学家都缺乏观念，充满了恐惧，旨在产生某种无价值的结果，使他们卷入空洞的论文洪流之中，如今在许多领域内，这些空洞的论文构成了"科学的进步"。除此之外，更重要的是什么呢？是引导人们心明如镜地选择生活吗？或者，是让人们花时间紧张地企图避免一些不那么聪明的人所谓的"僵局"吗？科学家的人数将不会减少到最终没有人管理我们的宝贵实验室吗？我不这么认为。如果有一种选择的话，许多人可以选择科学，因为由自由的行动者管理的科学，看起来比由奴隶（体制的奴隶和"理由"的奴隶）管理的科学，更具有吸引力。而且，如果出现了暂时缺少科学家的情况，总会有各种鼓励措施来加以补救。当然，科学家在我设想的社会里起不到任何主导作用。他们将受到巫师或神父或占星家很大的制衡。对于许多人（老人和年轻人，右派和左派）来说，这样一种情况是无法忍受的。几乎你们大家都有的坚定信念是：至少已经发现了某种真理，它必须被坚持，而我拥护的教学法和我保卫的社会形式会弱化真理，直至最后使之消失。你们有这种坚定的信念；你们中的许多人甚至有各种理由。但你不得不考虑的是，缺乏很好地唱反调的理由实属历史意外；这不在于问题的本质。建立我推荐的这种社会，确立你现在蔑视的观点（当然还不知道这些观点），将返回到这样一种壮观的场面：你将不得不努力工作，坚持你自己的立场，而且也许将完全不能这样做。你们不相信我吗？那么，看一下历史吧。科学天文学坚定地建立在托勒密和亚里士多德体系的基础上，他们是西方思想史上最伟大的两个聪明人。是谁推翻了他们的有说服力的论证呢？是经验的适当性和精确表述的体系吗？是狂妄古老的毕达哥拉斯学派的菲罗劳斯（Philolaos）。菲罗劳斯如何能这样东山再起呢？因为他发现了一个有能力的辩护者：哥白尼。当然，像我凭我的直觉一样，你们也可以凭你们的直觉。但记住，你们的直觉是你们的"科学"训练的结果，在这里，我意指的科学也是卡尔·

马克思的科学。我受到的训练，或更确切地说，我没有受到的训练是对奇闻怪事感兴趣的新闻记者的训练。最后，在当前的世界局势中，由于成千上万的人还在挨饿，还有人受奴役、受压迫，肉体与心灵处于悲惨的贫穷之中，那么，思考诸如此类的奢侈思想，不是完全不负责任吗？在这些情况下，自由选择不是一种奢侈吗？在这些情况下，我希望看到的轻率和情绪，不是自由选择的一个组成部分吗？我们一定不要放弃所有的自我纵容和行动吗？联合起来然后行动吗？这是今天提出的对我推荐的这条进路的最重要的反对。它大声呼吁，呼吁无私奉献。无私奉献——奉献什么呢？让我们来看看吧！

我们假定抛弃我们的自私倾向，奉献于解放受压迫者。那么，自私倾向是什么呢？它们是我们在现在生活的社会里最大限度地解放思想的愿望，最大限度地解放不仅是一个抽象类，而且体现在适当的教学制度和教学方法中。在我们自己的环境中具体地解放智力和肉体的这种愿望，暂时不予考虑。首先，这假设，我们不需要这种解放来完成我们的任务。它假设，我们以坚决禁止某些替代选择的思想来完成我们的任务。它假设，我们历来能发现解放他人的正确方式，所需要的一切是付诸实践。我很抱歉，在这些极其重要的问题上，我无法接受这样一种教条的自信。这意味着我们根本不能行动吗？不是。它不过意味着，我们在行动时，不得不努力实现我所推荐的自由，以使我们可以根据我们在增加自由时获得的观念纠正我们的行动。这无疑将会使我们放慢脚步，但假如我们拼搏进取，只是因为有些人告诉我们，他们已经对所有的苦难作出说明，并找到摆脱苦难的最佳出路了吗？我们也希望解放人民，使他们不要成为新型的奴隶，而是实现自己的愿望，然而，这些愿望可能不同于我们自己的愿望。自以为是和思想狭隘的解放者做不到这一点。作为一个规则，他们很快强加一种束缚，这种束缚由于更系统，比他们已经排除的很草率的束缚更加糟糕。而且，至于情绪和玩世不恭，答案应该是显而易见的。为什么任何一个人都希望解放其他人呢？当然不是因为解放的某种抽象的优势，而是因为解放是自由发展从而达到幸福的最好方式。我们希望解放人民，使他们能微笑。如果我们自己已经忘记如何微笑，反对仍然牢记微笑的那些人，我们能做这一点吗？与我们希望排除的疾病（清教徒式的自以为是的疾病）相比，我们就不会传播另一种疾病吗？不要反对奉献精神和情绪不能共存——苏格拉底是一个极好的反例。最艰难的任务需要最巧妙的手段，否则，任务的完成将不会通向自由，而是通向比它所替代的东西更加糟糕的暴政。

（成素梅 译）

Does *The Structure of Scientific Revolutions* Permit a Feminist Revolution in Science?

Helen E. Longino

1

Kuhn's influence on feminist science studies and feminist theory of knowledge might well be understood as an example of the principle of unintended consequences. Kuhn's notions of theory-laden meaning and observation and of revolutionary science were embraced by feminist thinkers, who applied them in ways that seem their natural and logical extensions. Judging from remarks in later essays such as "The Trouble with the Historical Philosophy of Science", Kuhn would have had serious reservations about these applications, as he had about many of those in science studies who took his views as a mandate to inquire into the social nature of scientific inquiry. [1] Nevertheless, the power of his challenge to logical empiricist philosophy of science provided a philosophical basis for a wide range of critical approaches to the sciences.

When *The Structure of Scientific Revolutions* burst upon the academic scene in the early 1960s, the second wave of feminism was in its earliest stages: identifying the forms of legal discrimination against women, challenging the cultural expectations of femininity, agitating for access to contraception and abortion, and rebelling against the second-class status accorded to women in the civil rights and antiwar movements. By the early 1970s, feminists in the academy had expanded the reach of feminism to analysis and critique of the research and scholarship that supported the discriminatory legal and social treatment of women. They argued that the traditional academic disciplines were guilty not only of professional discrimination in university admissions, hiring, and promotion, but also of scholarly discrimination. History, literary studies, sociology, and anthropology were characterized by an exclusionary focus on men's activities and accomplishments and a minimizing account of women, women's activities, and gender relations. Psychology and biology seemed to rationalize this imbalance by supporting views of male and female nature that coincided with the *Kinder*, *Kuche*, *und Kirche* view of women's roles in the social world. Nowadays we would call the focus on masculine activities "androcentrism" and the minimizing of women's activities "sexist" or "gender-biased"; in the beginning, there was no language with which to identify and diagnose the (mis)representation and neglect that were the lot of women.

Feminist researchers uncovered the activities of women ignored in conventional scholarship and challenged the values that privileged men's contributions to social and cultural life over women's contributions. Feminist historians began to reveal the shifts in consciousness of women's situation, the long but forgotten legacy of feminist activity in the past, and to make clear the shifts in gender ideology and relations over time. Feminist literary scholars analyzed the sexual politics of canonical works of literature. They reclaimed writers such as Sappho, Jane Austen, and Emily Dickinson and reconsidered the literary values that consigned them to the margins of literary history. While disciplines such as history and literary studies were obviously susceptible to charges of bias and more responsive to new directions, those disciplines that cloaked themselves in the garb of scientific objectivity and neutrality posed a quite different problem. [2] If hypotheses prejudicial to women passed the standards of scientific scrutiny, not only the content but the forms of validation of that content required challenge.

Kuhn's *Structure* offered a vocabulary for articulating the complex critique of science and of its ideology that feminist scientists sought to develop, and many feminist biologists and psychologists referred to Kuhn in their work. In spite of Kuhn's animating and legitimating role in the initial stages, however, a number of his ideas are in considerable tension with the aspirations of feminist scientists and philosophers of science. This essay will describe the landscape opened to feminists by Kuhn's work, showing how Kuhn's ideas made possible an increasingly sophisticated and far-reaching understanding of gender ideology in science. I will then discuss the limitations of those ideas from a feminist perspective and indicate how feminists have modified them to support a more transformative agenda.

2

It is difficult in 2002 to credit the kinds of ideas about women and gender relations that commanded scientific respectability in the 1950s and 1960s. As though the suffrage movement had never happened, these exhibited a remarkable continuity with ideas current in the nineteenth century. In keeping with a legal system that subordinated married women to their husbands and a culture that saw motherhood as the ultimate female accomplishment, psychologists attributed docility, dependence, and nurturance to women and assertiveness, independence, and competitiveness to men. Women who exhibited the latter rather than the former were deemed unhealthy, but the masculine traits were the signs of human psychological well-being. [3] Psychological sexism began to give way under assault from empirical researchers who rejected the stereotypes and sought to establish unbiased standards for mental health. [4] These challenges were developed in the name of science and objectivity.

The biological sciences, although harboring just as much sexism, however, seemed impervious to feminist critique. This was due in part to their more secure position in the scientific hierarchy and in part to the embeddedness of biological sexism in more extensive theoretical structures. Sociobiologists held that several biological factors accounted for most social behavior. For example, the different patterns of courting and sexual behavior were explained by differential parental investment on the part of males and females—males having a minor investment in each offspring and females, whose eggs represented a greater investment of resources, having a major investment. Males found reproductive advantage in frequent mating; females found reproductive advantage in careful selection of mates. To make a long story short, the patterns of male dominance and female subordination observed in just about every human society had their basis in biological differences. Ethologists obligingly found male dominance everywhere, not just in human societies, but also in other primates, in birds, in mountain sheep. And for each instance there was a biological basis. Sociobiology was a solution to the problem of self-sacrificing behavior in a variety of species. Evolutionary theory held that variations that conferred survival advantages were inherited, but how could a behavior that conferred disadvantage be inherited? The sociobiological answer to that question was known as "kin selection" . The genes of relatives of the self-sacrificing individual were passed on to offspring, and because relatives share genes, the self-sacrificing individual's genes, some of them, too, found their way into successive generations. The account of sex differences was just part of a much bigger theoretical picture. Views about human evolution located the selection pressures favoring distinctively human anatomical adaptations in male behavior: Not only were men dominant by nature in contemporary societies, but it was male variability in the past that provoked evolutionary change.

Biologist Ruth Hubbard, writing about evolutionary theory, ethology, and sociobiology, took an approach whose broad outlines she attributes to Kuhn. "Every theory is a self-fulfilling prophecy that orders experience into the framework it provides.... There is no such thing as objective value-free science"[5], she says before documenting the multitude of ways in which students of animal behavior ascribed stereotypical feminine characteristics to female organisms (or organisms identified as female) from algae to apes. What they claim to observe in nature are the very codes of behavior prescribed for human societies by Victorian mores. Scientists are constrained by the language available to them to describe behavior; their vocabulary as well as their gender ideology produce androcentric and sexist accounts of social behavior. Even when their descriptions indicate female activity rather than passivity, they read them as conforming to gender stereotypes. Hubbard quotes passages from Darwin such as the following from *The Descent of Man*: "Man is more courageous, pugnacious and energetic than woman, and has more inventive genius."[6] She summarizes his view as follows:

So here it is in a nutshell: men's mental and physical qualities were constantly improved through competition for women and hunting, while women's minds would have become vestigial if it were not for the fortunate circumstance that in each generation daughters inherit brains from their fathers.[7]

Darwin's ideas about male and female roles in evolution are not restricted to the 1870s but, as Hubbard shows, are repeated in the work of ethologists and physical anthropologists in the 1970s.

The approach that focused on the vocabulary of a theory could also work for neuroendocrinological approaches to behavior. The language used to describe the behavior of girls and boys was saturated with gender theory masquerading as common knowledge. For example, girls who engaged in activities stereotypically associated with boys were referred to colloquially as "tomboys", while boys who engaged in the less strenuous activities stereotypically associated with girls were described as afflicted with the "sissy syndrome". Research on children who had been exposed to anomalous levels of gonadal hormones in utero purported to show an increased incidence of tomboyism in girls exposed to excess levels of androgenic hormones in utero and an increased incidence of sissy syndrome in boys exposed to insufficient levels of androgenic hormones in utero. The gender loading extended from the labels for behavior to the very identification of the gonadal hormones themselves. Male and female gonads secrete a set of chemically very similar steroidal hormones. The gonads differ in the relative proportion of these steroids that they produce, but some quantity of all these hormones is necessary for proper physiological function in male and female mammalian organisms. A number of feminist scholars have examined the discussions in the 1930s concerning appropriate no- menclature for the gonadal hormones. In spite of their chemical similarity and their phys- iological roles in both male and female organisms, the researchers who wished to identify them by gender prevailed over those who preferred a more neutral form of identi- fication. Even more striking was the asymmetry in labeling. Those hormones secreted in greater quantities by male gonads were called "androgenic" (male-producing), while those secreted in greater quantities by female gonads were called "estrogenic" (frenzy- producing). This labeling had consequences for subsequent research: the multiple functions of the hormones were not recognized for decades. [8]

The story of gonadal hormone research thus provided an excellent case for application of Kuhn's notions of theory-laden observation and theory-laden meaning. Theory operated at several levels. In the first instance, theory about gender difference informed the ways in which the hormones were identified and labeled. In the second, that identification and labeling determined what physiological effects of hormone secretion were observed and recorded. Kuhn's ideas of theory-ladenness gave feminist scientists and scholars a language in which to express their perception that even methodologically impeccable science could nevertheless incorporate social biases. Its very impeccability, in turn, gave those biases intellectual respectability. Other critics have argued that plain old empiricism has the resources to support critical examination of scientific sexism[9], but Kuhn's way of putting things was preferred by those feminists who thought that the problems for women posed by the sciences ran deeper than sloppy observation practices. There was a connection between the content of theories in sociobiology and neuroendocrinology with the institutional exclusion of women from scientific education and careers. Kuhn's larger picture of scientific change, which emphasized the sociological factors in scientific revolutions, offered a means of articulating and examining that connection.

Kuhn's notions of theory-ladenness could explain how two researchers could look at a pride of wild horses, one seeing a male with his harem of attendant females and the other seeing a group of females tolerating the presence of their stallion in exchange for his services. They seemed to offer ways to make sense of the perpetuation of gender stereotypes in an area allegedly characterized by objective empirical methods. This is why even scholars who seemed only to be calling into question the empirical adequacy of biological descriptions of females and female-identified behavior invoked Kuhn, or Kuhnian ideas, in elaborating their critiques. The problem wasn't individual biased scientists but a shared gender ideology. The difficulty, however, is that feminists also wanted to say that one description of the horses is right.

Although Hubbard cites Kuhn at the beginning of the essay, she makes clear that the observations reported in support of views about the centrality of males are just wrong. The philosophical views that launch the essay are left behind. She concludes her essay by stating that the gender-biased "paradigm of evolution" requires that women "rethink our evolutionary history"[10]. Part of this rethinking requires getting close to the "raw data". Donna Haraway took Hubbard and other feminists arguing as she had to task for employing the analytic framework of Kuhn for critique and the empiricist framework he had criticized in putting forward a positive program for research.[11] Surely the problem was not Hubbard's but the poverty of the philosophical frameworks she was given to work with. Kuhn gave feminists a platform from which to reject the idea that reports from the field of submissive females and dominant males were just the facts. While he gave feminist scientists a way to talk about the ways in which a socially shared gender ideology had colored observation of males and females of all biological species, his analysis of scientific revolutions did not give them a language or rubric for describing the kinds of changes they wished to recommend. This point becomes clearer in thinking about the work of feminist philosophers of science.

3

Feminist philosophers of science, too, made use of Kuhn's ideas. They focused on different themes in Kuhn's work, extending and modifying his claims about the character of scientific knowledge and about scientific change to address a series of epistemological concerns.

Kathryn Addelson invoked Kuhn both in ethical and in epistemological contexts. The women's movements of the nineteenth and twentieth centuries, Addelson argued, were simultaneously enactments of and calls for moral revolution.[12] Women participating in those movements were stepping out of the prescribed behavior for women by speaking in public in assembly halls and in the streets, thus forcing themselves into public affairs, and the content of their message was a demand for change in those prescriptions that they violated. In the nineteenth century the demands were for the right to vote, to higher education, to property, to divorce, to participation in the civic life of society. In the twentieth century the demands included reproductive rights, an equal rights amendment, equal opportunity, and comparable worth. A moral revolution does not consist in bringing a society's behavior or lower-level commitments into conformity with its higher-level principles, but rather in a deeper change in those constitutive principles, indeed in the very concept of morality. In this way, said Addelson, a moral revolution was similar to a Kuhnian scientific revolution, which involved a change not only in descriptions and explanations of nature, but in the very criteria by which descriptions and explanations were evaluated.

Where Hubbard had been concerned to show the seepage of stereotypes of social life into scientific ideas, Addelson elaborated the connection between the content of ideas and forms of social life. She also appealed to a Kuhnian framework to raise issues about scientific knowledge. Here, she was interested in Kuhn's notion of a paradigm. This enabled him to understand scientific knowledge as consisting not just in " theories and laws, but also metaphysical commitments, exemplars, puzzles, anomalies, and various other features "[13] . Paradigms guided practice, and it was practice that was central to scientific inquiry, not the doctrinal results of practice.

From Addelson's point of view, Kuhn's construal of science as activity, as practice, and his documentation of the rise and fall of theories in the course of scientific change helped to focus attention on the social dynamics among scientists, those that contributed to the persistence of a theory and those that contributed to its replacement by a new one. Kuhn's remarks concerning the adherents of old paradigms and the champions of the new are well known: long-time adherents of paradigms are not converted to the new but retire, leaving the field for others, while champions of new ones tend to be young, uninvested in the success of the old ideas and likely to benefit from the adoption of new ones. Addelson saw Kuhn's work as opening up questions of cognitive authority. Debates during paradigm shifts were contests for cognitive authority, and because of the conceptually pervasive character of paradigms, the outcome of such contests included shifts in the authority to define the fundamental structures of our common world. The victory of an atomistic physical theory in early modern Europe was the triumph of a metaphysical view that extended beyond physics into social life (as individualism) .

Kuhn's acknowledgment of the multiple factors influencing individuals' theoretical, experimental, or practical preferences during paradigm shifts suggested that the boundaries between scientific and extrascientific considerations were fuzzy and/or porous. What made intuitive sense to an individual was an important factor in the judgments of plausibility that cumulatively tipped the balance to one or another of the contestants. But what made intuitive sense, argued Addelson, was largely influenced by one's social experience. To the extent that the metaphysical outlook of those who par-ticipated in and gained authority in the resolution of scientific controversy reflected their social experience, to that extent the metaphysical outlook thus legitimated for the culture at large reflected and thus reinforced a particular social reality.

Addelson focused on sociology, showing how a functionalist metaphysics was expressed in the research agenda and results of 1950s American sociology. She cited anthropological research from the 1970s documenting the different social realities of women and men and of members of different socioeconomic classes. These claims were then integrated in her claim that a system in which only some kinds of people were absorbed into the locus of cognitive authority, the scientific professions and leadership positions in that profession, was a system in which the metaphysical outlook of that class shaped that of the rest of society.

> The leading physicists, biologists, and philosophers of science ... live in societies marked by dominance of group over group. As specialists, they compete for positions at the top of their professional hierarchies that allow them to exercise cognitive authority more widely. Out of such cultural understandings, it is no wonder ... that our specialists present us with metaphysical descriptions of the world in terms of hierarchy, dominance, and competition. [14]

Addelson recognized the power of scientific inquiry not only to shape a society's worldview, but also to represent the world in ways that worked. Her point was that by paying attention to its social structure and correcting the disproportionate privileging of one social group, we could eliminate continuing irrationalities. Of course, with this claim, she lays herself open to the same kind of challenge Haraway raised for Hubbard's vision of a feminist approach to evolution. From within what paradigm are these alleged irrationalities identified as such? Why not call for a completely different science or for the social conditions that might result in such?

Sandra Harding discussed both *The Structure of Scientific Revolutions* and *The Copernican Revolution* in her book *The Science Question in Feminism*. [15] She hailed the former for its demonstration that the rational reconstructions of scientific judgment offered by logical empiricists were misrepresentations of the historical situations whose logic they sought to elucidate. Like Addelson, she saw this volume as legitimating a naturalized approach in science studies, one that looked particularly at the role of social relations to provide explanations of scientific outcomes. This should, in principle, extend to the study of the role of gender relations in the production of science. Harding writes that the Kuhnians and post-Kuhnians, by persisting in treating gender as a biological rather than as a social relation, failed to take the Kuhnian program to its logical extension. That is, they felt free to ignore the effects of gender relations and hence missed an important aspect of the development of modern science.

In a discussion of the historiography of the Scientific Revolution, however, Harding is more critical of Kuhn. *The Copernican Revolution*, in her reading, participates in the treatment of the Scientific Revolution as an instance of the triumph of intellect over superstition. Harding draws especially on passages likening the medieval mind, the mind committed to an Aristotelian worldview, to the minds of children and primitives, and reads Kuhn as celebrating the release of science from morality and politics effected in the sixteenth and seventeenth centuries. This strikes me as not quite fair. *The Copernican Revolution* contributed to the power of the later *Structure* by demonstrating the coherence, plausibility, and empirical adequacy of the Aristotelian physics and worldview. That Kuhn was less able to see into the worldview that promoted and was promoted by the intellectual and technical accomplishments of Copernicus, Galileo, Newton, and their fellow natural scientists does not diminish the implication that modern science, too, is in a similar relation of mutual support with a larger worldview that includes moral and political views as well as metaphysical ones.

This mutual support is precisely what Harding wishes to demonstrate. Rejecting the standard picture of the Scientific Revolution as an origin myth, she fills in the framework provided by *Structure* with work by Marxist historians of science to show the social and political dimensions of the Scientific Revolution. The Scientific Revolution was associated with the end of the feudal order and the emergence of a new middle class, with antiauthoritarianism, with belief in progress, with humanitarian ideals, and with a division of labor that separates the methodical investigation of nature (science) from the maintenance of the institutions that support that investigation (politics) . These social and political developments coemerge with a cosmology characterized by atomism (the view that nature is constituted of ultimately uniform and least bits of inert matter), value neutrality (the impersonal universe that replaces the teleological universe of the medievals), and faith in method (as guarantor of impartiality and independence from political and religious authority). However insightful the Marxist and post-Kuhnian social and historical studies of science were, they neglected the gender relations that were part of the new European world order. Harding cites the work of feminist scholars Ludmilla Jordanova, Carolyn Merchant, and Evelyn Keller to suggest the gendered dimensions of the new world order—its restriction of property rights to male members of the new middle class and the reconstruction of masculinity to harmonize with the new values of early modernism.

Harding's treatment of the possibility of feminist science shows her deepest debt to Kuhn. Feminists are like the seventeenth-century radicals in challenging contemporary structures of authority, in believing in progress insofar as that includes overcoming gender, race, and class hierarchies, stressing educational reform and humanitarianism. And Harding claims that feminists seek knowledge that unifies empirical with moral and political understanding. Harding's view is that this "successor" would be unrecognizable from within the categories of current mainstream science. Thus she treated the relationship between current science and science acceptable to feminists as, like that between Aristotelian science and cosmology and the New Science of the seventeenth century, one of incommensurability. The new science of the twenty-first century would be a unified science but not, as envisioned by the Vienna Circle, a unified science that took physics as its foundation. The new science would be directed by moral and political beliefs, and thus, according to Harding, would take social science, not physics, as its foundation. "Science and theorizing itself" must be reinvented. [16]

Another feminist scholar took a somewhat different but no less radical lesson from Kuhn. Evelyn Fox Keller has been the most visible of feminists concerned with the sciences. She has consistently urged the viability of models of complex interaction in contrast with the reductionist and linear analysis she sees as characterizing contemporary science. Her concern has not been the description of females and gender relations, but the ways in which gender ideologies have been expressed in areas of science having nothing to do with gender or social behavior. Keller writes as a scientist as much as a historian and philosopher, and reports being struck by the resistance or indifference of scientists themselves to Kuhn's claims. [17] Keller noted that while Kuhn's views of scientific change had laid the groundwork for research that investigated the social dimensions of scientific practice and judgment, he had not himself pursued such investigations, thus leaving open the exact nature and role of social and cultural factors in scientific practice. But she noted, "the direct implication of [Kuhn's claims] is that not only different collections of facts, different focal points of scientific attention, but also different organizations of knowledge, different interpretations of the world, are both possible and consistent with what we call science"[18]. Where Harding had linked Kuhnian with Marxist historiography, Keller proposed instead to employ the tools of psychoanalysis to explain simultaneously the scientific community's resistance to Kuhnian ideas, the gendering of past and contemporary science, and the way out of what she saw as a scientific dead end.

Keller relied primarily on what is known as "object relations theory". According to this theory, one of the maturational tasks of infants and children is the development of individual identity. The relations the child has with its closest adults profoundly affect this process and have lasting effects in its overall outlook on and behavior in the world. Male and female infants in typical Western families faced distinctive challenges the psychic resolution of which shaped their orientation to social and physical reality, an orientation that expressed itself cognitively, affectively, and practically. Boys had to achieve their individual identities in a context in which their primary adult figure was their mother and from which their father was largely absent. In a sex-differentiated social and domestic world, their task was then to become something about which they knew very little. Their developmental energies were therefore directed to not becoming that which they knew—their mother. As a consequence, the identity of boys and the men they became was fragile and needed constant reinforcement. One psychic strategy for coping with this need was to develop exaggerated psychic detachment from others. Little girls, on the other hand, because their task was to become female, tended to be over-attached.

Keller applied this analysis of psychological development to explain features of the sciences. The strategy of distancing and disidentification expressed itself affectively in what Keller called a "stance of static autonomy". Because the conditions of masculine individuation induced deep anxieties, it required continual confirmation, provided most vividly and reassuringly by domination of that which one needed not to be. To this point, the analysis is in keeping with that pursued by other object relations feminists. [19] Keller's innovation was to extend it to conceptions of knowledge and of science. In parallel with static autonomy, a cognitive attitude dubbed by Keller "static objectivity" emerged as an aspect of personal development. Static objectivity was characterized by its equation of knowledge with emotional detachment from and control over the objects of knowledge and by its treatment of the pursuit of knowledge, scientific inquiry, as an adversarial process. Static objectivity was contrasted with dynamic objectivity, which aimed at a reliable understanding of the world that granted to its elements their independent integrity and affirmed the connectivity of subjects and objects of knowledge. Knowledge is understood neither as detachment and control nor as loss of identity, but as flexible connection and relationship that acknowledges the autonomy of objects. The normal developmental processes of boys and girls led them to identify the behaviors associated with static autonomy and static objectivity with masculinity. Femininity, by contrast, was characterized by overidentification and the submergence of individuality. Dynamic autonomy and objectivity were (by implication) orientations that could be achieved only through struggle against the prevailing social, including gender, norms.

Keller supported her thesis that modern science was constituted by static objectivity by quoting from writers from Bacon to Simmel to contemporary scientists, and by showing how certain research programs in the sciences were driven by a goal of dominating nature. She supported the feasibility of her alternative—dynamic objectivity—by citing researchers such as Michel Polanyi and Barbara McClintock, who advocated and practiced approaches to science characterized by that attitude. Feminists, including Ruth Hubbard and Sandra Harding, had deplored the reductionism that seemed to characterize modern science. Harding attributed this to the social and economic conditions that permitted the development of modern science. Keller argued that that reductionism was part of a worldview whose tenacity was due to its psychic roots.

Keller could then explain the resistance to Kuhn in the scientific community as a reaction to the threat to cognitive autonomy posed by Kuhn's in-principle acceptance of the role that extrascientific social or subjective factors could play in determining scientific judgment. Where she differed from Harding was not only in stressing the psychological dimensions of scientific cognition, but in affirming that static objectivity was primarily a feature of the ideology of science rather than of its actual practice. Where the ideology of science stressed a unity of purpose in emotional detachment from and practical domination of nature, study of the practices of science revealed a greater variety and richness of ideas. Keller is famous, of course, for her exposition and advocacy of the work of geneticist Barbara McClintock, but other scientists, too, exemplified the ideal of dynamic objectivity, the ability to move in and out of intimate closeness with the objects of knowledge, to employ empathy rather than distance in seeking understanding. These cognitive capacities were associated for Keller with representation of the natural world as complex and heterogeneous, as contrasted with its representation as reducible to one basic level and ultimately explicable by simple one-way causal models. The ideology of science and its emotional connection to an ideology of masculinity explained why interactionist approaches such as McClintock's were consistently marginalized in favor of approaches that pursued forms of knowledge congruent with domination of rather than coexistence with the known. An alternative form of science suitable to feminist purposes did not need to be reinvented. We needed only to look more closely at forms of practice currently relegated to the margins of science for our models.

By stressing the availability throughout science's history of models of natural processes that employed representational or explanatory principles out of step with the mainstream, Keller departed from Kuhn's picture of scientific growth and change. Kuhn held that normal science in a field was characterized by a single explanatory approach and that the presence of multiple approaches signaled its immaturity. In addition, as noted in the discussion of Harding, Kuhn argued that successive (or contesting) theories of the same subject area were incommensurable. Keller does not treat McClintock's views about the mutability of the genome as involving theory-laden observations that researchers committed to different theories could not share. She understands the mainstream rejection of McClintock's views as a function of the mainstream's attachment to a conception of scientific knowledge and an associated metaphysics of nature to which McClintock's views simply did not conform. And while Harding seems to embrace incommensurability in her description of the relation between mainstream science and the science that will replace it, in other ways her conception, too, is at odds with the Kuhnian prescription. The feminist scientific revolution advocated by Harding will come about not because empirical anomalies accumulate and throw the current paradigm into crisis, but because changes in social values and relationships require a different way of knowing the natural world. Kuhn's conceptions of scientific knowledge and scientific change are of value to both of these thinkers because of the challenge he articulates to the then regnant logical empiricist philosophy of science. Science was either a battle between contesting paradigms (revolutionary science) or puzzle solving within a paradigm (normal science). But this characterization of science offered no tools for thinking about how to effect change. These feminists, however, were interested not just in understanding science but also in changing it.

4

It is telling that feminists appealed to Kuhn to legitimate their rejection of positivism (whether its philosophical expression in logical empiricism or its popular expression as scientism) but left Kuhn entirely or partially behind when talking about alternative forms of scientific knowledge. Kuhn's views are a hindrance to that project. There is, however, a way of thinking about scientific inquiry and scientific knowledge that, while indebted to Kuhn and reasonably seen as a product of the Kuhnian revolution in philosophy of science, does not impose the same constraints on the feminist project. Epistemological pluralism, that is, pluralism about knowledge, is grounded in broadly speaking Kuhnian insights about the history of science, but it employs some different philosophical principles. These enable explanation of androcentric or sexist science as something more than just empirically inadequate science, without undermining the case for an alternative. To see this, I recapitulate the relevant basic philosophical ideas of *The Structure of Scientific Revolutions*.

Kuhn claimed that successive theories about the same subject matter, say bodies in motion, were both in contradiction with one another and incommensurable, by which he meant that such theories could not be empirically tested vis-a′-vis one another in the mode envisaged by empiricists. That is, their relationship was not such that they could be comparatively evaluated against a common set of data or facts. Kuhn's explanation for this incommensurability is the theory dependence of meaning and of observation. Logical empiricists held that observation was independent of theory and that the meaning of observation terms and statements was independent of theory. Meaningfulness and confirmation (evidential support) flowed from observation to theory. Kuhn, by contrast, argued that the meaning of observation terms was determined by theory and that the meaning of theoretical terms, too, shifted when their theoretical context changed. For example, "mass" in classical physics refers to a quantity that is conserved, while "mass" in relativity physics refers to a quantity that is (under some conditions) convertible to energy. So what might seem to be common observation terms affording a shared point of contact with the observable world turned out to be, on this theory of meaning, no more than homonyms. [20] Kuhn also held that observation itself was theory-determined, and he supported this contention by citing a variety of psychological experiments that demonstrated the dependence of perception on expectation. One consequence of these views about meaning and observation is that genuine communication between scientists holding different theories, in the grip of different paradigms, is impossible. They may use language that sounds similar, they may point to the same phenomena in their sensory range, but the terms they use are different in meaning, and what they observe when looking at the same phenomena is also different.

Kuhn's views about meaning and observation were most unsettling to philosophers of science. Combined with the view that evidence for scientific hypotheses and theories lies in what can be observed, which Kuhn did not deny, evidential reasoning seemed to be circular: a theory was supported by observations whose content and description were determined by the theory. To counter the charge that this made theory choice entirely subjective (a charge made plausible by Kuhn's analogies with religious conversion), Kuhn claimed that theory choice in science was guided by a set of values. [21] These included accuracy, internal and external consistency, simplicity, breadth of scope, and fruitfulness. While the precise interpretation and relative priority or weight assigned to these values might differ in practice, they nevertheless provided a constant, a touchstone, by reference to which scientific judgment, theory choice, could be understood to be objective. [22] It is clear from Kuhn's discussion of these values, however, that they do not offer an independent yardstick, or as he put it, an algorithm, for the comparison of theories, nor could there be such a yardstick, given his interpretation of incommensurability. Kuhn sometimes elaborated the theory-ladenness of meaning and observation by implying that scientists who held different theories inhabited different worlds. There are different ways of understanding such a claim, but as an articulation of incommensurability, it does not serve feminist science scholars well.

Feminist science scholars want to affirm the (gender) value-ladenness of much contemporary science, the success of that science by conventional measures of success, and the need for (or desirability of) an alternative to the mainstream trends in science. They do not dispute all of the science to which they object merely on the grounds of empirical adequacy, but also on the grounds that it encodes and thus reinforces noxious (sexist, racist, capitalist) social values. They inhabit the same world as do the scientists to whose theories they object—they want to see different scientific accounts of that world. They are not content to ascribe differences to the inhabitation of different worlds or to semantic or cognitive incommensurabilities. They don't just want to do science in their other world; they want to change the way mainstream science is done in our common world. It is not that feminists are committed to a worldview from which it is impossible to understand the science they oppose. They would say they understand it only too well.

I would locate the source of the ultimate unsuitability of Kuhn's views to feminist projects in his notions of the theory dependence and theory-ladenness of meaning and observation. Since these doctrines are problematic on other grounds (both conceptual and empirical) as well, we should look elsewhere for philosophical support. The pluralism of contextual empiricism is based in Kuhnian insights about the nature of scientific change, but it relinquishes the theory-ladenness of meaning and observation as explanations of incommensurability.

Contextual empiricism treats the apparently equivalent empirical adequacy of different theories not as a matter of theory—ladenness, but as a function of differences in background assumptions facilitating inferences between data and hypotheses. [23] The language used to articulate or describe them is independently meaningful. Contextual empiricism sees the logical problem that gives rise to multiple theories as underdetermination (the gap between our evidential resources, whatever we can observe or measure, and our explanatory aspirations, the discovery of principles, capacities, or causal regularities underlying what we experience). This gap is bridged by background assumptions that constitute the context in which empirical, that is, observational, data acquire evidential relevance. A change in assumptions brings a change in evidential relevance. One of the advantages of contextual empiricism is that it offers an account of the historical phenomena that Kuhn sought to explain, that is, that two scientists could look at the same thing—a sealed jar with a dead mouse or a sunset or a pride of lions—but explain what they saw quite differently. Contextual empiricism is, therefore, compatible with pluralism: incompatible theories of the same phenomena can both offer adequate accounts. While pluralism holds that such theories can offer correct accounts, their correctness is judged from the perspective of different background assumptions and cognitive goals. Pluralism also holds that in such cases theories are partial or incomplete, unable to encompass all the aspects of a complex phenomenon in their range. Of course, contextual empiricism has its own philosophical difficulties, which I have addressed elsewhere. [24] Here I want to stress the advantage to feminist science studies of giving up theory-ladenness for contextualism. There are three points of contact.

4. 1　Regarding Incommensurability

The Kuhnian appeal, as we have seen, is to theory-ladenness, which as a general theory of meaning holds that the meanings of all terms in a theory are determined by the theory. Terms have no independent meaning outside of the theory. This approach to meaning, and its corollary concerning observation, while promising to explain how one researcher can see submission where another sees craftiness or how one sees dominance where another sees dependence or stress, is in the end not helpful to feminists because it leaves them unable to criticize the misrepresentations of gender as incorrect for anyone, regardless of their gender ideology. It disables empirical critique of sexist science.

Contextual empiricism permits a different approach to incommensurability. It does not take either theory or experience as a foundation of meaning. It thus departs from both Kuhnian semantic holism and logical empiricist semantic reductionism. While maintaining that the contents of meaning and observation are not theory-determined, it does agree that the categories of observation and measurement are theory— or context— relative. Apparent incommensurability arises when measurements are not separable from their context of measurement. That context provides the questions, the goals, and the standards of measurement. What counts as an observation in one context may not in another. Two theoretical approaches to a phenomenon might be incommensurable to the extent that they take different aspects of the phenomenon to be evidentially relevant or employ different measurement scales relevant to contrasting questions and cognitive goals. These are not incommunicable, as the semantic approach to incommensurability implies, and while neither of the theories provides a common standard, researchers employing different theoretical approaches may still share enough outside of their theories to engage critically with each other's ideas and observations. Treating incommensurability as a function of context avoids the undermining of the feminist empirical critique of sexist science deplored by Haraway.

4. 2　Regarding Paradigms and Normal Science

Kuhn stated that a main characteristic, almost a defining condition, of normal science is the organizing of research under a single paradigm. Contextual empiricism is compatible with the form of pluralism that holds that, in many cases, the phenomena to be explained are so complex that multiple approaches are necessary to provide a comprehensive account. Any single account, while correct, is in such cases incomplete. A clear example is organismic development. Both a gene-centered account and an environmental account of the development of some trait may be correct but partial. Their theoretical structures are such that they cannot, however, be combined into a single account. The existence of multiple approaches is not the sign of scientific immaturity or of preparadigmatic revolutionary science, but if required by the phenomena, it may be an unavoidable feature of normal science. While Keller herself might not endorse this pluralism, pluralism seems to offer a better account of the existence of multiple research traditions in the sciences such as she documents than do Kuhnian paradigms.

4.3 Regarding Values and Scientific Judgment

Kuhn held that the values determinant of (objective) scientific judgment were variably interpreted and variably weighed or prioritized, and even that individual scientists might interpret them in ways influenced by personal, subjective factors. Nevertheless, he also held that they were internal to the scientific community and that the ones he cited were in some way constitutive of scientificity. A contextualist holds that there can be no such in- principle circumscribing of scientific values. The values Kuhn listed are those conventionally recited by philosophers, but they reflect a particular intellectual tradition. There are other values that can be advocated that stand in complicated relations with the conventional ones. And, contra those who saw comfort for social studies of science in Kuhn's ideas, paradigm-governed science suggests that once a paradigm shift has occurred, a single set of values (with an implied prioritizing and interpretation) becomes normative, so that scientific judgment is once again fully internal.

Contextual empiricism, then, follows the Kuhnian approach in taking fidelity to the actual practice of science as a criterion of adequacy for a theory of scientific knowledge. It also stresses the complexity of scientific judgment, its dependence on factors not given in the immediate experimental or observational situation. The differences I have just cataloged, however, make it more amenable to the concerns of feminist science studies.

5

There is one point, however, where Kuhn offers a potentially significant jumping-off point. The last of the scientific values he discusses is fruitfulness. Now this could be understood as comparable to having empirical content, since one way a theory or paradigm can be fruitful is by generating empirical consequences. But Kuhn glosses this value in an interesting way. For him fruitfulness is a theory's or paradigm's capacity to generate interesting puzzles or problems to work on, that is, its capacity to direct research, to provide intellectual challenges. This introduces a note of pragmatism into an otherwise representational account of inquiry. Feminists have been concerned not only with the representation of gender and the use of gender in the representation of nature, but with the ways in which scientific ideas are deployed in the social world. Feminists are concerned to support forms of science that will distribute power throughout society rather than concentrating it in experts. Feminists have become concerned to support forms of inquiry that will preserve rather than consume natural resources. Feminists are concerned to encourage noninvasive and nondominating models of inquiry. Now it may well be that Kuhn's understanding of fruitfulness is entirely inward-looking, restricted to the puzzle-generating capacity. But there seems no in-principle reason not to extend the value of fruitfulness pragmatically understood to include these outward-looking ways in which a theory might be fruitful. If this licenses the evaluation of a theory by reference to the particular kinds of interventions it permits in the world outside the laboratory or seminar room, then we count ourselves indebted to a thinker who included pragmatic as well as representational concerns among the values that ought to guide scientific judgment. And if feminist values come to prevail in that evaluation, then *The Structure of Scientific Revolutions* will turn out, in spite of my earlier reservations, to have abetted a feminist revolution in science.

Notes

1. Thomas Kuhn, "The Trouble with the Historical Philosophy of Science", Robert and Maurine Rothschild Distinguished Lecture, November 19, 1991, Cambridge, MA: An Occasional Publication of the Department of History of Science, Harvard University.

2. I don't mean to minimize the resistance that feminist historians and literary scholars encountered in their efforts to correct scholarly bias, but rather to point to a qualitatively different problem facing feminist thinking about the sciences: a conception of science that to a large extent they themselves shared.

3. A study by Inge Broverman and her colleagues demonstrated the pervasiveness of this double bind for women in the thinking of clinical psychologists. See Inge Broverman, D. M. Broverman, F. E. Clarkson, P. S. Rosenkranz, and S. R. Vogel, "Sex Role Stereotypes and Clinical Judgments of Mental Health", *Journal of Consulting and Clinical Psychology* (1970) 34: 1-7.

4. See, for example, Naomi Weisstein, "Psychology Constructs the Female", in Vivian Gornick and Barbara Moran, eds., *Woman in Sexist Society* (New York: Basic Books, 1971), pp. 133-146; Eleanor Maccoby and Carol Jacklin, *The Psychology of Sex Differences* (Stanford, CA: Stanford University Press, 1974); Paula Caplan, Gael MacPherson, and Patricia Tobin, "Do Sex Differences in Spatial Ability Really Exist?" *American Psychologist* (1985) 40 (7): 786-798. In many ways, this work recapitulates the critique of sex differences that research and ideology performed decades earlier by psychologists such as Leta Hollingworth. The collective amnesia regarding such critique is part of the atmosphere that made a deep analysis such as Kuhn offered so attractive.

5. Ruth Hubbard, "Have Only Men Evolved?" in Ruth Hubbard, Mary Henifin, and Barbara Fried, eds., *Women Look at Biology Looking at Women* (Cambridge, MA: Schenkman, 1979), p. 9.

6. Charles Darwin, *The Origin of Species and the Descent of Man* (New York: Random House, Modern Library Edition, n. d.), as quoted in Hubbard, "Have Only Men Evolved?", p. 19.

7. Hubbard, "Have Only Men Evolved?", p. 20.

8. This story is told in a number of recent histories of reproductive endocrinology. See Nellie Oudshorn, *The Making of the Hormonal Body* (New York: Routledge, 1997), and Anne Fausto Sterling, *Sexing the Body: Gender Politics and the Construction of Sexuality* (New York: Basic Books, 2000). One ironic aspect of the ideological construction of hormonal identity is the labeling of equine estrogen, "premarin", suggesting that it is derived from mares, when it is actually derived from the urine of stallions.

9. See, for example, Noretta Koertge, "Methodology, Idealogy, and Feminist Critiques of Science", in Peter Asquith and Ronald Giere, eds., *PSA 1980*, vol. 2 (East Lansing, MI: Philosophy of Science Association, 1981), pp. 346-359.

10. Hubbard, "Have Only Men Evolved?", p. 31.

11. Donna Haraway, "In the Beginning Was the Word: The Genesis of Biological Theory",

Signs: Journal of Women in Culture and Society (1981) 6 (3): 469-482.

12. Kathryn Addelson, "Moral Revolution", in Julia A. Sherman and Evelyn Torton Beck, eds. , *The Prism of Sex* (Madison: Univer-sity of Wisconsin Press, 1979), pp. 189-227; reprinted in Addelson, *Impure Thoughts* (Philadelphia: Temple University Press, 1991), pp. 35-61.

13. Kathryn Addelson, "The Man of Professional Wisdom", in Sandra Harding and Merrill Hintikka, eds. , *Discovering Reality* (Dordrecht: Reidel, 1983), p. 166.

14. Ibid. , p. 184.

15. Sandra Harding, *The Science Question in Feminism* (Ithaca, NY: Cornell Univer sity Press, 1986), pp. 197-210.

16. Ibid. , p. 251.

17. Evelyn Fox Keller, *Reflections on Gender and Science* (New Haven, CT: Yale University Press, 1985) .

18. Ibid. , p. 5.

19. One of the best-known applications of object relations theory to gender relations is Nancy Chodorow, *The Reproduction of Mothering* (Berkeley: University of California Press, 1978) .

20. This theory can't be right, since "mass" refers to the same feature of bodies in the two theories. One problem with Kuhn's view of meaning is that he seems to have made no distinction between defining properties and contingent features. This is in some ways consonant with Quine's views about meaning, but to treat all contingent features as defining properties as Kuhn seems to have done leaves one no longer able to say of a statement that it is false (and uncertain of what it means to say that it is true) . Kuhn expended considerable effort over the years trying to make his notion of meaning incommensurability plausible, but he did not, to my knowledge, address this aspect of the view. Cf. his Presidential Address to the Philosophy of Science As-sociation: The Road since Structure," in Arthur Fine, Micky Forbes, and Linda Wessels, eds. , *PSA 1990*, vol. 2 (East Lansing, MI: Philosophy of Science As-sociation, 1991), pp. 3-13. Reprinted in Thomas Kuhn, *The Road since Structure*, edited by James Conant and John Haugeland. (Chicago: University of Chicago Press, 2001), pp. 91-104.

21. Thomas Kuhn, "Objectivity, Value Judgment, and Theory Choice", in *The Essential Tension* (Chicago: University of Chicago Press, 1977), pp. 320-339.

22. This aspect of Kuhn's views has been recently taken up by some feminist philoso-

phers of science, but as part of a discussion in general of the notion of epistemic of cognitive values in science. Feminists emphasize either the role of contextual values in determining the weight or application of epistemic values or the variety of possible cognitive values the choice among which can be seen to have social and/or cultural dimensions. See Alison Wylie and Lynn Nelson, "Coming to Terms with Value（s）of Science: Insights from Feminist Science Studies Scholarship", Workshop on Science and Values, Center for Philosophy of Science, University of Pittsburgh, October 9-11, 1998, on the first strategy, and Helen Longino, "Cognitive and Non-Cognitive Values in Science: Rethinking the Dichotomy", in Lynn Hankinson Nelson and Jack Nelson, eds. , *Feminism, Science and Philosophy of Science* （Boston: Kluwer, 1996）, pp. 39-58, on the second.

23. Equivalent empirical adequacy is not the same as empirical equivalence. The latter requires that two theories have the same empirical consequences; the former requires only that two theories be equally successful in predicting observations, but the set of observations relevant to the theories may not be the same.

24. Helen E. Longino, "Toward an Epistemology for Biological Pluralism", in Richard Creath and Jane Maienschein, eds. , *Biology and Epistemology* （Cambridge: Cambridge University Press, 2000）, pp. 262-286; *The Fate of Knowledge* （Princeton: Princeton University Press, 2002）.

《科学革命的结构》与科学中的女性主义革命

海伦·朗基诺[*]

一

库恩对女性主义科学研究和女性主义知识论的影响，很可能被理解为是造成意外结果（the principle of unintended consequences）的一个例子。库恩的意义、观察承载理论的概念、革命科学的概念受到了女性主义者的欢迎，她们应用这些概念的方式，似乎是其自然的逻辑的外延。从库恩后期的论文（如《历史的科学哲学出了问题》）中的评论来判断，他对这些应用会有很多保留意见，就像他对待科学研究领域内的许多人把他的观点看成是要求追究科学探索的社会本性的命令那样。[①] 然而，他对逻辑经验主义科学哲学的挑战力量，为广泛的科学批评进路提供了一种哲学基础。

当《科学革命的结构》在 20 世纪 60 年代早期突然出现在学术领域时，女性主义的第二次浪潮处于最初阶段：辨别法律上歧视妇女的形式，挑战对妇女（femininity）的文化预期，鼓动有权避孕和堕胎，反抗在民权运动和反战运动中赋予妇女的二等地位。到 20 世纪 70 年代初，学术界的女性主义者，把女性主义的范围扩大到分析和批评那些赞同在法律上、社会上歧视女性的研究和学术（scholarship）。她们认为，传统学术界在大学入学、聘用和职务晋升方面不仅存在职业歧视，而且存在学术歧视。历史学、文学研究、社会学和人类学的特点是，专门关注男性的活动和成就，而把对妇女、妇女的活动和性别关系的解释最小化。心理学和生物学支持男性和女性天性和认为妇女的社会角色是"孩子、厨房和教堂"（Kinder, Kuche, und Kirche）的说法相吻合的观点，似乎使这种不平衡合理化了。现在，我们将把只关注男性活动的做法称为"大男子主义"，把

---

* 海伦·朗基诺（Helen E. Longino），美国著名女性主义科学哲学家，斯坦福大学 Lewis 哲学教授。本文载于 *Thomas Kuhn*, edited by Thomas Nickles, Cambridge University Press, 2003. ——译者注

① Thomas Kuhn, "The Trouble with the Historical Philosophy of Science", Robert and Maurine Rothschild Distinguished Lecture, November 19, 1991, Cambridge, MA: An Occasional Publication of the Department of History of Science, Harvard University.

极力贬低女性活动的做法称为"男性至上主义"或"性别偏见"；在开始时，尚无语言可用于辨别和诊断对妇女命运的这种（错误）表征和忽视。

女性主义研究者发现了传统学术所忽视的女性活动，挑战了认为有优势的男性对社会和文化生活的贡献超过女性的价值观。女性主义历史学家开始揭示女性的处境意识（consciousness of women's situation）的转变（即过去长期被遗忘了的女性主义活动的遗产）和澄清性别意识形态与性别关系随时间的变化。女性主义文学研究者剖析了经典文学著作中的性别政治学。她们重提像萨福（Sappho）①、简·奥斯汀（Jane Austen）② 和艾米莉·狄金森（Emily Dickinson）③ 之类的作家，并且重新考虑置她们于文学史边缘的文学价值。当像历史和文学研究之类的学科，明显地易于被指控为是有偏见的并更多地响应新的趋势时，披着科学的客观性和中立性外衣的那些学科，提出了一个非常不同的问题。④ 如果对女性不利的假说通过了科学审查的标准，那么，不仅其内容，而且确证内容的形式都应受到挑战。

库恩的《科学革命的结构》为阐述女性主义科学家试图提出的对科学及其意识形态的复杂批评提供了语汇，而且，许多女性主义生物学家和心理学家在她们的著作中都提到了库恩。尽管库恩在开始阶段起到了激励和合法化的作用，然而，他的许多观点与女性主义科学家和女性主义科学哲学家的抱负是很不和谐的。本文将描述库恩的工作通向女性主义的前景，表明库恩的思想如何使得对科学中的性别意识形态的日益成熟的广泛理解成为可能。然后，我将从女性主义的视角讨论那些思想的局限性，简要地说明女性主义者如何修改这些思想来支持更有变化的议题。

二

到 2002 年，人们很难信任曾在 20 世纪 50～60 年代掌控科学名望（respectability）的关于女性和性别关系的那类思想。仿佛选举权运动从未发生过似的，这些思想表现出与 19 世纪的思潮明显的连续性。为了与已婚女性从属于丈夫的法律体系以及把成为母亲视为女性最终成就的文化保持一致，心理学家们把顺从、依赖和

---

① 萨福（Sappho，约公元前 630 或者 612～约前 592 或者 560），古希腊著名的女抒情诗人，一生写过不少情诗、婚歌、颂神诗、铭辞等。——译者注

② 简·奥斯汀（Jane Austen，1775～1817），英国著名女小说家。她在短暂的一生中创作了 6 部小说，其中《傲慢与偏见》为经典名著。——译者注

③ 艾米莉·狄金森（Emily Dickinson，1830～1886），美国抒情女诗人，意象派诗歌的先驱之一，被誉为美国现代派诗歌的鼻祖。——译者注

④ 我不是想要最小化女性主义历史学家和文学家在她们努力纠正学术偏见时遇到的阻力，而是要指出女性主义思考面临的不同性质的困难：她们自己在很大程度上共同拥有的一种科学观。

养育归于女性，把自信、独立和竞争归于男性。性格中展现了后者而不是前者的女性就被认为是不健康的，而男子汉的个性是人类心理健康的标志。① 在拒绝陈腐观念和寻求确立心理健康的公正标准的经验研究者的攻击下，心理学的男性至上主义开始作出让步。② 这些挑战是以科学和客观性的名义提出的。

然而，生物科学尽管也隐藏着同样多的男性至上主义，似乎没有受到女性主义批评的影响。一部分原因是它在科学等级中处于较安全的地位，另一部分原因是生物学的男性至上主义镶嵌在更加广泛的理论结构中。社会生物学家认为，几个生物学因素解释了大多数社会行为。例如，求偶和性行为的不同模式，可以通过雄性和雌性表现出来的不同的亲代投入来说明——雄性对每个子女投入都很少，雌性的卵子代表了较多的资源投入，因此，她们投入得更多。雄性在频繁的交配中找到繁殖优势；雌性经过对配偶的仔细挑选找到繁殖优势。长话短说，大概在每一个人类社会中都能观察到的男性支配而女性服从的模式，其基础在于生物学的差异。动物行为学家主动地发现，雄性不仅在人类社会中，而且也在其他灵长类动物、鸟类和山羊中，都占有支配地位。在每一个实例中都有生物学基础。社会生物学解答了不同物种的自我牺牲行为的问题。进化论认为，赋予幸存优势的变异能被遗传，但赋予劣势的行为如何能被遗传呢？社会生物学对此问题的解答被称为"亲缘选择"。自我牺牲的个体亲属的基因传递给后代，因为亲属享有共同的基因，某些自我牺牲个体的基因也有途径被世代连续传递。性别差异的解释只是更大理论图景的一个部分。人类进化观把明显有利于人体解剖学上的适应的选择性突变压力归于男性行为：男子不仅在当代社会中天生处于支配地位，而且正是男性在过去的变异引发了进化的改变。

撰写了进化理论、动物行为学（ethology）和社会生物学著作的生物学家露丝·哈伯德（Ruth Hubbard），采取了把其大致轮廓归功于库恩的一条进路。她在记录研究动物行为的学者把常见的雌性特征，归于从藻类（algae）到猿的雌

---

① 布罗沃曼（Inge Broverman）和她的同事的研究证明，在临床心理学家的思维中，对女性的这种双重约束是普遍的。参见 Inge Broverman, D. M. Broverman, F. E. Clarkson, P. S. Rosenkranz, and S. R. Vogel, "Sex Role Stereotypes and Clinical Judgments of Mental Health", *Journal of Consulting and Clinical Psychology* (1970) 34: 1-7.

② 例如，参看 Naomi Weisstein, "Psychology Constructs the Female", in Vivian Gornick and Barbara Moran, eds., *Woman in Sexist Society* (New York: Basic Books, 1971), pp. 133-146; Eleanor Maccoby and Carol Jacklin, *The Psychology of Sex Difference* (Stanford, CA: Stanford University Press, 1974); Paula Caplan, Gael MacPherson, and Patricia Tobin, "Do Sex Differences in Spatial Ability Really Exist?" *American Psychologist* (1985) 40 (7): 786-798. 在许多方面，这一工作重点概括了像霍林沃斯（Leta Hollingworth）之类的心理学家几十年前提供的研究和意识形态的性别差异的批评。关于这种批评的集体失忆在某种程度上造成了像库恩提出的深层分析那么吸引人的氛围。

性器官（或者确定为雌性的器官）的许多方式之前这样说，"每一理论都有把经验纳入它所提供的框架内的自我实现的前瞻能力（self-fulfilling prophecy）……根本没有像客观的与价值无关的科学之类的东西"①。她们声称在自然界中所观察到的，正是维多利亚道德观（Victorian mores）为人类社会规定的行为准则。科学家受到了他们用来描述行为的语言的制约；他们的词汇，还有他们的性别意识形态，产生了以男性为主（androcentric）和男性至上主义的社会行为的解释。即使当他们的描述展现了雌性的活动而不是被动性时，他们也将其解读为符合旧的性别成见。哈伯德引用了达尔文的几段话，例如，下面是《人类起源》中的一段："男性比女性更勇敢、好斗和精力充沛，而且有更多的创造天赋。"② 她把这样的观点概括如下：

> 因此，这里一言以蔽之：男性的心理和生理素质通常通过争夺女性和打猎得以提高，而女性的心智，如果不是每一代的女儿都继承了她们父亲的智慧这种幸运情况，就会是退化的（vestigial）。③

达尔文关于雄性和雌性在进化中所起作用的观念并不局限于19世纪70年代，而是正如哈伯德所表明的那样，它们不断出现在20世纪70年代的动物行为学家和生理人类学家（physical anthropologists）的著作中。

关注理论词汇的这条进路，对行为的神经内分泌进路，也可能是有效的。用来描述女孩和男孩行为的语言，充满了冒充为常识的性别理论。例如，女孩参加通常与男孩有关的活动，就被通俗地说成是"假小子"（tomboys），而男孩参加通常与女孩有关的不太激烈的活动，就被描述成患有"女性化综合征"（sissy syndrome）。据说对子宫内性腺激素水平异常的儿童的研究表明，子宫内雄性激素水平过高生出的女孩中，野丫头症（tomboyism）发生率会增加，而子宫内的雄性激素水平不足生出的男孩中，"女性化综合征"发生率会增加。性别的赋予从行为标签扩展到了性腺激素本身的确认。雄性和雌性的性腺分泌了一组在化学上非常相似的甾类激素（steroidal hormones）。性腺在它们产生的这些类固醇的相对比例上是不同的，但一定量的所有这些激素，是维持雄性和雌性哺乳动物器官的正常生理机能所必需的。许多女性主义学者考察了20世纪30年代关于性腺激素的恰当的专业术语的讨论。尽管它们有相似的化学作用，对雄性和雌性器官都起到了生理作用，但是，希望用性别来识别它们的研究者，超过了那些偏爱更中

① Ruth Hubbard, "Have Only Men Evolved?" in Ruth Hubbard, Mary Henifin, and Barbara Fried, eds., *Women Look at Biology Looking at Women*, Cambridge, MA: Schenkman, 1979, p. 9.

② Charles Darwin, *The Origin of Species and the Descent of Man*, New York: Random House, Modern Library Edition, n. d., as quoted in Hubbard, "Have Only Men Evolved?", p. 19.

③ Hubbard, "Have Only Men Evolved?", p. 20.

立的识别形式的研究者。甚至更加令人惊讶的是标注上的不对称。那些更多由雄性性腺分泌的激素被称为"雄激素"（androgenic）（雄性产生的），而那些更多由雌性性腺分泌的激素则被称为"雌激素"（estrogenic）［产生极度兴奋的（frenzy-producing）］。这种标注对后来的研究有重要的影响：几十年来，人们不承认激素有多种功能。①

因此，这个性腺激素的故事，为应用库恩的观察负载理论（theory-laden observation）和意义负载理论（theory-laden meaning）的观念，提供了极好的案例。理论在几个层次上起作用。在第一种情况下，关于性别差异的理论告知了辨别和标注激素的方式。在第二种情况下，辨别和标注决定了观察和记录什么样的激素分泌的生理效应。库恩的理论负载的思想，为女性主义科学家和学者提供了一种语言来表达他们的感知：即使方法论上无懈可击的科学，也仍然会包含社会偏见。它的这种无懈可击，转而为那些偏见带来了智力上的声望（respectability）。其他批评者认为，朴素的旧经验主义者具有的资源，支持了对科学的男性至上主义的批评考察②，但是库恩处理问题的方式深受那些女性主义者的偏爱：她们认为，由科学提出的女性问题比肤浅的观察实践更加深刻。社会学、神经内分泌学的理论内容与科学教育和职业对女性的体制化排斥之间有一定的联系。库恩关于科学变化的更大图景，强调了科学革命中的社会因素，这为阐述和考虑这种联系提供了一种手段。

库恩的理论负载概念说明了，两个研究者会如何看待一群野马，一个研究者看到，一匹"妻妾成群"的雄马被众多随从的雌马簇拥着，而另一个研究者看到，一群雌马容忍（tolerating）种马的存在，以换取它的服务。它们似乎为理解旧的性别成见在所谓由客观经验方法刻画的领域里经久不衰提供了方法。这就是为什么甚至那些仅仅质疑雌性和雌性识别行为的生物学描述的经验适当性的学者，在阐述他们的批评时，也会求助于库恩或库恩的观点之原因所在。问题不是个别科学家是有偏见的，而是共同的性别意识形态。然而，困难在于，女性主义者也想说，对那些马的某一种描述是正确的。

尽管哈伯德在这篇论文的开头引用了库恩的话，但她明确指出，支持以男性

---

① 所讲的这个故事在最近的生殖内分泌学史的许多版本中都有。参看 Nellie Oudshorn, *The Making of the Hormonal Body*, New York: Routledge, 1997 和 Anne Fausto Sterling, *Sexing the Body: Gender Politics and the Construction of Sexuality*, New York: Basic Books, 2000。这种激素认同的意识形态建构的一个讽刺意味的方面是马的雌激素的标示"马雌激素"（premarin），暗示了它是来自雌马，但实际上它是来自公马的尿。

② 参见，例如，Noretta Koertge, "Methodology, Idealogy, and Feminist Critiques of Science", in Peter Asquith and Ronald Giere, eds., PSA 1980, vol. 2, East lansing, MI: Philosophy of Science Association, 1981, pp. 346-359.

为中心的观点的观察报告恰好是错误的。引发这篇论文的哲学观点被抛到了一边。她在论文的结尾处声明，性别偏见的"进化范式"要求女性"重新思考我们的进化史"①。这种重新思考部分地要求接近"原始材料"。堂娜·哈拉维（Donna Haraway）②指责哈伯德和像她那样论证的女性主义者，同时运用了库恩用于批判的分析框架和他在提出积极的研究纲领时所批判的经验主义框架。③的确，问题不是哈伯德的观点，而是缺少她需要的哲学框架。库恩为女性主义者拒绝下列观念提供了一个平台：来自雌性顺从和雄性支配领域内的报告恰好是事实。社会上共有的性别意识形态歪曲了对所有生物种类的雄性和雌性的观察，当他为女性主义科学家提供了谈论这些问题的一种方法时，他对科学革命的分析并没有为描述她们希望推荐的某种变化提供一种语言或注释。在思考女性主义科学哲学家的工作时，这一点变得更加清楚。

三

女性主义科学哲学家也利用了库恩的思想。她们关注库恩工作中的不同主题，扩展并修改了库恩关于科学知识的特征和科学变化的主张，提出了一系列认识论的担忧。

凯瑟琳·阿德尔森（Kathryn Addelson）在伦理学语境和认识论语境中援引了库恩的思想。阿德尔森认为，19世纪和20世纪的妇女运动，同时是对道德革命的设定与呼吁。④参与那些运动的女性，通过在会场和街头的公开演讲，逐渐超越了为女性规定的行为规范，因而迫使她们自己参与公共事务，而且，她们表达的信息内容，就是要求改变她们违反的那些规定。19世纪的要求是有选举权、接受高等教育权、财产权、离婚权、参与市民的社会生活权。20世纪的要求包括生育权、平等权的改善、机会平等和同工同酬。道德革命并不在于使一个社会的行为或较低层次的承诺（commitments）与其较高层次的原则相符合，而在于深刻改变那些构成原则，即真正的道德观方面。阿德尔森说，这样的话，道德革命类似于库恩的科学革命，其中涉及的不仅是改变对自然界的描述和说明，而且

---

① Hubbard, "Have Only Men Evolved?", p. 31.

② 堂娜·哈拉维是美国著名的跨学科学者。她在生物学、灵长类动物学、科学史学、科学哲学、科学社会学、科幻文学等方面都有很深的造诣。——译者注

③ Donna Haraway, "In the Beginning Was the Word: The Genesis of Biological Theory", *Signs: Journal of Woman in Culture and Society* (1981) 6 (3): pp. 469-482.

④ Kathryn Addelson, "Moral Revolution", in Julia A. Sherman and Evelyn Torton Beck, eds., *The Prism of Sex*. Madison: University of Wisconsin Press, 1979, pp. 189-227; reprinted in Addelson, *Impure Thoughts*, Philadelphia: Temple University Press, 1991, pp. 35-61.

是改变评价那些描述和说明的特有标准。

在哈伯德对表明社会生活的陈旧观念渗透到科学思想中表示担忧的情况下，阿德尔森详细阐述了思想内容和社会生活形式之间的联系。她也求助于库恩的框架来提出有关科学知识的问题。在此，她对库恩的范式概念十分感兴趣。这使他能够把科学知识理解为不仅包括"理论和定律，而且还包括形而上学的承诺、范例、难题、反常等不同的特征"①。范式指导了实践，而且，对科学探索而言，重要的是实践，而不是教条的实践结果。

根据阿德尔森的观点，库恩把科学解释为活动、实践的说法，以及他提供的在科学变化过程中理论兴衰的证据，有助于把注意力集中于科学家之间的社会动力学（social dynamics），这些科学家包括有助于坚持一个理论的那些人和有助于用一个新理论替代一个旧理论的那些人。库恩对坚持旧范式和倡导新范式的评论是非常著名的：长期支持某些范式的人不会转向新的范式，而是退隐，把该领域让给他人，而新范式的倡导者往往都是年轻人，他们没有为旧思想的成功出力，很可能由于采纳新思想而受益匪浅。阿德尔森把库恩的工作看成揭露了认知权威的问题。范式转换期间的争论是在争夺认知权威，而且，由于从概念上看，范式的特征是普遍的，这些竞争的结果包括在确定我们的共同世界的基本结构时权威的转换。早期现代欧洲的原子物理学理论的胜利，是超越物理学扩展到社会生活（比如个人主义）的形而上学观点的胜利。

库恩承认，在范式转换期间，一个人的理论、实验或实践偏好会受到多重因素的影响，这表明，科学的考虑和科学之外的考虑之间的边界是模糊的，以及/或者是相互渗透的。在不断地偏向某一位竞争者的似乎合理的判断中，一个人具有的直觉感是一个重要因素。但是，阿德尔森认为，一个人具有的直觉感在很大程度上受到了他的社会经验的影响。参与科学争论的解决并从中获得权威的那些人的形而上学观点，反映了他们的社会经验，就此而言，使文化合法化的形而上学观点在很大程度上反映了进而强化了具体的社会实在。

阿德尔森致力于社会学，表明了功能主义的形而上学如何体现在20世纪50年代美国社会学的研究议题（research agenda）和结果中。她引用了20世纪70年代以来的人类学研究，这些研究记录了女性和男性以及不同社会经济阶层成员的不同的社会现实。然后，把这些断言纳入到她的下列断言中：只有某些类型的人才能进入认知权威的核心（locus）、科学行业和这些行业中的领导岗位，这样的体系是这个阶层的形而上学观点塑造社会中其他人观点的体系。

---

① Kathryn Addelson, "The Man of Professional Wisdom", in Sandra Harding and Merrill Hintika, eds., *Discovering Reality*, Dordrecht: Reidel, 1983, p. 166.

　　杰出的物理学家、生物学家和科学哲学家……生活在一个群体控制另一个群体的社会里。作为专家，他们为使自己能更广泛地行使认知权威的最高职业等级而竞争。出于这样的文化认识，毫不奇怪……我们的专家为我们提供了根据等级、优势和竞争对世界的形而上学描述。①

　　阿德尔森认识到，科学探索的力量，不仅仅是塑造一个社会的世界观，而且是以有效的方式表征世界。她的观点是，通过把注意力转向其社会结构，纠正给予某个社会群体不相称的特权，我们就能够消除持续的不合理性（irrationalities）。当然，由于这个主张，她很容易使自己受到挑战，其性质与哈拉维向哈伯德的进化进路的女性主义版本提出的挑战一样。从什么范式内部，这些所谓的不合理性可以如此确定？为什么不提倡一个完全不同的科学呢？或者，为什么不呼吁可能会导致如此结果的社会条件呢？

　　桑德拉·哈丁（Sandra Harding）在她的《女性主义的科学问题》一书中讨论了《科学革命的结构》和《哥白尼革命》。② 她欢迎前者，因为它证明了，逻辑经验主义提供的科学判断的理性重建，歪曲了他们设法阐明其逻辑的历史情境。像阿德尔森一样，她把该书看成是使科学研究（science studies）中的自然化进路合法化，特别是考虑了社会关系对提供科学结果说明的作用的力作。这在原则上应该扩展到研究性别关系对科学活动的作用。哈丁写道，库恩的追随者和后库恩的追随者，由于坚持把性别当做是生物的，而不是一种社会关系，因而未能获得库恩范式的逻辑外延。也就是说，他们随意地忽视了性别关系的影响，因而错过了现代科学发展的一个重要方面。

　　然而，在对科学革命的编年史的讨论中，哈丁对库恩持更加严厉的批判态度。在她的解读中，《哥白尼革命》是把科学革命看成智慧战胜迷信的一个实例。哈丁特别利用了把中世纪的思维、忠于亚里士多德世界观的思维比作孩子和原始人的思维的那些段落，并且把库恩解读成是赞扬把科学从16世纪和17世纪的道德和政治中解放出来。这给我留下的印象是很不公平的。《哥白尼革命》通过证明亚里士多德的物理学和世界观的一致性、可靠性（plausibility）和经验的适当性，增强了后来《科学革命的结构》的力量。库恩几乎未能预见到促进了哥白尼、伽利略、牛顿以及与他们共事的自然科学家的智力成就和技术成就的世界观和由他们的这些成就创立的世界观，这一点并没有减少这样的暗示：现代科学也处在受包括道德观、政治观和形而上学观点在内的更大的世界观相互支持的

---

① Ibid., p. 184.

② Sandra Harding, *The Science Question in Feminism*, Ithaca, NY：Cornell University Press, 1986, pp. 197-210.

类似关系中。

这种相互支持正好就是哈丁希望证明的。她拒绝了作为一种起源神话的科学革命的标准图像，把马克思主义的科学史学家的成果补充到《科学革命的结构》提供的框架内，来表明科学革命的社会与政治维度。科学革命伴随着封建秩序的终结和新的中产阶级的出现，伴随着反权威主义、对进步的信念、人道主义理想和把对自然的系统研究（科学）和支持这种研究的制度维持（政治）分离开来的劳动分工。这些社会和政治发展与下列宇宙学是共存的：由原子论（这种观点认为，自然界是由根本上一致的少量无生命的物质构成的）、价值中立（代替了中世纪的目的论宇宙的与人无关的宇宙），以及信任方法（作为公正性和独立于政治与宗教权威的保证）所刻画的宇宙学。不管马克思主义者和后库恩主义者对科学的社会历史研究是多么富有洞察力，他们都忽视了作为新的欧洲世界秩序一部分的性别关系。哈丁引用女性主义学者乔丹诺娃（Ludmilla Jordanova）、麦茜特（Carolyn Merchant）[1] 和凯勒（Evelyn Keller）[2] 的成果，来建议新世界秩序的性别维度——它限制新的中产阶级男性成员的财产权，重建与早期现代主义的新价值相协调的男性主义（masculinity）。

哈丁对女性主义科学的可能性的探讨表明，她最感激库恩。女性主义者像17 世纪的激进分子一样，挑战当代的权威结构，相信包括克服性别、种族和阶级等级方面的进步，强调教育改革和人道主义。而且哈丁声称，女性主义寻求把经验的理解与道德和政治的理解统一起来的知识。哈丁的观点是，从当前主流科学的范畴来看，这个"后继者"（successor）不会得到承认。因此，她把当前的科学和女性主义可接受的科学之间的关系看成是不可通约的，像亚里士多德的科学与宇宙学和 17 世纪的新科学之间关系一样。21 世纪的新科学将会是统一的科学，但不是像维也纳学派想象的那样，把物理学当做其基础的统一的科学。新科学将受到道德和政治信念的指导，因此，根据哈丁的观点，新科学将是把社会科学而不是物理学当做其基础。必须彻底改造"科学及其自身的理论化"[3]。

另一位女性主义学者从库恩那里汲取了稍微不同但同样激进的经验。凯勒（Evelyn Fox Keller）一直是最明显的关心科学的女性主义者。她一贯强调多元相互作用模型的可行性，相比之下，她把还原论的分析和线性分析看成是当代科学的特征。她关心的不是对女性和性别关系的描述，而是在与性别或社会行为无关的科学领域内表现出的性别意识形态的那些方面。凯勒作为科学家，也作为历史

---

[1] 卡罗琳·麦茜特（Carolyn Merchant，1936～），美国生态女性主义哲学家和科学史家。——译者注
[2] 伊芙琳·凯勒（Evelyn Fox Keller，1936～），美国女性主义哲学家和科学史家，麻省理工学院教授。——译者注
[3] Ibid.，p. 251.

学家和哲学家评论说，科学家对库恩的主张的抵制和冷漠是令人震惊的。① 凯勒注意到，虽然库恩关于科学变化的观点为审查科学实践和科学评价的社会维度的研究打下了基础，但是他本人并没有从事这样的审查，因此，他并没有解决科学实践中的社会和文化因素的确切性质和作用问题。不过，她注意到，"（库恩的主张）的直接含义是，不仅不同的数据收集、不同的科学关注焦点，而且不同的知识组织、不同的世界解释，都是可能的，也都与我们所说的科学相一致"②。在哈丁把库恩的编年史和马克思的编年史联系起来的地方，凯勒提议，运用心理分析的方法来同时说明科学共同体对库恩思想的抵制，过去和当代科学的性别化（gendering）以及如何走出她认定的科学终结困境。

凯勒主要依赖于所谓的"对象关系理论"。根据这个理论，婴儿和儿童成长的任务之一是形成个人认同。儿童和他最亲近的成年人的关系，深刻地影响了这个过程，而且持续地影响着他对世界的总体看法和他在世界中的行为。典型的西方家庭里的男性婴儿和女性婴儿，面临着独特的挑战，对这些挑战的心理应对形成他们对社会现实和物理实在的取向（即在认知、情感和实践意义上体现出来的一种取向）。男孩不得不在家中主要的成年人是母亲而父亲多半缺席的环境中，获得他们的个人认同。在一个强调性别差异的社会和家庭环境里，他们的任务就是成为某种他们所知甚少的角色。因此他们成长的潜力不是指向成为他们所熟悉的角色——他们的母亲。结果，男孩对他们成为的那种男人的认同（identity）是脆弱的，需要不断强化。满足这种需要的一个心理策略是形成与他人夸张的心理分离。另一方面，由于小姑娘们的任务是成为女性，她们往往变得过度依恋（overattached）。

凯勒运用对心理发展的这种分析来说明科学的特征。这种疏远和不认同的策略，本身在情感上表达了凯勒所说的一种"静态自主的立场"（stance of static autonomy）。因为男性个性化（masculine individuation）的条件导致了深度焦虑，所以，这种策略需要不断地确证，即最生动和最令人信服地由人们不必要的那种控制来提供。就这一点而言，这种分析与其他强调对象关系的女性主义者的追求相一致。③凯勒的创新就是把它扩展到知识观和科学观。与静态的自主性相并行，由被凯勒称为"静态的客观性"（static objectivity）的一种认知态度作为个人发展的一个方面脱颖而出。静态的客观性被刻画为把知识等同于在情感上从知识对象中分离出来，然后控制知识对象，并且，把追求知识、科学探索看成一个

① Evelyn Fox Keller, *Reflections on Gender and Science*, New Haven, CT: Yale University Press, 1985.

② Ibid., p. 5.

③ 把对象关系理论应用于性别关系的名著之一是南希·乔多罗（Nancy Chodorow）的 *The Reproduction of Mothering*, Berkeley: University of California Press, 1987.

对抗性过程。静态的客观性和动态的客观性形成鲜明对比，其目的在于，可靠地理解赋予其要素独立的完整性的世界，确认知识的主体和客体的连通性。知识被理解为既不是分离和控制，也不是丧失同一性（identity），而是承认对象自主的灵活的联系和关系。男孩和女孩的正常成长过程使得他们认为与静态的自主性和静态的客观性相联系的行为等同于男性（masculinity）。相比之下，女性（femininity）的特点是过度认同（overidentification）和淹没个性。动态的自主性和客观性是只有通过与包括性别在内的流行的社会规范作斗争才能达到的（含蓄的）取向。

凯勒通过引用从培根到西美尔（Simmel）① 再到当代科学家这些作者的观点，以及通过表明科学中的某些研究纲领是以控制自然为目标来推动的，来支持她的论题：现代科学是由静态的客观性构成的。她通过引证那些提倡和实践以这种态度走进科学的研究者 [比如，迈克尔·波兰尼（Michel Polanyi）② 和芭芭拉·麦克林托克（Barbara McClintock）③] 的观点来支持她的替代选择——动态的客观性——的可行性。包括露丝·哈伯德和桑德拉·哈丁在内的女性主义者，强烈反对似乎充斥近代科学的还原论。哈丁把此归于允许现代科学发展的社会经济条件。凯勒认为，还原论是由于其心理根源（psychic roots）而得以坚持的世界观的一部分。

因此，凯勒可能把科学共同体对库恩的抵制解释为是对下列观点的一种回应：这种观点是，库恩原则上认可，科学之外的社会因素或主观因素在决定科学判断时发挥了作用，这对认知的自主性构成了威胁。她与哈丁的不同之处是，不仅强调科学认知的心理维度，而且断言，静态的客观性基本上是科学意识形态的特征，而不是其具有实践的特征。在科学的意识形态强调在情感上脱离自然和在实践中统治自然的统一目标的情况下，科学实践研究揭示了更加丰富多彩的思想。当然，凯勒之所有出名，是因为她阐述与支持了遗传学家芭芭拉·麦克林托克的工作，但是，其他科学家也例证了动态的客观性的理想、有能力密切接近知识对象，也有能力脱离知识对象，运用共鸣而不是疏远追求理解。对凯勒来说，这些认知能力与把自然界表征为复杂的和异质的观点联系在一起，完全不同于把自然界表征为可还原到某一基本层面，并且最终可用简单的单向因果模型解释。科学的意识形态及其与男性意识形态的情感联系，说明了为什么我们一直把像麦克林托克等人的互动主义进路（interactionist approaches）边缘化，支持追求符合

① 齐奥尔格·西美尔（Georg Simmel，1858~1918），德国哲学家和社会学家。——译者注
② 迈克尔·波兰尼（Michel Polanyi，1891~1976），英国学者，提出了显性知识（explicit knowledge）和隐性知识（tacit knowledge）的概念。——译者注
③ 芭芭拉·麦克林托克（Barbara McClintock，1902~1992），美国著名生物学家。——译者注

控制的知识形式进路，而不是与现有认识共存的知识形式进路。我们不需要重新发明适合女性主义目的的替代的科学形式，只需要更加密切地注视当前归于科学边缘的那些实践形式来支持我们的模式。

通过强调在整个科学史中有许多自然过程的例子（models），它们运用了非主流的表征或解释原则，凯勒偏离了库恩的科学增长、变化的图景。库恩认为，一个领域的常规科学是由一种解释进路来描述的，如果有多种进路并存，则表明它还不成熟。此外，正如在讨论哈丁时所注意到的，库恩认为，同一学科领域的前后相继的（或竞争的）理论是不可通约的。凯勒并没有把麦克林托克的基因组变异观点，看成涉及负载理论的观察，对于这些观察，承诺不同理论的研究者是不能共享的。她把主流对麦克林托克观点的拒绝，理解成主流坚持某种科学知识观以及相应的和麦克林托克观点完全不符的有关自然的形而上学的结果（function）。而且，尽管哈丁在描述主流科学和要替代它的科学之间的关系时，似乎接受了不可通约性，但在其他方面，她的概念也和库恩的建议相冲突。哈丁提倡的女性主义的科学革命之所以会发生，不是因为反常经验不断积累，并使当前的范式陷入危机，而是因为社会价值和关系的变化，要求不同的认识自然界的方法。库恩的科学知识观和科学变化观对这两位思想家都有价值，因为他的阐述对当时占优势的逻辑经验主义的科学哲学提出了挑战。科学要么是竞争范式（革命的科学）之间的较量，要么是解决一个范式（常规科学）之内的难题。但是，科学的这种特点并没有为思考如何影响变化提供工具。然而，这些女性主义者不仅对理解科学感兴趣，而且对改变科学也感兴趣。

四

据说女性主义者求助于库恩，是为了使她们对实证主义（无论其哲学表述是逻辑经验主义还是流行的科学主义）的拒绝合理化，但是，她们在讨论替代的科学知识形式时，却完全或部分地抛弃了库恩。库恩的观点成了这个计划的障碍。然而，有种思考科学探索和科学知识的方式，尽管它受惠于库恩，可以合理地看成科学哲学中库恩革命的产物，但它并没有对女性主义的计划施加同样的限制。认识论的多元主义，即关于知识的多元主义，从广义上说，是以库恩关于科学史的见解为基础的，但是，它也运用了某些不同的哲学原理。这些使得她们能够把以男性为中心的科学或男性至上主义的科学，不仅仅解释为经验上不充分的科学，还不破坏对替代的追求。为了明白这一点，我扼要重述《科学革命的结构》的相关的基本哲学思想。

库恩声称，关于相同主题（比如说，运动的物体）的前后相继的理论，既是相互矛盾的，又是不可通约的，他这样说的意思是，这些理论在面对另一个理论时，不可能以经验主义者设想的模式在经验上得到检验。也就是说，它们的关系不是这样的：能根据一组共同的数据或事实，对它们作出可比较的评价。库恩对这种不可通约性的解释是，理论依赖于意义和观察。逻辑经验主义者认为，观察独立于理论，观察术语和陈述的意义也独立于理论。富有意义（meaningfulness）和确证（证据的支持）是从观察到理论。相反，库恩认为，观察术语的意义是由理论决定的，而且，当它们的理论语境发生变化时，理论术语的意义也发生转变。例如，经典物理学中的"质量"是指一个守恒量，而相对论物理学中的"质量"是指一个（在某些条件下）可转变为能量的量。因此，似乎共同的观察术语（提供和可观察世界接触的共同点），根据这个意义理论，结果只是同音异义词。① 库恩也认为，观察本身是由理论决定的，而且他引用了证明知觉依赖于期望的种种心理学实验来支持这个论点。关于意义和观察的这些观点的一个结果是，在不同的范式控制下，拥有不同理论的科学家之间的真正交流是不可能的。他们运用的语言，也许听起来相同，他们也许指的是感觉范围内的相同现象，但是，他们使用的术语有不同的意义，而且，当他们看到相同的现象时，他们所观察的东西也是不同的。

库恩关于意义和观察的观点是最令科学哲学家感到不安的。库恩承认，科学假说和理论的证据在于能够观察到什么，与这种观点相结合，证据推理似乎是循环论证：一个理论受到观察的支持，而观察的内容和描述又由该理论来决定。这使得理论选择完全成为主观的。为了反对这种指责（库恩用宗教信仰的转变作比喻，使得这种指责看似有理），库恩声称，科学中的理论选择是由一系列价值引导的。② 这些价值包括精确性、内部和外部的一致性、简单性、范围的广度和富有成效性（fruitfulness）。虽然在实践中，赋予这些价值的精确解释和相对优先权或权重，有可能不同，但是，它们提供了一个不变的标准，参照这个标准，科学

---

① 这个理论不可能正确，因为"质量"是指在两个理论中物体的共同特征。库恩的意义观的一个问题是，他似乎没有在定义的特性和可能的特征之间作出区分。这在某些方面和奎因关于意义的观点一致，但像库恩所做的那样，把所有可能的特征看成定义的特性，使得人们不再能讨论一个错误的陈述（也不能肯定说出正确的陈述是什么）。多年来库恩花费了大量的精力，试图使他的意义不可通约性概念更为可靠，但是，据我所知，他没有讨论该观点的这一方面。参见，他的科学哲学学会主席的任职讲演《结构之后的道路》，Arthur Fine, Micky Forbes and Linda Wessels, eds., *PSA 1990*, vol. 2, East Lansing, MI: Philosophy of Science Association, 1991, pp. 3-13. 重印于 Thomas Kuhn, *The Road since Structure*, edited by James Conant and John Haugeland, Chicago: University of Chicago Press, 2001, pp. 91-104.

② Thomas Kuhn, "Objectivity, Value Judgment, and Theory Choice", in *The Essential Tension*, Chicago: University of Chicago Press, 1977, pp. 320-339.

判断、理论选择能够理解为客观的。① 然而，从库恩对这些价值的讨论可以看出，它们并没有为理论的比较提供一个独立的准绳（yardstick），或者如他所说，提供一个运算法则（algorithm）。考虑到他对不可通约性的解释，也不可能有这样的一个准绳。库恩有时通过暗示持有不同理论的科学家生活在不同的世界，来详细阐述意义和观察的理论承载。对这样一个主张有不同的理解方式，但是作为不可通约性的一种表述，它不能很好地满足女性主义科学学者的需要。

女性主义科学学者希望确认：许多当代科学都负载有（性别）价值；那种科学的成功是从衡量成功的传统标准来看的；有必要（或希求）替代科学中的主流趋势。她们不仅是在经验充分性的基础上抵制所有科学，而且是在它们包含（encodes）并强化有害的（男性至上主义的、种族主义的、资本主义的）社会价值的基础上抵制它们。她们和反对其理论的科学家生活在同一个世界——她们希望看到对那个世界的不同的科学解释。她们不满足于把差异归于生活在不同的世界，或者说归于语义的或认知的不可通约性。她们只是不希望在她们的另一个世界里做科学；她们希望改变在我们共同的世界里从事科学的主流方式。这不是说，女性主义者拥有一种不可能理解她们反对的科学的世界观。她们会说，她们只是太了解科学了。

我将把库恩的观点最终对女性主义事业的不适合性，定位于他的理论依赖和意义与观察的理论承载。既然这些学说在（概念的和经验的）其他基础上也是有问题的，那么，我们应该在别处寻找哲学支持。语境经验主义（contextual empiricism）的多元论（pluralism）建立在库恩关于科学变化的本质的洞见之基础上，但是，它放弃了作为说明不可通约性的意义和观察的理论承载。

语境经验主义没有把不同理论之间明显等价的经验充分性看成理论承载的问题，而是看成有助于数据和假说之间推理的背景假设中的差异函数②，用来阐述或描述它们的语言具有独立的意义。语境经验主义把产生多种理论的逻辑问题看成为是非充分决定的（underdetermination）［在我们所能观察的或测量的证据来源和我们的说明抱负（即发现作为我们经验基础的原理、能力或因果规则）之

---

① 某些女性主义科学哲学家最近吸收了库恩观点的这个方面，但是作为通常讨论科学中认知价值的认识概念的一部分，女性主义者要么强调语境价值在决定认识价值的权重或应用时的作用，要么强调在能够被看成具有社会维度和/或文化维度的范围内作出选择时可能的认知价值的变化。参见 Alison Wylie and Lynn Nelson, "Coming to Terms with Value（s）of Science: Insights from Feminist Science Studies Schlarship", Workshop on Science and Values, Center for Philosophy of Science, University of Pittsburgh, October 9-11, 1998, on the first strategy, and Helen Longino, "Cognitive and Non-Cognitive Values in Science: Rethinking the Dichotomy", in Lynn Hankinson Nelson and Jack Nelson, eds. , *Feminism*, *Science and Philosophy of Science*, Boston: Kluwer, 1996, pp. 39-58, on the second.

② 等价的经验适当性并不同于经验的等价。后者要求两个理论具有相同的经验结果；前者仅要求两个理论在预言观察时是同样成功的，但与这些理论相关的一组观察不可能是相同的。

间的差距]。构成语境的背景假设弥补了这个差距，在这种语境中，经验数据，即观察数据，获得了证据相关性。假设的改变，带来了证据相关性的改变。语境经验主义的优势之一就是，它提供了对库恩设法说明的历史现象的一个解释，即两位科学家可能注视着相同的东西——装有一只死老鼠的密闭容器或落日或一群狮子——但却解释说，他们看出了相当不同的东西。因此，语境经验主义和多元论（即不一致的理论都能对相同的现象提供充分的解释）是相融的。虽然多元论认为，这样的理论能提供正确的解释，但是它们的正确性是从不同的背景假设和认知目标的视角来评价的。多元论也认为，在这样的情况下，理论是部分的或不完备的，不能包括其范围内的复杂现象的所有方面。当然，语境经验主义有自己的哲学困难，这一点我已经在别处讨论过。① 这里我想强调支持语境主义放弃理论承载的女性主义科学研究的优势。这里有三个切入点。

1. 关于不可通约性

正如我们所看到的，库恩求助的是理论承载作为意义的一般理论，它认为一个理论中的所有术语的意义都是由该理论决定的。在该理论之外，任何术语都没有独立的意义。这条意义进路及其关于观察的推论，在承诺说明，一位研究者看成是奸诈的情况，另一位研究者为何能看成是顺从，或者，一位研究者看成是依赖和压力的情况，另一位研究者为何看成是优势时，最终对女性主义者来说是没有帮助的，因为它使得女性主义者，无论其性别意识形态如何，都不能指责对任何人来说都是不正确的性别的歪曲。它没有能力对男性至上主义者的科学作出经验批评。

语境经验主义准许不可通约性的不同进路。它没有把理论或经验看成是意义的基础。因此，它脱离了库恩的语义整体论和逻辑经验主义的语义还原论。当它坚持意义内容和观察不是由理论决定的时，它确实同意，观察和测量的范畴是理论相关的或语境相关的。当测量离不开其测量语境时，就产生了明显的不可通约性。这个语境提供了测量的问题、目标和标准。在一个语境中算作是观察的东西，也许在另一个语境中不能算作是观察。一种现象的两条理论进路，在把该现象的不同方面看成是证据相关的，或者，使用与对比问题和认知目标相关的不同测量尺度的程度上，可能是不可通约的。这两条进路是可以交流的，正如不可通约性的语义进路所暗示的那样，而且，当两个理论都不能提供共同的标准时，运

---

① Helen E. Longino, "Toward an Epistemology for Biological Pluralism", in Richard Creath and Jane Maienschein, eds., *Biology and Epistemology*, Cambridge: Cambridge University Press, 2000, pp. 262-286; *The Fate of Knowledge*, Princeton: Princeton University Press, 2002.

用不同理论进路的研究者，仍然可能足以在他们的理论外部分享对彼此思想与观察的批评交锋。把不可通约性看成语境的一个功能，避免了破坏女性主义者对哈拉维强烈反对的男性至上主义者的科学的经验批评。

2. 关于范式和常规科学

库恩声明，常规科学的主要特征（差不多是一个界定条件）是在单一范式下组织研究。语境经验主义与下列多元论的形式是相容的，即认为在许多情况下，被说明的现象是如此之复杂，以至于要提供全面解释，需要有多重进路。任何一个解释，即使是正确的，在这些情况下，也是不完备的。一个明显的例子是有机体的发育。对某个特征发育的以基因为中心的解释和环境解释也许是对的，但却是片面的。无论如何，它们的理论结构不可能被合并到一个解释中。存在多重进路，不是科学不成熟的标志或范式形成之前处于革命时期的科学的标志，而是假如根据现象的要求来看的话，它可能是常规科学的一个不可避免的特征。尽管凯勒本人也许不赞同这种多元论，但与库恩的范式相比，多元论更好地解释了她所引证的学科中多重研究传统的存在。

3. 关于价值和科学判断

库恩认为，对（客观的）科学判断的价值决定因素的解释、权重或优先性是多变的，乃至个别科学家对它们的解释，在某些方面会受到个人主观因素的影响。尽管如此，他也认为，它们是内在于科学共同体的，而且，他所列举的那些因素在某方面是科学性的构成要素。语境主义者认为，可能不存在对科学价值的如此原则性的限定。库恩列举的价值是哲学家习惯性地列举的那些价值，但它们反映了一种特殊的学术传统（intellectual tradition）。还有能够提倡的其他的价值，这些价值与传统价值之间的关系是错综复杂的。于是，与把库恩的思想看成是支持了科学的社会研究的那些人相反，由范式支配的科学意味着，一旦发生范式转移，一系列价值（隐含着优先性和解释）就成为规范的，因此，科学判断再一次完全是内在的。

这样，语境经验主义在把信守实际的科学实践，看成是科学知识论的充分性的一个标准时，遵循了库恩的进路。它也强调了科学判断的复杂性，即科学判断依赖于特定的直接实验或观察情境之外的因素。然而，我刚才分类出的差异使得它更易于受到女性主义科学研究的关注。

五

然而，在库恩提供了具有潜在意义的出发点的地方，还有一点，他讨论的科学价值中的最后一个价值是富有成效性。现在，我们能够把这理解为算得上具有经验内容，因为判断一个理论或范式是否富有成效的一个方面是根据产生的经验结果。但是库恩以一种有趣的方式注解这个价值。对他而言，富有成效性是指一个理论或范式有能力产生有趣的难题或继续研究的问题，也就是说，它有能力指导研究、提出智力挑战。这就把实用主义的调子引入科学探索的其他有代表性的解释中。女性主义者不仅关心性别表征和运用性别表征自然界，而且关心把科学思想应用到社会领域的方式。女性主义者关心支持在全社会而不是集中在专家范围内分配权力的科学形式。女性主义者关心支持保护而不是消耗自然资源的探索形式。女性主义者关心鼓励非侵害性的和非支配性探索模式。现在，很可能是，库恩对富有成效的理解完全是内视型的，限于产生难题的能力。但似乎原则上没有理由不能把实用地理解的富有成效性的价值，扩展到包括一个理论可能是富有成效的那些外视型的方面。如果这准许参照实验室或研究室之外的世界中允许的特殊类型的干预来评价理论，那么，我们就认为自己受惠于这些思想家：他们在应该引导科学判断的价值中包括了实用和表征的关注。于是，如果女性主义的价值在那种评价中开始流行起来，那么《科学革命的结构》的结果就是促进了科学中的女性主义革命，尽管以前我有保留意见。

（魏洪钟 译，成素梅 校）

Experts: Which Ones Should You Trust?

Alvin I. Goldman

1. Expertise and Testimony

Mainstream epistemology is a highly theoretical and abstract enterprise. Traditional epistemologists rarely present their deliberations as critical to the practical problems of life, unless one supposes—as Hume, for example, did not—that skeptical worries should trouble us in our everyday affairs. But some issues in epistemology are both theoretically interesting and practically quite pressing. That holds of the problem to be discussed here: how laypersons should evaluate the testimony of experts and decide which of two or more rival experts is most credible. It is of practical importance because in a complex, highly specialized world people are constantly confronted with situations in which, as comparative novices (or even ignoramuses), they must turn to putative experts for intellectual guidance or assistance. It is of theoretical interest because the appropriate epistemic considerations are far from transparent; and it is not clear how far the problems lead to insurmountable skeptical quandaries. This paper does not argue for fiat-out skepticism in this domain; nor, on the other hand, does it purport to resolve all pressures in the direction of skepticism. It is an exploratory paper, which tries to identify problems and examine some possible solutions, not to establish those solutions definitively.

The present topic departs from traditional epistemology and philosophy of science in another respect as well. These fields typically consider the prospects for knowledge acquisition in "ideal" situations. For example, epistemic agents are often examined who have unlimited logical competence and no significant limits on their investigational resources. In the present problem, by contrast, we focus on agents with stipulated epistemic constraints and ask what they might attain while subject to those constraints.

Although the problem of assessing experts is non-traditional in some respects, it is by no means a new problem. It was squarely formulated and addressed by Plato in some of his early dialogues, especially the *Charmides*. In this dialogue Socrates asks whether a man is able to examine another man who claims to know something to see whether he does or not; Socrates wonders whether a man can distinguish someone who pretends to be a doctor from someone who really and truly is one (*Charmides* 170d-e). Plato's term for posing the problem is *techne*, often translated as "knowledge" but perhaps better translated as "expertise" (see Gentzler 1995, LaBarge 1997). [1]

In the recent literature the novice/expert problem is formulated in stark terms by John Hardwig (1985, 1991). When a layperson relies on an expert, that reliance, says Hardwig, is necessarily *blind*. [2] Hardwig is intent on denying full-fledged skepticism; he holds that the receiver of testimony can acquire "knowledge" from a source. But by characterizing the receiver's knowledge as "blind", Hardwig seems to give us a skepticism of sorts. The term "blind" seems to imply that a layperson (or a scientist in a different field) cannot be *rationally justified* in trusting an expert. So his approach would leave us with testimonial skepticism concerning rational justification, if not knowledge.

There are other approaches to the epistemology of testimony that lurk in Hardwig's neighborhood. The authors I have in mind do not explicitly urge any form of skepticism about testimonial belief; like Hardwig, they wish to expel the specter of skepticism from the domain of testimony. Nonetheless, their solution to the problem of testimonial justification appeals to a minimum of *reasons* that a hearer might have in trusting the assertions of a source. Let me explain who and what I mean.

The view in question is represented by TylerBurge (1993) and Richard Foley (1994), who hold that the bare assertion of a claim by a speaker gives a hearer prima facie reason to accept it, quite independently of anything the hearer might know or justifiably believe about the speaker's abilities, circumstances, or opportunities to have acquired the claimed piece of knowledge. Nor does it depend on empirically acquired evidence by the hearer, for example, evidence that speakers generally make claims only when they are in a position to know whereof they speak. Burge, for example, endorses the following Acceptance Principle: "A person is entitled to accept as true something that is presented as true and that is intelligible to him, unless there are stronger reasons not to do so. " (1993: 467) He insists that this principle is not an empirical one; the "justificational force of the entitlement described by this justification is not constituted or enhanced by sense experiences or perceptual beliefs" (1993: 469). Similarly, although Foley does not stress the a priori status of such principles, he agrees that it is reasonable of people to grant *fundamental* authority to the opinions of others, where this means that it is "reasonable for us to be influenced by others even when we have no special information indicating that they are reliable" (1994: 55). Fundamental authority is contrasted with *derivative* authority, where the latter is generated from the hearer's *reasons for thinking* that the source's "information, abilities, or circumstances put [him] in an especially good position" to make an accurate claim (1994: 55). So, on Foley's view, a hearer need not have such reasons about a source to get prima facie grounds for trusting that source. Moreover, a person does not need to acquire empirical reasons for thinking that people generally make claims about a subject only when they are in a position to know about that subject. Foley grants people a fundamental (though prima facie) epistemic right to trust others even in the absence of any such empirical evidence. [3] It is in this sense that Burge's and Foley's views seem to license "blind" trust.

I think that Burge, Foley, and others are driven to these sorts of views in part by the apparent hopelessness of reductionist or inductivist alternatives. Neither adults nor children, it appears, have enough evidence from their personal perceptions and memories to make cogent inductive inferences to the reliability of testimony (cf. Coady 1992). So Burge, Foley, Coady and others propose their "fundamental" principles of testimonial trustworthiness to stem the potential tide of testimonial skepticism. I am not altogether convinced that this move is necessary. A case might be made that children are in a position to get good inductive evidence that people usually make claims about things they are in a position to know about.

A young child's earliest evidence of factual reports is from face-to-face speech. The child usually sees what the speaker is talking about and sees that the speaker also sees what she is talking about, e. g. , the furry cat, the toy under the piano, and so forth. Indeed, according to one account of cognitive development (Baron-Cohen 1995), there is a special module or mechanism, the "eye-direction detector", that attends to other people's eyes, detects their direction of gaze, and interprets them as "seeing" whatever is in the line of sight. [4] Since seeing commonly gives rise to knowing, the young child can determine a certain range of phenomena within the ken of speakers. Since the earliest utterances the child encounters are presumably about these *speaker-known* objects or events, the child might easily conclude that speakers usually make assertions about things within their ken. Of course, the child later encounters many utterances where it is unclear to the child whether the matters reported are, or ever were, within the speaker's ken. Nonetheless, a child's early experience is of speakers who talk about what they apparently know about, and this may well be a decisive body of empirical evidence available to the child.

I don't want to press this suggestion very hard. [5] I shall not myself be offering a full-scale theory about the justification of testimonial belief. In particular, I do *not* mean to be advancing a sustained defense of the reductionist or inductivist position. Of greater concern to me is the recognition that a heater's evidence about a source's reliability or unreliability can often *bolster* or *defeat* the hearer's justifiedness in accepting testimony from that source. This can be illustrated with two examples.

As you pass someone on the street, he assertively utters a sophisticated mathematical proposition, which you understand but have never previously assessed for plausibility. Are you justified in accepting it from this stranger? Surely it depends partly on whether the speaker turns out to be a mathematics professor of your acquaintance or, say, a nine-year-old child. You have prior evidence for thinking that the former is in a position to know such a proposition, whereas the latter is not. Whether or not there is an a priori principle of default entitlement of the sort endorsed by Burke and Foley, your empirical evidence about the identity of the speaker is clearly relevant. I do not claim that Burge and Foley (etc.) cannot handle these cases. They might say that your recognition that the speaker is a math professor *bolsters* your *overall* entitlement to accept the proposition (though not your prima facie entitlement); recognizing that it is a child *defeats* your prima facie entitlement to accept the proposition. My point is, however, that your evidence about the properties of the speaker is crucial evidence for your overall entitlement to accept the speaker's assertion. A similar point holds in the following example. As you relax behind the wheel of your parked car, with your eyes closed, you hear someone nearby describing the make and color of the passing cars. Plausibly, you have prirna facie justification in accepting those descriptions as true, whether this prima facie entitlement has an a priori or inductivist basis. But if you then open your eyes and discover that the speaker is himself blindfolded and not even looking in the direction of the passing traffic, this prima facie justification is certainly defeated. So what you empirically determine about a speaker can make a massive difference to your overall justifiedness in accepting his utterances.

The same obviously holds about two putative experts, who make conflicting claims about a given subject-matter. Which claim you should accept (if either) can certainly be massively affected by your empirical discoveries about their respective abilities and opportunities to know the truth of the matter (and to speak sincerely about it) . Indeed, in this kind of case, default principles of the sort advanced by Burge and Foley are of no help whatever. Although a hearer may be prima facie entitled to believe each of the speakers, he cannot be entitled *all things considered* to believe both of them; for the propositions they assert, we are supposing, are incompatible (and transparently incompatible to the hearer) . So the hearer's all-things-considered justifiedness vis-à-vis their claims will depend on what he empirically learns about each speaker, or about the opinions of other speakers. In the rest of this paper I shall investigate the kinds of empirical evidence that a novice hearer might have or be able to obtain for believing one putative expert rather than her rival. I do not believe that we need to settle the "foundational" issues in the general theory of testimony before addressing this issue. This is the working assumption, at any rate, on which I shall proceed. [6]

2. The Novice/Expert Problem Versus the Expert/Expert Problem

There are, of course, degrees of both expertise and novicehood. Some novices might not be so much less knowledgeable than some experts. Moreover, a novice might in principle be able to turn himself into an expert, by improving his epistemic position vis-à-vis the target subject-matter, e. g. , by acquiring more formal training in the field. This is not a scenario to be considered in this paper, however. I assume that some sorts of limiting factors—whether they be time, cost, ability, or what have you—will keep our novices from becoming experts, at least prior to the time by which they need to make their judgment. So the question is: Can novices, while remaining novices, make justified judgments about the relative credibility of rival experts? When and how is this possible?

There is a significant difference between the novice/expert problem and another type of problem, the expert/expert problem. The latter problem is one in which experts seek to appraise the authority or credibility of other experts. Philip Kitcher (1993) addresses this problem in analyzing how scientists ascribe authority to their peers. A crucial segment of such authority ascription involves what Kitcher calls "calibration" (1993: 314-322.) . In *direct* calibration a scientist uses his own opinions about the subject-matter in question to evaluate a target scientist's degree of authority. In *indirect* calibration, he uses the opinions of still other scientists, whose opinions he has previously evaluated by direct calibration, to evaluate the target's authority. So here too he starts from his own opinions about the subject-matter in question.

By contrast, in what I am calling the novice/expert problem (more specifically, the novice/2-expert problem), the novice is not in a position to evaluate the target experts by using his own opinion; at least he does not think he is in such a position. The novice either has no opinions in the target domain, or does not have enough confidence in his opinions in this domain to use them in adjudicating or evaluating the disagreement between the rival experts. He thinks of the domain as properly requiring a certain expertise, and he does not view himself as possessing this expertise. Thus, he cannot use opinions of his own in the domain of expertise—call it the *E-domain*—to choose between conflicting experts' judgments or reports.

We can clarify the nature of the novice/expert problem by comparing it to the analogous listener/eyewitness problem. (Indeed, if we use the term "expert" loosely, the latter problem may just be a species of the novice/expert problem.) Two putative eyewitnesses claim to have witnessed a certain crime. A listener—for example, a juror—did not himself witness the crime, and has no prior beliefs about who committed it or how it was committed. In other words, he has no personal knowledge of the event. He wants to learn what transpired by listening to the testimonies of the eyewitnesses. The question is how he should adjudicate between their testimonies if and when they conflict. In this case, the E-domain is the domain of propositions concerning the actions and circumstances involved in the crime. This E-domain is what the listener (the "novice") has no prior opinions about, or no opinions to which he feels he can legitimately appeal. (He regards his opinions, if any, as mere speculation, hunch, or what have you.)

It may be possible, at least in principle, for a listener to make a reasonable assessment of which eyewitness is more credible, even without having or appealing to prior opinions of his own concerning the E-domain. For example, he might obtain evidence from others as to whether each putative witness was really present at the crime scene, or, alternatively, known to be elsewhere at the time of the crime. Second, the listener could learn of tests of each witness's visual acuity, which would bear on the accuracy or reliability of their reports. So in this kind of case, the credibility of a putative "expert's" report can be checked by such methods as independent verification of whether he had the opportunity and ability to see what he claims to have seen. Are analogous methods available to someone who seeks to assess the credibility of a "cognitive" expert as opposed to an eyewitness expert?

Before addressing this question, we should say more about the nature of expertise and the sorts of experts we are concerned with here. Some kinds of experts are unusually accomplished at certain skills, including violinists, billiards players, textile designers, and so forth. These are not the kinds of experts with which epistemology is most naturally concerned. For epistemological purposes we shall mainly focus on cognitive or intellectual experts: people who have (or claim to have) a superior quantity or level of knowledge in some domain and an ability to generate new knowledge in answer to questions within the domain. Admittedly, there are elements of skill or know-how in intellectual matters too, so the boundary between skill expertise and cognitive expertise is not a sharp one. Nonetheless, I shall try to work on only one side of this rough divide, the intellectual side.

How shall we define expertise in the cognitive sense? What distinguishes an expert from a layperson, in a given cognitive domain? I'll begin by specifying an objective sense of expertise, what it is to *be* an expert, not what it is to have a reputation for expertise. Once the objective sense is specified, the reputational sense readily follows: a reputational expert is someone widely believed to be an expert (in the objective sense), whether or not he really is one.

Turning to objective expertise, then, I first propose that cognitive expertise be defined in "veritistic" (truth-linked) terms. As a first pass, experts in a given domain (the E-domain) have more beliefs (or high degrees of belief) in true propositions and/ or fewer beliefs in false propositions within that domain than most people do (or better: than the vast majority of people do). According to this proposal, expertise is largely a comparative matter. However, I do not think it is wholly comparative. If the vast majority of people are full of false beliefs in a domain and Jones exceeds them slightly by not succumbing to a few falsehoods that are widely shared, that still does not make him an "expert" (from a God's-eye point of view). To qualify as a cognitive expert, a person must possess a substantial body of truths in the target domain. Being an expert is not simply a matter of veritistic superiority to most of the community. Some non-comparative threshold of veritistic attainment must be reached, though there is great vagueness in setting this threshold.

Expertise is not all a matter of possessing accurate information. It includes a capacity or disposition to deploy or exploit this fund of information to form beliefs in true answers to new questions that may be posed in the domain. This arises from some set of skills or techniques that constitute part of what it is to be an expert. An expert has the (cognitive) know-how, when presented with a new question in the domain, to go to the right sectors of his information-bank and perform appropriate operations on this information; or to deploy some external apparatus or data-banks to disclose relevant material. So expertise features a propensity element as well as an element of actual attainment.

A third possible feature of expertise may require a little modification in what we said earlier. To discuss this feature, let us distinguish the *primary* and *secondary* questions in a domain. Primary questions are the principal questions of interest to the researchers or students of the subject-matter. Secondary questions concern the existing evidence or arguments that bear on the primary questions, and the assessments of the evidence made by prominent researchers. In general, an expert in a field is someone who has (comparatively) extensive knowledge (in the weak sense of knowledge, i. e., true belief) of the state of the evidence, and knowledge of the opinions and reactions to that evidence by prominent workers in the field. In the central sense of "expert" (a strong sense), an expert is someone with an unusually extensive body of knowledge on both primary and secondary questions in the domain. However, there may also be a weak sense of "expert", in which it includes someone who merely has extensive knowledge on the secondary questions in the domain. Consider two people with strongly divergent views on the primary questions in the domain, so that one of them is largely right and the other is largely wrong. By the original, strong criterion, the one who is largely wrong would not qualify as an expert. People might disagree with this as the final word on the matter. They might hold that anyone with a thorough knowledge of the existing evidence and the differing views held by the workers in the field deserves to be called an expert. I concede this by acknowledging the weak sense of "expert".

Applying what has been said above, we can say that an expert (in the strong sense) in domain D is someone who possesses an extensive fund of knowledge (true belief) and a set of skills or methods for apt and successful deployment of this knowledge to new questions in the domain. Anyone purporting to be a (cognitive) expert in a given domain will claim to have such a fund and set of methods, and will claim to have true answers to the question (s) under dispute because he has applied his fund and his methods to the question (s). The task for the layperson who is consulting putative experts, and who hopes thereby to learn a true answer to the target question, is to decide who has superior expertise, or who has better deployed his expertise to the question at hand. The novice/2-experts problem is whether a layperson can *justifiably* choose one putative expert as more credible or trustworthy than the other with respect to the question at hand, and what might be the epistemic basis for such a choice?[7]

3. Argument-Based Evidence

To address these issues, I shall begin by listing five possible sources of evidence that a novice might have, in a novice/2-experts situation, for trusting one putative expert more than another. I'll then explore the prospects for utilizing such sources, depending on their availability and the novice's exact circumstance. The five sources I shall discuss are:

(A) Arguments presented by the contending experts to support their own views and critique their rivals' views.

(B) Agreement from additional putative experts on one side or other of the subject in question.

(C) Appraisals by "meta-experts" of the experts' expertise (including appraisals reflected in formal credentials earned by the experts).

(D) Evidence of the experts' interests and biases vis-à-vis the question at issue.

(E) Evidence of the experts' past "track-records".

In the remainder of the paper, I shall examinethese five possible sources, beginning, in this section, with source (A).[8]

There are two types of communications that a novice, N, might receive from his two experts, El and E2.[9] First, each expert might baldly state her view (conclusion), without supporting it with any evidence or argument whatever. More commonly, an expert may give detailed support to her view in some public or professional context, but this detailed defense might only appear in a restricted venue (e. g., a professional conference or journal) that does not reach N's attention. So N might not encounter the two experts' defenses, or might encounter only very truncated versions of them. For example, N might hear about the experts' views and their support from a second-hand account in the popular press that does not go into many details. At the opposite end of the communicational spectrum, the two experts might engage in a full-scale debate that N witnesses (or reads a detailed reconstruction of). Each expert might there present fairly developed arguments in support of her view and against that of her opponent. Clearly, only when N somehow encounters the experts' evidence or arguments can he have evidence of type (A). So let us consider this scenario.

We may initially suppose that if N can gain (greater) justification for believing one expert's view as compared with the other by means of their arguments, the novice must at least understand the evidence cited in the experts' arguments. For some domains of expertise and some novices, however, even a mere grasp of the evidence may be out of reach. These are cases where N is an "ignoramus" vis-à-vis the E-domain. "This is not the universal plight of novices. " Sometimes they can understand the evidence (in some measure) but aren't in a position, from personal knowledge, to give it any credence. Assessing an expert's evidence may be especially difficult when it is disputed by an opposing expert.

Not every statement that appears in an expert's argument need be epistemically inaccessible to the novice. Let us distinguish here between *esoteric* and *exoteric* statements within an expert's discourse. Esoteric statements belong to the relevant sphere of expertise, and their truth-values are inaccessible to N—in terms of his personal knowledge, at any rate. Exoteric statements are outside the domain of expertise; their truth-values may be accessible to N—either at the time of their assertion or later[10], I presume that esoteric statements comprise a hefty portion of the premises and "lemmas" in an expert's argument. That's what makes it difficult for a novice to become justified in believing any expert's view on the basis of arguments per se. Not only are novices commonly unable to assess the truth-values of the esoteric propositions, but they also are ill-placed to assess the support relations between the cited evidence and the proffered conclusion. Of course, the proponent expert will claim that the support relation is strong between her evidence and the conclusion she defends; but her opponent will commonly dispute this. The novice will be ill-placed to assess which expert is in the right.

At this point I wish to distinguish *direct* and *indirect argumentative justification*. In direct argumentative justification, a hearer becomes justified in believing an argument's conclusion by becoming justified in believing the argument's premises and their (strong) support relation to the conclusion. If a speaker's endorsement of an argument helps bring it about that the hearer has such justificational status vis-à-vis its premises and support relation, then the hearer may acquire "direct" justification for the conclusion via that speaker's argument. [11] As we have said, however, it is difficult for an expert's argument to produce direct justification in the hearer in the novice/2-expert situation. Precisely because many of these matters are esoteric, N will have a hard time adjudicating between E_1's and E_2's claims, and will therefore have a hard time becoming justified vis-à-vis either of their conclusions. He will even have a hard time becoming justified in trusting one conclusion more than the other.

The idea of indirect argumentative justification arises from the idea that one speaker in a debate may demonstrate dialectical superiority over the other, and this dialectical superiority might be a plausible *indicator*[12] for N of greater expertise, even if it doesn't render N directly justified in believing the superior speaker's conclusion. By dialectical superiority, I do not mean merely greater debating skill. Here is an example of what I do mean.

Whenever expert E_2 offers evidence for her conclusion, expert E_1 presents an ostensible rebuttal or defeater of that evidence. On the other hand, when E_1 offers evidence for her conclusion, E_2 never manages to offer a rebuttal or defeater to E_1's evidence. Now N is not in a position to assess the truth-value of E_2's defeaters against E_2, nor to evaluate the truth-value or strength of support that E_1's (undefeated) evidence gives to E_1's conclusion. For these reasons, E_1's evidence (or arguments) are not directly justificatory for N. Nonetheless, in "formal" dialectical terms, E_1 seems to be doing better in the dispute. Furthermore, I suggest, this dialectical superiority may reasonably be taken as an indicator of E_1's having superior expertise on the question at issue. It is a (non-conclusive) indicator that E_1 has a superior fund of information in the domain, or a superior method for manipulating her information, or both.

Additional signs of superior expertise may come from other aspects of the debate, though these are far more tenuous. For example, the comparative quickness and smoothness with which E1 responds to E_2's evidence may suggest that E_1 is already well familiar with E_2's "points" and has already thought out counterarguments. If E_2's responsiveness to E_1's arguments displays less quickness and smoothness, that may suggest that E_1's prior mastery of the relevant information and support considerations exceeds that of E_2. Of course, quickness and smoothness are problematic indicators of informational mastery. Skilled debaters and well-coached witnesses can appear better-informed because of their stylistic polish, which is not a true indicator of superior expertise. This makes the proper use of indirect argumentative justification a very delicate matter. [13]

To clarify the direct/indirect distinction being drawn here, consider two different things a hearer might say to articulate these different bases of justification. In the case of direct argumentative justifiedness, he might say: "In light of this expert's argument, that is, in light of the truth of its premises and the support they confer on the conclusion (both of which are epistemically accessible to me), I am now justified in believing the conclusion. " In indirect argumentative justifiedness, the hearer might say: "In light of the way this expert has argued—her argumentative *performance*, as it were—I can infer that she has more expertise than her opponent; so I am justified in inferring that her conclusion is probably the correct one. "

Here is another way to explain the direct/indirect distinction. Indirect argumentative justification essentially involves an *inference to the best explanation*, an inference that N might make from the performances of the two speakers to their respective levels of expertise. From their performances, N makes an inference as to which expert has superior expertise in the target domain. Then he makes an inference from greater expertise to a higher probability of endorsing a true conclusion. Whereas *indirect* argumentative justification essentially involves inference to the best explanation, direct argumentative justification need involve no such inference. Of course, it *might* involve such inference; but if so, the topic of the explanatory inference will only concern the objects, systems, or states of affairs under dispute, not the relative expertise of the contending experts. By contrast, in indirect argumentative justifiedness, it is precisely the experts' relative expertise that constitutes the target of the inference to the best explanation.

Hardwig (1985) makes much of the fact that in the novice/expert situation, the novice lacks the expert's reasons for believing her conclusion. This is correct. Usually, a novice (1) lacks all or some of the premises from which an expert reasons to her conclusion, (2) is in an inferior position to assess the support relation between the expert's premises and conclusions, and (3) is ignorant of many or most of the defeaters (and "defeater-defeaters") that might bear on an expert's arguments. However, although novice N may lack (all or some of) an expert's reasons R for believing a conclusion p, N *might* have reasons R^* for believing *that* the expert has good reasons for believing p; and N might have reasons R^* for believing that one expert has *better* reasons for believing her conclusion than her opponent has for hers. Indirect argumentative justification is one means by which N might acquire reasons R^* without sharing (all or any) of either experts' reasons R. [14] It is this possibility to which Hardwig gives short shrift. I don't say that a novice in a novice/2-expert situation invariably has such reasons R^*; nor do I say that it is easy for a novice to acquire such reasons. But it does seem to be possible.

4. Agreement from Other Experts:
the Question of Numbers

An additional possible strategy for the novice is to appeal to further experts. This brings us to categories (B) and (C) on our list. Category (B) invites N to consider whether other experts agree with E_1 or with E_2. What proportion of these experts agree with E_1 and what proportion with E_2? In other words, to the extent that it is feasible, N should consult the numbers, or degree of consensus, among all relevant (putative) experts. Won't N be fully justified in trusting E_1 over E_2 if almost all other experts on the subject agree with E_1, or if even a preponderance of the other experts agree with E_1?

Another possible source of evidence, cited under category (C), also appeals to other experts but in a slightly different vein. Under category (C), N should seek evidence about the two rival experts' relative degrees of expertise by consulting third parties' assessments of their expertise. If "meta-experts" give E_1, higher "ratings" or "scores" than E_2, shouldn't N rely more on E_1 than E_2? Credentials can be viewed as a special case of this same process. Academic degrees, professional accreditations, work experience, and so forth (all from specific institutions with distinct reputations) reflect certifications by other experts of E_1's and E_2's demonstrated training or competence. The relative strengths or weights of these indicators might be utilized by N to distill appropriate levels of trust for E_1 and E_2 respectively. [15]

I treat ratings and credentials as signaling "agreement" by other experts because I assume that established authorities certify trainees as competent when they are satisfied that the latter demonstrate (1) a mastery of the same methods that the certifiers deem fundamental to the field, and (2) knowledge of (or belief in) propositions that certifiers deem to be fundamental facts or laws of the discipline. In this fashion, ratings and conferred credentials ultimately rest on basic agreement with the meta-experts and certifying authorities.

When it comes to evaluating specific experts, there is precedent in the American legal system for inquiring into the degree to which other experts agree with those being evaluated. [16] But precedented or not, just how good is this appeal to consensus? If a putative expert's opinion is joined by the consensual opinions of other putative experts, how much warrant does that give a hearer for trusting the original opinion? How much evidential worth does consensus or agreement deserve in the doxastic decision-making of a hearer?

If one holds that a person's opinion deserves prima facie credence, despite the absence of any evidence of their reliability on the subject, then numbers would seem to be very weighty, at least in the absence of additional evidence. Each new testifier or opinion-holder on one side of the issue should add weight to that side. So a novice who is otherwise in the dark about the reliability of the various opinion-holders would seem driven to agree with the more numerous body of experts. Is that right?

Here are two examples that pose doubts for "using the numbers" to judge the relative credibility of opposing positions. First is the case of a guru with slavish followers. Whatever the guru believes is slavishly believed by his followers. They fix their opinions wholly and exclusively on the basis of their leader's views. Intellectually speaking, they are merely his clones. Or consider a group of followers who are not led by a single leader but by a small elite of opinion-makers. When the opinion-makers agree, the mass of followers concur in their opinion. Shouldn't a novice consider this kind of scenario as a possibility? Perhaps (putative) expert Et belongs to a doctrinal community whose members devoutly and uncritically agree with the opinions of some single leader or leadership cabal. Should the numerosity of the community make their opinion more credible than that of a less numerous group of experts? Another example, which also challenges the probity of greater numbers, is the example of rumors. Rumors are stories that are widely circulated and accepted though few of the believers have access to the rumored facts. If someone hears a rumor from one source, is that source's credibility enhanced when the same rumor is repeated by a second, third, and fourth source? Presumably not, especially if the hearer knows (or justifiably believes) that these sources are all uncritical recipients of the same rumor.

It will be objected that additional rumor spreaders do not add credibility to an initial rumor monger because the additional ones have no established reliability. The hearer has no reason to think that any of their opinions is worthy of trust. Furthermore, the rumor case doesn't seem to involve "expert" opinions at all and thereby contrasts with the original case. In the original case the hearer has at least some prior reason to think that each new speaker who concurs with one of the original pair has *some* credibility (reliability). Under that scenario, don't additional concurring experts increase the total believability of the one with whom they agree?

It appears, then, that greater numbers should add further credibility, at least when each added opinion-holder has positive initial credibility. This view is certainly presupposed by some approaches to the subject. In the Lehrer-Wagner (1981) model, for example, each new person to whom a subject assigns "respect" or "weight" will provide an extra vector that should push the subject in the direction of that individual's opinion. [17] Unfortunately, this approach has a problem. If two or more opinion-holders are totally *non-independent* of one another, and if the subject knows or is justified in believing this, then the subject's opinion should not be swayed—even a little—by more than one of these opinion-holders. As in the case of a guru and his blind followers, a follower's opinion does not provide any additional grounds for accepting the guru's view (and a second follower does not provide additional grounds for accepting a first follower's view) even if all followers are precisely as reliable as the guru himself (or as one another) —which followers must be, of course, if they believe exactly the same things as the guru (and one another) on the topics in question. Let me demonstrate this through a Bayesian analysis.

Under a simple Bayesian approach, an agent who receives new evidence should update his degree of belief in a hypothesis H by conditioning on that evidence. This means that he should use the ratio (or quotient) of two likelihoods: the likelihood of the evidence occurring if H is true and the likelihood of the evidence occurring if H is false. In the present case the evidence in question is the belief in H on the part of one or more putative experts. More precisely, we are interested in comparing (A) the result of conditioning *on* the evidence of a single putative expert's belief with (B) the result of conditioning on the evidence of concurring beliefs by two putative experts. Call the two putative experts X and Y, and let X (H) be X's believing H and Y (H) be Y's believing H. What we wish to compare, then, is the magnitude of the likelihood quotient expressed in (1) with the magnitude of the likelihood quotient expressed in (2) .

(1) $\dfrac{P (X (H) /H)}{P (X (H) / \sim H)}$

(2) $\dfrac{P (X (H) \& Y (H) /H)}{P (X (H) \& Y (H) / \sim H)}$

The principle we are interested in is the principle that the likelihood ratio given in (2) is always larger than the likelihood ratio given in (1), so that an agent who learns that X and Y both believe H will always have grounds for a larger upward revision of his degree of belief in H than if he learns only that X believes H. At least this is so when X and Y are each somewhat credible (reliable) . More precisely, such comparative revisions are in order if the agent is *justified* in believing these things in the different scenarios. I am going to show that such comparative revisions are not always in order. Sometimes (2) is not larger than (1); so the agent—if he knows or justifiably believes this—is not justified in making a larger upward revision from the evidence of two concurring believers than from one believer.

First let us note that according to the probability calculus, (2) is equivalent to (3).

$$(3) \quad \frac{P (X (H) /H) P (Y (H) /X (H) \&H)}{P (X (H) / {\sim}H) P (Y (H) /X (H) \& {\sim}H)}$$

While looking at (3), return to the case of blind followers. If Y is a blind follower of X, then anything believed by X (including H) will also be believed by Y. And this will hold whether or not H is true. So,

(4) P (Y (H) / X (H) & H) = 1,

and

(5) P (Y (H) / X (H) & ~H) = 1.

Substituting these two values into expression (3), (3) reduces to (1). Thus, in the case of a blind follower, (2) (which is equivalent to (3)) is the same as (1), and no larger revision is warranted in the two-concurring-believers case than in the single-believer case.

Suppose that the second concurring believer, Y, is not a *blind* follower of X. Suppose he would sometimes agree with X but not in all circumstances. Under that scenario, does the addition of Y's concurring belief always provide the agent (who possesses this information) with more grounds for believing H? Again the answer is no. The appropriate question is whether Y is more likely to believe H when X believes H and H is true than when X believes H and H is false. If Y is just as likely to follow X's opinion whether H is true or false, then Y's concurring belief adds nothing to the agent's evidential grounds for H (driven by the likelihood quotient). Let us see why this is so.

If Y is just as likely to follow X's opinion when H is false as when it's true, then (6) holds:

(6) P (Y (H) / X (H) & H) = P (Y (H) / X (H) & ~H)

But if (6) holds, then (3) again reduces to (1), because the right-hand sides of both numerator and denominator in (3) are equal and cancel each other out. Since (3) reduces to (1), the agent still gets no extra evidential boost from Y's agreement with X concerning H. Here it is not required that Y is certain to follow X's opinion; the likelihood of his following X might only be 0.80, or 0.40, or whatever. As long as Y is just as likely to follow X's opinion when H is true as when it's false, we get the same result.

Let us describe this last case by saying that Y is a *non-discriminating reflector* of X （with respect to H）. When Y is a non-discriminating reflector of X, Y's opinion has no extra evidential worth for the agent above and beyond X's opinion. What is necessary for the novice to get an extra evidential boost from Y's belief in H is that he （the novice）be justified in believing （6′）:

(6′) P （Y （H）/ X （H）& H）> P （Y （H）/ X （H）& ~H）

If （6′）is satisfied, then Y's belief is at least partly *conditionally independent* of X's belief. Full conditional independence is a situation in which any dependency between X and Y's beliefs is accounted for by the dependency of each upon H. Although full conditional independence is not required to boost N's evidence, *partial* conditional independence is required. [18]

We may now identify the trouble with the （unqualified）numbers principle. The trouble is that a novice cannot automatically count on his putative experts being （even partially）conditionally independent of one another. He cannot automatically count on the truth of （6′）. Y may be a non-discriminating reflector of X, or X may be a non-discriminating reflector of Y, or both may be non-discriminating reflectors of some third party or parties. The same point applies no matter how many additional putative experts share an initial expert's opinion. If they are all non-discriminating reflectors of someone whose opinion has already been taken into account, they add no further weight to the novice's evidence.

What type of evidence call the novice have to justify his acceptance of （or high level of credence in）（6′）? N call have reason to believe that Y's *route* to belief in H was such that even in possible cases where X fails to recognize H's falsity （and hence believes it）, Y *would* recognize its falsity. There are two types of causal routes to Y's belief of the right sort. First, Y's route to belief in H might entirely *bypass* X's route. This would be exemplified by cases in which X and Y are causally independent eyewitnesses of the occurrence or non-occurrence of H; or by cases in which X and Y base their respective beliefs on independent experiments that bear on H. In the eyewitness scenario X might falsely believe H through misperception of the actual event, whereas Y might perceive the event correctly and avoid belief in H. A second possible route to Y's belief in H might *go partly through X* but not involve uncritical reflection of X's belief. For example, Y might listen to X's reasons for believing H, consider a variety of possible defeaters of these reasons that X never considered, but finally rebut the cogency of these defeaters and concur in accepting H. In either of these scenarios Y's partly "autonomous" causal route made him poised to avoid belief in H even though X believes it (possibly falsely). If N has reason to think that Y used one of these more-or-less autonomous causal routes to belief, rather than a causal route that guarantees agreement with X, then N has reason to accept (6'). In this fashion, N would have good reason to rate Y's belief as increasing his evidence for H even after taking account of X's belief.

Presumably, novices could well be in such an epistemic situation vis-à-vis a group of concurring (putative) experts. Certainly in the case of concurring *scientists*, where a novice might have reason to expect them to be critical of one another's viewpoints, a presumption of partial independence might well be in order. If so, a novice might be warranted in giving greater evidential weight to larger numbers of concurring opinion-holders. According to some theories of scientific opinion formation, however, this warrant could not be sustained. Consider the view that scientists' beliefs are produced entirely by negotiation with other scientists, and in no way reflect reality (or Nature). This view is apparently held by some social constructionists about science, e. g., Bruno Latour and Steve Woolgar (1979/1986); at least this is Kitcher's (1993: 165-166) interpretation of their view. [19] Now if the social constructionists are right, so interpreted, then nobody (at least nobody knowledgeable of this fact) would be warranted in believing anything like (6'). There would never be reason to think that any scientist is more likely to believe a scientific hypothesis H when it's true (and some other scientist believes it) than when it's false (and the other scientist believes it). Since causal routes to scientific belief never reflect "real" facts—they only reflect the opinions, interests, and so forth of the community of scientists- (6') will never be true. Anybody who accepts or inclines toward the indicated social-constructionist thesis would never be justified in believing (6'). [20]

Setting such extreme views aside, won't a novice normally have reason to expect that different putative experts will have some causal independence or autonomy from one another in their routes to belief? If so, then if a novice is also justified in believing that each putative expert has some slight level of reliability (greater than chance), then won't he be justified in using the numbers of concurring experts to tilt toward one of two initial rivals as opposed to the other? This conclusion might be right when *all* or *almost all* supplementary experts agree with one of the two initial rivals. But this is rarely the case. Vastly more common are scenarios in which the numbers are more evenly balanced, though not exactly equal. What can a novice conclude in those circumstances? Can he legitimately let the greater numbers decide the issue?

This would be unwarranted, especially if we continue to apply the Bayesian approach. The appropriate change in the novice's belief in H should be based on two sets of concurring opinions (one in favor of H and one against it), and it should depend on *how reliable* the members of each set are and on *how (conditionally) independent* of one another they are. If the members of the smaller group are more reliable and more (conditionally) independent of one another than the members of the larger group, that might imply that the evidential weight of the smaller group exceeds that of the larger one. More precisely, it depends on what the novice is *justified* in believing about these matters. Since the novice's justifiedness on these matters may be very weak, there will be many situations in which he has no distinct or robust justification for going by the relative numbers of like-minded opinion-holders.

This conclusion seems perfectly in order. Here is an example that, by my own lights, sits well with this conclusion. If scientific creationists are more numerous than evolutionary scientists, that would not incline me to say that a novice is warranted in putting more credence in the views of the former than in the views of the latter (on the core issues on which they disagree). At least I am not so inclined on the assumption that the novice has roughly comparable information as most philosophers currently have about the methods of belief formation by evolutionists and creationists respectively. [21] Certainly the numbers do not *necessarily* outweigh considerations of individual reliability and mutual conditional independence. The latter factors seem more probative, in the present case, than the weight of sheer numbers. [22]

5. Evidence from Interests and Biases

I turn now to the fourth source of possible evidence on our original list: evidence of distorting interests and biases that might lie behind a putative expert's claims. If N has excellent evidence for such bias in one expert and no evidence for such bias in her rival, and if N has no other basis for preferential trust, then N is justified in placing greater trust in the unbiased expert. This proposal comes directly from common sense and experience. If two people give contradictory reports, and exactly one of them has a reason to lie, the relative credibility of the latter is seriously compromised.

Lying, of course, is not the only way that interests and biases can reduce an expert's trustworthiness. Interests and biases can exert more subtle distorting influences on experts' opinions, so that their opinions are less likely to be accurate even if sincere. Someone who is regularly hired as an expert witness for the defense in certain types of civil suits has an economic interest in delivering strong testimony in any current trial, because her reputation as a defense witness depends on her present performance.

As a test of expert performance in situations of conflict of interest, consider the results of a study published in the *Journal of American Medical Association* (Friedberg et al., 1999). The study explored the relationship between published research reports on new ontology drugs that had been sponsored by pharmaceutical companies versus those that had been sponsored by nonprofit organizations. It found a statistically significant relationship between the funding source and the qualitative conclusions in the reports. Unfavorable conclusions were reached by 38% of nonprofit-sponsored studies but by only 5% of pharmaceutical company-sponsored studies.

From a practical point of view, information bearing on an expert's interests is often one of the more accessible pieces of relevant information that a novice can glean about an expert. Of course, it often transpires that *both* members of a pair of testifying experts have interests that compromise their credibility. But when there is a non-negligible difference on this dimension, it is certainly legitimate information for a novice to employ.

Pecuniary interests are familiar types of potential distorters of an individual's claims or opinions. Of greater significance, partly because of its greater opacity to the novice, is a bias that might infect a whole discipline, sub-discipline, or research group. If all or most members of a given field are infected by the same bias, the novice will have a difficult time telling the real worth of corroborating testimony from other experts and meta-experts. This makes the numbers game, discussed in the previous section, even trickier for the novice to negotiate.

One class of biases emphasized by feminist epistemologists involves the exclusion or underrepresentation of certain viewpoints or standpoints within a discipline or expert community. This might result in the failure of a community to gather or appreciate the significance of certain types of relevant evidence. A second type of community-wide bias arises from the economics or politics of a sub-discipline, or research community. To advance its funding prospects, practitioners might habitually exaggerate the probativeness of the evidence that allegedly supports their findings, especially to outsiders. In competition with neighboring sciences and research enterprises for both resources and recognition, a given research community might apply comparatively lax standards in reporting its results. Novices will have a difficult time detecting this, or weighing the merit of such an allegation by rival experts outside the field. [23]

6. Using Past Track Records

The final category in our list may provide the novice's best source of evidence for making credibility choices. This is the use of putative experts' past track records of cognitive success to assess the likelihoods of their having correct answers to the current question. But how can a novice assess past track records? There are several theoretical problems here, harking back to matters discussed earlier.

First, doesn't using past track records amount to using the method of (direct) "calibration" to assess a candidate expert's expertise? Using a past track record means looking at the candidate's past success rate for previous questions in the E-domain to which she offered answers. But in our earlier discussion (section 2), I said that it's in the nature of a novice that he has no opinions, or no confidence in his own opinions, about matters falling within the E-domain. So how can the novice have any (usable) beliefs about past answers in the E-domain by which to assess the candidate's expertise? In other words, how can a novice, *qua* novice, have any opinions at all about past track records of candidate experts?

A possible response to this problem is to revisit the distinction between *esoteric* and *exoteric* statements. Perhaps not every statement in the E-domain is esoteric. There may also be a body of exoteric statements in the E-domain, and they are the statements for which a novice might assess a candidate's expertise. But does this really make sense? If a statement is an exoteric statement, i. e. , one that is epistemically accessible to novices, then why should it even be included in the E-domain? One would have thought that the E-domain is precisely the domain of propositions accessible only to experts.

The solution to the problem begins by sharpening our esoteric/exoteric distinction. It is natural to think that statements are categorically either esoteric or exoteric, but that is a mistake. A given (timeless) statement is esoteric or exoteric only *relative* to an epistemic standpoint or position. It might be esoteric relative to one epistemic position but exoteric relative to a different position. For example, consider the statement, "There will be an eclipse of the sun on April 22, 2130, in Santa Fe, New Mexico". Relative to the present epistemic standpoint, i. e. , the standpoint of people living in the year 2000, this is an esoteric statement. Ordinary people in the year 2000 will not be able to answer this question correctly, except by guessing. On the other hand, on the very day in question, April 22, 2130, ordinary people on the street in Santa Fe, New Mexico will easily be able to answer the question correctly. In that different epistemic position, the question will be an exoteric one, not an esoteric one. [24] You won't need specialized training or knowledge to determine the answer to the question. In this way, the epistemic status of a statement can change from one time to another.

There is a significant application of this simple fact to the expert/novice problem. A novice might easily be able to determine the truth-value of a statement after it has become exoteric. He might be able to tell *then* that it is indeed true. Moreover, he might learn that at an earlier time, when the statement was esoteric for the likes of him, another individual managed to believe it and say that it is (or would be) true. Furthermore, the same individual might repeatedly display the capacity to assert statements that are esoteric at the time of assertion but become exoteric later, and she might repeatedly turn out to have been right, as determined under the subsequently exoteric circumstances. When this transpires, novices can infer that this unusual knower must possess some special manner of knowing—some distinctive expertise—that is not available to them. They presumably will not know exactly what this distinctive manner of knowing involves, but presumably it involves some proprietary fund of information and some methodology for deploying that information. In this Fashion, a novice can verify somebody else's expertise in a certain domain by verifying their impressive track record within that domain. And this can be done without the novice himself somehow being transformed into an expert.

The astronomical example is just one of many, which are easily proliferated. If an automobile, an air-conditioning system, or an organic system is suffering some malfunction or impairment, untrained people will often be unable to specify any true proposition of the form, "If you apply treatment X to system Y, the system will return to proper functioning". However, there may be people who can repeatedly specify true propositions precisely of this sort. [25] Moreover, that these propositions are true can be verified by novices, because novices might be able to "watch" the treatment being applied to the malfunctioning system and see that the system returns to proper functioning (faster than untreated systems do). Although the truth of the proposition is an exoteric matter once the treatment works, it was an esoteric matter before the treatment was applied and produced its result. In such a case the expert has knowledge, and can be determined to have had knowledge, at a time when it was esoteric. [26]

It should be emphasized that many questions to which experts provide answers, at times when they are esoteric, are not merely yes/no questions that might be answered correctly by lucky guesses. Many of them are questions that admit of innumerable possible answers, sometimes indefinitely many answers. Simplifying for purposes of illustration, we might say that when a patient with an ailment sees a doctor, he is asking her the question: "Which medicine, among the tens of thousands of available medicines, will cure or alleviate this ailment?" Such a question is unlikely to be answered correctly by mere guesswork. Similarly, when rocket scientists were first trying to land a spaceship on the moon, there were indefinitely many possible answers to the question: "Which series of steps will succeed in landing this (or some) spaceship on the moon?" Choosing a correct answer from among the infinite list of possible answers is unlikely to be a lucky guess. It is feats like this, often involving technological applications, that rightly persuade novices that the people who get the correct answers have a special fund of information and a special methodology for deploying it that jointly yield a superior capacity to get right answers. In this fashion, novices can indeed determine that others are experts in a domain in which they themselves are not.

Of course, this provides no algorithm by which novices call resolve all their two-expert problems. Only occasionally will a novice know, or be able to determine, the track records of the putative experts that dispute an issue before him. A juror in a civil trial has no opportunity to run out and obtain track record information about rival expert witnesses who testify before him. Nonetheless, the fact that novices can verify track records and use them to test a candidate's claims to expertise, at least in principle and in some cases, goes some distance toward dispelling utter skepticism for the novice/2-expert situation. Moreover, the possibility of "directly" determining the expertise of a few experts makes it possible to draw plausible inferences about a much wider class of candidate experts. If certain individuals are shown, by the methods presented above, to have substantial expertise, and if those individuals train others, then it is a plausible inference that the trainees will themselves have comparable funds of information and methodologies, of the same sort that yielded cognitive success for the original experts. [27] Furthermore, to the extent that the verified experts are then consulted as "meta-experts" about the expertise of others (even if they didn't train or credential them), the latter can again be inferred to have comparable expertise. Thus, some of the earlier skepticism engendered by the novice/2-expert problem might be mitigated once the foundation of expert verification provided in this section has been established.

7. Conclusion

My story's ending is decidedly mixed, a cause for neither elation nor gloom. Skeptical clouds loom over many a novice's epistemic horizons when confronted with rival experts bearing competing messages. There are a few silver linings, however. Establishing experts' track-records is not beyond the pale of possibility, or even feasibility. This in turn can bolster the credibility of a wider class of experts, thereby laying the foundation for a legitimate use of numbers when trying to choose between experts. There is no denying, however, that the epistemic situations facing novices are often daunting. There are interesting theoretical questions in the analysis of such situations, and they pose interesting practical challenges for "applied" social epistemology. What kinds of education, for example, could substantially improve the ability of novices to appraise expertise, and what kinds of communicational intermediaries might help make the novice-expert relationship more one of justified credence than blind trust? [28]

Notes

1. Thanks to Scott La Barge for calling Plato's treatment of this subject to my attention.
2. In his 1991 paper, Hardwig at first says that trust must be "at least partially blind" (p. 693). He then proceeds to talk about knowledge resting on trust and therefore being blind (pp. 693, 699) without using the qualifier "partially".
3. However, there is some question whether Foley can consistently call the epistemic right he posits a "fundamental" one, since he also says that it rest on (A) my justified *self* trust, and (B) the *similarity* of others to me—presumably the *evidence* I have of their similarity to me (see pp. 63-64). Another question for Foley is how the fundamentality thesis fits with his view that in cases of conflict I have more reason (prima facie) to trust myself than to trust someone else (see p. 66). If my justified trust in others is really fundamental, why does it take a backseat to self-trust?
4. Moreover, according to Baron-Cohen, there is a separate module called the "shared attention mechanism", which seeks to determine when another person is attending to the same object as the self is attending to.
5. For one thing, it may be argued that babies' interpretations of what people say is, in

the first instance, constrained by the assumption that the contents concern matters within the speakers' perceptual ken. This is not an empirical finding, it might be argued, but an a priori posit that is used to fix speakers' meanings.

6. Some theorists of testimony, Burge included, maintain that a heater's justificational status vis-à-vis a claim received from a source depends partly on the justificational status of the source's own belief in that claim. This is a *transpersonal*, *preservationist*, or *transmissional* conception of justifiedness, under which a recipient is not justified in believing p unless the speaker has a justification and entitlement that he *transmits* to the hearer. For purposes of this paper, however, I shall not consider this transmissional conception of justification. First, Burge himself recognizes that there is such a thing as the recipient's "proprietary" justification for believing an interlocutor's claim, justification localized "in" the recipient, which isn't affected by the source's justification (1993: 485-486). I think it is appropriate to concentrate on this "proprietary" justification (of the recipient) for present purposes. When a hearer is trying to "choose" between the conflicting claims of rival speakers, he cannot appeal to any inaccessible justification lodged in the heads of the speakers. He can only appeal to his *own* justificational resources. (Of course, these might include things *said* by the two speakers by way of defense of their contentions, things which also are relevant to *their own* justifications.) For other types of (plausible) objections to Burge's preservationism about testimony, see Bezuidenhout (1998).

7. In posing the question ofjustifiedness, I mean to stay as neutral as possible between different approaches to the concept of justifiedness, e. g., between internalist versus externalist approaches to justifiedness. Notice, moreover, that I am not merely asking whether and how the novice can justifiably decide to accept one (candidate) expert's view *outright*, but whether and how he can justifiably decide to give *greater* credence to one than to the other.

8. I do not mean to be committed to the exhaustiveness of this list. The list just includes some salient categories.

9. In what follows I shall for brevity speak about two experts, but I shall normally mean two *putative* experts, because from the novice's epistemic perspective it is problematic whether each, or either, of the self-proclaimed experts really is one.

10. It might be helpful to distinguish *semantically* esoteric statements and *epistemically* esoteric statements. (Thanks to Carol Caraway for this suggestion.) Semantically esoteric statements are ones that a novice cannot assess because he does not even

understand them; typically, they utilize a technical vocabulary he has not mastered. Epistemically esoteric statements are statements the novice understands but still cannot assess for truth-value.

11. By "direct" justification I do not, of course, mean anything having to do with the basicness of the conclusion in question, in the foundationalist sense of basicness. The distinction I am after is entirely different, as will shortly emerge.

12. Edward Craig (1990: 135) similarly speaks of "indicator properties" as what an inquirer seeks to identify in an informant as a guide to his/her truth-telling ability.

13. Scott Brewer (1998) discusses many of the same issues about novices and experts can vassed here. He treats the present topic under the heading of novices' using experts' " demeaner " to assess their expertise. Demeanor is an especially untrustworthy guide, he points out, where there is a lucrative " market " for demeanor itself—where demeanor is "traded" at high prices (1998: 1622) . This practice was prominent in the days of the sophists and is a robust business in adversarial legal systems.

14. Of course, in indirect argumentative justification the novice must at least *hear* some of the expert's premises—or intermediate steps between "ultimate" premises and conclusion. But the novice will not share the expert's *justifiedness* in believing those premises.

15. These items fall under Kitcher's category of "unearned authority" (1993: 315) .

16. Appealing to other experts to validate or underwrite a putative expert's opinion—or, more precisely, the *basis* for his opinion—has a precedent in the legal system's procedures for deciding the admissibility of scientific expert testimony. Under the governing test for admitting or excluding such testimony that was applicable from 1923 to 1993, the scientific principle (or methodology) on which a proffered piece of testimony is based must have "gained general acceptance in the particular field in which it belongs" (*Frye v. United States*, 292 F. 1013 D. C. Cir. (1923)) . In other words, appeal was made to the scientific community's opinion to decide whether the basis of an expert's testimony is sound enough to allow that testimony into court. This test has been superseded as the uniquely appropriate test in a more recent decision of the Supreme Court (Daubert v. Merrell Dow Pharmaceuticals, 509 U. S. 579 (1993)) ; but the latter decision also appeals to the opinions of other experts. It recommends that judges use a combination of four criteria (none of them necessary or sufficient) in deciding whether proffered scientific expert testimony is

admissible. One criterion is the old general acceptance criterion and another is whether the proffered evidence has been subjected to peer review and publication. Peer review, obviously, also introduces the opinions of other experts. Of course, the admissibility of a piece of expert testimony is not the same question as how heavily a hearer—e. g. , a juror—should trust such testimony if he hears it. But the two are closely intertwined, since courts make admissibility decisions on the assumption that jurors are likely to be influenced by any expert testimony they hear. Courts do not wish to admit scientific evidence unless it is quite trustworthy. Thus, the idea of ultimately going to the opinions of other experts to assess the trustworthiness of a given expert's proffered testimony is certainly a well-precedented procedure for trying to validate an expert's trustworthiness.

17. Lehrer and Wagner say (p. 20) that one should assign somebody else a positive weight if one does not regard his opinion as "worthless" on the topic in question— i. e. , if one regards him as better than a random device. So it looks as if every clone of a leader should be given positive weight—arguably, the same weight as the leader himself, since their beliefs always coincide—as long as the leader receives positive weight. In the Lehrer-Wagner model, then, each clone will exert a positive force over one's own revisions of opinion just as a leader's opinion will exert such force; and the more clones there are, the more force in the direction of their collective opinion will be exerted.

18. I am indebted here to Richard Jeffrey. (1992: 109-110), He points out that it is only conditional independence that is relevant in these kinds of cases, not "simple independence" defined by the condition: $P (Y (H) / X (H)) = P (Y (H))$. If X and Y are even slightly reliable independent sources of information about H, they won't satisfy this latter condition.

19. I myself interpret Latour and Woolgar as holding a more radical view, viz. , that there is no reality that could causally interact, even indirectly, with scientists' beliefs.

20. This is equally so under themore radical view that there are no truths at all (of a scientific sort) about reality or Nature.

21. More specifically, I am assuming that believers in creation science have greater (conditional) dependence on the opinion leaders of their general viewpoint than do believers in evolutionary theory.

22. John Pollock (in a personal communication) suggests a way to bolster support for the

use of "the numbers". He says that if one can argue that P (X (H) / Y (H) & H) = P (X (H) / H) , then one can cumulate testimony on each side of an issue by counting experts. He further suggests that, in the absence of countervailing evidence, we should believe that P (X (H) / Y (H) & H) = P (X (H) / H) . He proposes a general principle of probabilistic reasoning, which he calls "the principle of nonclassical direct inference", to the effect that we are defeasibly justified in regarding additional factors about which we know nothing to be irrelevant to the probabilities. In Pollock (2000) (also see Pollock 1990) he formulates the idea as follows. If factor C is irrelevant (presumably he means *probabilistically* irrelevant) to the causal relation between properties B and A, then conjoining C to B should not affect the probability of something's being A. Thus, if we have no reason to think that C is relevant, we can assume defeasibly that P (Ax / Bx & Cx) = P (Ax / Bx) . This principle can be applied, he suggests, to the case of a concurring (putative) expert. But, I ask, is it generally reasonable for us—or for a novice—to assume that the opinion of one expert is probabilistically irrelevant to another expert's holding the same view? I would argue in the negative. Even if neither expert directly influences the opinion of the other, it is extremely common for two people who work in the same intellectual domain to be influenced, directly or indirectly, by some common third expert or group of experts. Interdependence of this sort is widespread, and could be justifiably believed by novices. Thus, probabilistic irrelevance of the sort Pollock postulates as the default case is highly questionable.

23. In a devastating critique of the mental health profession, Robyn Dawes (1994) shows that the real expertise of such professionals is, scientifically, very much in doubt, despite the high level of credentialism in that professional community.

24. In the present discussion only *epistemic* esotericness, not *semantic* esotericness, is in question (see note 10) .

25. They can not only recognize such propositions as true when others offer them; they can also produce such propositions on their own when asked the question, "What can be done to repair this system?"

26. I have discussed such cases in earlier writings: Goldman 1991 and Goldman 1999 (p. 269) .

27. Of course, some experts may be better than others at transmitting their expertise. Some may devote more effort to it, be more skilled at it, or exercise

stricter standards in credentialing their trainees. This is why good information about training programs is certainly relevant to judgments of expertise.

28. For helpful comments on earlier drafts, I am indebted to Holly Smith, Don Fallis, Peter Graham, Patrick Rysiew, Alison Wylie, and numerous participants at the 2000 Rutgers Epistemology Conference, the philosophy of social science roundtable in St. Louis, and my 2000 NEH Summer Seminar on "Philosophical Foundations of Social Epistemology".

References

Baron-Cohen, Simon (1995). *Mindblindness.* Cambridge, MA: MIT Press.

Bezuidenhout, Anne (1998). "Is Verbal Communication a Purely Preservative Process?" *Philosophical Review* 107: 261-288.

Brewer, Scott (1998). "Scientific Expert Testimony and Intellectual Due Process", *Yale Law Journal* 107: 1535-1681.

Burge, Tyler (1993). "Content Preservation", *Philosophical Review* 102: 457-488.

Coady, C. A. J. (1992). *Testimony.* Oxford: Clarendon Press.

Craig, Edward (1990). *Knowledge and the State of Nature—An Essay in Conceptual Synthesis.* Oxford: Clarendon Press.

Dawes, Robyn (1994). *House of Cards: Psychology and Psychotherapy Built on Myth.* New York: Free Press.

Foley, Richard (1994). "Egoism in Epistemology", in F. Schmitt, ed., *Socializing Epistemology* Lanham, MD: Rowman & Littlefield.

Friedberg, Mark et al. (1999). "Evaluation of Conflict of Interest in Economic Analyses of New Drugs Used in Ontology", *Journal of the American Medical Association* 282: 1453-1457.

Gentzler, J. (1995). "How to Discriminate between Experts and Frauds: Some Problems for Socratic Peirastic", *History of Philosophy Quarterly* 3: 227-246.

Goldman, Alvin (1991). "Epistemic Paternalism: Communication Control in Law and Society", *Journal of Philosophy* 88: 113-131.

Goldman, Alvin (1999). *Knowledge in a Social World.* Oxford: Clarendon Press.

Hardwig, John (1985). "Epistemic Dependence", *Journal of Philosophy* 82: 335-349.

Hardwig, John (1991) . "The Role of Trust in Knowledge", *Journal of Philosophy* 88: 693-708.

Jeffrey, Richard (1992) . *Probability and the Art of Judgment.* New York: Cambridge University Press.

Kitcher, Philip (1993) . *The Advancement of Science.* New York: Oxford University Press.

LaBarge, Scott (1997) . "Socrates and the Recognition of Experts", in M. McPherran, ed. , *Wisdom, Ignorance and Virtue: New Essays in Socratic Studies.* Edmonton: Academic Printing and Publishing.

Latour, Bruno and Woolgar, Steve (1979/1986) . *Laboratory Life: The Construction of Scientific Facts.* Princeton: Princeton University Press.

Lehrer, Kcith and Wagner, Carl (1981) . *Rational Consensus in Science and Society.* Dordrecht: Reidel.

Pollock, John (1990) . *Nomic Probability and the Foundations of Induction.* New York: Oxford University Press.

Pollock, John (2000) . "A Theory of Rational Action", Unpublished manuscript, University of Arizona.

专家：哪些是你应该信任的？

阿尔文·戈德曼[*]

一、专长与证言

主流的认识论是一项极其理论和抽象的事业。传统的认识论者很少提出，他们的深思熟虑对实际生活问题很关键，除非人们假定——比如，像休谟就没有假定——怀疑论的担忧使我们对日常琐事感到烦恼。但有些认识论问题，既在理论上令人注目，在实践中也相当紧迫。这就提出了这里讨论的问题：外行应该如何评价专家的证言，以及如何在两个或更多的相互竞争的专家中确定哪一位专家的证言最可信。这有现实意义，因为在一个复杂的高度专业化的社会里，人们常常面对这样一些情境：作为相对的新手（乃至无知者），他们为了获得智力上的引导或帮助，必须求助于社会公认的专家（putative expert）。这有理论意义，因为适当的认识考虑很不明显，也不清楚这些问题离导致不可逾越的怀疑论的困惑还有多远。本文不为这个领域内的直接的怀疑论而争辩，也不自称解决了怀疑论方面的所有困扰。这是一篇说明文，试图辨别问题和考察某些可能的解决方案，而不是明确地确立这些解决方案。

当前的主题从传统认识论出发，在其他方面，也从科学哲学出发。这些领域典型地考虑了在"理想"情况下知识获得的前景。例如，通常考察这样的认识能动者：他们有无限的逻辑能力，并且，他们的研究资源没有重大限制。相比之下，在当前的问题中，我们关注受到规定认识约束的能动者，并质问，当他们受到这些约束时，可能得到什么。

尽管评价专家的问题在某些方面是非传统的，但这绝不是一个新问题。柏拉图在他的早期对话中，特别是在《论节制》（*Charmides*）中，曾明确地阐述和讨论过这个问题。在这本对话集中，苏格拉底问，一个人是否能审查一下，声称知

* 阿尔文·戈德曼（Alvin I. Goldman），美国新泽西州立大学认知科学中心和哲学系教授。本文收录于 *The Philosophy of Expertise*，New York：Columbia University Press，2006.——译者注

道某件事的另一个人，看一看他是否真的知道此事；苏格拉底想弄明白，一个人是否能区分出真假医生（*Charmides* 170d-e）。柏拉图在提出这个问题时用的术语是 techne，通常翻译为"知识"（knowledge），但也许译为"专长"（expertise）更好。①

在最近的文献中，约翰·哈德威格②用呆板的术语阐述了新手/专家问题。哈德威格说，当外行依靠专家时，那种信赖必定是盲目的。③ 哈德威格旨在否认很成熟的怀疑论（fledged skepticism）；他认为，证言的接受者能够从证人那里获得"知识"。但由于把接受者的知识描述为是"盲目的"，因此，哈德威格似乎向我们提供了各种不同类型的怀疑论。术语"盲目的"似乎意味着，外行（或不同领域的科学家）不可能在理性意义上有充分的理由信任专家。因此，就理性辩护（如果不是知识的话）而言，他的进路使我们对证言产生怀疑。

在哈德威格的邻域内还潜藏着证言认识论（epistemology of testimony）的其他进路。我想到的这些作者，没有明显地强迫对证言式信念（testimonial belief）产生任何形式的怀疑，像哈德威格一样，他们希望从证言域内排除怀疑论的幽灵。然而，他们对证言辩护问题的解决方案求助于下列最低限度的理由：听者可能信任证人的断言。我来说明我指的是谁，我是什么意思。

所讨论的这种观点的代表人是泰勒·伯格④和里查德·弗雷⑤，他们认为，说话者对一种主张的坦率认定，向听者提供了接受这个主张的表面理由（prima facie reason），与听者可能所知的一切完全无关，或者，无可非议地相信说话者的能力、环境、或者，有机会获得所声称的知识。这也不依赖于听者获得的经验证据，例如，只有当说话者能够知道他们所谈论的事情时，他们通常才提出主张的证据。例如，伯格赞成下列**接受原则**（acceptance principle）："如果某件事情被作为真的来介绍，那么，一个人有权把它接受为是真的，对他来说，这是可理

---

① 感谢斯柯特·拉巴格（Scott LaBarge）使我注意到柏拉图对这个问题的论述。Gentzler, J., "How to Discriminate between Experts and Frauds: Some Probliem for Socratic Peirastic", *History of Philosophy Quarterly* Vol. 3, 1995, pp. 227-246; LaBarge, Scott, "Socrates and the Recognition of Expters", in M. McPherran, ed., *Wisdom, Ignorance and Virtue: New Essays in Socratic Studies*, Edmonton: Acadmic Printing and Publishing, 1997.

② Hardwig, John, "Epistemic Dependence", *Journal of Philosophy*, Vol. 82, 1985, pp. 693-708; Hardwig, John, "The Role of Trust in Knowledge", *Journal of Philosophy*, Vol. 88, 1991, pp. 693-708.

③ 在1991年的这篇文章中，哈德威格第一次说，信任一定是"至少部分盲目的"（第693页）。然后，他在不用限定词"部分"的前提下，继续讨论与信任相关因而是盲目的知识（第693、699页）。

④ Burge, Tyler, "Content Preservation", *Philosophical Review*, Vol. 102, 1993, pp. 457-488.

⑤ Foley, Richard, "Egoism in Epistemology", in *Socializing Epistemology*, edited by F. Schmitt, Lanham, M D: Rowman & Littlefield, 1994.

解的，除非有很充分的理由不能这样做。"① 他坚持认为，这个原则不是一个经验原则；"由这种辩护所描述的这种资格（entitlement）的辩护力，不是通过感觉经验或知觉信念来建构的或强化的"②。同样，尽管弗雷没有强调这些原则的先验地位，但在这意味着"对我们而言，受别人的影响是有理由的，即使我们没有特殊的信息来表明，他们是可信赖的"情况下，他同意，人们有理由承认，别人的看法有基本的权威性。③ 在认为证人的"信息、能力或环境使［他］特别能够很好地"作出准确主张的理由，产生了派生权威性的地方，基本的权威性与派生的权威性形成了对比。④ 因此，根据弗雷的观点，听者不需要有关于证人的那些理由，来得到信任证人的表面根据（prima facie grounds）。此外，一个人不需要获得经验的理由，就能认为，只有当人们能够了解一位受试者时，他们通常才相信这位受试者。即使在缺乏任何这样的经验证据时，弗雷也赋予人们基本的（尽管是表面上的）认识权利来信任他人。⑤ 正是在这种意义上，伯格的观点与弗雷的观点似乎准许"盲目"信任。

我认为，伯格、弗雷等人之所以提出这些类型的观点，在某种程度上，是由于对还原论者或归纳论者的替代选择感到明确的绝望。不管是成年人还是小孩子，似乎都没有足够的证据，根据他们的个人感知和记忆，对证言的可靠性作出有说服力的归纳推理⑥。因此，伯格、弗雷、考迪（Coady）等人提出，他们的证言的可信赖性的"基本"原则，阻止了证言怀疑论的潜在潮流。我完全不相信，这种措施是必要的。举一个可能的例子来说，小孩子能够得到有用的归纳证据是：人们通常对他们能够知道的事情作出断言。

一个小孩子最早对事实报告的证据来自面对面的交谈。这个孩子通常明白，说话者正在谈论什么，并明白，这位说话者也明白她正在谈论什么，比如，毛茸茸的猫、钢琴下面的玩具等。的确，根据认知发展的一种说明（Baron-Cohen），有一个特殊的模块或机制，即注意别人眼睛的视觉方向的觉测器（detector），来

---

① Burge, Tyler, " Content Preservation", p. 467.

② Foley, Richard, "Egoism in Epistemology", p. 479.

③ Ibid. , p. 55.

④ Ibid.

⑤ 然而，弗雷是否能一致地把他假定的"基本"权利称为认识权利，是有问题的，因为他也说：这依赖于（A）我对自信的辩护，以及（B）他人与我的相似性——大概我有他们类似于我的证据（参见第63～34页）。弗雷的另外一个问题是，基础性论点如何与他们的下列观点相符合：在有冲突的情况下，与信任他人相比，我有更多的理由（表面的）信任我自己（参见第66页）。如果我对信任他人的辩护，确实是基本的，为什么它使自信处于次要地位呢？

⑥ 参见 Coady, C. A. J. , *Testimony*, Oxford：Clarendon Press, 1992.

检测他们凝视的方向，并把他们解释为是"看见了"视线内的一切。① 既然看见（seeing）通常导致知道（knowing），这个小孩子就能把现象的特定范围确定在说话者的视野内。因为这个小孩子遇到的最早表达大概是关于说话者已知的对象或事件的表达，所以，他可能很容易得出结论说，说话者通常作出他们能看见的东西的断言。当然，在小孩子不清楚所传说的事情现在乃至过去是否都在说话者的视野内的情况下，他后来会遇到许多种说法。尽管如此，一个孩子的早期经验是这样的说话者：他们谈论自己明确知道的事情，而且，这很可能是对孩子有用的一组决定性的经验证据。

我不希望很艰难地强调这个建议。② 我自己也没有提出辩护证言式信念的完备理论。特别是，那并不意味着，提出支持捍卫还原论者的立场或归纳论者的立场。对我而言，最关注的是承认，在接受证人的证言时，听者关于证人的可靠性或不可靠性的证据，通常能支持或击败（bolster or defeat）听者的正当性（justifiedness）。下面举两个例子来说明。

当你在街上遇见某个人时，他武断地说出一道复杂的数学题，你理解这道题，但以前从来没有作出可信的评价。你有理由接受这位陌生人的数学题吗？无疑，这在某种程度上取决于，这位说话者被证明是一位你熟悉的数学家，还是比如说是一位9岁的孩子。你事先有证据认为，前者能够知道这道题，而后者则不可能知道。不管是否存在着由伯格和弗雷赞同的这类默认资格（default entitlement）的先验原则，你关于说话者身份的经验证据显然是切题的。我没有断言，伯格和弗雷（等人）不能处理这些情况。他们可能说，你承认这位说话者是一位数学教授，支持了你接受这道题的总的资格（overall entitlement）（尽管不是你的表面资格）；承认他是个孩子，击败了你接受这道题的表面资格。然而，我的观点是，你关于说话者的性质的证据，是你接受这位说话者的断言的总的资格的关键证据。同样的观点适用于下面的例子。在停放的汽车的轮胎后面，当你闭上眼、放松时，你听到附近有人描述经过的车牌和颜色。似乎合理的是，你在把那些描述接受为是真的时，你有表面辩护，不管这种表面资格是否有先验的或归纳论者的根据。但如果你睁开眼后发现，这位说话者自己也被蒙着眼睛，甚至看不见过路车辆的方向，那么，这就肯定击败了这种表面辩护。那么，在接受他的意见时，你在经验上对一位说话者所确定的情况，对于你在接受他的表达时的

---

① 此外，根据拜伦-柯恩（Baron-Cohen）的观点，有一个独立的称为"共享注意机制"的模块设法确定，另一个人正在注意的对象何时与自己正在注意的对象一样。

② 首先，可能有人认为，在第一个事例中，小孩子对大人所说的话的解释受制于这样的假设：在说话者的感知范围内，内容与问题相关。也许有人认为，这一点并不是一个经验发现，而是用来确定说话者意义的一个先验的假定。

总的正当性来说，有很大的不同。

这显然也适用于对一个特定问题作出矛盾断言的两位社会公认的专家。你应该接受哪一个人的断言（如果是二选一的话），必然要受到你关于他们各自了解事情真相（和诚实地谈论此事）的能力和机会的经验证据的极大影响。的确，在这种情况下，伯格和弗雷提出的这类击败原则没有任何帮助。尽管听者在表面上有权相信每一位说话者，但如果全面考虑，他就无权同时相信俩人；因为我们假定，他们断言的命题是矛盾的（而且对这位听者来说显然是矛盾的）。因此，这位听者的全面考虑的正当性与他们的主张相比，将取决于他在经验上了解到的每一位说话者的情况，或者，其他说话者的看法。在本文的其余部分，我将研究，一位新的听者，为了相信一位社会公认的专家，而不是她的竞争者，具有的或能够获得的这种经验证据。我不认为，在提出这个问题之前，我们需要在一般的证言理论中解决这些"基本"问题。这无论如何是我将继续进行下去的一个工作假设。①

二、新手/专家问题与专家/专家问题

当然，有不同程度的专家与新手。一些新手的知识渊博程度可能并不逊色于专家。此外，一位新手通过提高他关于目标问题的认知立场，比如，通过在本领域内获得更正式的训练，原则上能够使自己变成一位专家。然而，这不是本文要考虑的场景。我假定，某些类型的限制因素——比如，时间、费用、能力诸如此类的因素——将阻止我们的新手至少在他们需要作出自己的判断之前成为专家。因此，问题是，新手（在依然是新手时）能够对相互竞争的专家的相对可信性的判断作出辩护吗？何时可能？如何可能？

在新手/专家问题和其他类型的问题（专家/专家问题）之间有着很大的不同。后面的问题是专家寻找评价其他专家的权威性或可信性的问题。菲利普·基切尔②

---

① 包括伯格在内的有些证言理论家坚持认为，听者关于证人的主张的辩护状态部分地依赖于证人自己相信这个主张的辩护。这是一种超越个人的、保护传统的或可传递的辩护观，根据这种辩护观，一个接受者不能合理地相信 P，除非说话者拥有他传递给听者的一种辩护和资格。然而，为了达到本文的目标，我将不考虑这种可传递的辩护观。首先伯格本人承认，为了相信对话者的主张，存在着像接受者的"专用"辩护之类的事情，即定位"于"接受者，不受证人辩护影响的辩护（1993 年，485-486 页）。我认为，就当前目标而言，集中于这个（接受者的）"专用"辩护，是适当的。当一位听者正在试图在竞争的说话者的矛盾主张中作出"选择"时，他不可能求助于加在说话者头上的不懂的辩护。他只能求助于他自己的辩护资源。（当然，这些包括两位说话者通过他们争论的抗辩方式所说的事，即也是与他们自己的辩护相关的事。）对伯格关于证言的保守主义的另一种（可信的）反对，参见 Bezuidenhout, Anne, "Is Verbal Communication a Purely Preservative Process?" *Philosophical Review*, Vol. 107, 1998, pp. 261-288.

② Kitcher, Philip, *The Advancement of Science*, New York: Oxford University Press, 1993.

在分析科学家如何把权威性赋予他们的同行时提出了这个问题。这种权威归属的关键部分包括基切尔所称的"校准"（calibration）[①]。在直接校准时，科学家利用他自己对所研究问题的看法来评价目标科学家的权威度。在间接校准时，他仍然是利用其他科学家的看法来评价目标的权威性，这些科学家的看法是他事先已经通过直接校准进行过评价的。因此这里他也是从他自己关于所研究问题的看法出发的。

相比之下，在我们所说的新手/专家问题（更具体地说，是新手/2-专家问题）上，新手不能通过他自己的看法来评价目标专家；至少他不认为，他能这么做。这位新手，要么在这个目标域内没有任何看法，要么对他在这个目标域内的看法没有足够的自信，用它们来对相互竞争的专家之间的分歧作出裁定或评价。他把这个目标域看成是恰好需要特定专长的领域，并且，他认为他自己没有这种专长。这样，他们在这个专长域内——称之为 E-域——不能运用他自己的看法在矛盾的专家判断或报告之间作出选择。

我们通过把新手/专家问题比作类似于倾听者/目击者问题，能够澄清新手/专家问题的本性（的确，如果我们不严格地使用"专家"这个术语的话，后面的问题可能恰好是一个新手/专家问题）。两位被公认的目击者都声称看到了某个犯罪。一位倾听者——例如，一位陪审员——他自己并没有目击到这种犯罪，也没有关于这是谁干的或犯罪过程的先验信念。换言之，他对这个事件没有个人知识。他希望通过聆听目击者的证言来了解所发生的情况。问题是，当两位目击者的说法是矛盾的时，他应该如何在他们的证言之间作出裁定呢？在这个案件中，E-域是与犯罪相关的行动和情况的命题域。这个 E-域是，倾听者（这位"新手"）对他觉得能够合法地求助于哪一位目击者，事先没有看法。（如果有的话，他把他的看法只当作是推测、预感等。）

至少在原则上，对于一位倾听者来说，即使没有或不求助于他自己关于这个 E-域的先入之见，也有可能合理地评价哪一位目击者更可信。例如，第一，他可以通过别人获得关于每一位被公认的目击者是否真的在案发现场的证据，或者，在案发期间，知道在别的地方的证据。第二，这位倾听者可能获悉测试每位目击者的视力，这与他们报告的精确度或可靠性有关。那么，在这类案件中，通过诸如分别证实他是否有机会和有能力看到他声称所看到的情况之类的方法，能够核实一位社会公认的"专家的"报告的可信性。当某人试图评价一位"认知"专家而不是目击者专家的可信性时，类似的方法也有用吗？

在提出这个问题之前，我们应该更多地讨论一下专家的本性和在这里我们关注的专家类型。某些类型的专家与众不同地擅长某些技能，包括小提琴家、台球

---

① Ibid.，pp. 314-322.

运动员、纺织设计师等。这些人不是最本能地关注认识论的专家。对于认识论的目标来说，我们将主要关注认知专家或智力专家：这些人在某些领域内有（或者声称有）较高的素质或知识水平，而且，在回答这个领域内的问题时，有（或声称有）能力产生新的知识。诚然，在智力问题上也有技能元素或技能知识（know-how），因此，技能专长和认知专长之间的边界是不明确的。尽管如此，我将试图只致力于这种大致划分的一个方面，即智力方面。

在认知意义上，我们将如何定义专长呢？在一个特定的认知领域内，专家与外行的区别是什么呢？我首先具体说明专长的客观意义，即成为一名专家是怎么回事，而不是享有专长的声誉是怎么回事。一旦阐明了这种客观意义，声誉的意义就容易理解了：一位有声誉的专家是指，某人被广泛地认为是一位专家（在客观意义上），不管他实际上是不是一位专家。

那么，在转向阐述客观的专长时，我首先提议，用"求真"（与真理相关）的术语来定义认知专长。作为第一步，一个特定领域（E-域）内的专家比大多数人（或更恰当地说，比绝大多数人）更相信（或高度相信）这个领域内的真命题和/或更不相信这个领域内的假命题。根据这个提议，专长在很大程度上是一个比较问题。然而，我认为，它不是完全可比较的。如果在一个领域内，绝大多数人充满了错误的信念，而琼斯由于不屈服于被广泛共享的少数谎言而胜过他们，这仍然不会使他成为一名"专家"（从上帝之眼的观点看）。一个人为了有资格成为一名认知专家，他必须拥有目标域中的大多数真理。成为一名专家不只是优越于这个共同体的大多数人的一个求真问题。尽管在设置这个阈值时有很大的模糊性，但是，必须达到求真获得（veritistic attainment）的某个非比较的阈值。

专长并不完全是拥有准确信息的问题。它包括针对这个领域内可能提出的新问题，调用或探索这些信息储存，形成相信正确答案的一种能力或倾向。这源于成为一名专家需要具备的某一组技能或技巧。一名专家当面对本领域内的一个新问题时，他有（认知的）技能知识诉诸他的信息库的正确部分，并对这些信息进行适当的处理；或者，调用某些外在设置或数据库来揭示相关内容。因此，专长以倾向性元素和实际获得的元素为特征。

专长的第三个可能特征也许要求对我前面的说法作出一点修改。为了讨论这个特征，让我们在一个领域内区分出主要问题和次要问题。主要问题是本问题的研究者或学者主要感兴趣的问题。次要问题关系到与主要问题有关的已有证据或论证，以及对杰出研究者提供的证据的评价。一般情况下，一个领域内的专家是这样的人：他（在可比较的意义上）拥有证据状态的广泛知识（在弱的知识意义上，即真信念的意义上），也拥有对本领域内的杰出工作者提供的证据作出反映和提出意见的知识。在"专家"的这个核心意义（强的意义）上，一名专家是在本领域内的主要问题和次要问题上都拥有非常渊博知识的人。然而，也存在着一种弱的"专家"意义，在这种意义上，包括只在本领域的次要问题上拥有广泛知识的人。考虑在本领域的次要问题上持有强烈分歧观点的两个人，以使一个人的观点在很大程度上是正确的，另一个人的观点在很大程度上是错误的。根据最初的强标准，在很大程度上是错误的那个人没有资格成为专家。人们可能不同意把这一点作为问题的最终结论。他们可能认为，全面了解现有证据和由本领域内的工作者提供的不同观点的任何一个人，都值得称为一名专家。我由于承认弱的"专家"意义而认可这一点。

应用上面所说的观点，我们能说，D 域内的一名专家（在强的意义上）是这样的人：在本领域内，他拥有广泛的知识（真信念）储存和聪明而成功地利用这些知识解答新问题的一组技能或方法。声称成为特定领域内的一名（认知）专家的任何一个人，都要求具有这样一种储存和一组方法，也要求能正确回答所争论的问题，因为他已经把他的储存和方法应用于这个（这些）问题。对于咨询一般认定的专家和希望因此而了解对目标问题的正确回答的外行来说，这项任务是确定谁有较好的专长或者谁能更好地利用他的专长解决眼前的问题。新手/2-专家问题是，外行是否能合理地把一名被公认的专家选择为比关于眼前问题的另一名专家更可信或更值得信赖吗？这样一种选择的认知基础可能是什么呢？①

三、基于论证的证据

为了提出这些问题，我首先列出了一位新手，在新手/2-专家的情形中，为了在两位被公认的专家中更信任其中的一位，所拥有的五种可能的证据来源。然后，我将根据它们的可用性和新手的准确情况，探索利用这些来源的视角。我讨论的这五种来源是：

（A）支持他们自己观点和批评对方观点的相竞争的专家所提供的论证；

（B）讨论问题的这一方或那一方得到了被公认的其他的专家的认同；

（C）对专家们的专长的"元专家"的评价（包括由专家们获得的正式证书反映出的评价）；

（D）专家们关于所讨论问题的利益和偏见的证据；

---

① 在提出正当性问题时，我的意思是说，在正当性概念的不同进路之间，例如，在内在论者与外在论者的正当性之间，尽可能地保持中立。此外，请注意，我不只是质问新手是否和如何能正当地决定完全接受一位（候选）专家的观点，而且还质问，他是否和如何能正当地决定对一个人的信任大于另一个人。

（E）专家过去"记录"的证据。

在本文的其余部分，我将考察这五种可能的来源，在这一部分，从来源（A）开始。①

新手 N 可能从他的两位专家 E_1 和 E_2 获得两种交流。② 首先，每一位专家都可能大胆地陈述她的观点（结论），不用任何证据或论证来支持她的观点。更一般地说，一位专家可能在某个公共语境或专业语境中对她的观点提供了详细的支持，但这种详细的辩护只可能出现在 N 注意不到的受限制的地方（例如，专业会议或期刊），因此，N 不可能看到这两位专家的辩护，或者，可能只看到他们恰好删减过的版本。例如，N 可能从很不详细的大众读物的二手描述中听说过两位专家的观点和他们的支持。在交流范围的另一端，这两位专家可能进行着 N 目击到的（或者，读到一个详细重构的）全部争论。在支持她的观点和反对对方的观点时，每位专家都可能提供相当完备的论证。显然，只有当 N 以某种方式看到专家的证据或论证时，他才能有类型（A）的证据。这样，让我们考虑这种情节。

我们可能开始假定，如果 N 根据两位专家的论证，经过比较后，能够获得对信任专家的观点的（更大的）辩护，那么，这位新手至少必须理解两位专家论证中所引证的证据。然而，对于某些专长领域和某些新手来说，即使只是掌握了证据，也可能力所不及。在有些情况下，N 关于 E-域是一个"无知者"。这不是新手的普遍困境。有时，他们能够理解证据（多少的问题），但不能根据个人知识为它提供任何凭证。当一位专家的证据受到另一位反对专家的质疑时，评价他的证据可能是很困难的。

对于新手来说，并不是专家论证中出现的每个陈述在认识论意义上都一定是不可理解的。这里，让我们在专家的论述中区分出深奥的陈述和通俗的陈述。深奥的陈述属于与专长相关的领域，它们的真值对 N 来说——至少根据他的个人知识——是不可理解的。通俗的陈述是外在于专长领域的；它们的真值对 N 来说——或者是它们当时的断言，或者是后来的断言——是可理解的。③ 我假定，在专家的论证中，深奥的陈述是由大量的前提和"引理"构成的。那才是一位新手基于论证本身对所相信的任何一位专家的观点作出辩护的困难。新手不仅通

---

① 我不意味着承诺要穷尽这个列表。这个列表只包括某些重要的类型。

② 接下来，我将简洁地谈论两位专家，不过，我通常是意指两位被公认的专家，因为从新手的认识视角来看，每一位自封的专家或两者之一实际上是否是专家，是有问题的。

③ 区分出在语义意义上深奥的陈述和在认知意义上深奥的陈述，可能是有益的。[感谢卡洛尔·卡拉威（Carol Caraway）的这个建议。] 在语义意义上深奥的陈述是新手不可能评价的陈述，因为他甚至都不理解这些陈述；在典型意义上，它们用了他没有掌握的技术词汇。在认知意义上深奥的陈述是新手能理解但仍然不能评价真值的陈述。

常不能评价这些深奥陈述的真值，而且他们也不能令人满意地评价所引证的证据和所提出的结论之间的支持关系。当然，支持者专家将声称，在她的证据与她辩护的结论之间，支持关系是有说服力的；但她的反对者通常会抗议这一点。新手不能令人满意地评价哪位专家是对的。

在这一点上，我愿意区分出直接论证的辩护和间接论证的辩护。在直接论证的辩护中，听者根据有充分的理由相信论证前提及其对结论的（有说服力的）支持关系，完全可以相信一个论证的结论。如果说话者对一个论证的认可有助于使得听者相对于论证的前提和支持关系拥有这样的辩护状态，那么，听者通过说话者的论证获得了对这个结论的"直接"辩护。[1] 然而，正如我们所说的那样，在新手/2-专家情况下的听者当中，一位专家的论证很难产生直接辩护。恰好是因为这些问题有许多是深奥的，所以，N 在 E_1 的主张和 E_2 的主张之间作出裁定，困难重重，因而很难对他们的结论中的任一结论作出辩护。他甚至很难有充分理由信任这个结论胜过信任另一个结论。

间接论证辩护的观点源于这样的观点：在一场争论中，一位说话者可能显示出超越于另一位说话者的辩证优势，而且，对于 N 来说，这种辩证优势可能是有更好专长的一个可信的标志[2]，即使这没有使 N 对相信优胜说话者的结论作出直接辩护。我说的辩证优势不只是指更多的争论技能。下面的一个例子说明了我的意思。

只要专家 E_2 为她的结论提供证据，专家 E_1 就对这个证据提出一个明显的反驳或驳斥。另一方面，当 E_1 提供证据支持她的结论时，E_2 从来不对 E_1 的证据提出反驳或驳斥。现在，N 不能评价驳斥 E_2 的 E_2 证据的真值，也不能评价 E_1（获胜方）的证据导致 E_1 结论的支持力度或真值。由于这些原因，E_1 的证据（或论证）对于 N 来说不是直接可辩护的。尽管如此，用"形式的"辩证术语来说，在这场争论中，E_1 似乎做得更好。此外，我建议，就所争论的问题而言，这种辩证优势可以被合理地看成是 E_1 有优势专长（superior expertise）的一个标志。这是一个（非结论性的）标志，即 E_1 在本领域内具有占优势的信息储存，或者，具有操纵她的信息的优势方法，或者，同时拥有两者。

优势专长的另外的标志可能来自这场争论的其他方面，尽管这些方面很不可靠。例如，E_1 对 E_2 的证据的回应比较敏捷和圆滑，可能意味着，E_1 已经很熟悉 E_2 的"要点"，并且，已经想出了反证。如果 E_2 对 E_1 的论证的回应，显得不太敏

---

[1] 当然，在基础主义的基础性（basicness）意义上，我的"直接"辩护的意思不是指与所讨论的结论的基础性相关。我后面的这种区分是完全不同的，不久将会出现。

[2] 爱德华·克雷格同样把"标志特性"说成是，一个咨询者从作为影响他/她的真理论述能力的一位信息提供者身上设法辨别出来的。Craig, Edward, *Knowledge and the State of Nature-An Essay in Conceptual Synthesis*, Oxford：Clarendon Press，1990.

捷和圆滑，那可能意味着，E_1 事先掌握的相关信息和支持的考虑超过了 E_2。当然，敏捷和圆滑作为信息掌握的标志是成问题的。有技能的争论者和受过良好训练的目击者，由于他们的文体修饰，可能看起来有更详细的信息，这并不是优势专长的一个真标志。这使得正确运用间接论证的辩护成为一件很微妙的事情。①

为了澄清这里划分的直接/间接的区分，考虑听者可能说清楚地表达了辩护的这些不同基础的两种不同情况。在直接论证辩护的情况下，他可能说："根据专家的论证，也就是说，根据论证前提的真实性及其对结论的支持（对我而言，这两者在认识论意义上都是可达到的），我现在有充分的理由相信这个结论。"在间接论证辩护的情况下，这位听者可能说："根据这位专家论证的方式——可以说是她论证的表现——我能够推出，她比她的对手有更好的专长；因此，我有理由推出，她的结论可能是正确的结论。"

下面是说明这种直接/间接区分的另一种方式。间接论证的辩护在基本意义上包括最佳说明推理，即 N 从两位说话者的表现到他们各自的专长水平作出的一种推理。根据他们的表现，N 对哪位专家在这个目标域内拥有优势专长作出一种推理。然后，他从拥有的专长水平越高推出支持真结论的概率越大。间接论证的辩护在基本意义上包括了最佳说明推理，而直接论证的辩护不需要包括这样的推理。当然，它可能包括这样的推理；但即使如此，说明推理的主题也将只涉及争论中的对象、系统，或事态。不涉及竞争的专家的相对专长。相比之下，在间接论证的辩护中，恰好是专家的相对专长构成了最佳说明的目标。

哈德威格（1985）提供了许多这样的事实：在新手/专家情境中，新手缺乏专家的理由相信她的结论。这是对的。通常，新手缺乏专家推出她的结论的所有前提或某些前提；在评价专家的前提和结论之间的支持关系时处于劣势；以及对可能与专家论证相关的许多或大多数反驳（和对反驳的反驳）是无知的。然而，尽管新手 N 可能缺乏（所有或某些）专家的理由相信结论 P，但是，N 可能有理由 R^* 相信这一点：专家有好的理由相信 P；N 可能有理由 R^* 相信，一位专家相信自己结论的理由，比她的对手相信其结论的理由更充分。间接论证的辩护是，N 在不共享任何一位专家的（所有或任何）理由 R^* 的前提下，可能据此获得理

---

① 斯科特·布鲁尔讨论了关于这里详细讨论的新手/专家的许多相同问题。他在新手用专家的"行为举止"（demeanor）评价他们的专家的标题下，讨论了当前的话题。他指出，在行为举止本身有赚钱"市场"的情况下——在行为举止被以高价"交易"的情况下，行为举止是一个特别不值得信任的向导。这种实践在诡辩论者的时代是突出的，而且，在对抗的司法系统中是一项健全的商业。Brewer, Scott, "Scientific Expert Testimony and Intellectual Due Process", *Yale Law Journal*, Vol. 107, 1998, pp. 1535-1681.

由 R* 的一种方法。① 哈德威格漠不关心的正是这种可能性。我没有说，在新手/专家情境中的新手总是有这些理由 R*；我也没有说，新手很容易获得这些理由。但这似乎是可能的。

四、来自其他专家的论证：人数问题

对于新手来说，另一个可能的策略是进一步求助于专家。这就把我们带到我们列出的类型（B）和（C）。类型（B）要求 N 考虑其他专家是同意 E_1，还是同意 E_2。同意 E_1 的专家比例有多大？同意 E_2 的专家比例有多大？换言之，在切实可行的程度上，在所有相关的（被公认的）专家中，N 应该咨询的人数或共识程度。如果几乎所有的其他专家在此问题上都同意 E_1，或者，如果同意 E_1 的其他专家恰好在数量上占有优势，那么，N 有充分的理由信任 E_1 超过信任 E_2 吗？

在类型（C）条件下引证的证据的另一种可能来源也求助于其他专家，但思路稍有不同。在类型（C）条件下，N 应该通过咨询第三方对竞争的两位专家的专长的评价，寻求证明他们的相对专业水平的证据。如果"元专家"支持 E_1 的"比率"或"分数"大于支持 E_2 的"比率"或"分数"，那么，E_1 和 E_2 相比，难到 N 不应该更加信赖 E_1 吗？证书能被看成是这个同样过程的一个特例。学位、专业评审、工作经历等（一切都来自具有独特荣誉的特殊制度）反映了其他专家对 E_1 和 E_2 表现出的训练或胜任能力的鉴定。N 可能利用这些标志的相对优势或权重，提炼出分别信任 E_1 和 E_2 的适当层次。②

我把比率和证书看成是由其他专家发出的"同意"的信号，因为我假定，当受训者相信下列事实时，已确立的权威证明了他们有胜任能力。这些事实是，证书证实了：①精通相同的方法，认证机构认为这些方法对本领域是基本的；②命题（或相信）的知识，认证机构认为这些命题知识是本学科的基本事实或定律。以这种方式，比率和授予的证书最终依赖于元专家和发证机构的基本认同。

当提到评价特殊专家时，在美国的司法体系中，调查其他专家在多大程度上

---

① 当然，在间接论证辩护中，新手一定至少听到某些专家的前提——或"最终的"前提与结论之间的中间步骤。但是，新手在相信那些前提时，并不共享专家的正当性。

② 这些术语属于基彻尔的"自然权威"的范畴（1993：315）。

认同被评价的那些专家有例在先。① 但不管有无先例，这种求助于共识的做法究竟有多好呢？如果把一位被公认专家的看法结合到其他被公认专家的共识看法中，那么，这为听者相信原来的看法提供了多大的保证呢？在听者作出信念的决策时，共识或认同有多大的证据价值呢？

如果人们认为，个人看法在表面上是值得信任的，尽管缺乏任何证据证明它们在这个问题上的可靠性，那么，至少在缺乏额外证据的情况下，人数似乎是很重要的。问题一方的每一个新证人或看法持有者都应该增加了这一方的权重。因此，在别的方面对各种不同看法持有者的可靠性一无所知的一位新手，似乎被迫同意绝大多数专家的看法。这对吗？

下面举两个例子，对"用人数"来判断对方立场的相对可信性的做法提出质疑。第一个例子是，具有奴性追随者的领袖。凡是领袖相信的，他的追随者都奴隶般地相信。他们把自己的看法完全和排他性地建立在他们的领导的观点之基础上。从智力上说，他们只是他的克隆。或者，考虑这样一组追随者：他们不是受一位领导者的领导，而是受舆论制造者的少数社会精英的领导。当舆论制造者持有相同看法时，众多的追随者就赞成他们的看法。难道新手不应该把这种场景看成是一种可能性吗？也许（被公认的）专家 E_1 属于这样一个教条的共同体：这个共同体的成员衷心地和不加鉴别地与某个领导的看法或领导层的阴谋小集团的看法相一致。难道人多的专家共同体应该使他们的看法比人少的专家组的看法更可信吗？另一个例子是谣传的例子，这个例子也对更多人数的诚实性提出挑战。谣言是被广泛流传或公认的故事，尽管没有几个信徒能理解传说的事实。如果某人从某个传播者那里听到一则谣言，当同样的谣言被第二个传播者、第三个传播者和第四个传播者重复时，这就强化了第一个传播者的可信性吗？想必没

---

① 求助于其他专家来证实或同意一位被公认专家的看法——或者，更确切地说，他的看法的基础——在决定科学的专家证言的可采性（admissibility）的司法系统的程序中，有一个先例。在对承认或排斥从1923年到1993年适用的这种证言的调节检验的过程中，提供证言基础的科学原理（或方法论）必须"在它所属的特殊领域内得到一般的接受"［Fry v. United States，292 F. 1013 D. C. Cir. (1932)］。换言之，求助于科学共同体的看法来决定，专家证言的基础是否牢固，足以允许那种证言提交到法院。在更新近的最高法院的裁决中，这种检验被替代成为唯一适当的检验［Daubert v. Merrell Dow Pharmaceuticals，599 U. S. 579 (1993)］；但后者的裁决也求助于其他专家的看法。由此建议，法官在确定所提供的科学的专家证言是否可采纳时，把四个标准结合起来使用（四个标准中，没有一个标准是必要的或充分的）。一个标准是旧的一般可接受标准，另一个标准是，所提供的证据是否受同行审查和公开。同行审查显然也引入了其他专家的看法。当然，一个专家证言的可采性的问题，不同于一位关键的听者——例如，一位陪审员——应该如何信任他听到的这种证言的问题。但这两者紧密地纠缠在一起，因为法庭基于下列假设作出可采性的决定：陪审员可能受到他们听到的任何一位专家证言的影响。法庭不愿意承认科学证据，除非它是相当值得信任的。因此，最终诉诸其他专家的看法来评价一位给定专家所提供的证言的可信任性的观点，肯定是设法证实一位专家的可信任性的一个好的有先例可循的步骤。

有，特别是，如果听者知道（或有理由相信），这些传播者全都是同样谣言的盲目接受者。

有人反对说，另外的谣言传播者没有增加最初的谣言传播者的可信性，是因为另外这些人没有确立可靠性。听者没有理由认为，他们的任何看法都是值得信任的。此外，谣言的例子似乎根本没有包括"专家"的看法，因而与最初的例子形成鲜明的对比。在最初的例子中，听者至少有一些先验的理由认为，赞成最初的两位专家之一的每个新的说话者都有某种可信性（可靠性）。在那种场景中，难道另外赞成的专家没有增加他们认同的那位专家的总的可信度吗？

于是，似乎是至少当每个新增的看法持有者最初都有肯定的可信性时，人数越多，应该越能增加可信性。这个受试者的某些进路肯定预先假定了这种观点。例如，在莱尔－瓦格纳（Lehrer-Wagner）[1] 模型中，如果受试者把"尊敬"或"权重"赋予每一个新人，那么，他就提供了应该使这个受试者向着这个人的看法方向推进的一个额外矢量。[2] 不幸的是，这条进路有一个问题。如果两个或更多的看法持有者完全是相互依赖的，如果受试者知道或有理由相信这一点，那么，受试者的看法不应该被一个以上的这些看法的持有者所动摇——一点都不动摇。像在领袖及其盲从者的例子中那样，一位追随者的看法没有提供接受领袖观点的任何额外的根据（而且，第二位追随者没有为接受第一位追随者的观点提供额外根据），即使所有的追随者都正好与领袖本人（或与另一个人）一样可靠——当然，如果在所讨论的问题上，追随者恰好相信与领袖（和另一个人）一样的看法，那么，他们一定是如此。让我通过贝叶斯分析来证实这一点。

在简单的贝叶斯进路中，接受新证据的一位行动者，应该以那个证据为条件，修正他相信一个假设 H 的程度。这意味着，他应该用两种可能性之比（或商）：如果 H 是为真的证据发生的可能性，以及如果 H 是为假的证据发生的可能性。在当前的例子中，所讨论的证据是，站在一个或多个被公认专家的立场上，相信 H。更精确地说，我们感兴趣的是，比较下列两个结果：（A）以一位社会公认专家的信念的证据为条件的结果；（B）以两位社会公认专家都赞成的信念

① Lehrer, Keith and Wagner, Carl, *Rational Consensus in Science and Society*, Dordrecht: Reidel, 1981.

② 莱尔和瓦格纳说（第20页），如果人们不把他的看法看成对所讨论的问题是"无价值的"——即，如果人们认为他比一台随机装置更好，那么，人们应该为别人赋予一个肯定的权重。因此，只要一位领导接受肯定的权重，看起来好像对这位领导的每一次克隆，都应该是给予肯定的权重——可以认为，与领导自己的权重一样，因为他们的信念总是一致的。于是，在莱尔-瓦格纳模型中，每一次克隆都将把正向力施加给人们对自己看法的修改，正好像一位领导的看法将施加这样一种正向力一样；克隆越多，在他们集体看法方向所施加的力越大。

的证据为条件的结果。把这两位社会公认的专家称为 X 和 Y，设 X（H）是 X 相信 H；设 Y（H）是 Y 相信 H。那么，我们所希望的比较是，把（1）中所表示的商的可能性的大小与（2）中所表示的商的可能性的大小相比较。

(1) $\dfrac{P(X(H)/H)}{P(X(H)/\sim H)}$

(2) $\dfrac{P(X(H)\&Y(H)/H)}{P(X(H)\&Y(H)/\sim H)}$

我们感兴趣的是这样的原则：（2）中给出的可能性之比总是大于（1）给出的可能性之比，所以，了解到 X 和 Y 都相信 H 的一位能动者，总是有根据比如果他只了解到 X 相信 H，更加提高了他对 H 的信任度。当 X 和 Y 每个人都有点是可信的（可靠的）时，至少情况是如此。更精确地说，如果这位能动者在不同场景中都有充分的理由相信这些事情，那么，这些经过比较的修改是妥当的。我将表明，这些经过比较的修改并非总是妥当的。有时，（2）不大于（1）；因此，这位能动者——如果他知道或合理地相信这一点——就没有理由根据两位赞成的信徒的证据，对 H 的信任度的提高，大于根据一位信徒的证据，对 H 的信任度的提高。

首先让我们注意，根据概率计算，（2）等价于（3）。

(3) $\dfrac{P(X(H)/H)\,P(Y(H)/X(H)\&H)}{P(X(H)/\sim H)\,P(Y(H)/X(H)\&\sim H)}$

当接受（3）时，回到了盲从者的例子。如果 Y 是 X 的盲从者，那么，凡是 X 相信的（包括 H），也是 Y 所相信的。并且，不管 H 是否为真，这都成立。因此，

(4) P（Y（H）/X（H）&H）=1，

和

(5) P（Y（H）/X（H）&～H）=1。

把这两个值代入（3），（3）简化为（1）。这样，在盲从者的例子中，（2）[它等价于（3）]与（1）是一样的，并且，根本不能保证在两位赞成信徒的情况下作出的修改，大于在一位信徒的情况下作出的修改。

假设第二位赞成的信徒 Y 不是 X 的盲目随从者。假设他有时与 X 相一致，但不是在所有情况下都一致。在这种场景中，附加 Y 的赞同信念，总是为行动者（他拥这些信息）提供了相信 H 的更多依据吗？回答还是否定的。适当的问题是，当 X 相信 H 和 H 是真的时，Y 相信 H 的可能性是否大于，当 X 相信 H 和 H 是假的时，Y 相信 H 的可能性。如果不管 H 是真，还是假，Y 恰好可能追随 X 的看法，那么，Y 的赞同信念没有增加行动者对 H 的证据基础（由可能性的商导致的）。让我们来看一下为什么会是这种情况。

如果当 H 为假时，Y 可能追随 X 的看法，与当 H 为真时，Y 追随 X 的看法，

恰好一样可能，那么，（6）成立：

(6) P（Y（H）/X（H）&H）= P（Y（H）/X（H）& ~H）

但如果（6）成立，那么，（3）再一次简化为（1），因为（3）中的分子和分母的右边相等，并且可彼此相消。既然（3）简化为（1），所以，就 H 而言，能动者从 Y 与 X 的一致，仍然没有得到额外的证据提升。这里不是要求，Y 肯定追随 X 的看法；他追随 X 的可能性只可能是 0.8 或 0.4，等等。只要 Y 可能追随 X 的看法，在 H 为真时恰好与 H 为假时一样，我们就得到同样的结果。

让我们通过说 Y 是 X（相对于 H）的非歧视的反映者（non-discrimination reflector）来描述后一种情况。当 Y 是 X 的非歧视的反映者时，Y 的看法，对上面的能动者来说，没有额外的证据价值，并且超越了 X 的看法。对于从 Y 对 H 的相信获得额外证据提升的新手来说，所必要的是，他（新手）合理地相信（6′）：

(6′) P（Y（H）/X（H）&H）> P（Y（H）/X（H）& ~H）

如果（6′）被满足，那么，Y 的信念至少与 X 的信念是部分条件无关的。完全条件无关是这种情境：X 的信念和 Y 的信念之间的任何依赖性，都通过各自对 H 的依赖性，加以说明。尽管完全条件无关不要求提升 N 的证据，但部分条件无关有这个要求。①

我们现在可能认为这种困惑等同于（不合格的）人数原理。这种困惑是，新手不可能自动地指望他的被公认的专家，是（甚至部分是）有条件地独立于另一位专家。他不可能自动地指望（6′）的真实性。Y 可能是对 X 的非歧视没有辨别能力的反映者，或者，X 可能是对 Y 的非歧视的反映者，或者，两者可能是对某个第三方或第三组的非歧视的反映者。不管有多少另外被公认的专家共享原来专家的看法，都适用于同样的观点。如果某人的看法已经被考虑，他们全是这个人的非歧视的反映者，那么，他们对新手的证据没有增加进一步的权重。

新手能有什么类型的证据为他接受（或很信任）的（6′）辩护呢？N 能有理由相信，Y 相信 H 的路径是：即使在 X 不承认 H 为假（并因此相信 H）的可能情况下，Y 也将承认 H 为假。Y 的正确类型的信念有两种类型的因果路径。第一，Y 相信 H 的路径完全不顾 X 的路径。下面两个例子说明了这一点：X 和 Y 对 H 的发生或不发生在因果意义上是独立的目击者；或者，X 和 Y 把他们各自的信念建立在与 H 有关的独立实验的基础上。在目击者的场景中，X 由于对实际事件的错觉，可能错误地相信 H，而 Y 可能正确地感知了事件，避免了相信 H。

---

① 我在这里要感谢理查德·杰弗里。他指出，在这几种情况下，只有条件独立是相关的，而不是由下列条件定义的"简单独立"：P（Y（H）/X（H））= P（Y（H））。如果 X 和 Y 恰好是关于 H 的信息的稍微可靠的独立来源，他们就不满足后面这个条件。Jeffrey, Richard, *Probability and the Art of Judgment*, New York: Cambridge University Press, 1992, pp. 109-110.

Y 相信 H 的第二种可能路径可能部分地接受 X，但不包括对 X 信念的不加鉴别的反映。例如，Y 可以听取 X 相信 H 的理由，考虑对 X 从来没有考虑到的那些理由的各种可能的反驳，但最终击败这些反驳的说服力，赞成接受 H。在这两种场景的任何一种情况下，Y 对因果路径的部分"自主性"使他沉着地避免相信 H，即使 X 相信 H（可能在虚假意义上）。如果 N 有理由认为，Y 运用信念的这些多少自主的因果路径之一，而不是确保与 X 一致的一种因果路径，那么，N 有理由接受（6'）。照这样，即使在考虑了 X 的信念之后，N 也有好的理由把 Y 的信念算作是增加了他对 H 的证据。

据推测，相对于赞成（被公认的）专家组而言，新手很可能处于这样一种认知境况。无疑，在赞成科学家的例子中，即在新手有理由期待科学家批评另一个人的观点的情况下，部分独立的推测可能是很妥当的。如果是这样，新手可以保证，赞成看法的持有者越多，提供的证据权重越大。然而，根据科学看法形成的某些理论，这种保证不可能持久。我们考虑这样的观点：科学家的信念完全是通过与其他科学家的协商产生的，不可能反映实在（或自然界）。这种观点显然是关于科学的某些社会建构论者所持有的，例如，布鲁诺·拉图尔（Bruno Latour）和史蒂夫·沃尔伽（Steve Woolgar）[1]；至少这是基彻尔[2]对他们观点的解释。[3] 现在，如果社会建构论者是正确的，这种解释也是正确的，那么，无人（至少没有人知晓这种事实）保证相信像（6'）那样的条件。根本没有理由认为，任何一位科学家都在一个科学假设为真（和某些其他科学家相信它）时，比它为假（和另一些科学家相信它）时，更有可能相信这个假设。既然科学信念的因果路径没有反映"真的"事实——它们只反映科学共同体的看法、利益等——（6'）将绝对不是真的。接受或倾向于表明社会建构论者的论点的任何一个人，绝对没有理由相信（6'）。[4]

抛开这两种极端的观点，难道新手通常将没有理由预期，彼此不同的被公认的专家，在他们的信念路径中，将有某种因果独立性或自主性吗？如果是这样，那么，假如一位新手也有充分的理由相信，每一位被公认的专家都有一点可靠度（大于碰运气），那么，在最初竞争的两位专家中，难道他没有充分的理由用赞成专家的人数偏向一方反对另一方吗？当所有的或几乎所有的增补专家都同意两

① Latour, Bruno and Woolgar, Steve, *Laboratory Life: The Construction of Scientific Facts*, Princeton University Press, 1979/1986.

② Kitcher, Philip, *The Advancement of Science*, New York: Oxford University Press, 1993.

③ 我自己把莱尔和瓦格纳解释为拥有更激进的观点，即根本没有能在因果意义上（即使是间接地）与科学家的信念相互作用的实在。

④ 根本没有关于实在或自然界的（科学类型的）真理，在这种更激进观点的情况下，这同样如此。

个竞争者之一时，这个结论可能是正确的。但很少有这种情况。更普遍的情况是这样的场景：人数大致相当，尽管不是完全相等。在这些情况下，新手能得出什么样的结论呢？他能合法地准许根据人数较多来决定问题吗？

特别是，如果我们继续应用贝叶斯进路，这一点将是没有根据的。新手相信H的适当变化，应该以两组赞成专家（一组赞成H，另一组反对H）为基础，而且，这种变化应该依赖于每一组成员有多么可靠，他们相互之间是如何（有条件地）独立。如果人数较少的小组比人数较多的大组更可靠和更（有条件地）相互独立，这就意味着小组的证据权重超过了大组吗？更确切地说，这取决于新手有充分的理由相信这些问题。既然新手在这些问题上的正当性可能很弱，所以，有许多这样的情况：他根据观点相同的看法持有者的相对人数，无法对进展情况作出明确的或有说服力的辩护。

这个结论似乎是完全妥当的。根据我自己的观点，下面是接受这个结论的一个例子。如果科学创世者的人数大于进化论的科学家，这不使我倾向于说（在他们不一致的核心问题上）前者的观点与后者的观点相比，保证新手更信任前者的观点。至少我并不倾向于这样的假设，以至于新手大致拥有可比较的信息，就像当前的大多数哲学家几乎拥有进化论者和创世论者分别形成的信念方法一样。①无疑，人数不一定优于考虑个人的可靠性和互为条件的独立性。在当前的例子中，后面的因素比纯粹人数的权重似乎更有证据。②

---

① 更具体地说，我正在假设，创世说的信徒比进化论的信徒更多地依赖于他们的一般观点的意见领袖。

② 约翰·波洛克（John Pollock）（在一次个人通信中）提出了一种强化支持运用"人数"的方法。他说，如果人们能认为，P（X（H）/Y（H）&H）＝P（X（H）/H），那么，人们就能通过数出问题每一方的专家人数积累证言。他进一步建议说，在缺乏补偿证据的情况下，我们应该相信，P（X（H）/Y（H）&H）＝P（X（H）/H）。他提出了一个概率推理的一般原理，称之为"非经典的直接推理原理"，大意是，我们有理由取消我们一无所知的与概率无关的附加因素。波洛克在2000年的文章中（也参见Pollock, 1990）阐述的观点如下：如果因素C与特性B和A之间的因果关系无关［大概，他意指在概率意义上无关］，那么，C与B的联合不应影响某物是A的可能性。因此，如果我们没有理由认为，C是相关的，我们就能假定取消P（Ax/Bx&Cx）＝P（Ax/Bx）。他建议说，能把这个原理应用于赞成（被公认的）专家的事件中。但我问道，假设一位专家的看法与另一位专家持有的同样观点在概率意义上是无关的，对我们来说——或对新手来说——这通常是合理的吗？我认为是不合理的。即使两位专家都不会直接影响对方的看法，对于工作在同一个学术领域的两个人来说，最普遍的情况是，会直接或间接地受到共同的第三位专家或专家组的影响。这类相互依赖性是普遍的，能合理地得到新手的信任。因此，波洛克假定为是默认情况的概率无关是很可疑的。Pollock, John, *Nomic Probability and the Foundations of Induction*, New York: Oxford University Press, 1990; Pollock, John, "A Theory of Rational Action", Unpublished manuscript, University of Arizona, 2000.

五、利益与偏见的证据

我现在转向我们当初列出的第四种可能来源：扭曲的利益和偏见的证据，这可能支持被公认的专家的主张。如果 N 有极好的证据对一位专家产生这样的偏见，而没有证据对她的竞争者产生这样的偏见，如果 N 没有优先信任（preferential trust）的其他基础，那么，N 有理由更信任无偏见的专家。这种建议直接来自常识和经验。如果两个人提供了相矛盾的报告，恰好其中一人有一种理由撒谎，就会严重地危及后者的相对可信性。

当然，撒谎不是利益与偏见能降低专家的可信赖性的唯一方式。利益与偏见能对专家的看法产生更微妙的扭曲影响，所以，他们的看法更不可能是准确的，即使是真诚的。在特定类型的民事诉讼中，某人经常被雇用为被告的专家证人，他在当前的任何审判中，对提供有力的证词有利可图，因为作为被告的证人，她的信誉取决于她当前的表现。

作为对利益冲突情境中的专家表现的一种检验，考虑在《美国医协学会杂志》①上发表的一项研究结果。该研究探索了关于新的肿瘤药品的两份公开的研究报告之间的关系，两项研制分别受到制药公司的资助与非营利组织的资助。在这两份报告中发现，在基金来源与定性结论之间有一种统计意义的关系。非营利组织资助研究的 38% 得出不利结论，但制药公司资助研究的只有 5% 得出不利结论。

从实践的观点来看，与专家利益相关的信息，通常是新手能够收集的有关专家的更易接近的相关信息之一。当然，通常透露说，一对作证专家的两位成员具有危及他们的可信性的利益。但当在这个维度上有不可忽视的分歧时，新手肯定会利用合法的信息。

金钱利益是潜在地扭曲个人的主张或看法的常见类型。更重要的是可能影响整个学科、子学科或研究小组的一种偏见，部分原因是，它对新手来说更费解。如果一个特定领域的所有或大多数成员都受到相同偏见的影响，那么，新手将很难把确证的证言的真正价值和其他专家与元专家区分开来。对于协商的新手来说，这使得前面讨论过的人数游戏变得更加复杂。

女性主义认识论者强调的一类偏见包括，在一门学科或专家共同体中，对在一门学科或专家共同体中的某些观点或立场的排斥或欠表达。这可能导致一个共

---

① Friedberg, Mark et al., "Evaluation of Conflict of Interest in Economic Analyses of New Drugs Used in Oncology", *Journal of the American Medical Association*, Vol. 282, 1999, pp. 1453-1457.

同体不能收集或重视特定类型的相关证据的重要性。共同体范围内的偏见的第二种类型源于子学科或研究共同体的经济因素或权术。为了增加基金资助的希望，从业人员可能习惯性地夸大所谓支持他们成果的证据的可检验性，外行尤其如此。为了获得资源与承认，在与相近学科和研究计划的竞争中，一个特定的研究共同体在报告其成果时，可能应用相对不太严格的标准。新手很难发觉这一点，或者，很难权衡场外相竞争的专家的这样一种辩解的价值。①

六、运用过去的记录

我们列表中的最后一个类型可能提供了新手作出可信性选择的最佳证据来源。这是运用被公认专家的过去认知成功的记录，评价他们对当前问题给出正确回答的可能性。但新手如何能评价过去的记录呢？这里有几个理论问题，可追溯到前面讨论的问题。

首先，运用过去的记录不等于用（直接）"校准"方法来评价一位候选专家的专长吗？运用过去的记录意味着，考虑候选者对前面在 E-域内的问题提供回答的成功率。但在我们前面的讨论中（第二部分），我说过，新手的本性正是，他对 E-域内的问题没有看法或对他自己的看法没有自信。因此，新手如何能有关于用来评价候选者专长的 E-域内的过去回答的任何（不可能的）信念呢？换言之，新手作为新手，如何能有关于候选专家的过去记录的任何看法呢？

对这个问题的一个可能回应是，重新讨论深奥的陈述和通俗的陈述之间的区别。也许并不是 E-域内的任何一个陈述都是深奥的。在 E-域内也有许多通俗的陈述，而且，它们是新手可能评价候选者专长的陈述。但是，这实际上有意义吗？如果一个陈述是通俗的陈述，即在认识论意义上新手易于接近的陈述，那么，它为什么正好应该被包括在 E-域内呢？有人会认为，E-域恰好是只有专家才易于接近的命题域。

对这个问题的解决办法首先使我们的深奥/通俗的区分变得更加明显。认为陈述在类型上或者是深奥的，或者是通俗的，这很自然，但这是一种误解。一个给定的（无时间性的）陈述是深奥的或通俗的，只是相对于一种认知立场或形势而言的。它可能是深奥的，例如，考虑陈述："2130 年 4 月 22 日，在新墨西哥的圣达菲，将会有日食。"相对于当前的认知立场，即生活在 2000 年的人的立

---

① 在对心理健康职业的一种毁灭性批评中，洛宾·道斯表明，这些职业的真正专长，在科学意义上，是非常可疑的，尽管在那个职业共同体中，文凭主义的层次很高。Dawes, Robyn, *House of Cards: Psychology and Psychotherapy Built on Myth*, New York: Free Press, 1994.

场，这是一个深奥的陈述。一方面，2000 年的普通人不能正确地回答这个问题，除非他们对此进行猜测。另一方面，在所讨论的 2130 年 4 月 22 日当天，在新墨西哥州圣达菲大街上的普通人将能轻易地正确回答这个问题。在这种不同的认知形势下，这个问题将是一个通俗的问题，而不是一个深奥的问题。①你不需要专门训练或知识来确定问题的答案。这样，一种陈述的认知地位是随时间的变化而变化的。

把这个简单的事实应用于专家/新手问题是有意义的。当一个陈述变成通俗的时，新手能轻易确定它的真值。他能确定，那时它的确是真的。此外，他可能了解到，在早期，当这个陈述对于像他那样的人来说是深奥的时，另一个人设法相信这个陈述，并说它是（或将是）真的。此外，这个人可能反复地显示出评价下列陈述的能力：这个陈述在断言时是深奥的，但后来变成通俗的，而且，他可能再三证明在后来通俗情况下被确定为是对的。当证明在后来通俗情况下被确定为是对的时，新手们能推出，这位不寻常的认知者一定拥有某种特殊的认知方式——某种独特的专长——那是他们没有的。他们大概恰好不知道，这种独特的认知方式，除了大概包括某些专用的信息储存和调用那些信息的某种方法论之外，还包括什么。照这样，新手们通过证实某些人在一个特定领域内的令人印象深刻的记录，能证实他们在该领域内的专长。在新手本人不能以某种方式转变为专家的前提下，也能做到这一点。

天文学的例子恰好是容易被传播的许多事例之一。如果一辆车、一个空调系统或一个机体出现了功能异常或故障，未受过训练的人通常不能详述下列形式的真命题："如果你把治疗方案 X 应用于机体 Y，该机体将恢复正常功能。"然而，可能有人正好反复地详述这类真命题。② 此外，新手能证实这些命题是真的，因为新手能"看到"对异常功能机体进行的治疗，并看到这个机体恢复了正常功能（比不进行这种治疗的机体恢复得更快）。尽管一旦治疗起作用，这个命题的真实性就是一个通俗的问题，但是，在实施治疗和产生治疗结果之前，它是一个深奥的问题。在这样一个例子中，当它是深奥的时，专家拥有知识并能够被确定为已经拥有知识。③

① 在当前的讨论中，只是认知的深奥性，而不是语义的深奥性，是成问题的（参见注释 166 页注释①）。

② 当别人提出命题时，他不可能只承认这样的命题为真；当质问"如何修复这个机体？"这样的问题时，他们也能独立地提出这样的命题。

③ 我在早期的著作中已经讨论过这些例子：戈德曼 1991 年和戈德曼 1999 年（第 269 页）的研究。Goldman, Alvin, "Epistemic Paternalism: Communication Control in Law and Society", *Journal of Philosophy*, Vol. 88, 1991, pp. 113-131; Goldman, Alvin, *Knowledge in a Social World*, Oxford: Clarendon Press, 1999.

应该强调的是，专家提供答案的许多问题，当它们是深奥的时，不只是通过幸运的猜测就能正确回答的是/否问题。其中的许多问题是允许有无数答案，有时是许多模糊答案的问题。为达到简化举例说明的目标，我们可以说，当一位病人看医生时，他问医生的问题是："在成千上万的有效药物中，哪一种药将治愈或缓解这种病？"这样一个问题，只通过猜测，不可能得到正确的回答。同样，当研究火箭的科学家首先试图让一艘宇宙飞船着陆在月球上时，对下列问题有尚不确定的许多可能的回答："哪些步骤系列将使这一艘（或一些）宇宙飞船成功地着陆在月球上？"从可能答案的无限列表中选择出一个正确答案，不可能是一种幸运的猜测。正是像这种通常包括技术应用在内的技艺，正确地劝告新手说，得到正确答案的人，具有联合产生超常能力获得正确答案的特殊的信息储存和利用信息的特殊的方法论。照这样，新手的确能确定，在他们自己不是专家的这个领域内，其他人是专家。

当然，这没有为新手提供能全部解决他们的2-专家问题的一种算法。新手只是偶尔知道或能确定，在他面前争论一个问题的两位被公认专家的记录。民事诉讼中的一位陪审员根本没有机会用尽和获得在他面前作证的竞争专家证人的记录信息。虽然如此，新手至少在原则上和某些情况下，能证实记录并用它们检验一位候选者应有的专长，这个事实离完全消除对新手/2-专家情境的怀疑还有一段距离。此外，"直接"确定少数专家专长的可能性，使得有可能对更广泛阶层的候选专家作出可信的推理。如果某些人通过上面提供的方法体现了有重要的专长，如果这些人能够培训另一些人，那么，受训者本人将有可比较的信息储存和产生最初专家的认知成功的同样类型的方法论，这个推理是可信的。[①] 此外，已被证实的专家就被作为关于他人（即使他们没有接受训练或不信任他们）专长的元"专家"来咨询，在这种程度上，能再一次推出后者具有可比较的专长。因此，一旦本部分提供的专家证实的基础得以建立，就减轻了对新手/2-专家问题产生的某些早期怀疑。

七、结　　论

我的故事的结束无疑是混合的，一个理由是，它既不会令人兴奋，也不会令人失望。当面对带有竞争信息的相竞争的专家时，怀疑的乌云隐约呈现在新手的

---

① 当然，在传播他们的专长时，某些专家可能比其他专家更好。有些专家可能更努力热衷于传播，或更擅长于传播，或在对他们的学员进行资格审查时，执行更严格的标准。这就是为什么关于培训项目的好的信息肯定与专长的判断相关的原因。

很多认知范围之内。然而，这里仍有一线曙光。建立专家记录没有超越可能性乃至可行性的范围。这能依次强化更广泛层次的专家的可信性，因而，当试图在专家之间作出选择时，奠定了合理运用人数的基础。然而，这并没有否认这种认知情境面对新手时通常是不容乐观的。在分析这样的情境时，有一些令人感兴趣的理论问题，而且，这些问题对"应用的"社会认识论提出了引人感兴趣的实际挑战。例如，什么类型的教育能实质性地提高新手评价专家的能力；什么类型的互动调节有助于使新手/专家关系成为比盲目信任更能得到辩护的信任之一。①

（成素梅 译）

---

① 关于早期草稿的有益评论，我要感谢史密斯（Holly Smith）、费利斯（Don Fallis）、格拉汉姆（Pter Graham）、赖肖（Patrick Rysiew）、威利（Alison Wylie）和 2000 年罗格斯认识论会议的许多参加者，圣路易斯的社会科学哲学圆桌会议以及我关于"社会认识论的哲学基础"的 2000 年 NEH 暑期班。

Dreyfus on Expertise: The Limits of Phenomenological Analysis

Even Selinger and Robert P. Crease

Introduction

Expertise is of central importance to contemporary life, in which many economic, political, scientific, and technological decisions are routinely delegated to experts (Barbour, 1993, pp. 213-223). Citizens deter to the authority of experts not only in circumstances involving technical dimensions, but also in "all sorts of common decisions" (Walton, 1997, p. 24). On the one hand, routine deference to experts has political consequences and some scholars even suggest that it undermines rational democratic procedures and communicative action by allowing ideology to substitute for critical discussion (Turner, 2001). [1] On the other hand, volatile controversies over issues with a scientific-technical dimension can result in the suspension of routine deference and increased suspicion towards experts. It is hardly surprising, therefore, that the nature and the proper criteria for identifying expertise have been hotly debated in political and legal contexts. In legal contexts, for instance, the question of the proper criteria for expertise regularly arises in connection with developing the appropriate criteria for certifying expert witnesses. [2]

Expertise is an issue for which philosophical clarification seems appropriate and even essential. The 1993 landmark decision by the U. S. Supreme Court concerning the use of expert witnesses, *Daubert v. Merrell Dow Pharmaceuticals*, appealed to several concepts of philosophical origin, with particular attention given to Karl Popper's notion of "falsification" (Huber and Foster, 1999, pp. 37-68). Philosophers themselves occasionally have been called upon to serve as expert legal witnesses while medical ethicists have advised hospital boards, politicians, and U. S. Presidents concerning such politically sensitive programs as the Human Genome Project and stem cell research (Nussbaum, 2001; Ruse, 1996). [3]

Aside from its social implications, the issue of expertise is also philosophically important for several reasons. One is that it bears on the philosophy of mind. The classical locus of expertise, or "expert knower", is the subject, and the way experts understand and are attuned to the world bears on the nature of subjectivity, intentionality, and rationalist and representational notions of consciousness (Pappas, 1994). The nature of expertise, for instance, is the focal point around which turns the debate over whether intelligence can be successfully disembodied in artificial intelligence (AI) schemes, expert systems, and computer-based distance learning programs (Collins, 1995).

A second reason for the philosophical importance of the question of expertise is that it crystallizes the conflict between two traditions, classical philosophy of science and science and technology studies (henceforth STS). Classical philosophy of science takes expertise for granted and assumes the legitimacy of an expert-lay divide, while STS takes a skeptical approach towards experts for granted, presupposing the need to expose their illegitimacy (Mialet, 1999, pp. 552-553).

Finally, addressing expertise stimulates the interface between phenomenology and the sciences. Though some have argued that phenomenological description is only capable of capturing subjective dimensions of experience, and hence is inappropriate to use when trying to understand science (Latour, 1999, p. 9), scientific practice requires the development, exercise, and coordination of a variety of expert skills that are open to phenomenological clarification.

Nevertheless, philosophers have rarely addressed the subject explicitly, though implicit and unexamined notions of expertise often lurk under rubrics such as "authority", "colonization", "power", and "rational debate" (Turner, 2001). Hubert Dreyfus is one of the few to have overtly addressed the concept, and this essay is devoted to critically appraising his account. This analysis of Dreyfus will proceed in five steps. We shall: (1) place his model of expertise in an embodied context, (2) outline his general conception of skills, (3) summarize his descriptive model of expertise, (4) present his normative theory regarding which expectations about experts are justifiable, and (5) point out certain problems with his account. We argue that Dreyfus, by proposing that fundamental expert characteristics can be specified independently of cultural and historical considerations, demonstrates the importance of phenomenology to the subject by showing persuasively that expertise cannot be examined exhaustively by sociological, historical, and anthropological analyses. But we also identify certain descriptive and normative problems in Dreyfus's account. While Dreyfus shows phenomenologically that experts cannot be reduced to ideologues and artifacts of social networking, he also lacks hermeneutical sensitivity by overstating the independence of the expert and expert decision making from cultural embeddedness.

1. Expertise and the Body

The significance of Dreyfus's account of expertise call be highlighted by recalling just how the two traditions mentioned above, classical philosophy of science and STS, avoid addressing the issue, Each in effect treats creative expert performance as something extraordinary that "just happens", and which poses no special philosophical problems. The goal of traditional philosophy of science, for instance, is the rational reconstruction of the organizational dimensions at the root of science's efficacy and objectivity, with particular emphasis on how its operation is only temporarily disrupted by anomalies, it approaches creativity as a predominantly mental act for which true philosophical discussion can take place only about its products. Where creative ideas come from is not considered a proper epistemological question and is relegated to psychology or history in the framework of the distinction between context of discovery and context of justification (Mialet, 1999, p. 552).

Contemporary STS, on the other hand, treats expertise as "distributed", externalized into particular settings such as laboratory and social networks, and standardized in technologies, criteria of scientificity, protocols for evaluating proof, and the rhetorical means of recruiting allies (Mialet, 1999, p. 552). To do otherwise, according to STS proponents, would risk "naturalizing" expertise, conferring undue authority on experts and leading to the repression of lay knowledge, values, and interests. By focusing on how experts become overly exalted through processes of mediation, STS proponents place a non-distributed sense of expertise into a "black box" (Mialet, 1999, p. 553). [4] While STS theorists occasionally appeal to tacit knowledge in their work, these appeals are not tied to any theory of embodiment. [5]

Although traditional philosophy of science and STS have different motives for wanting to demystify expertise, they produce similar results. For all their sharp disagreements, both traditional philosophy of science and STS refrain from discussing the relation between expertise and the body. They both agree that to demystify expertise and provide an accurate account of science, invariant features of bodily praxis need to be ignored.

The neglect of embodiment broaches the traditional phenomenological theme that practical involvements of living bodies ultimately ground any knowledge that they have about the world, including abstract-scientific knowledge.[6] For phenomenologists, all practical and theoretical activities, no matter how abstract their outcomes, need to be understood on a continuum with basic lifeworld practice. Dreyfus follows in this tradition by rigorously treating expert judgment and behavior as an instance of embodied human performance. Like classical philosophers of science and STS proponents, he seeks to demystify expertise—but he does so by placing expertise on a continuum of lifeworld activities, rather than isolating it from them. Experts merely act the way each of us does when performing mundane tasks: "We are all experts at many tasks and our everyday coping skills function smoothly and transparently so as to free us to be aware of other aspects of our lives where we are not so skillful. " (Dreyfus and Dreyfus, 1990, p. 243)

Dreyfus explicitly links his skill model to phenomenological tradition by calling it an explicit development of Merleau-Ponty's notion of the lived body (*le corps vecu*) and the concepts "intentional arc" and "maximal grip" .[7] By closely attending to "how our relation to the world is transformed as we acquire a skill", Dreyfus claims that he intends to "lay out more fully than Merleau-Ponty does" how skills are acquired, improved, and used (Dreyfus, 1999a, p. 1) . Dreyfus aims to study experts as *embodied*, *situated subjects*, seeking to note "under what conditions deliberation and choice appear" in order to avoid "making the typical philosophical mistake of reading the structure of deliberation and choice into [his/her] account of everyday coping" (Dreyfus and Dreyfus, 1990, p. 239) . But Dreyfus also contends that his model of how experts act is empirically verified by neural network researchers, specifically Walter Freeman in his studies of the brain dynamics underlying perception (1999a; 1999b, pp. 6-10; 1998) .[8]

Dreyfus's account offers what might be called a metaphysics of expertise (though he prefers the term "ontological" to "metaphysical") . This is in sharp contrast to both traditional philosophy of science and STS. For the former, there is not enough subject matter for a metaphysical account; for the latter, expertise is a culturally dependent phenomenon whose definition changes in relation to the historical transformations that govern how it is perceived.

2. Expertise and Skills

Dreyfus first developed the basis for his descriptive account of expertise with his brother Stuart during the 1960s, when hired by the RAND corporation as a consultant to evaluate their work on artificial intelligence. His research culminated in a paper called "Alchemy and Artificial Intelligence" (Dreyfus, 1967) and the book *What Computers Can't Do* (Dreyfus, 1992) .[9] In *Mind Over Machine: The Power of Human intuition and Expertise in the Era of the Computer*, the two brothers developed a model of expert skill acquisition whose scope was claimed to be universal (Dreyfus and Dreyfus, 1986) . They aimed to provide a phenomenological account of how adults acquire skills by instruction in all fields involving skilled performance[10], whether of the intellectual or motor kind. [11]

This model of expert skill acquisition serves as a touchstone in Dreyfus's career. He uses it, for instance, as the basis to: (1) demythologize the hype associated with artificial intelligence projects, in particular, "expert computer systems" designed to simulate human expertise (Dreyfus and Dreyfus, 1986; Dreyfus, 1992); (2) judge social biases that "endanger" professional experts (such as nurses, doctors, teachers, and scientists) by imposing "rationalization" constraints (Dreyfus and Dreyfus, 1986); (3) explain what is wrong with dominant tendencies in American styles of business management (Dreyfus and Dreyfus, 1986); (4) defend the accuracy of Merleau-Ponty's non-representational account of intentionality and action (Dreyfus, 1998); (5) expose the practical limits of Jürgen Habermas's neo-Kantian conception of ethics (Dreyfus and Dreyfus, 1990); (6) explain the expertise of political action groups (Dreyfus, Spinosa and Flores, 1997); (7) clarify what *techne* and *phronesis* mean for Martin Heidegger and Aristotle, and in doing so, correct his acclaimed, book-length interpretation of Heidegger (Dreyfus, 2000); (8) explain the relevance of Kierkegaard's normative stages of the subject's development to evaluations of the Internet's value as a medium for communication (Dreyfus, 2001); and (9) critique the educational potential of "distance learning" programs (Dreyfus, 2001) .

A first key element of Dreyfus's account is his rejection of the common tendency to define experts as sources of information. Expert skills are principally a matter of practical reasoning, of "knowing how" rather than "knowing that" (Ryle, 1984) . "Knowing that" is prepositional knowledge of and about things, obtainable through reflection and conscious appreciation. "Knowing how", such as the ability to walk, talk, and drive, involves practical knowledge that is mostly experienced as a "thoughtless mastery of the everyday" and does not require conscious deliberation for successful execution (Dreyfus and Dreyfus, 1990, p. 244). In many instances, knowing how involves the exercise of inarticulatable skills of which one cannot fully give an account, though one should not confuse this with Polanyi's concept of "tacit knowledge" (Dreyfus and Dreyfus, 1986, p. 16). [12]

It is possible to suggest hints and maxims that approximate elements of smooth performance, but seeing the point of these hints, and being capable of following these maxims, to a large extent presuppose the skill that the hints and maxims are supposed to account for. Moreover, Dreyfus holds that once skill is acquired, one tends not to follow the maxims used during the initial stages of learning.

Following Heidegger, Dreyfus argues that practical agents tend to reflect only on how their experience is organized during "breakdown" scenarios when pre-thematic styles of environmental coping prove insufficient for accomplishing ordinary goals; people only reflect on how they do what they do when what they do fails to work effectively. The justification for this derivative use of reflection is also practical, based on the agent's experience that to reflect upon what one does and the rules for doing it usually leads to practical problems doing what one wants to do. Going from pre-conscious behavior to a conscious appreciation of the rules followed during particular actions marks the agent's transition from practical to theoretical reasoning, from "knowing how" to "knowing that" (Dreyfus and Dreyfus, 1986, p. 7). For Dreyfus, the most basic tasks requiring skilled performance are best accomplished if one does not consciously focus on what one is doing.

Finally, Dreyfus contends that skills are flexible ranges of response and even physical skills cannot be reduced to a repetitive series of kinesthetic movements. He contends, "A skill, unlike a fixed response or set of responses can be brought to bear in all indefinite number of ways" (Dreyfus, 1992, p. 249). Joseph Rouse elaborates this point in a discussion of Dreyfus. In learning to throw, he writes, what is involved is not a series of repetitive motions, "but a range of responses to throwable things. Learning to throw overhand means that one can also throw sidearm, though the movement is different. Having learned to imitate a fairly limited number of sentences, I can produce an unlimited variety of different ones" (Rouse, 1987, p. 61). Thus to learn a flexible range of response is not to memorize an "actual movement" or "thought pattern" with the

intention of repeating them, but to grasp of a field of possibilities (Rouse, 1987, p. 61).
Dreyfus, however, no doubt would add that while skills are flexible, there are limits,
borderline areas, and marginal spaces; knowing how to throw a baseball overhand does
not enable one to throw a javelin, to throw underhand, or to throw a softball sidearm.

After *Mind Over Machine*, Dreyfus's model of expertise appeared nearly verbatim in
numerous articles. In them, he never challenges the core theses about, nor descriptions
of, expertise articulated in *Mind Over Machine*, but expands the range of what the
model can be applied to, and comments on the epistemological and metaphysical
features of expert performance that he previously applied but did not fully articulate. [13]

3. The Descriptive Model

Dreyfus's descriptive model of expertise has several key features. A first is that it is
supposed to have *phenomenological justification.* Traditionally, the concept of
justification relates propositions to a public sphere that can demonstratively verify or
refute the content of; or logical or inferential connections in, statements. From a
phenomenological perspective, propositions are justified when they detail experiential
invariants that all subjects are capable of recognizing as according with their own
experience. This is signaled in the following invitation Dreyfus extends to his readers:
"You need not merely accept our word but should check to see if the process by which
you yourself acquired various skills reveals a similar pattern." (Dreyfus and Dreyfus,
1986, p. 20) Dreyfus claims to have based his model of expertise on invariant patterns
found in descriptions of skill acquisition relayed by first-person testimonies of "airplane
pilots, chess players, automobile drivers, and adult learners of a second language"
who discussed how they learned to make "unstructured" decisions (Dreyfus and
Dreyfus, 1986, p. 20). [14] Thus, his account is not supposed to be vulnerable to
accusations of mischaracterizing idiosyncratic experts as paradigmatic or begging the
question of how one knows what an exemplary expert in a field is. Because of adhering to
the phenomenological method and its justificatory mechanisms, his account is also not
meant to be susceptible to counterexamples that could disprove its applicability to adults
who deliberately seek to acquire skills. It is putatively immune to these types of criticism
because his skill model is expected to be meaningful in such a way that all experts can
rediscover and verify its essential elements for themselves.

Another key feature is that his model is *developmental*, and envisions skill acquisition as occurring sequentially through five ascending stages: (1) an initial "beginner" phase, (2) an "advanced beginner" phase, (3) a "competent" phase, (4) a "proficient" phase, and (5) finally a culminating "expert" stage. In the first stage, the beginner, who "wants to do a good job", learns a "context free" set of "rules for determining action" and tends to act slowly in remembering how to apply them (Dreyfus and Dreyfus, 1986, p. 21). In the advanced beginner stage, the student who now has more "practical experience in concrete situations" begins to "marginally" improve by recognizing "meaningful additional aspects" of the situation that are not codified by rules (Dreyfus and Dreyfus, 1986, pp. 22-23).

What is particularly interesting about Dreyfus's account is that the learner undergoes not just cognitive and practical transformations but affective ones as well. Beginners and advanced beginners, he claims, typically experience their commitment to a practice as "detached" while a competent performer feels "involved" in the outcome of his or her performance (Dreyfus and Dreyfus, 1986, p. 26). In the competent stage the learner frequently feels "overwhelmed", as if he or she is "on an emotional roller coaster", having to cope with "nerve-wracking and exhausting" aspects of the practice, and feels "overloaded" due to facing too many potentially relevant elements to remember (Dreyfus, 2001, p. 35). Consequently, the competent learner narrows down those elements, devises a "plan" and chooses a "perspective" in order to selectively address "relevant features and aspects" of the situation (Dreyfus and Dreyfus, 1986, pp. 26-27). By making these changes, the competent performer experiences "a kind of elation unknown to the beginner", including "pride" and "fright" (Dreyfus and Dreyfus, 1985, pp. 117-118). Accepting responsibility means "mistakes are felt in the pit of the stomach" (Dreyfus and Dreyfus, 1985, p. 118).

In the proficient stage, the student transcends what Dreyfus calls the "Hamlet model of decision making", the "detached, deliberative, and sometimes agonizing selection of alternatives", which typifies the first three stages of skill acquisition (Dreyfus and Dreyfus, 1986, p. 28). Here the performer's reliance on rules and principles for seeing what goals need to be achieved is largely replaced by "know-how", an "*arational*", grasp of the situation that Dreyfus calls "intuitive behavior"; although the proficient performer must still contemplate and deliberate about what to do to achieve his or her goals (Dreyfus and Dreyfus, 1986, pp.: 27-36). [15] "Action becomes easier and less stressful" as the proficient performer "simply sees what needs to be done rather than using a calculative procedure to select one of several possible alternatives"

(Dreyfus, 2001, p. 40) .

In the final stage, the expert not only sees what needs to be done, but also how to achieve it without deliberation, immediately, yet "unconsciously", recognizing "new situations as similar to whole remembered ones" and intuiting "what to do without recourse to rules" (Dreyfus and Dreyfus, 1986, p. 35) —though, interestingly, the Dreyfus's removed all references to remembered cases in the 1988 paperback edition. Thus, the expert, like masters in the "long Zen tradition" or Luke Skywalker when responding to Obi-Wan Kenobi's advice to "use the force", transcends "trying" or "efforting" and "just responds" (Dreyfus, 1999b, p. 22, n. 13) . Dreyfus summarizes the "fluid performance" of expertise as: "*When things are proceeding normally, experts don't solve problems and don't make decisions; they do what normally works.* " (Dreyfus and Dreyfus, 1986, pp. 30-31) He even claims that at the expert level the ability to distinguish between subject and object disappears: "The expert driver becomes one with his car, and he experiences himself simply as driving, rather than as driving a car. " (Dreyfus and Dreyfus, 1986, p. 30) When the expert experiences the "flow" of peak performance he or she "does not worry about the future and devise plans" (Dreyfus and Dreyfus, 1986, p. 30) . By being immersed in the moment, the expert can experience "euphoria", which athletes describe as playing "out of your head" (Dreyfus and Dreyfus, 1986, p. 40) .

Yet another key feature of Dreyfus's model has to do with the demarcations between beginners and experts. Whereas some STS researchers argue that it is untenable to decisively distinguish the expert from the nonexpert because their apparent differences are social illusions, Dreyfus contends that beginners and experts can be demarcated in three different ways. [16] A first is based on the expert's "immersion into experience and contextual sensitivity" . The expert differs from the beginner because he or she no longer relates principally to a practice analytically through context-free features that are recognizable without experience. Instead, through skillful behavior rooted in experience, he or she recognizes important features as contextually sensitive (Dreyfus and Dreyfus, 1986, p. 35) . This situational engagement leads to a change in the expert's judgment: " [The] novice and advanced beginner exercise no judgment... and those who are proficient or expert make judgments based upon their prior concrete experience in a manner that defies explanation. " (Dreyfus and Dreyfus, 1986, p. 36) A second demarcation centers on the temporal connection between action and decision making. While slowly following rules and deliberating characterize the beginner's actions, the expert's actions are immediate and intuitive situational responses. A final way of demarcation concerns affective transformation. In passing through developmental stages the expert's subjectivity and relation to the world are transformed in a manner that qualitatively differs, and can be demarcated, from a beginner's relation to the world. While in early stages the learner is "frustrated" and "overwhelmed", in the last stage the expert, who learned that the outcome of a situation matters by making risky commitments, sheds those affects and enjoys "fluid" and "smooth" performance. In his critique of John Searle's account of the background and its relation to intentionality, Dreyfus describes an expert tennis player as being able to become so absorbed in the "flow" of the game that he or she no longer feels the pressure to win, and only responds to the gestalt tensions on the court (1999b, pp. 4-5) . [17] Additionally, the change in affect from beginner to expert corresponds with a change in meaning. While the beginner's attitude is essentially one of interest or curiosity, the expert is fully invested in his or her own being. Dreyfus quotes chess grandmaster Bobby Fischer—chess being the paradigmatic intellectual skill for Dreyfus—to the effect that "chess is life" (Dreyfus and Dreyfus, 1986, p. 33) .

Despite these demarcations Dreyfus's account also involves *continuities*. Even the most abstract practice maintains an essential, though sometimes hidden, relation to the lifeworld. This is why he avoids using the contrasting and antonymic terms "nonexpert" and "layperson" and instead uses "beginner", suggesting someone on one end of a spectrum of potentialities. [18]

Dreyfus's model has *foundational implications*. Different fields are organized around different types of essential skills; still, his account is a model of skill acquisition that can be formally specified without reference to any particular field. Of course diverse fields define experts in different ways due to the type of content that typifies the field. For example, a relation to winning defines an expert chess player, whereas a relation to arriving at a destination defines an expert driver. Nevertheless, a fundamental definition of expert as intuitive, committed, rule-transcending subject "whose skill has become so much a part of him that he need be no more aware of it than he is of his own body" applies to both of these (Dreyfus and Dreyfus, 1986, p. 30) . At the phenomenological level, the expert chess player and car driver are functionally equivalent *qua* expert (Dreyfus and Dreyfus, 1986, pp. 21-35) .

Finally, Dreyfus's account provides a *practical expert point of view*. Based upon the universal scope of his model, he attempts to describe a common cognitive and affective relation to the world that all experts share, which can be evoked as the basis of justifying the expert's commitments. When other theorists have attempted to delineate the "expert point of view", they have mostly done so from a theoretically holistic perspective, pointing to the expert's synoptic command of a field. For example, Scott Brewer argues that while nonexperts can know particular, true things about the methods and facts pertinent to a field, only experts know how the relevant features of a field, such as "enterprise" and "axiological" characteristics relate to provide a shared sense of what, how, and why practitioners in that field can claim to be true (1998, pp. 1568-1593) .[19] Brewer notes that his focus on the organization of theoretical knowledge is somewhat Platonic (1998, p. 1591) . By contrast, Dreyfus often refers to his skill model as anti-Platonic. Rather than providing a theoretically holistic account of the "expert point of view", he presents a description of the common features of practical understanding. The expert's practical understanding does not come from beliefs or theoretical commitments, but from acquired and embodied skills. Hence, Dreyfus writes, "The moral of the five-stage model is there is more to intelligent behavior than calculative rationality" (Dreyfus and Dreyfus, 1986, p. 36) .

4. Normative Implications

Dreyfus's account describes which expectations about expert services are justifiable, and what can and cannot be legitimately asked of them. Dreyfus's account thus has normative implications, which he discusses empirically in connection with, among others, ballistics examiners, chicken sexers, citizens, judges, nurses, paramedics, physicians, science advisors, and teachers (Dreyfus and Dreyfus, 1986, pp. 196-201; Dreyfus, Spinosa and Flores, 1997, p. 106) . A phenomenological understanding of the nature of expert decision making, Dreyfus suggests, needs to be the basis for identifying the means by which we may achieve some of the social and political goals that involve experts, such as consulting experts for personal, institutional, and legal reasons. Although directly deducing normative obligations from a descriptive foundation might suggest that Dreyfus commits the naturalistic fallacy of deducing obligations about what ought to be done from premises that only state what is the case, he contends that the relation between phenomenology and normativity is an issue of "priority" .[20]

These normative implications are significant for several reasons. A first is the renowned difficulty experts have communicating with others. Although this is sometimes attributed to the disparity between technical and ordinary language, and sometimes to psychological factors such as arrogance, Dreyfus's account suggests deeper causes, an issue to which we shall return. Another reason the normative implications of his account are significant is the frequency with which experts serving as personal, institutional, and legal advisors are challenged as to their motives and biases, which may arise in connection with: (1) their professional training, which can influence how experts conceptualize and bound the problems they deal with; (2) their employers, which can influence politically the conclusions experts arrive at; (3) economic interests, which are capable of turning experts into hired guns whose recommendations can be purchased for the right price; and (4) the desire for recognition and reputation—another issue to which we shall return. By contrast, Dreyfus suggests that essential normative restrictions surrounding expert performance can be determined without considering the social forces that have sometimes influenced what claims experts make—another issue to which we shall return in our critique.

The normative implications arise from the claim that novices cannot transcend rules through developed intuition in the way that experts can in any field, no matter how dire or pressing the social circumstances might be. [21] It is therefore illegitimate, according to Dreyfus, to expect them to describe their process of decision making in prepositional statements, because their decisions are made on the basis of tacitly operating intuition. Not only does the chess grandmaster act on intuition, so too does the ballistics expert, who cannot propositionally express in a truthful manner how he or she determined whether or not a particular bullet originated in a particular gun (Dreyfus and Dreyfus, 1986, p. 199). Yet in assuming the role of expert legal witness the ballistics expert will be expected to do so, correlating conclusions to rules, and only in so doing will he or she potentially be convincing to a jury. But Dreyfus argues this persuasive power comes at the expense of prioritizing the "form" over the "content" of the explanation (Dreyfus and Dreyfus, 1986, p. 199). Indeed, the expert even "forfeits" expertise when explaining his or her decision making process based on rules, for that involves abandoning the intuitive experience that guided the decision making process (Dreyfus and Dreyfus, 1986, p. 196). Rational reconstruction of expert decision making, Dreyfus argues, inaccurately represents a process that is in principle unrepresentable. When nonexperts demand that experts walk them through their decision making process step by step so that they can follow the expert's chain of deductions and inferences (perhaps hoping to make this chain of deductions and inferences for themselves) , they are, according to Dreyfus, no longer allowing the expert to function as expert, but instead, are making the expert produce derivative, and ultimately false, representations of his or her expertise. Hence Dreyfus argues that too much pressure should not be placed upon experts to "rationalize" their "intuitive" process of decision making to nonexperts. [22]

Dreyfus's normative position thus amounts to what Douglas Walton calls "a strong form of the inaccessibility thesis" (henceforth IT) : "that expert conclusions cannot be tracked back to some set of premises and inference rules (known facts and rules) that yield the basis of expert judgment. " (Walton, 1997, p. 110) According to IT, when experts render a verdict in matters of their expertise, their judgments are inaccessible to nonexperts in the sense that experts cannot propositionally express "a set of laws and initial conditions (principles and facts) that would exhibit an implication of the conclusion [i. e. what the expert concludes] by deductive (or even inductive) steps of logical inference " (Walton 1997, p. 109) . Due to the expert's reliance on inexpressible dimensions of intuition, IT suggests that "the most advanced expert in a field, who has achieved outstanding mastery of the skills of her field, may be the least able to communicate her knowledge" (Walton 1997, p. 113) . It is not merely that experts will be unable to communicate how their knowledge was achieved to nonexperts; their knowledge will be equally opaque to themselves and other experts.

Dreyfus's chief normative concern, therefore, involves the ways nonexperts can possibly jeopardize experts, not how experts can jeopardize nonexperts. "Experts are an endangered species. " (Dreyfus and Dreyfus, 1986, p. 206) He characterizes this as a problem confronting the U. S. in particular: "Demanding that its experts be able to explain how they do their job can seriously penalize a rational culture like ours, in competition with an intuitive culture like Japan's. " (Dreyfus and Dreyfus, 1986, p. 196) Furthermore, Dreyfus argues that historical changes have precipitated the expert's current crisis. He contends that recent advances in computer technology and bureaucratic social organization exacerbate this cultural problem: "The desire to rationalize society would have remained but a dream were it not for the invention of the modern digital computer. The increasingly bureaucratic nature of society is heightening the danger that in the future skill and expertise will be lost through overreliance on rationality. " (Dreyfus and Dreyfus, 1986, p. 195) In order to solve the problem of the disappearance of expertise Dreyfus seeks to reeducate people on the difference between "knowing how" and "knowing that". The future that Dreyfus speaks of is not distant. He suggests that "within one generation" we may lose "our professional experts" (Dreyfus and Dreyfus, 1986, p. 206). The goal is to get people to appreciate the value of intuition and the limits of rational deliberation. [23]

Dreyfus analyzes expertise from both an *asocial* as well as a *social* perspective. At the *asocial* level, he phenomenologically investigates how all human beings acquire skills regardless of who they are, what field they are apprenticing in, and when and where they learn their skills. At the *social* level, he presents a normative position that dictates how experts should be treated when they are asked to serve as consultants. The arrangement of this particular combination of *social* and *asocial* arguments is predicated upon Dreyfus's foundational assumption that social expectations of what experts *ought* to do should correspond with the embodied limitations that circumscribe what they can *in fact do*. This assumption is the justificatory mechanism behind his views that possessing expert level skills is both a necessary and sufficient condition for being an expert; and it is erroneous to take the ability to rationalize as a sign of expertise.

5. Problems with Dreyfus's Descriptive Account

Dreyfus's account, however, admits certain categories of people as experts which do not belong, and omits several which do. Dreyfus's claim, for instance, that adults are "experts" in walking and talking—which as we have seen is essential to his account for it grounds them in the same lifeworld spectrum as conventionally expert behavior—collides with ordinary usage. We do not call people who are merely ambulatory or verbal "expert" walkers or talkers, but reserve the adjective for those who undergo special training, give professional advice, etc. We do not call licensed drivers "experts" — nor even driving enthusiasts or competitive amateurs—even when they have an intuitive relation to operating their vehicles. Rather, we reserve the word for drivers who belong to professional driving organizations, participate in certain kinds of competitions, and so forth.

Meanwhile, Dreyfus's account also excludes certain classes of people from being experts who do belong. One of us, for instance, has drawn a distinction between "expert x", which is an adjectival use of "expert" that stems from the Latin *expertus*, and "expert in x", which substantively treats "expert" as a noun (Selinger, forthcoming). In Dreyfus's terms, "expert x" corresponds to "knowing how" while "expert in x" corresponds to "knowing that". Whereas an "expert x" could be an "expert farmer," an "expert in x" could be an expert "in farming". An expert "in farming" could effectively communicate, coordinate, and synthesize accurate propositional information about farming—could become Secretary of Agriculture—even if terrified of plows and tractors. An "expert in sports", who correlates the past behavior of athletes to current situations, could be crippled and lack physical capacity to play the sport; an "expert in music" could be a terrible musician. Nevertheless, Dreyfus denies that the propositional knowledge definitive of an "expert in x" is expert knowledge.

Listening to. . . commentators, who take up at least half the time on erudite talk shows, is like listening to articulate chess kibitzers, who have an opinion on every move, and an array of principles to invoke, but who have not committed themselves to the stress and risks of tournament chess and so have no expertise. (Dreyfus, Spinosa, and Flores, 1997, p. 87). Dreyfus associates the knowledge of "experts in x" ("commentators" or "kibitzers") with the "idle talk" definitive of the "public sphere" that "undermines commitment stemming from practical rationality" (1997, p. 86). The reasoning of an "expert in x" is "not grounded in local practices" but rather in "abstract solutions" and "anonymous principles" that fail to display "wisdom" (1997, p. 87). Moreover, Dreyfus argues, an "expert in x" lacks the bodily commitment a genuine "expert x" possesses; only an "expert x" affectively cares about the outcome of a situation and experiences the "risk" of performance. But this is an inaccurate characterization. An "expert in art history" presumably does not spout abstractions and generalities, but helps guide others, and sometimes even artists themselves, to appreciate the meaning in works and the artistic process itself. George Sterner once proposed what amounts to an effective refutation by reduction of Dreyfus's position here by trying to imagine a society without critics and commentators, in which "all talk *about* the arts, music, and literature is prohibited. . . held to be illicit verbiage" (Steiner 1989, pp. 4-5). While Steiner found himself yearning for such a "counter-Platonic republic", like Plato he also recognized the impossibility of his thought-experiment. Ultimately, Steiner realized, the result would be to stifle not only the arts and the creative imagination, but the lifeworld itself. Dreyfus could presumably agree, to the extent that his first three stages require verbalization and thus expertise could not be acquired in such a world—but the spirit of Steiner's thought-experiment is that even artists, musicians, and writers can only flourish in a sea of high-level discourse about what they do.

Another category of expertise passed over by Dreyfus's account of skill acquisition is the "coach". Dreyfus's account posits the instructor as delivering rules and commands in a standardized way to masses of people at once—and the instructor disappears after the initial stages of learning. Even Dreyfus's slightly revised, seven-stage model that includes copying from a "master", the master is no coach, but simply offers an example of a developed "style" which, in some unspecified way, helps the learner in developing his or her own (2001, pp. 43-46). But even though linguistically mediated, coaching is mischaracterized as a dispensing of rules in a standardized way; rather, it involves modeling and demonstration and is addressed to the specific performances of specific embodied learners. Since these performances differ from body to body, the knowledge embodied in coaching cannot be reduced to rules of thumb. Nor does it disappear after the beginning stages; if anything, it grows in importance. Many professional athletes, musicians, actors, and performers of various kinds—including Dreyfus's paradigmatic expert Bobby Fischer—would find it unthinkable to perform without coaches to help evaluate and strengthen their playing. For a coach does more than correct bad performance; a coach also facilitates good performance to become better.

The expert performer and the coach or "expert in x", to be sure, are functionally different. Nevertheless, they are closely related enough to suggest that Dreyfus is relying on a false dichotomy between those who "do" and those who comment, kibitz, or at best instruct as a propaedeutic to "doing". (One wonders, perversely, whether the prejudice against teaching evident here isn't an occupational hazard of academics.) A lived, embodied subject lacks an objective purchase on its own performing process. Though Dreyfus would clearly disagree, the performing of that subject is always, therefore, in principle open to being better disclosed to outsiders, whose intercessions may then help the subject to perform better. This is not to say, of course, that the outsider's insights are achieved because of an objective stance; rather, it is because the outsider is situated differently. Moreover, coaching involves not only understanding what the performer is doing and what it is to perform well, but also how to intercede best in order to transform the performing of the pupil. Coaching, therefore, is a first person process requiring different types of contextual sensitivity; it has its own techniques, its own intentional arc, and its own quest for maximal grip (which in this context means the ability to optimize the conditions to allow the pupil to go beyond not only his or her present performing ability, but the coach's as well). Coaching, in short, is not just a practice in which one can be expert, but a form of expertise itself—even in Dreyfus's own terms. The expertise is not identical with that of the expert performer, but it is closely related and shares a common end: the achievement of good performance.

The false dichotomy not only exposes a problem with Dreyfus's invocation of the difference between "knowing how" and "knowing that", but a deeper flaw in his account as well. Dreyfus assumes that the body which acquires skill has no relevant biography, gender, race, or age (Young, 1998; Sheets-Johnstone, 2000). He does acknowledge that "cultural styles" affect how skills are learned, noting for example that differences exist between how American and Japanese mothers "handle" their babies (2000, pp. 46-47). However, this notion of "cultural style" is not developed beyond unsubstantiated generalities, and assumes as well the insignificance of any biographic differences existing within individuals sharing a single culture. From Dreyfus's perspective, one develops the affective comportment and intuitive capacity of an expert solely by immersion into a practice; the skill-acquiring body is assumed to be able, in principle at least, to become the locus of intuition without influence by forces external to the practice in which one is apprenticed.

Sheets-Johnstone has sharply critiqued this "adultist" stance. She argues that it is a methodological mistake to take the adult human body as the phenomenological ground zero, for it overlooks "the originating ground of our knowledge, our capacities, our being" (Sheets-Johnstone, 1999, p. 232). The behavior of the adult human body does not simply happen but is itself the outcome of an "apprenticeship" carried out in infancy, an apprenticeship which leaves adult bodies full of meanings and experiences that are essential to all of their future behavior, including even the skills that they go on to learn and their application. "Whatever the particular adult skill-learning situation playing the piano, driving a car, playing chess, making trousers—it is a compound of experiences sedimented with skills and concepts accruing from our history. " (Sheets-Johnstone, 2000, p. 359)

To be sure, what Sheets-Johnstone means by apprenticeship (as when she calls an infant apprenticed to its body) is different from what Dreyfus means. For her, apprenticeship involves constant experimenting in the dark, and rules—even if only rules of thumb—flow from rather than precede it. For Dreyfus, on the other hand, apprenticeship is an explicit and deliberate process, and follows the formulation of rules. Nevertheless, she points out something overlooked in Dreyfus's account: that the learning body is always already *embedded*, and does not disembed itself in skill acquisition. In order to learn a skill such as dancing, for instance, I must *already* be able to move my body. Each body that takes up dancing is already embedded in a way of moving, and learning the dance is not a matter of learning to move *tout court*—as learning a new language is much different from learning to speak. Adult human bodies *never* move in a generic way but always in *this* way or *that*, and in a way that reflects how one has been brought up, and has brought oneself up, to move.

Learning dancing, therefore, cannot be said to exemplify the Dreyfusian progression of skill acquisition in which the learner begins by following rules, learns to make exceptions to the rules in the advanced beginner stage, learns to enact spontaneously and fluidly in the higher stages until "called" by the situation, and so forth. Learning dancing as a skill almost always requires coaching by someone who is sensitive to the differences between performances of individual bodies and to ways of adjusting such performances by modeling and demonstration. The learner does not leave this embeddedness behind, departing from what one already has and becoming delivered to a new way of moving, but rather carries it forward and transforms it (Crease, 2002a). Even the affective demeanor and intuitive capacity of a practice are not therefore separable from broader social influences that first do not appear directly connected with the skilled practice. The claims sometimes made in biographies of Bobby Fischer, for instance, that his distinctive aggressive chess playing style was not solely developed by playing many games of American style chess but shaped in part by personal childhood experiences are not implausible and even persuasive; if true, his "expert responses", the choices of specific moves, were not fully and exhaustively forged by contextual sensitivity plus experience, and chess is even more life than Dreyfus suspects.

Oddly, given Dreyfus's philosophical commitments, his account lacks hermeneutical sensitivity. The flaw in his assumption that skilled behavior crystallizes out of contextual sensitivity plus experience without contribution from individual or cultural biography can be traced to a failure to take into account the fact that the embodied subject, *even when behaving expertly*, brings to the situation what has been historically and culturally transmitted to it, and in a way such that the subject can never grasp cognitively all at once. The individual expert performer, as a consequence, does not have a complete purchase on his or her own expert behavior. Therefore, *contra* Dreyfus, it will always be possible in principle for an expert performer to learn about one's own performance from another, contextually sensitive person—though, again, this is not because the other has managed to obtain an objective position, but on the contrary, because the other is differently situated.

6. Problems with Dreyfus's Normative Account

Lack of hermeneutical sensitivity also affects Dreyfus's normative account, for his assumption of the autonomy of expert training suggests a naïveté in his counsel to "trust experts". While a beginner might have entered a training program *culturally* or *situationally embedded* with prejudices, ideologies, or hidden agendas, these would all be left behind by the time one reached the expert stage; an expert's knowledge for Dreyfus, as we saw, crystallizes out of contextual sensitivity plus experience. But if one can never leave the hermeneutic circle, the best that one can do is transform one's embeddedness, rather than extricate oneself from it. The acquisition of expertise is not a transcending of embeddedness and context, but a deepening and extension of one's relationship to it (Crease, 1997). This also means that experts will never be able to free themselves a priori from the suspicion that prejudices, ideologies, or hidden agendas might lurk in the pre-reflective relation that characterizes expertise. [24] Not only is this suspicion to be expected; its absence would be socially dangerous. This gives rise to a *recognition problem* that is such a prominent part of many actual controversies involving a technical dimension where expert advice is required. But there is no room for this recognition problem in Dreyfus's account. He leaves no grounds for understanding how an expert might be legitimately challenged (or instructed, for that matter, as in the case of sensitivity training, nonexpert review panels, etc.). One would never imagine, from Dreyfus's account, that society could possibly be endangered by experts, only how society's expectations and actions could endanger experts. The stories of actual controversies, we will argue, not only shows things do not work the way Dreyfus says they do, but also that it would be less salutary if they did. Such stories, we claim, amount to a counterexample to Dreyfus's normative claims, and point to serious shortcomings in his arguments.

Dreyfus, let us recall, assumes from the start that the people who possess expert level skills are the same people who *should* be socially recognized as experts. It is unproblematic, for him, who "counts as" an expert in a given social situation; he assumes the absence of a recognition problem. This is clear from his examples. On the one hand, he refers to people who are socially recognized as experts, such as airplane pilots, surgeons, and chess masters, to illustrate how embodied expert performance functions (Dreyfus and Dreyfus, 1986, pp. 30-35). These references are descriptively loaded because they use *socially acknowledged* experts as data for *asocial* phenomenological descriptions of expertise. On the other hand, he portrays mundane examples of everyday action, such as driving a car, walking, talking, and carrying on a conversation, as paradigmatic instances of how experts behave, even though these activities would not normally be socially recognized as being performed by experts (Dreyfus and Dreyfus, 1986, p. 30). By arguing that the *same type* of expertise exists in both extraordinary performances of skill that are socially recognized as occurring at the expert level and mundane performances of skill that are not recognized as occurring at the expert level, Dreyfus advances his end of demystifying the seemingly magical quality of expertise and establish continuity between expert and everyday lifeworld activity. In both instances, for him, expertise is a pre-reflexive relation between skill and environment: "An expert's skill has become so much a part of him that he need be no more aware of it than his body." (Dreyfus and Dreyfus, 1986, p. 30)

From Dreyfus's perspective, social problems can arise when social agents are unable to recognize the essential qualities of expertise that are rendered explicit by the *phenomenological* investigator. These social problems arise from a different source than expertise itself; they are brought about by the *interference* of political, social, and cultural involvements. But the hermeneutic circle would turn this around; these involvements *give rise to* recognizing and trusting experts in the first place. Even if one defers to the traditional "experts", that deference, too, has been brought about as a result of one's particular background and lifeworld involvements. For expertise is a two-way relation: the *claim to* expertise itself involves a social demand; it is not merely a neutral identification label but a declaration that others should defer to the expert's judgments. The phenomenon of expertise, therefore, is ultimately and inextricably tied to its social utility; an expert is not only "in" a field but "for" an audience.

An obvious example is the "selection" problem of experts. In many instances experts who endorse different conclusions within the same field can be pitted against one another as "counter experts". A common strategy for "counter experts" consists in claiming that the judgment proffered by the expert is tainted due to the presence of prejudices, ideologies, or hidden agendas. "Counter experts" are particularly prevalent in the legal arena where expert witnesses function in accord with the logic of an adversarial system by proffering testimony for both the prosecution and defense. [25]

Stephen Turner's account of expertise addresses the recognition issue by pointing out that, in order for someone to be an expert, he or she needs not only to be skilled, but also to have an audience that socially recognizes his or her type of skills as skilled expertise (Turner, 2001, p. 138). [26] Turner contends that although what Merton calls "cognitive authority" is neither an "object that can be distributed" nor something that can be "simply granted", it nevertheless is "open to resistance and submission" based upon the evaluations of different audiences (Turner, 2001, p. 128). [27] By contrast, Dreyfus, who defines an expert solely on the basis of skill acquisition and use, methodologically excludes the audience from the description of experts and expertise, defining an expert as one with the right affective comportment and intuitive response to the situation.

This methodological move has many advantages over Dreyfus's phenomenology. It explains why: (1) expertise is not a stable property, but can be gained and lost; (2) why discussions of expertise tend not only to focus on epistemological, but also political issues; and (3) why perceptions of expertise can be based on historical transformations. In connecting expertise with these, Turner does not relinquish the philosophical aim of revealing general structures and transcending particularity (i. e. expertise can only be discussed with reference to local features). For example, he provides the following "taxonomy" that accounts for five general kinds of distinguishable experts:

Experts who are members of groups whose expertise is generally acknowledged, such as physicists; experts whose personal expertise is tested and accepted by individuals, such as the authors of self-help books; members of groups whose expertise is accepted only by particular groups, like theologians whose authority is only accepted by their section; experts whose audience is the public but who derive their support from subsidies from parties interested in the acceptance of their opinions as authoritative; and experts whose audience is bureaucrats with discretionary power, such as experts in public administration whose views are accepted as authoritative by public administrators (Turner, 2001, p. 140).

Turner's account thus relates the phenomenon of expertise to changing historical and social perceptions. Dreyfus might well retort that there is no reason why his phenomenological model should be understood as incompatible with Turner's—that

Turner simply expands on Dreyfus's model, with Dreyfus describing what expert skills are and Turner describing how different audiences come to recognize these skills as occurring at the expert level. But the critical methodological difference is that Turner does not treat the possession of skill as a necessary *and* sufficient condition of expertise, omitting reference to an audience, while Dreyfus does. [28]

Dreyfus's failure to address the recognition problem is highlighted by the following paradox. On the one hand, he argues that only an expert can recognize another person as a genuine expert, for nonexperts do not know what to look for when evaluating whether someone is skilled. [29] It takes an expert—and only one—to know one. This might be called a *difference* claim (DC): experts are "not like us". DC is fairly innocuous, and even Turner adheres to a version: "[It] is the character of expertise that only other experts may be persuaded by argument of the truth of the claims of experts; the rest of us must accept them as true on different grounds than other experts do." (2001, p. 129) On the other hand, Dreyfus also advances a *similarity* claim (SC), according to which "they" —the experts—are very much "like us". They behave in a similar way the rest of us do in our everyday activities. At the most basic level of everyday coping, everybody deserves to be characterized as an expert: "Citizens will be speaking in terms of their expertise, whether they are university professors who have expertise in foreign cultures doing business with their state or farmers or small-store owners speaking about concrete problems that need legislative solution." (Dreyfus, Spinosa and Flores, 1997, p. 107) Thus we can projectively identify with experts, and understand the kind of knowledge they use in their judgments. The problem is not only that Dreyfus needs to advance both, it is that, in the end, he needs SC to trump DC. Otherwise, nonexperts would lack any basis to recognize, accept, and trust the kind of knowledge that experts possess.

This point can be exposed most directly by reference to the situation depicted in Ibsen's play *Enemy of the People*. The central character, Thomas Stockmann, is a doctor at a spa on which the livelihood of his town depends. He thinks an invisible poison is polluting the spa's water, and confirms it by sending water samples to an expert at the University for analysis. He seeks to inform members of the community, thinking they will thank him for bringing the danger to their attention. But a friend warns him not to be so sure how they will respond: "You're a doctor and a man of science, and to you this business of the water is something to be considered in isolation. I think you don't perhaps realize how it's tied up with a lot of other things." Stockmann, citing the expert, insists that the "shrewd and intelligent" people will be "forced" to accept the news. The mayor points out that, for citizens, the matter is more economic than scientific; and indeed, the citizens condemn Stockmann as an "enemy of the people". The basic conflict thus involves a scientist who accepts expert technical advice as authoritative, versus citizens who do not find that expert advice authoritative, who find *it* threatening to their world, and who seek guidance from others.

Ibsen's invented situation is like a model which strips away inessential details to clarify the essential forces of a situation. The situation involves a volatile controversy with a scientific-technical dimension in which people have lost confidence in traditional "experts". From the audience position, we see that two kinds of stories can be told about such a situation from two different perspectives. In the *expert's* perspective, that of Stockmann and the University chemist, there is no uncertainty or grounds for contestation regarding whom the expert is, what kind of training that person requires, and what kind of information (the technical issue of the quality of the water) is relevant to maximizing the good of that particular social situation. The experts are like the citizens in that they want to maximize the good of the town; they do so with their special knowledge. Failure of the citizens to recognize the expert, and defer to the expert's advice, is due to the peripheral, and even deleterious and corrupting, influence of economic and political motives and the involvements of the citizens with politicians, the media, and other nonexpert authorities. In the *citizen's* perspective, on the other hand, things are much more confused. The background and lifeworld involvements of citizens mean that economic and political motives loom much larger than they do for Stockmann and the chemist, and suggest different people to whose advice they should defer in seeking to advance their welfare. In this perspective, the purported "experts" — Stockmann and the chemist—are precisely *not* like the other citizens for they have different agendas (and the chemist is even literally an outsider).

We see the difference between these two situated perspectives clearly from a *third* perspective, sitting in the audience. This third perspective is also not neutral, and its effect is to dramatize the similarity claim; we recognize the common humanity between Stockmann and the citizens, and the relevance of his knowledge to their welfare. One realizes that the welfare of the citizens requires that they defer to the experts, even as one fully appreciates the severity of the conflict and the impossibility of its resolution. But we in the audience have no doubt about who the real experts are, and the difference between essential and peripheral involvements. The person in the audience is not standing anywhere, not situated with respect to this aspect; this third position, in short, is not a hermeneutically sensitive one.

In any real controversy, however, no one occupies the audience position; everyone is as it were "on stage" in a situated position, standing someplace with *particular* involvements which give rise to a *particular* understanding of the situation and, with it, an inclination to accept some people rather than others as authoritative. It is not a priori clear which involvements are essential and which particular; whose actions are in the grip of prejudices, ideologies, and hidden agendas, and whose are in the best interests of society. Meanwhile, experts are also in a particular position, standing someplace and with particular involvements, and the claim to expertise is a charge, a valence, a demand that one should be deferred to. Each real life controversy involving expertise takes the form of a jockeying between those who advance claims of expertise to advance their authority and those seeking the right authorities to whom to defer. This is the hermeneutic predicament: there is no escape from the particular involvements of a given situation. There is and can be no talisman for expertise.

This problem is most visible in connection with volatile public controversies involving a technical dimension—and especially involving public safety—where traditional sources of authority have become distrusted. In these cases, the question of who speaks authoritatively, of who is an expert, is contested. Each citizen, and each person proposed as an expert, has a particular set of involvements, and there is no safe audience position from which to sort out the essential and inessential involvements in an expert's judgment.

Consider, for instance, the case of the closing of the National Tritium Labeling Facility (NTLF) —where, as it happened, Dreyfus's wife Genevieve worked. The facility created unique tagged molecules by putting tritium into specific molecular positions, creating tracers that are used to study mechanisms of biochemical transformation in basic and applied research. But antinuclear activists objected to the fact that some tritium was released to the environment in the process. Scientists and local, state, and federal public health experts, after carefully examining the situation, said the emissions were safe, a fraction of the Environmental Protection Agency suggested limit, which in turn was a fraction of the background level. But the anti-nuclear activists sought to discredit those claims, saying that they were either made by those who worked at the facility, or from institutions connected in some way with that facility, or by scientists who knew too much about tritium to be disinterested. The activists were effective at disrupting the normal socially negotiated procedures for who speaks authoritatively about safety, and the facility was closed (Crease, 2002a).

Or again, consider the controversy surrounding the shipment of spent fuel rods from Brookhaven National Laboratory's High Flux Beam Reactor in 1976 (Crease, 1999). The controversy pitted activists, who associated research reactors with power reactors/nuclear weapons/the military, versus scientists for whom such associations made no sense at all. The situation spawned a spectrum of experts of the sort described by Turner's taxonomy. In yet another controversy involving the Brookhaven lab, a program it ran studying the health of Marshall Islanders accidentally exposed to fallout from a nuclear weapons test came under attack, with one complicating factor being that it involved the classic colonialist situation of U. S. scientists working in a third-world country whose language and customs they were not familiar with (Crease, 2002c).

Each of these controversies was complex, and involved high stakes. Each turned on a scientific-technical issue, and thus necessary recourse to expertise. Yet who the "real" experts were deemed to be depended on who one was and where one stood. To understand such controversies involving experts and expertise requires moving beyond the practical expert's point of view.

7. Conclusion

Dreyfus's model of expert skill acquisition is philosophically important because it shifts the focus on expertise away from its social and technical externalization in STS, and its relegation to the historical and psychological context of discovery in the classical philosophy of science, to universal structures of embodied cognition and affect. In doing so he explains why experts are not best described as ideologues and why their authority is not exclusively based on social networking. Moreover, by phenomenologically analyzing expertise from a first person perspective, he reveals the limitations of, and sometimes superficial treatment that comes from, investigating expertise from a third person perspective. Thus, he shows that expertise is a prime example of a subject that is essential to science but can only be fully elaborated with the aid of phenomenological tools.

However, both Dreyfus's descriptive model and his normative claims are flawed due to the lack of hermeneutical sensitivity. He assumes, that is, that an expert's knowledge has crystallized out of contextual sensitivity plus experience, and that an expert has shed, during the training process, whatever prejudices, ideologies, or hidden agendas that person might have begun with. This assumption not only flaws in Dreyfus's descriptive account but his normative account as well.

The phenomenological goal is to expose presuppositions that lurk unapprehended in the natural attitude. The phenomenological experience reveals that the most difficult presuppositions to expose and get a grip on are those closest to home. In this spirit, though possibly with a trace of perversity, one might expose this limitation by posing the following question: "Why does Genevieve Dreyfus no longer work at the National Tritium Labeling Facility?"

From Dreyfus's own account, it is because of a breakdown, a corruption of a legitimate and phenomenologically justifiable authority. A group of anti-nuclear crusaders managed to hijack a socially negotiated process and persuade the administrative authorities to overlook robust science and expert advice, and make a purely political decision. But from another, more all-too-common, highly important, and potentially powerful point of view, one could, with hermeneutic sensitivity, tell quite a different story. Individuals do not take in an item of information, even scientific information, nakedly. It matters who conveys that piece of information and in what context. The Berkeley anti-nuclear activists live in a different interpretive world than the scientists, with a different set of supporting behaviors, values, and institutions; their meaning-generating process by which they interpret facts, principles, and their application is very different (Crease, 1999, p. 498). The difference between the scientists and activists thus cannot be regarded as one between knowers and those who corrupt or betray that knowledge, but more like the relation between the members of one culture and another. Surely a description of this hermeneutic predicament, missing from Dreyfus's own, belongs in any account of expertise.

Finally, we believe that when evaluated from an immanent perspective, Dreyfus's account of expertise fails according to his own philosophical standards. In many publications Dreyfus champions Heidegger's hermeneutic approach to phenomenology over Husserl's. He even argues that Heidegger's account of authentic Dasein in Division II of *Being and Time* entails that Dasein is an "expert" (Dreyfus, 2000). However, by *approaching trust as a static scenario* in which a nonexpert solicits advice from an expert, Dreyfus resorts to an implicit Husserlian schema of intersubjectivity. The essence of SC is an analogy: just as I do not expect myself to be able to articulate rules for how to drive or ride a bicycle, I should not expect experts, such as ballistics examiners, nurses, and ecologists, to justify their decisions by referring to rules. Dreyfus implicitly argues that even though trust is established at the social level on the basis of an expert's track record, at the phenomenological level, it is established through what Husserl calls "intersubjective pairing". By making a pre-reflexive analogy from my behavior to the behavior of an expert, I recognize that an expert's behavior is essentially similar to my own. The expert is an expert "alter ego". In the Dreyfusian scheme, I should trust experts because: (1) I trust myself to make decisions in a similarly intuitive manner, and (2) I trust my decisions to be good ones, even though I, like an expert, cannot propositionally justify them according to rules. Even though an expert has more training than I do, our cognitive similarities outweigh our technical differences.

Dreyfus's account of intersubjectivity here, with its lack of hermeneutic sensitivity, recalls Husserl more than Heidegger. Indeed, his portrait of the expert, who masters his or her own relation to expertise by making it through all of the developmental stages, and who feels no need to seek external ratification of his or her own abilities, evokes the caricature of Heidegger according to which authentic Dasein is accorded too much heroic freedom. But Heideggerian sensitivity to the hermeneutic dimensions of worldhood would have to depict an expert as engaged in a much more fragile and vulnerable process, in which expertise does not appear as a destination in which the individual surpasses one's embeddedness in that world. Expertise is always a process of becoming, and, in principle at least, it will always be possible for coaches, commentators, and others whose own expertise overlaps with the expert's to disclose aspects of expert performance which escape the grasp even of that performer.

Notes

1. Turner discusses the bases of Stanley Fish, Jürgen Habermas and Michel Foucault's concerns about why expertise undermines liberal democracy. He argues that these concerns "depend upon a kind of utopianism about the character of knowledge that social constructionism undermines" (Turner, 2001, p. 147, n. 7).

2. The problem of certifying expert witnesses is frequently discussed in relation to the issue of "junk science" (See Black, Ayala and Saffran-Brinks, 1994; Jasanoff, 1992; Caudill and Redding, 2000; Huber, 1991).

3. When theorists assume an expert social role based upon the prescriptive dimensions of

their research then the problem of what Winner calls the "values expert" emerges (Winner, 1995, pp. 65-67). Winner argues that applied theorists not only provide counsel in "interminable" debates, but tend to misconstrue the audiences they address as an overly generalized "we", fail to recognize how social change is instituted, and occasionally help legitimate already made political decisions.

4. "Black box" is an engineering term used in science studies analogously to the Marxist concept of hegemony. "Black boxing", like hegemony, refers to background assumptions that are generally regarded as self-evidently true and not requiring further investigation (Feenberg, 1995, p. 7). To open the "black box" of scientific expertise is to show, through close empirical and conceptual analysis, that what appears to be self-evidently true, culturally sanctioned, and not requiring further investigation about experts is false, hidden from cultural scrutiny, and in dire need of critical analysis. When the "black box" of expertise is opened, STS theorists contend that experts do not emerge as self-sufficient geniuses whose knowledge is infallible, certain, and objective. Scientific experts are revealed rather to be non-extraordinary, biased people whose successes and failures emerge from working within a competitive network of distributed knowledge and prestige. From the STS perspective, it is principally because the network's operation is rarely explicitly described and theoretically examined that nonexperts mistakenly perceive scientific experts as more knowledgeable, authoritative, and trustworthy than is appropriate.

5. STS has occasionally focused on the tacit knowledge involved in the production of scientific results. For example, Harry Collins recently sought to expand Michael Polanyi's classification of tacit knowledge by expanding its range of application beyond skills to the production of scientific results in empirically verifiable cases (2001). Proposed as a general thesis, Polanyi claims that in order to understand how to use a machine one needs to know how its components fulfill their function together (1974). On the basis of field work in the United States and observations of experimental work in Glasgow University, Collins claims that the problem of tacit knowledge accounts for why twenty year old Russian measurements of the quality factors (Q) of sapphire have only just been repeated in the U. S. Collins's focus on tacit knowledge allows him to stress the importance of personal contact and trust between scientists. It also allows him to suggest that information that is not currently contained in experimental reports should be added to increase the likelihood of reproducing scientific findings. Despite some overlap, the examples that Collins and Dreyfus focus on differ in one key

respect. Collins focuses on cases in which the tacit knower produces intersubjectively accepted results; his emphasis is on the routine production of these results, and the impediments that prevent them from being easily reproduced. What Collins does not present is a theory of embodiment that explains how knowers are capable of tacitly mastering equipment.

6. Abstract contexts alone do not allow scientists to mathematize, model, and formalize the world. Body oriented skills are used to operate the technological instruments that stabilize phenomena in order for scientists to manipulate and interpret them. Ihde repeatedly argues that as with mundane uses of technology, the technological instruments used in scientific settings extend and transform bodily praxes through "embodiment relations"; they are absorbed and incorporated into bodily experience of the world like Heidegger's hammer or Merleau-Ponty's blind person's cane, and the phenomena that scientists can produce change as the forms of embodiment change (Ihde, 1998, pp. 42-43).

7. Dreyfus defines "intentional arc" and "maximal grip" as follows: "The *intentional arc* names the tight connection between the agent and world, viz. that, as the agent acquires skills, those skills are 'stored', not as representations in the mind, but as dispositions to respond to the solicitations of the world. *Maximal grip* names the body's tendency to respond to these solicitations in such a way as to bring the current situation closer to the agent's sense of an optimal gestalt. Neither of these abilities requires mental or brain operations." (1998; 1999a).

8. Although we believe that Dreyfus turns to the theme of "brain topics" simply to establish another perspective that can be made compatible with his phenomenological approach, Sheets-Johnston argues that the reference to neural networks and brain functions contradicts his phenomenological aspirations: "[We] find that any erstwhile sense we might have had of a phenomenological subject has given way to a neurological one, while at the same time we find a phenomenological subject to be ostensibly fully present but transformed in ways utterly foreign to our experience in that it 'presumably senses' its own 'brain dynamics'." (2000, p. 357) Moreover, she argues that, "Not only can they [neural nets] not distinguish between formal and informal learning, but the very vocabulary by which they operate is not simulation-friendly to the latter kind of learning and is thus inappropriate to its description and deflective to its understanding" (Sheets-Johnstone, 2000, pp. 357-358). We believe that what Sheets-Johnston forgets is that some still consider phenomenology to be a subjective

form of analysis and that the reason why Dreyfus probably gestures towards neural nets is to demonstrate that phenomenology can reveal objective structures of bodily praxis.

9. During the 1960s, when Dreyfus first formulated his critique of artificial intelligence and its failed hype, the intellectual atmosphere at the Artificial Intelligence Laboratory at the Massachusetts Institute of Technology was overly hostile to recognizing the implications of what he said, and as a result, he almost lost his job. By contrast, Winograd notes that today, "some of the work being done at that laboratory seems to have been affected by... Dreyfus" (1995, p. 110).

10. According to Dreyfus, the primary desideratum of phenomenology has always been to adequately describe expertise (even if historical phenomenologists such as Husserl, Heidegger, and Merleau-Ponty did not use the term "expertise" in their analyses), because at bottom, phenomenology aims at getting behind the prejudices that impede how human experience is understood (Dreyfus and Dreyfus, 1986, pp. 2-5).

11. Sheets-Johnston, though, argues that by analytically separating intellectual from bodily skills Dreyfus assumes a "pernicious" "Cartesian split" between mind and body by forgetting how bodily skills are foundational for intellectual skills (2000, pp. 355-356).

12. Although there are many surface similarities between Dreyfus and Polanyi (Polanyi, 1974) on the issue of tacit knowledge, there also is a notable difference. Dreyfus argues that while Polanyi recognizes that formalisms cannot account for the tacit performance of riding a bicycle, he still believes that such performance is governed by "hidden rules": "The reference to hidden rules shows that Polanyi, like Plato, fails to distinguish between performance and competence, between explanation and understanding, between the rule one is following and the rule which can be used to describe what is happening." (1992, pp. 330-331) Dreyfus also differs from Kuhn's analysis of tacit knowledge since this analysis primarily situates tacit dimensions of knowledge in the general structure of a paradigm (Stengers, 2000, p. 6). For other writings on tacit knowledge see Rawls, 1968; Reber, 1995; Searle, 1983; 1992.

13. The only time Dreyfus alters his five-stage model is when he adds two additional stages—mastery and wisdom—in his recent reflections on the Internet (2001).

14. Dreyfus takes his target group involved in making "unstructured' decisions to be paradigmatic of the "typical learner" and classifies "a common pattern" observable in their behavior (Dreyfus and Dreyfus, 1986, p. 20). The adjective "typical" denotes

a class of learners who "possesses innate ability" and also have "the opportunity to acquire sufficient experience " (Dreyfus and Dreyfus, 1986, p. 20) . "Unstructured" and " structured" are standard terms discussed in theories of information management. They are used to classify organizational differences in the range of decision making, usually with "semi-structured" appearing as a mid-point in this range. When Dreyfus references "structured" decisions, he means the type of decisions that are made when "the goal and what information is relevant are clear, the effects of the decisions are known, and verifiable solutions can be reasoned out" (Dreyfus, 1986, p. 20) . In other words, "structured" decisions involve well-understood situations that have a common procedure for handling them. "Structured" decisions can be found in situations that are repetitive, routine, and have the pertinent pieces of evidence remaining stable as time passes. Examples of "structured" decision making can be found in situations where "context-free" deliberation dominates, such as "mathematical manipulations, puzzles, and, in the real world, delivery truck routing and petroleum blending" (Dreyfus and Dreyfus, 1986, p. 20) . In contrast with " structured " decisions, Dreyfus characterizes "unstructured" ones as: intuitive, commonsensical, heuristic, and involving trial and error approaches. He states they have a tendency to be ad hoc, are not programmable, and "contain a potentially unlimited number of possibly relevant facts and features, and the ways those elements interrelate and determine other events is unclear" (Dreyfus and Dreyfus, 1986, p. 20) . The situations that require "unstructured" decisions typically are elusive ones: where it is not possible to specify in advance most of the decision procedures to follow; where the decision maker must provide judgment, evaluation, and insights into the problem definition whose parameters cannot be precisely identified; where unquantifiable factor are central; and where there is no agreed-upon procedure for making decisions. These types of decision are routinely made by people in "management, nursing, economic forecasting, teaching, and all social interactions " and require " considerable concrete experience with real situations" (Dreyfus and Dreyfus, 1986, p. 20) .

15. Dreyfus writes: "Although irrational behavior. . . , should generally be avoided, it does not follow that behaving rationally should be regarded as the ultimate goal. A vast area exists between irrational and rational that might be called *arational*. The word *rational*, deriving from the Latin word *ratio*, meaning to reckon or calculate, has come to be equivalent to calculative thought and so carries with it the connotation

of 'combining component parts to obtain a whole'; arational behavior, then, refers to action without conscious analytic decomposition and recombination. *Competent performance is rational*; *proficiency is transitional*; *experts act arationally.*" (1986, p. 36)

16. As Michael Callon puts this point: "Researchers in the wild participate in the subversion of modern institutional framing by challenging the oppositions that we had come to take for granted, yet that are crucial, such as the distinction between expert and layperson." (Callon, 2001)

17. This emphasis on affect is important for Dreyfus's criticism of so-called "expert" computer systems, which he argues can approximate competent human performance. Skill acquisition for Dreyfus involves not only a cognitive acquisition by the subject but also an affective transformation that computers cannot experience. This point is especially significant in serving to highlight Dreyfus's phenomenological background. Phenomenologists have long argued that to be a subject is to have an intentional relation to the world such that changes made to the subject correlate to changes made to the subject's world. The subject who goes through the developmental process of expert apprenticeship is not the same subject as the one who began the process, and the world is not meaningful for the subject in the same way; experts and nonexperts are indeed different subjects. They are different types of *people* who deliberate and feel differently, and to whom the world responds differently. Not only can experts do more things than beginners, but their whole affective demeanor changes. The way that experts care about their activities in a practice changes from when they were beginners, progressing from relative detachment to engaged commitment. This is why Dreyfus characterizes the five developmental stages as "qualitatively different perceptions" of what a task is and what mode of decision making is appropriate to handling that task (Dreyfus and Dreyfus, 1986, p. 19).

18. Although not every practice provides every beginner with an opportunity to achieve expert level mastery, many do: "Not all people achieve an expert level in their skills. Some areas of skill have the characteristics that only a very small fraction of beginners can ever master the domain of." (Dreyfus, 1986, p. 21)

19. Brewer, like many other philosophers and legal theorists, appeals to the "point of view" as an analytical device that captures how perspective and justification relate (1998, PP. 1568-1570). He turns to the point of view in order to articulate a

common theoretical perspective that applies to all scientific experts—and that is arguably generalizable to all experts in all fields of expertise—regardless of which particular scientific field the practitioner is an expert in. He writes: "One invokes a point of view to justify some claim. To serve this justificatory function, the point of view is assumed to be a reliable method for achieving the (explicit or implicit) aims of some rational enterprise" (Brewer, 1998, p. 1575). Brewer's line of reasoning is that one turns to a point of view to rationally justify either a theoretical claim about what ought to be believed or a practical claim about what action ought to occur. What is distinctive about this type of validation is that it relates the justification of a claim to a distinct, yet "reliable" method, which is chosen to achieve a specific cognitive aim. The "reliable method" common to all "rational" points of view is defined in terms of two characteristics: "enterprise" and "axiological" conceptions. An "enterprise" is defined as the choice and use of particular methods of analysis in order to serve specific cognitive goals. He acknowledges that even within the "same generic enterprise" practitioners can disagree about the "proper specific aims of the enterprise" (Brewer, 1998, p. 1571). But such a disagreement will take place within a "holistic" network involving an "axiological" component. When Brewer discusses the "axiological" dimensions of justification, he does so in Larry Laudan's sense of the term. Where Laudan analyzes scientific reasoning, he distinguishes between "factual", "methodological", and "axiological" levels of analysis. Factual analysis focuses on what exists in the world, including theoretical and unobservable entities for scientists. At the methodological level, practitioners in a given field share precise as well as vague rules. For scientists, this can include vague rules, such as avoid ad hoc explanations, to precise rules, such as calibrate instrument 'x' to standard 'y'. The axiological level, which is often explicable in the form of rules, designates cognitive aims. Brewer, like Laudan, argues that the relation between facts, methods, and axiological aims should not be understood as a "simple linear hierarchy", but as "reticular" "constraints [that] are multidirectional within the holistic network of aims, methods, and beliefs" (Brewer, 1998, p. 1575). Facts, methods, and axiological aims relate in a "multidirectional" sense because each can "constrain" the other, without any one of the three being a priori valued more than the others. Facts, methods, and axiological aims relate in a "holistic" sense since the point of view they collectively constitute is comprised of all three features relating together. In other words, a point of view is a complete and systematic perspective, irreducible to isolated observations. Finally, due to the holistic nature of a point of view, Brewer describes its epistemic status as "understanding" and not "knowledge." "The important difference between knowledge and understanding", Miles Burnyeat claims and Brewer repeats, "is that knowledge can be piecemeal, can grasp isolated truths one by one, whereas understanding always involves seeing connections and relations between the items known" (Brewer, 1998, p. 1591).

20. Dreyfus argues that the relation between phenomenology and normativity concerns "priority", in the sense that normative obligations are ascertained by phenomenologists "prioritizing" how agents do in fact respond to concrete situations. This "priority" is also, according to Dreyfus, held by those in the Hegelian tradition of *Sittlichkeit*, such as Bernard Williams, Charles Taylor, and Carol Gilligan. By contrast, Dreyfus argues, a group of theorists in the Kantian tradition of *Moralität*, such as Jürgen Habermas, John Rawls, and Lawrence Kohlberg, "prioritize" detached principles that detail what the right thing to do is over an understanding of the empirical conditions that allow for certain decisions to be made (1990, p. 237).

21. This is an important corrective to Paul Feyerabend, who for example, misses this point. He argues that when the prestige of science does not demand excessive complication, and social circumstances are such that experts cannot be overly esteemed, such as during wartime, medicine is capable of being effectively simplified. He claims that evidence for this can be found in army recruits who have historically proven themselves capable of being instructed as physicians with only half a year of training (Feyerabend, 1987, p. 307). The point that Feyerabend misses is that army recruits may quickly become instructed to be competent at various aspects of medicine, such as triage, but this training does not produce expert physicians or undermine how normal training allows expert physicians to do more than quickly trained ones.

22. It is often desirable that experts defend their recommendations against other experts, or in some way be cross-examined so that those affected can question their presuppositions. If this is taken to mean that the expert must articulate his values, rules, and factual assumptions, examining becomes a futile exercise in rationalization in which expertise is forfeited and time is wasted (Dreyfus and Dreyfus, 1986, p. 196).

23. Dreyfus writes: "Society... must encourage its children to cultivate their intuitive

capacities in order that they might achieve expertise, not encourage them to become human logic machines. And once expertise has been attained, it must be recognized and valued for what it is. To confuse the common sense, wisdom, and mature judgment of the expert with today's artificial intelligence, or to value them less highly, would be genuine stupidity" (Dreyfus and Dreyfus, 1986, p. 201).

24. At one point, Dreyfus suggests that trust can be obtained legitimately if experts provide a certain type of narrative. When he makes this suggestion, Dreyfus seems to mitigate his IT thesis. He suggests experts can be effective communicators, so long as they do not need to provide deductive accounts of rules they followed when making decisions: "The cross-examination of competing experts in an intuitive culture might take the form of a conflict of interpretation in which each expert is required to produce and defend a coherent narrative which leads naturally to the acceptance of his point of view" (Dreyfus and Dreyfus, 1986; p. 196). This passage is interesting because it suggests that experts are capable not merely of producing, but "defending" something called "coherent narrative" and that this narrative may be evaluated in an intuitive culture without endangering expertise. The passage is problematic because Dreyfus fails to explain what a "coherent narrative" is and why it is so efficacious as to "lead naturally" (another undefined phrase) to acceptance of a "point of view". The implication is that somehow experts can avoid the problems of IT, which are connected with what I called a "practical point of view", by presenting another kind of viewpoint in their "coherent narrative". The word "coherent" can be taken to suggest holism, and it is possible that Dreyfus has in mind something like Brewer's "theoretically holistic" expert point of view. But if Dreyfus is evoking theoretical holism, then he fails to explain how he can expect that experts, who according to IT "forfeit" their expertise when producing prepositional content, are capable of providing such a narrative *qua* experts. The guiding presupposition is that whatever this "coherent narrative" is, it is only unproblematic for experts to produce in an "intuitive culture", such as Japan. Not only does this presupposition seem to be predicated upon unsupported cultural essentialism, but it also calls RT into question. Nonexperts may not be able to recognize experts on the basis of deductive procedural steps, but Dreyfus indicates recognition can occur on the basis of a "natural" response to the content of a "coherent narrative". Without explaining what these are, Dreyfus begs the question as to why nonexperts should trust expert intuition.

25. As Shelia Jasanoff points out, the issue of selecting real experts is so difficult in the courtroom that the original 1923 "general acceptance" criteria employed in *Frye v. United States* had to be refined because it "did not provide guidance as to how much agreement was enough, or among who" (Jasanoff, 1995, p. 62) . General acceptance, which is an implicit presupposition of RT, did not legally work because it failed to: (1) clarify the degree of consensus required for establishing general acceptance, (2) set guidelines for how the contradictory results produced by variations in boundary work should be resolved, and (3) determine how to weigh results provided by frontier as opposed to established science (Jasanoff, 1995, p. 62) . Since general acceptance proved to be a vague and ineffective standard for judging expert consensus in the legal arena, by extension it is problematic for Dreyfus to implicitly connect IT with RT. They are also routinely used in journalism, where sensationalism is generated by showing that a problem is so complicated that experts cannot agree on how to solve it. But for Dreyfus, this is not an issue. Dreyfus overestimates the overall trustworthiness of experts because he analyzes them as a general category and therein fails to recognize that expert consensus is field specific. Since the level of expert consensus is field specific it follows that the trustworthiness of expert intuition is not something that can be addressed in Dreyfus's general terms.

26. This does not mean that an audience of comparable scale recognizes every field of expertise. For example, experts in physics are more widely recognized as possessing expert skills than are theologians, who are only recognized by a particular sect, which Turner calls a "restricted audience" (Turner, 2001, p. 131) .

27. For example, when discussing the relation between massage therapists and recognition, Turner notes that some people feel they benefit from massage therapy, whereas others do not find the promise of massage therapy to be fulfilled. The massage therapist is thus only considered to be (or more strongly put, only is) an expert for the former audience: "So massage therapists have . . . a created audience, a set of followers for whom they are expert because they have proven themselves to this audience by their actions. " (Tuner, 2001, p. 131)

28. One may nevertheless be concerned that by insisting upon recognition as an essential dimension of expertise Turner undermines the objectivity of expertise. History is replete with examples of people who at one point are acknowledged to be experts and at other moments are denounced as charlatans. Turner acknowledges, " [What]

counts as 'expert' is conventional, mutable, and shifting, and that people are persuaded of claims of expertise through mutable, shifting conventions" (2001, p. 145). Nevertheless, he claims that the prerogative to revise fallible judgments is a crucial part of democratic life and that to insist on a higher standard is "Utopian" (Turner, 2001, p. 146).

29. Again, it is helpful to consider Dreyfus's examples. An example of DC can be found in Dreyfus's description of an experiment in which students, experienced paramedics, and CPR instructors watched videotapes of six exemplars (five students and one experienced paramedic) giving CPR to patients. This target group was then asked which of the exemplars he or she would choose to save his own life. Dreyfus writes: "The results were revealing. In the paramedic group, nine out of ten selected an experienced paramedic. The students chose the paramedic five times out of ten. The instructors, attempting to find a paramedic by looking for the individual closely following the rules they were taught, failed to find the expert because an experienced paramedic has passed beyond the rule-following stage!" (Dreyfus and Dreyfus, 1986, p. 201)

Bibliography

Barbour, I. (1993). *Ethics in an Age of Technology*. San Francisco: Harper Collins.

Black, B., Ayala, F., and Saffran-Brinks, C. (1994). "Science and the Law in the Wake of Daubert." *Texas Law Review* 72: 715-802.

Brewer, S. (1998). "Scientific Expert Testimony and Intellectual Due Process." *The Yale Law Review* 107 (4): 1535-1681.

Callon, M. (2001). "Researchers in the Wild and the Rise of Tehnical Democracy." Paper presented at *Knowledge in Plural Contexts*, Science and Technology Studies, Université de Lausanne, Switzerland.

Caudill, D. and Redding, R. (2000). "Junk Philosophy of Science? The Paradox of Expertise and Interdisciplinarity in Federal Courts." *Washington and Lee Law Review* 57 (3): 685-766.

Collins, H. M. (1995). "Humans, Machines, and the Structure of Knowledge." *Stanford Humanities Review* 4 (2): 67-83.

Collins, H. (2001). "Tacit Knowledge, Trust, and the Q of Sapphire."

Social Studies of Science 31：71-86.

Crease，R. P. （1997）．" Hermeneutics and the Natural Science：Introduction. " In R. Crease（ed. ），*Hermeneutics and the Natural Sciences.* Dordrecht：Kluwer，PP. 1-12.

Crease，R. P. （1999）．" Conflicting Interpretations of Risk：The Case of Brookhaven's Spent Fuel Rods. " *Technology* 6：495-500.

Crease，R. P. （2001）．" Anxious History：The High Flux Beam Reactor and Brookhaven National Laboratory. " *Historical Studies in the Physical and Biological Sciences* 32（1）：41-56.

Crease，R. P. （2002a）．" Conrpromising Peer Review. " *Physics World*，January 2002，17.

Crease，R. P. （2002b）．" The Pleasure of Popular Dance. " *Journal of the Philosophy of Sport* 39（2）：106-120.

Crease，R. P. （2002c）．" Fallout：Issues in the Study，Treatment，and Reparations of Exposed Marshall Islanders. " In R. Figueroa and S. Harding（eds. ），*Exploring Diversity in the Philosophy of Science and Technology.* Routledge（forthcoming）．

Dreyfus，H. （1967）．" Alchemy and Artificial Intelligence. " *Rand*，Paper P. 3244.

Dreyfus，H. and Dreyfus，S. （1985）．" From Socrates to Expert Systems：The Limits of Calculative Rationality. " In C. Mitcham and A. Huning（eds. ），*Philosophy and Technology 11：Information Technology and Computers in Theory and Practice.* Boston：D. Reidel Publishing Company，PP. 111-130.

Dreyfus，H. and Dreyfus，S. （1986）．*Mind Over Machine：The Power of Human Intuition and Expertise in the Era of the Computer.* New York：Free Press.

Dreyfus，H. and Dreyfus，S. （1990）．" What is Morality? A Phenomenological Account of the Development of Ethical Expertise. " In D. Rasmussen（ed. ），*Universalism vs. Communitarianism：Contemporary Debates in Ethics.* Cambridge：MIT Press，PP. 237-264.

Dreyfus，H. （1991）．*Being-in-the-World：A Commentary on Heidegger's Being and Time.* Cambridge：MIT Press.

Dreyfus，H. （1992）．*What Computers Still Can't Do：A Critique of Artificial Reason.* Cambridge：MIT Press.

Dreyfus，H. ，Spinosa，C. ，and Flores，F. （1997）．*Disclosing Worlds：*

Entrepreneurship, *Democratic Action*, *and the Cultivation of Solidarity*. Cambridge, MIT Press.

Dreyfus, H. (1998) . "Intelligence Without Representation. " *Network for Non-Scholastic Working Paper*, Department of Philosophy, Aarhus University, Denmark.

Dreyfus, H. (1999a) . "How Neuroscience Supports Merleau-Ponty's Account of Learning. " Paper presented at the *Network for Non-Scholastic Learning* Conference, Sonderborg, Denmark.

Dreyfus, H. (1999b) . " The Primacy of Phenomenology over Logical Analysis. " *Philosophical Topics* 27 (2): 3-24.

Dreyfus, H. (2000) . " Could Anything be More Intelligible than Everyday Intelligibility? Reinterpreting Division I of *Being and Time* in the Light of Division II. " In J. Faulconer and M. Wrathall (eds.), *Appropriating Heidegger*. Cambridge: Cambridge University Press, PP. 155-170.

Dreyfus, H. (2001) . *On the Internet*. New York: Routledge.

Feyerabend, P. (1987) . *Science in a Free Society*. London: Verso.

Feenberg, A. and Hannay, A. (eds.) (1995) . *Technology and the Politics of Knowledge*. Bloomington: Indiana University Press.

Huber, P. (1991) . *Galileo's Revenge: Junk Science in the Courtroom*. New York: Basic Books.

Huber, P. and Foster, K. (1999) . *Judging Science: Scientific Knowledge and the Federal Courts*. Cambridge: MIT Press.

Husserl, E. (1973) *Cartesian Meditations and the Paris Lectures*, ed. S. Strasser. Dordrecht: Kluwer.

Ibsen, H. (1988) . *Ibsen: The Complete Major Prose Plays*, trans. R. Fjelde. New York: New American Library.

Ihde, D. (1998) . *Expanding: Hermeneutics: Visualism in Science*. Evanston: Northwestern University Press.

Jasanoff, S. (1995) . *Science at the Bar: Law, Science, and Technolgy in America*. Cambridge: Harvard University Press.

Latour, B. (1999) . *Pandora's Hope: Essays on the Reality of Science Studies*. Cambridge: Harvard University Press.

MacKenzie, D. (1996) . *Knowing Machines: Eassays on Technological Change*. Cambridge: MIT Press.

Mialet, H. (1999). "Do Angels have Bodies? Two Stories about Subjectivity in Science: The Cases of William X and Mister H." *Social Studies of Science* 29 (4): 551-582.

Pappas, G. (1994). "Experts." *Acta Analytica* 9 (12).

Polanyi, M. (1974). *Personal Knowledge: Towards a Post-Critical Philosophy*. Chicago: University of Chicago Press.

Rawls, J. (1968). "Two Concepts of Rules." In J. Thomson and G. Dworkin (eds.), *Ethics*. New York: Harper & Row, PP. 104-135.

Reber, A. (1995). *Implicit Learning and Tacit Knowledge: An Essay on the Cognitive Unconscious*. New York: Oxford University Press.

Rouse, J. (1987). *Knowledge and Power: Toward a Political Philosophy of Science*. New York: Cornell University Press.

Ryle, G. (1984). *The Concept of Mind*. Chicago: University of Chicago Press.

Schon, D. (1983). *The Reflective Practitioner*. New York: Basic Books.

Searle, J. (1983). *Intentionality: An Essay in the Philosophy of Mind*. New York: Cambridge University Press.

Searle, J. (1992). *The Rediscovery of the Mind*. Cambridge: MIT Press.

Selinger, E. *The Paradox of Expertise*. Ph. D. dissertation, Stony Brook University (forthcoming).

Sheets-Johnstone, M. (1999). *The Primacy of Movement*. Philadelphia: John Benjamins.

Sheets-Johnstone, M. (2000). "Kinetic Tactile-Kinesthetic Bodies: Ontogenetical Foundations of Apprenticeship Learning." *Human Studies* 23: 343-370.

Stengers, I. (2000). *The Invention of Modern Science*, trans. D. Smith. Minneapolis: University of Minnesota Press.

Turner, S. (2001) "What Is the Problem With Experts?" *Social Studies of Science* 31: 123-149.

Walton, D. (1997). *Appeal to Expert Opinion: Arguments from Authority*. University Park: Pennsylvania State University Press.

Williams, R. (1976). *Keywords: A Vocabulary of Culture and Society*. New York: Oxford University Press.

Winner, L. (1995). *Citizen Virtues in a Technological Order*. In A. Feenberg and A. Hannay, (Eds.), *Technology and the Politics of Knowledge*. Bloomington: Indiana University Press, pp. 65-84.

Winograd, T. (1995) . *Heidegger and the Design of Computer Systems.* In A. Feenberg and A. Hannay (Eds.), *Technology and the Politics of Knowledge.* Bloomington: Indiana University Press, pp. 108-127.

Young, I. (1998) . "Throwing Like a Girl. " In D. Welton (ed.), *Body and Flesh: A Philosophical Reader.* Oxford and Maiden, MA: Blackwell Publishers, 259-273.

德雷福斯论专长：现象学分析的局限性

埃文·赛林格　罗伯特·克里斯*

引　言

专长（expertise）① 对当代生活来说是极其重要的，在当代生活中，许多经济、政治、科学和技术的决策都常规性地托付给专家。②市民不仅在涉及技术的问题上，而且在"所有类型的普通决策"中，都听从权威专家的意见。③一方面，常规性的听从专家会产生一些政治后果，而且，有些学者甚至建议说，由于允许意识形态代替批评讨论，因而阻碍了合理的民主程序，破坏了交往行为。④另一方面，有关科学技术方面的论题的反复争论，可能导致把常规性的听从悬置起来，增加对专家的怀疑。因此，在政治语境和法律语境中，就辨别专家评价的本质和正确标准展开激烈争论，就不足为奇了。比如，在法律语境中，要提供证明专家证人的正确标准，就经常会提出判断专家评价的恰当标准是什么的问题。⑤

从哲学上澄清有关专家评价的论题，似乎是适当的，乃至是基本的。美国最

---

* 埃文·赛林格（Evan Selinger），美国纽约州罗切斯特理工学院副教授。罗伯特·克里斯（Robert P. Crease），纽约州立大学石溪分校哲学系系主任、教授。本文在作者的授权下译自 *The Philosophy of Expertise*, edited by Evan Selinger and Robert P. Crease, New York：Columbia University Press，2006. ——译者注

① "expertise" 这个词的中文意思并不唯一，本译文将根据作者的具体使用语境分别译为"专长"或"专家评价"或"专家的看法"。——译者注

② I. Barbour，*Ethics in an Age of Technology*，San Francisco：Harper Collins，1993，pp. 213-223.

③ D. Walton，*Appeal to Expert Opinion：Arguments from Authority*，University Park：Pennsylvania State University Press，1997，p. 24.

④ 特纳（S. Turner）讨论了斯坦利·费什（Stanley Fish）、尤根·哈贝马斯和米歇尔·福柯凭什么担心专家的看法会破坏自由民主的根据。他认为，这些担心"取决于社会建构论破坏了的关于知识特性的一种乌托邦的理想"（S. Turner，"What Is the Problem With Experts？"，*Social Studies of Science*，No. 31，2001，p. 147，n. 7）。

⑤ 讨论证实专家级证人的问题经常关系到"垃圾科学"（junk science）的论题（参见 Black，B.，Ayala，F.，and Saffran-Brink，C.，"Science and the Law in the Wake of Daubert"，*Texas Law Review*，No. 72，1994；Jasanoff，1992；Caudill，D.，and Redding，R.，"Junk Philosophy of Science？The Paradox of Experties and Interdisciplinarity in Federal Courts"，*Washington and Lee Law Review*，Vol. 57，No. 3，2000；Huber，P.，*Galileo's Revenge：Junk Science in the Courtroom*，New York：Basic Books，1991）。

高法院关于使用专家证人的里程碑式的裁决，即达伯特诉麦热里·杜药品公司
（Daubert v. Merrell Dow Pharmaceuticals）案件，诉诸了一些具有哲学渊源的概念，
还特别关注卡尔·波普尔的"证伪"概念①。当医学伦理学家就人类基因组计划
和干细胞研究之类的政治敏感项目向医院董事会、政治家和美国总统提出忠告
时，他们号召哲学家本人充当专家级的法律鉴定人。②

抛开其社会意义不说，关于专家评价的论题由于几个原因在哲学上也是重要
的。第一个原因是，它与心灵哲学有关。专长的传统载体（locus）或"专家级
的认知者"（expert knower）是主体，并且，专家理解和适应世界的方式与主观
性、意向性和理性主义者的本性以及意识的表征概念相关。例如，专长的本性是
转变下列争论的焦点：在人工智能（AI）的框架内，即在专家系统和以计算机
为基础的远程学习项目中，能否成功地使智能（intelligence）脱离肉体
（disembodied）的争论③。

专家评价问题具有哲学重要性的第二个原因是，它使经典科学哲学与科学技
术的人文社会科学研究（science and technology studies，STS）④ 这两种传统之间
的矛盾具体化。经典的科学哲学把专家的看法看成是理所当然的，并假定，区分
专家与外行是合法的；而 STS 则把怀疑专家看法的进路视为理所当然的，这预示
着有必要探讨专家评价的非法性问题⑤。

最后一个原因是，讨论专家评价的问题促进了现象学与各门学科之间的联
系。尽管有人争辩说，现象学的描述只能捕获到体验（experience）的主观维度，
因而用来试图理解科学是不适当的⑥，但是，科学实践需要开发、训练和协调值
得作出现象学澄清的各项专业技能。

---

① Huber, H. and Foser, K. *Judging Science*: *Scientific Knowledge and the Federal Courts*, Cambridge：MIT
Press，1999，pp. 37-68.

② 当理论家们假定，专家的社会作用是建立在他们的研究视角之基础上时，就出现了温纳（Winner）
所说的"评价专家"的问题［L. Winner， "Citizen Virtues in a Technological Order"，In A. Feenberg and
A. Hannay（eds.），*Technology and the Politics of Knowledge*，Bloomington：Indiana University Press，1995，pp. 65-
67］。温纳认为，从事应用研究的理论家不仅对"无休止的"争论提出忠告，而且往往把他们称呼的群众误
解为过分概括的"我们"，茫然不知社会变化是如何开始的有时也不利于制定合法的政治决策。

③ H. M. Collins， "Humans，Machines，and the Structure of Knowledge"，*Stanford Humanities Review*，
Vol. 4，No. 2，1995.

④ "science and technology studies" 国内有多种译法，现在通常译为"科学技术论"，这里译为"科学
技术的人文社会科学研究" 是为了把科学技术的哲学研究分离出来，即"科学技术论"的译法中可以包括
科学技术的哲学研究在内，而"科学技术的人文社会科学研究"的译法中则不包括这种研究。这种译法在
意思上较为准确，但用词上有点烦琐。——译者注

⑤ H. Mialet， "Do Angels Have Bodies? Two Stories about Subjectivity in Science：The Cases of William X
and Mister H"，*Social Studies of Science*，Vol. 29，No. 4，1999，pp. 552-553.

⑥ B. Latour，*Pandora's Hope*：*Essays on the Reality of Science Studies*. Cambridge：MIT Press，1999，p. 9.

然而，哲学家很少明确地讨论这个主题，尽管潜在的没有经过考察的专长概念通常蕴涵在像"权威性"、"殖民化"、"权力"和"理性的争论"之类的标题中。① 赫伯特·德雷福斯（Hubert Dreyfus）是明确讨论这个概念的少数人之一，本文致力于对他的解释作出批判性的评论。对德雷福斯解释的这种分析分五步进行。我们将：①把他的专长模型（the model of expertise）置于体知型语境（embodied context）中；②勾勒出他的一般的技能观；③总结他描述的专长模型；④提出他期望专家的言行是可辩护的规范性理论；⑤提出他的解释中存在的某些问题。德雷福斯建议，不依赖文化与历史的考虑，也能具体说明专家的基本特征。我们认为，他根据这个建议，通过令人信服地表明社会学分析、历史分析和人类学分析不可能详尽地考察专长的问题，证明了现象学对这个主题的重要性。但我们在德雷福斯的解释中也发现了某些描述性问题和规范性问题。当德雷福斯在现象学的意义上表明，不能把专家降低为意识形态的拥护者和社交网络的人造物时，从文化嵌入性的视角来看，他由于高估了专家和专家决策的独立性，因而也缺乏解释学的敏感性。

一、专长和身体

只有回溯上面提到的经典科学哲学和 STS 这两个传统如何回避提出这个论题，才能突出德雷福斯的专长解释的意义。每一种传统实际上都把创造性的专家级的行为表现（expert performance）看成是某种特别的东西："就这么发生了"，没有提出任何哲学问题。例如，传统科学哲学的目标是，从科学的有效性和客观性的根源上，对科学组织的特征进行理性重建，特别强调反常如何只是暂时扰乱其运行。传统科学哲学把创造性当做是占主导地位的精神活动，为此，只对其产品进行真正的哲学讨论。在区分发现的语境和辩护的语境的框架内，创造性的观念从哪里来的问题，没有被认为是一个恰当的认识论问题，而是被归属于心理学或历史学。②

另一方面，当代的 STS 把专长看成是分配式的外化到像实验室和社交网络之类的特殊设置中，是标准化的技术、判断科学性的标准、评价证明的协议，以及吸纳同盟者的修辞手段。③ 按照 STS 的拥护者的观点，其他做法所冒的风险是，使专长"自

---

① S. Turner, "What Is the Problem With Experts?" *Social Studies of Science*, No. 31, 2001.

② H. Mialet, "Do Angels Have Bodies? Two Stories about Subjectivity in Science: The Cases of William X and Mister H", *Social Studies of Science*, Vol. 29, No. 4, 1999, p. 552.

③ H. Mialet, "Do Angels Have Bodies? Two Stories about Subjectivity in Science: The Cases of William X and Mister H", *Social Studies of Science*, Vol. 29, No. 4, 1999, p. 552.

然化"，赋予专家不当的权威性，导致抑制外行的知识、价值与兴趣。STS 的拥护者通过关注专家在整个仲裁过程中如何变得过分尊贵的问题，把专长的非分配式的意义置于"黑箱"① 之中。② 当 STS 的理论家在他们的工作中偶尔求助于意会知识时，这些求助未必就是任何体知合一的理论（theory of embodiment）。③

尽管传统科学哲学和 STS 希望去除专长的神秘性的动机有所不同，但它们产生了相同的结果。就它们所有的明显分歧而言，传统科学哲学和 STS 都避免讨论专长与身体之间的关系。双方都同意，为了去除专长的神秘性，对科学提供一种精确的解释，需要忽略身体实践的不变特征。

这种对体知合一的忽视建议讨论传统的现象学主题：鲜活的身体的实践参与最终确立了他们具有的关于世界的知识，包括抽象的科学知识在内。④对于现象

---

① 在科学的人文社会科学研究中所用的"黑箱"是一个工程术语，类似于马克思主义的霸权（hegemony）概念。"黑箱"，像霸权一样，涉及通常被认为是自身显然为真并且不需要进一步研究的背景假设 [Feenberg, A. and A. Hannay（eds.），*Technology and Politics of Knowledge*，Bloomington：Indiana University Press，1995，p. 7]。打开科学的专家评价的"黑箱"，是透过严密的经验分析和理论分析表明，专家评价看起来本身显然为真、得到了文化上的认可并且不需要进一步研究的观点，是错误的，潜藏了文化的审查，也迫切需要批评分析。当这种专家评价的"黑箱"被打开之后，STS 理论家主张，专家并不是作为他们的知识是不可错的、确定的和客观的自立性天才出现的，反而揭示出科学专家是平凡的、有偏见的人，他们的成败来自他们工作在被分配知识和声望的竞争性网络中。从 STS 的视角来看，正是主要因为很少明确地描述和从理论上考察这个网络的运行机制，非专家才错误地把科学专家视为比实际情况更博学、更有权威和更值得信赖。

② H. Mialet，"Do Angels Have Bodies? Two Stories about Subjectivity in Science：The Cases of William X and Mister H"，*Social Studies of Science*，Vol. 29，No. 4，1999，p. 553.

③ STS 偶尔关注产生科学结果时涉及的意会知识。例如，哈里·柯林斯（Harry Collins）最近在可证实的经验案例中，超越技能，把波兰尼的意会知识扩展应用到科学结果的产品中去，试图通过这种扩展，阐述波兰尼的意会知识的分类（H. Collins，"Tacit Knowledge，Trust，and the Q of Sapphire"，*Social Studies of Science*，No. 31，2001）。波兰尼断言，为了理解如何运用机器，人们需要知道它的零部件如何一起发挥功能，他把这举荐为是普遍的论点（M. Polanyi，*Personal Knowledge：Towards a Post-Critical Philosophy*，Chicago：University of Chicago Press，1974）。基于在美国的现场考察和在格拉斯哥大学进行的实验观察，柯林斯断言，意会知识的问题说明了，为什么 20 年前俄罗斯人对蓝宝石的品质因数的测量只能在美国得到重复。柯林斯对意会知识的关注使他强调科学家之间的个人接触和信任的重要性，也使他建议应该补充当前在实验报告中不包括的信息，来增加复制科学发现的可能性。柯林斯和德雷福斯关注的事例，尽管有些重叠，但也有一个关键的不同。在柯林斯关注的事例中，意会的认知者产生了主体间性意义上公认的结果；他强调了这些结果的常规产生和妨碍他们轻易复制的障碍。柯林斯并没有提出说明认知者如何能意会地掌握设备的体知合一的理论。

④ 只有抽象的语境，不可能使科学家对世界进行数学化、模型化和形式化。对于科学家来说，为了控制和解释现象，需要运用面向技能的身体，来操作稳定现象的技术仪器。伊德再三认为，就技术的日常用法而言，在科学实验中所用的技术仪器，通过"体知合一的关系"，扩大到和转变为身体实践；它们就像海德格尔的锤子或梅洛-庞蒂的盲人的拐杖一样被兼并或合并到对世界的身体体验中，科学家能够产生的现象随着体知合一的形式的变化而变化（D. Ihde，*Expanding Hermeneutics：Visualism in Science*，Evanston：Northwestern University Press，1998，pp. 42-43）。

学家来说，所有的实践活动和理论活动，不管它们的结果多么抽象，都需要基于基本的生活世界实践的连续体来理解，德雷福斯通过把专家的判断和行为严格地看成是体知合一的人类行为表现的一个实例，继承了这一传统。像经典的科学哲学家和 STS 的拥护者一样，他试图去除专长的神秘性——但他这么做的方法是，把专长置于生活世界活动的连续体中，而不是把专长从生活世界活动中分离出来。专家在完成世俗的任务时，只不过扮演了像我们每一个人那样的角色："我们大家在完成许多任务时都是专家，我们每天都顺利而明显地应对技能的作用，使我们自由地意识到我们生活中不太熟练的其他方面。"①

德雷福斯明显地把他的技能模型与现象学传统联系起来，称之为是明显地发展了梅洛-庞蒂的经验身体（le corps vécu）的概念和"意向弧"（intentional arc）与"极致掌握"（maximal grip）的观点。② 通过密切注意"我们与世界的关系问题如何被转化为我们获得一项技能的问题"，德雷福斯断言，他打算比梅洛-庞蒂更全面地提出如何获得、改进和使用技能的问题。③ 德雷福斯旨在把专家作为体知型的定位主体（embodied, situated subjects）来研究，即力图注意"在什么条件下，出现慎重考虑（deliberation）与选择"，以便避免"使得对慎重考虑与选择结构的典型的哲学误读成为［他或她］对每天应对问题的解释"④。但德雷福斯也争辩说，中立的网络研究者，特别是沃特·弗里曼（Walter Freeman）在

---

① Dreyfus, H. and Dreyfus, S., "What is Morality? A phenomenological Account of the Development of Ethical Expertise", In D. Rasmussen (ed.), *Universalism vs. Communitarianism: Contemporary Debates in Ethics*, Cambridge: MIT Press, 1990, p. 243.

② 德雷福斯对"意向弧"和"极致掌握"的定义如下："意向弧确定了能动者（agent）和世界之间的密切联系"，即当能动者获得技能时，这些技能是"被存储起来的"，不是被看成内心的表征，而是被看成倾向于对世界的诱惑作出回应。极致掌握确定了身体趋向于对这些诱惑的回应，以这样一种方式，使当前的情境越来越接近能动者最理想的格式塔的意义。这两种能力都不需要心理或大脑操作"（H. Dreyfus, "Intelligence Without Representation", *Network for Non-Scholastic Working Paper*, Department of philosophy, Aarhus University, Denmark, 1998; "How Neuroscience Supports Merleau-Ponty's Account of Learning", Paper presented at *the Network for Non-Scholastic Learning* Conference, Sonderborg, Dennark, 1999a）。

③ H. Dreyfus, "How Neuroscience Supports Merleau-Ponty's Account of Learning", Paper presented at *the Network for Non-Scholastic Learning* Conference, Sonderborg, Dennark, 1999a, p. 1.

④ Dreyfus, H. and Dreyfus, S., "What is Morality? A phenomenological Account of the Development of Ethical Expertise", In D. Rasmussen (ed.), *Universalism vs. Communitarianism: Contemporary Debates in Ethics*, Cambridge: MIT Press, 1990, p. 239.

他的潜在知觉的脑动力学研究中，从经验上证实了他的专家是如何活动的模型。①

德雷福斯的解释提供了所谓的专长的形而上学（尽管他喜欢"本体论的"术语胜过喜欢"形而上学的"术语）。这与传统科学哲学和 STS 形成了强烈的对比。对于前者而言，缺乏形而上学解释的内容；对于后者而言，专长是一种依赖于文化的现象，其定义的改变是相对于历史变迁的，这种历史变迁支配着如何感知专家的看法。

二、专长与技能

当德雷福斯被兰德公司聘为顾问来评价他们在人工智能方面的工作时，德雷福斯与他的弟弟斯图亚特（Stuart Dreyfus）一起于 20 世纪 60 年代首先提出了他对专长的描述性解释。他的研究在《炼金术与人工智能》②的文章中和《计算机不能干什么？》③的著作中达到了高潮。④ 在《心智高于机器：计算机时代人的直

---

① H. Dreyfus, "How Neuroscience Supports Merleau-Ponty's Account of Learning ", Paper presented at *the Network for Non-Scholastic Learning* Conference, Sonderborg, Dennark, 1999a；"The Primacy of Phenomenology over Logical Analysis ", Philosophical Topics, Vol. 27, No. 2, 1999b, pp. 6-10； "Intelligence Without Representation ", *Network for Non-Scholastic Working Paper*, Department of philosophy, Aarhus University, Denmark, 1998. 尽管我们相信，德雷福斯转向"大脑话题"的主题，简单地确立了能与他的现象学进路相一致的另一种视角，但席斯-约翰斯顿（Sheets-Johnston）认为，所提到的神经网络和大脑功能与他的现象学的抱负相矛盾："［我们］发现，我们具有的现象学主题，已经从过去的理解让位于神经学的理解，然而同时，我们也发现，现象学的主题表面上完全在场，但已经变得与我们的体验完全无关，我们的体验'大概理解了'它自己的'脑动力学'。"（M. Sheets-Johnston, "Kinetic Tactile-Kinesthetic Bodies：Ontogenetical Foundations of Apprenticeship Learning ", *Human Studies*, No. 23, 2000, p. 357）此外，她认为，"它们［神经网络］不仅能区分开形式的学习与非形式的学习，而且，它们据此运行的词汇不是对后一种学习的有用模拟，因而对后一种学习的描述是不适当的，对它的理解也是有偏差的"（M. Sheets-Johnston, "Kinetic Tactile-Kinesthetic Bodies：Ontogenetical Foundations of Apprenticeship Learning ", *Human Studies*, No. 23, 2000, pp. 357-358）。我们相信，席斯-约翰斯顿忘记了，有些人仍然把现象学看成是一种主观的分析形式，也忘记了，德雷福斯对待神经网络的可能态度，为什么是证实现象学能揭示身体实践的客观结构的理由。

② H. Dreyfus, "Alchemy and Artificial Intelligence", *Rand*, Paper , 1967.

③ H. Dreyfus, *What Computer Still Can't Do：A Critique of Artificial Reason*, Cambridge：MIT Press, 1992.

④ 在 20 世纪 60 年代，当德雷福斯第一个表述他对人工智能及其虚假广告宣传的批评时，在麻省理工学院的人工智能实验室里的学术氛围，显然是对承认他所说的意义很有敌意，结果，他差一点失去自己的工作。相比之下，威诺格拉德（Winograd）注意到，今天，"实验室里做的某些工作似乎受到了德雷福斯的影响"［T. Winograd, "Heidegger and Design of Computer Systens", In A. Feenberg and A. Hannay（eds.）, *Technology and Politics of Knowledge*, Bloomington：Indiana University Press, 1995, p. 110］。

觉与专长的力量》① 一书中，兄弟二人提出了一个专家级的技能获得（expert skill acquisition）模型，他们声称，这个模型的范围是普遍的。② 他们的目的是根据涉及熟练操作的所有领域的使用说明书，对成年人如何获得技能提供一种现象学的解释③，不管是智力型的熟练操作，还是运动型的熟练操作。④

在德雷福斯的生涯中，这种专家级的技能获得模型相当于一块试金石。比如，他以此作为下列工作的基础：①去除与人工智能项目相关的渲染式广告的神秘色彩，特别是去除为模拟人类的专长而设计的"专家计算机系统"的神秘色彩⑤；②判断通过强加"合理化"约束"危及"职业专家（比如，护士、医生、教师和科学家）的社会偏见⑥；③说明在美国式的企业管理中的主导趋势错在哪里⑦；④捍卫梅洛-庞蒂对意向性和行动的非表征解释的准确性⑧；⑤揭示尤根·哈贝马斯的新康德主义伦理观的实践界限⑨；⑥说明政治行动小组的专家评价⑩；⑦澄清马丁·海德格尔和亚里士多德的技艺与实践智慧的意义，并且在这么做

---

① 原书名是 Mind Over Machine：The Power of Human Intuition and Expertise in the Era of the Computer，国内有人译为《机器心智》或《机器思维》，这些译法都不符合原著的论点，本书论证的观点是，人的心智能力是机器无法模仿的。这里根据本书讨论的核心观点和这两个替换词的意思，认为译为《心智高于机器》较恰当。——译者注

② Dreyfus, H. and Dreyfus, S., Mind Over Machine：The Power of Human Intuition and Expertise in the Era of the Computer, New York：Free Press, 1986.

③ 根据德雷福斯的观点，现象学首先迫切需要的永远是适当地描述专长（即使像胡塞尔、海德格尔和梅洛-庞蒂之类的历史现象学家也没有在他们的分析中运用"专长"这一术语），因为实际上，现象学旨在识破妨碍如何理解人类经验的偏见（Dreyfus, H. and Dreyfus, S., Mind Over Machine：The Power of Human Intuition and Expertise in the Era of the Computer, New York：Free Press, 1986, pp. 2-5）。

④ 席斯-约翰斯顿完全认为，德雷福斯由于忘记了，对于智力技能（intellectual skill）来说，身体技能是多么的基本，所以，他在分析上把智力技能与身体技能分离开来，假定了身心之间的一种"有害的""笛卡儿式的分裂"（M. Sheets-Johnston, "Kinetic Tactile-Kinesthetic Bodies：Ontogenetical Foundations of Apprenticeship Learning", Human Studies, No. 23, 2000, pp. 355-356）。

⑤ Dreyfus, H. and Dreyfus, S., Mind Over Machine：The Power of Human Intuition and Expertise in the Era of the Computer, New York：Free Press, 1986；Dreyfus, H., What Computer Still Can't Do：A Critique of Artificial Reason, Cambridge：MIT Press, 1992.

⑥ Dreyfus, H. and Dreyfus, S., Mind Over Machine：The Power of Human Intuition and Expertise in the Era of the Computer, New York：Free Press, 1986.

⑦ Dreyfus, H. and Dreyfus, S., Mind Over Machine：The Power of Human Intuition and Expertise in the Era of the Computer, New York：Free Press, 1986.

⑧ H. Dreyfus, "Intelligence Without Representation", Network for Non-Scholastic Working Paper, Department of Philosophy, Aarhus University, Denmark, 1998.

⑨ Dreyfus, H. and Dreyfus, S., "What is Morality? A phenomenological Account of the Development of Ethical Expertise", In D. Rasmussen (ed.), Universalism vs. Communitarianism：Contemporary Debates in Ethics, Cambridge：MIT Press, 1990.

⑩ Dreyfus, H., Spinosa, C., and Flores, F., Disclosing Worlds：Entrepreneurship, Democratic Action, and the Cultivation of Solidarity, Cambridge：MIT Press, 1997.

时，纠正了他在单行本中对海德格尔解释的称赞①；⑧说明了齐克果（Kierkegaard）的主体发展的规范阶段与把互联网的价值评价为交流媒介的相关性②；⑨批判了"远程学习"项目的教育潜力③。

德雷福斯的解释的一个最关键要素是，他拒绝接受把专家定义为信念源的普遍倾向。专家级的技能主要是"知道如何去做"（know how）而不是"知道是什么"（know that）的一个实践推理问题。"知道是什么"是通过反思和有意识的判断获得的关于问题的命题性知识。"知道如何去做"涉及实践知识，比如，走路、说话和开车的能力，实践知识是对日常事情的不假思索的掌握，不需要对成功实施进行有意识的慎重考虑。④ 在许多事例中，知道如何去做，涉及人们无法完全给予解释的无言表达的技能训练，尽管人们不应该把这与波浪尼的"意会知识"概念相混淆。⑤

有人可能会提出暗示和准则，即平稳操作的近似要素，但明白这些暗示点和有能力遵守这些准则，在很大程度上，预设了应该解释这些暗示与准则的技能。此外，德雷福斯坚持认为，人们一旦获得技能，就往往不再遵守初学时用到的那些准则。

德雷福斯追随海德格尔认为，当应对环境能力的前主题风格（pre-thematic styles）证明不足以实现普通目标时，实践的能动者（practical agent）往往只反思在"分解"方案时如何整理他们的体验；当人们的所作所为不能有效进行时，

---

① Dreyfus, H., "Could Anything be More Intelligible than Everyday Intelligibility? "Reinterpreting Diviion Ⅰ of *Being and Time* in the Light of Division Ⅱ ", In J. Faulconer and M. Wrathall (eds.), *Appropriating Heidegger*, Cambridge：Cambridge University Press, 2000.

② H. Dreyfus, *On the Internet*, New York：Routledge, 2001.

③ H. Dreyfus, *On the Internet*, New York：Routledge, 2001.

④ Dreyfus, H. and Dreyfus, S., "What is Morality? A phenomenological Account of the Development of Ethical Expertise ", In D. Rasmussen (ed.), *Universalism vs. Communitarianism：Contemporary Debates in Ethics*, Cambridge：MIT Press, 1990, p. 244.

⑤ Dreyfus, H. and Dreyfus, S., *Mind Over Machine：The Power of Human Intuition and Expertise in the Era of the Computer*, New York：Free Press, 1986, p. 16. 尽管从表面上看，德雷福斯与波兰尼在意会知识的问题上有许多相似之处，但也存在着重要的差别。德雷福斯认为，当波兰尼承认拘泥于形式不能说明骑自行车的意会的行为表现时他仍然相信，这样的行为表现是受"潜规则"支配的："有关潜规则的参考文献表明，波兰尼像柏拉图一样没有在行为表现和胜任能力之间、说明与理解之间、人们遵守的规则与被用来描述正在发生的情况的规则之间作出区分。"（Dreyfus, H., *What Computer Still Can't Do：A Critique of Artificial Reason*, Cambridge：MIT Press, 1992, pp. 330-331）德雷福斯也不同于库恩对意会知识的分析，因为这种分析主要是使知识的意会维度处于一般的范式结构中（I. Stengers, *The Invention of Modern Science*, trans. D. Smith. Minneapolis：University of Minnesota Press, 2000, p. 6）。关于意会知识的其他文献参见 J. Rawls, "Two Concepts of Rules", In J. Thomson and G. Dworkin (eds.), *Ethics*, New York：Parper & Row, 1968; A. Reber, *Implicit Learning and Tacit Knowledge：An Essay on the Cognitive Unconscious*, New York：Oxford University Press, 1995; J. Searle, *Intentionality：An Essay in the Philosophy of Mind*, New York：Cambridge University Press, 1983; The Rediscovery of the Mind, Cambridge：MIT Press, 1992.

他们才对自己付诸实践的方式作出反思。对反思的这种派生用法的辩护也是在能动者反思人们的所作所为基础上的实践，而且，做事的规则通常导致了人们随心所欲的实践问题。在特殊的行动中，从对遵循规则的前意识行为到有意识的鉴赏，标志着能动者从实践推理转向了理论推理，从"知道如何去做"转向"知道是什么"①。对于德雷福斯来说，完成要求熟练操作的大多数基本任务的最高标准是，一个人不用有意识地关注他当下的所作所为。

最后，德雷福斯争辩说，技能就是作出各种灵活的回应，乃至不能把身体技能降低为重复一系列的动觉运动。他主张，"技能，与一种固定回应或一组回应不同，能够以许多不确定的方式加以传承"②。约瑟夫·劳斯（Joseph Rouse）在讨论德雷福斯的观点时阐述了这一点。他写道，在学习传球时，所涉及的不是一系列重复运动，"而是能对传来的球作出各种反应。学习投球意味着，人们也能侧投球，尽管这项运动是不同的。我学会模仿很有限的句子，就能写出无数个各种不同的句子"③。因此，学习作出各种灵活反应不是熟记"实际动作"或打算重复的"思考模式"，而是掌握运动场上的各种可能性④。然而，德雷福斯无疑还会说，尽管技能是灵活的，但也有极限、边界区域和边缘空间；知道如何上投球不能使人掷标枪、下投球或侧投垒球。

在《心智高于机器》之后，德雷福斯的专长模型几乎一字不差地出现在大量的文章中。在这些文章中，他没有对《心智高于机器》一书中阐明专长的描述和核心论点提出挑战，而是扩展了这个模型所能应用的范围，评价了专家级的行为表现的认识论特征和形而上学特征，在此之前，他只是应用但没有充分阐述专家级的行为表现。⑤

三、描述的模型

德雷福斯描述的专长模型有几个关键特征。第一个特征应该是具有现象学的

---

① Dreyfus, H. and Dreyfus, S., *Mind Over Machine: The Power of Human Intuition and Expertise in the Era of the Computer*, New York: Free Press, 1986, p. 7.

② H. Dreyfus, *What Computer Still Can't Do: A Critique of Artificial Reason*, Cambridge: MIT Press, 1992, p. 249.

③ J. Rouse, *Knowledge and Power: Toward a Political Philosophy of Science*, New York: Cornell University Press, 1987, p. 61.

④ J. Rouse, *Knowledge and Power: Toward a Political Philosophy of Science*, New York: Cornell University Press, 1987, p. 61.

⑤ 德雷福斯在最近反思互联网时追加了——驾驭和智慧——两个阶段，这是他对五阶段模型的唯一一次修改（H. Dreyfus, *On the Internet*, New York: Routledge, 2001）。

辩护。在传统意义上，辩护概念把命题与这样一个公共领域联系起来：它能够在可论证的意义上证实或拒绝陈述的内容，或者陈述中的逻辑或推理关系。从现象学的视角来看，当命题详述了所有学科都能确定与其经验相一致的经验不变性时，这些命题就得到了辩护。在德雷福斯向他的读者提出的下列请求中发出了这个信号："你不必只接受我们的世界，而是应该检查一下看看你自己获得各种技能的过程是否揭示了类似的模式。"① 德雷福斯断言，他的专长模型是建立在不变模式之基础上的，这种不变模式是在以第一人称证言的形式转述"飞行员、下棋高手、驾驶员和学习第二语言的成年人"在讨论他们如何学会做出"非结构化"决策（"unstructured" decisions）时对技能获得的描述中发现的。②因此，他的解释不应该轻易受到这样的指控：把异质专家们（idiosyncratic experts）错误地刻画为示范性的，或者回避了人们如何知道什么是一个领域内的模范专家的问

---

① Dreyfus, H. and Dreyfus, S., *Mind Over Machine：The Power of Human Intuition and Expertise in the Era of the Computer*, New York：Free Press, 1986, p. 20.

② Dreyfus, H. and Dreyfus, S., *Mind Over Machine：The Power of Human Intuition and Expertise in the Era of the Computer*, New York：Free Press, 1986, p. 20. 德雷福斯把专心于制定"非结构化"决策的目标群看成是"典型的学习者"的典范，并且，他把"共同的模式"界定为在他们的行为中是可观察的（Dreyfus, H. and Dreyfus, S., *Mind Over Machine：The Power of Human Intuition and Expertise in the Era of the Computer*, New York：Free Press, 1986, p. 20）。"典型的"这个形容词意指"很有天赋"也有"机会获到足够经验"的一类学习者（Dreyfus, H. and Dreyfus, S., *Mind Over Machine：The Power of Human Intuition and Expertise in the Era of the Computer*, New York：Free Press, 1986, p. 20）。"非结构化"与"结构化"在信息管理理论中是讨论的标准术语。它们被用来分类决策范围内的组织差异，通常伴随有看起来像是这个范围内的中点的"半结构化"。当德雷福斯提到"结构化"决策时，他指的是作出的决策的类型，即当"明确了目标和相关信息时，就会知道这些决策的效果，并能推论出可证实的解决方案"（Dreyfus, H. and Dreyfus, S., *Mind Over Machine：The Power of Human Intuition and Expertise in the Era of the Computer*, New York：Free Press, 1986, p. 20）。换言之，"结构化"的决策需要很好地了解情况，为应对这些情况提出标准的程序。"结构化"的决策能够在下列情况下获得：这些情况是可重复的、常规的以及拥有非常可靠的相关证据。在"语境无关"占有绝对优势的情况下，能够找到"结构化"决策的例子，比如，"数学运算、猜谜语以及在现实世界中送货车的路径和石油调和"（Dreyfus, H. and Dreyfus, S., *Mind Over Machine：The Power of Human Intuition and Expertise in the Era of the Computer*, New York：Free Press, 1986, p. 20）。与"结构化"决策相比，德雷福斯把"非结构化"决策描述为直觉的、常识的、试探性的，并涉及试错法。他声明，"非结构化"决策往往是特设性的，不是程序化的，"含有无数个可能相关的潜在事实与特征，而且，这些要素相互联系和决定其他事件的方式是不明确的"（Dreyfus, H. and Dreyfus, S., *Mind Over Machine：The Power of Human Intuition and Expertise in the Era of the Computer*, New York：Free Press, 1986, p. 20）。需要"非结构化"决策的情况在典型意义上是难以捉摸的情况：在该情况下，不可能事先指定接下来的大多数决策程序；决策制定者不可能准确地辨认出问题定义的参数，必须对这个问题定义作出判断、评估和洞察；无法量化的因素是最重要的；根本没有取得共识的制定决策的程序。这些类型的决策通常是由从事"管理、护理、经济预测、教学和所有社会交往"工作并且要求有"相当多的关于真实情况的具体经验"的人做出的（Dreyfus, H. and Dreyfus, S., *Mind Over Machine：The Power of Human Intuition and Expertise in the Era of the Computer*, New York：Free Press, 1986, p. 20）。

题。由于坚持现象学的方法及其辩护机制，所以，他的解释也不意味着受到下列反例的困扰：这些反例驳斥了他的解释对有意设法获得技能的那些成年人的适应性。可以推定，他的解释之所以不受这类批评的影响，是因为所有的专家都能为他们自己重新发现和证实他的技能模型的基本要素，这样，他的技能模型被预期为是有意义的。

另一个关键特征是，他的模型是发展的，而且把技能获得预想为通过五个上升阶段相继发生的：①最初的"初学者"阶段；②"高级初学者"阶段；③"胜任"阶段；④"精通"阶段；⑤最终达到"专家"阶段。在第一阶段，"想把事情做好"的初学者学到了一组"语境无关"的"决定行动的规则"，常常在记住如何应用这些规则的过程中缓慢地行动①。在高级初学者阶段，此刻有了更多"具体情境中的实践经验"的学生，开始通过辨认这种情境的"有意义的其他方面"，即不能用规则编码的那些方面进行"一点点地"改进②。

德雷福斯的解释特别有趣的是，学习者不仅经历认知转变和实践转变，而且也经历情感转变。他断言，初学者和高级初学者典型地感受到，他们投入的实践被看成是"分离的"，而能胜任的执行者则觉得"已经参与到"他或她的表现的结果中③。在胜任阶段，学习者频繁地感到"激动不已"，好像他或她的"情绪在激烈波动"，不得不应对"极度紧张和筋疲力尽的"实践问题，而且，由于面临着要记住太多潜在的相关要素而深感"超载"④。结果，能胜任的学习者缩小了那些要素的范围，即为了有选择地提出这种情境的"相关特征和问题"，设计了一个"计划"和选取了一个"视角"。⑤ 通过这些变化，能胜任的执行者体验到了"初学者无法懂得的一种得意"，包括"骄傲和惊吓"在内。⑥ 承担责任意

---

① Dreyfus, H. and Dreyfus, S., *Mind Over Machine：The Power of Human Intuition and Expertise in the Era of the Computer*, New York：Free Press, 1986, p. 21.

② Dreyfus, H. and Dreyfus, S., *Mind Over Machine：The Power of Human Intuition and Expertise in the Era of the Computer*, New York：Free Press, 1986, pp. 22-23.

③ Dreyfus, H. and Dreyfus, S., *Mind Over Machine：The Power of Human Intuition and Expertise in the Era of the Computer*, New York：Free Press, 1986, p. 26.

④ Dreyfus, H., *On the Internet*, New York：Routledge, 2001, p. 35.

⑤ Dreyfus, H. and Dreyfus, S., *Mind Over Machine：The Power of Human Intuition and Expertise in the Era of the Computer*, New York：Free Press, 1986, pp. 26-27.

⑥ Dreyfus, H. and Dreyfus, S., "From Socrates to Expert Systems：The Limit of Calculative Rationality", In C. Mitcham and A. Huning (eds.), *Philosophy and Technology* II：*Information Technology and Computers in Theory and Practice*, Boston：D. Reidel Publishing Company, 1985, pp. 117-118.

味着"打心里觉得有错"①。

在精通阶段，学生超越了德雷福斯所谓的"决策的哈姆雷特模型（Hamlet model）"，即"分离的、慎重考虑的、有时是令人烦恼的替代选择"，这代表了技能获得的头三个阶段。② 这里，履行者对需要达到的目标的规则和原理的依赖在很大程度上被"知道如何去做"（即对情境的一种"无理性的"把握，德雷福斯称之为"直觉行为"）所取代；尽管熟练的履行者必须仍然深思和慎重考虑如何实现他或她的目标。③ 当熟练的履行者"只明白需要做什么，而不是运用计算程序在几种可能的替代者中选择其中之一"时，"行动变得更容易和更从容"④。

在最后阶段，专家不仅明白需要做什么，而且明白如何实现，用不着马上慎重考虑，就能"无意识地"承认，"新情况"类似于所有能被记住的情况⑤——尽管有趣的是，德雷福斯在1988年的简装本中去除了所有涉及能被记住的案例。因此，专家，像"悠久的禅传统中"的大师一样，或者，像在对欧比旺·肯诺比（Obi-Wan Kenobi）"用武力"的忠告作出回应时的卢克·天行者（Luke Skywalker）一样，超越了"尝试"或"努力"，"只是作出回应"⑥。德雷福斯把专长的"易变的行为表现"总结为"当事情规范地进行时，专家没有解决问题，没有作出决策；他们做的工作是规范的"⑦。他甚至断言，在专家层，区分主体与客体的能力消失了：熟练的司机与他的车成为一体，他体验到自己只是在驾

① Dreyfus, H. and Dreyfus, S., "From Socrates to Expert Systems: The Limit of Calculative Rationality", In C. Mitcham and A. Huning (eds.), *Philosophy and Technology II: Information Technology and Computers in Theory and Practice*, Boston: D. Reidel Publishing Company, 1985, p. 118.

② Dreyfus, H. and Dreyfus, S., *Mind Over Machine: The Power of Human Intuition and Expertise in the Era of the Computer*, New York: Free Press, 1986, p. 28.

③ Dreyfus, H. and Dreyfus, S., *Mind Over Machine: The Power of Human Intuition and Expertise in the Era of the Computer*, New York: Free Press, 1986, pp. 27-36. 德雷福斯写道："尽管在一般情况下应该避免不合理的行为，但并不能由此推断出，应该把理性的行为看成是最终目标。大量的领域存在于理性的和非理性的之间，可以称之为无理性的（arational）。理性的这个词，来源于拉丁语 ratio，意思是估计或计算，相当于是计算思维，因此，含有'把部分结合起来得到一个整体'的内涵；那么，无理性的行为是指无意识地分解和重组的行为。能胜任的行为表现是无理性的；精通是过渡期；专家在无理性的意义上采取行动。"（Dreyfus, H. and Dreyfus, S., *Mind Over Machine: The Power of Human Intuition and Expertise in the Era of the Computer*, New York: Free Press, 1986, p. 36.）

④ H. Dreyfus, *On the Internet*, New York: Routledge, 2001, p. 40.

⑤ Dreyfus, H. and Dreyfus, S., *Mind Over Machine: The Power of Human Intuition and Expertise in the Era of the Computer*, New York: Free Press, 1986, p. 35.

⑥ H. Dreyfus, "The Primacy of Phenomenology over Logical Analysis", *Philosophical Topics*, Vol. 27, No. 2, 1999b, p. 22, n. 13.

⑦ Dreyfus, H. and Dreyfus, S., *Mind Over Machine: The Power of Human Intuition and Expertise in the Era of the Computer*, New York: Free Press, 1986, pp. 30-31.

驶，而不是驾驶一辆车。① 当专家体验到最佳表现的"流畅"（flow）时，他或她就不会担心未来，而是会制订各种计划。② 由于沉迷于这一时刻，专家能够体验到"精神愉悦"，运动员把这描述为"忘我"的玩耍。③

然而，德雷福斯模型的另外一个特征必须与初学者和专家之间的划界相关。某些 STS 的研究者认为，决定性地把专家与非专家区分开来是站不住脚的，因为他们之间的明显差别是社会错觉，相比之下，德雷福斯主张，能以三种不同的方式划出专家与初学者之间的界线。④ 第一种方式基于专家"沉浸于体验和语境的敏感性"。专家不同于初学者，因为他或她不再主要关系到通过无语境特征（即不需要体验就能承认的特征）的解析实践。相反，他或她通过植根于经验中的灵活行为，把重要特征辨别为是语境敏感的。⑤ 在专家的判断中，这种情境约定导致了一种变化："新手和高级初学者只练习，不判断……而精通者或专家基于他们先前的具体体验，以蔑视说明的方式，做判断。"⑥ 第二种划界方式集中于行动与决策之间的暂时联系。当用缓慢地遵守规则和慎重考虑刻画初学者的行动时，专家的行动是直接的和直觉的情境反应。最后一种划界方式是关于情感的转变。在经过整个发展阶段之后，专家的主观性及其与世界的关系发生了质的转变，能够同初学者与世界的关系区分开来。在早期阶段，学习者是"挫败的"和"不知所措的"，而在最后阶段，专家了解做有风险的事情会带来的后果，因此，他们打消这些情绪，享受"易变"和"顺利"的最后一种表现。德雷福斯批评了约翰·塞尔（John Searle）的背景解释及其与意向性的关系，在他的批评中，他把专家级的网球选手描述为能全神贯注地沉浸于"流畅"的比赛中，以

① Dreyfus, H. and Dreyfus, S., *Mind Over Machine: The Power of Human Intuition and Expertise in the Era of the Computer*, New York: Free Press, 1986, p. 30.

② Dreyfus, H. and Dreyfus, S., *Mind Over Machine: The Power of Human Intuition and Expertise in the Era of the Computer*, New York: Free Press, 1986, p. 30.

③ Dreyfus, H. and Dreyfus, S., *Mind Over Machine: The Power of Human Intuition and Expertise in the Era of the Computer*, New York: Free Press, 1986, p. 40.

④ 因为米歇尔·卡龙（Michael Callon）提出这种观点："从事事实考察工作的研究者（researchers in the wild）由于对我们视为是理所当然的甚至是关键的对比（比如，区分专家与外行）提出了挑战，从而参与颠覆了现代的制度框架。"（M. Callon, "Researchers in the wild and Rise of Technical Democracy", Paper presented at *Knowledge in Plural Contexts*, Science and Technology Studies, Université de Lausanne, Switzerland, 2001）

⑤ Dreyfus, H. and Dreyfus, S., *Mind Over Machine: The Power of Human Intuition and Expertise in the Era of the Computer*, New York: Free Press, 1986, p. 35.

⑥ Dreyfus, H. and Dreyfus, S., *Mind Over Machine: The Power of Human Intuition and Expertise in the Era of the Computer*, New York: Free Press, 1986, p. 36.

至于他或她不再觉得有获胜的压力，只是对球场上的整个紧张局势作出回应。①
除此以外，从初学者到专家的情绪变化符合意义的变化。初学者的态度基本上是
一种兴趣或好奇，而专家则完全是投身于他或她自己的存在。德雷福斯引用了象
棋大师鲍比·菲舍尔（Bobby Fischer）的话——大意是"下棋就是生活"——对
德雷福斯来说，下棋是作为范例的智力技能。②

　　尽管给出这些划界，德雷福斯的解释还涉及连续性问题。甚至最抽象的实践
也保持与生活世界的基本关系，尽管它有时是潜在的。这就是为什么他避免用对
比性的反义词"非专家"和"外行"，反而用术语"初学者"向人们提供了可能
性范围的一个端点。③

　　德雷福斯的模型具有基本的意义。按照不同类型的基本技能组建不同的领域；不
过，他的解释是一个技能获得模型，这个模型能从形式上加以阐述，不用提到任何一
个特殊的领域。当然，不同的领域以不同的方式定义专家，因为代表不同领域的内
容，有不同的类型。例如，用是否获胜来定义专家级的象棋选手，而用是否能达到目
的地来定义专家级的司机。尽管如此，专家基本上被定义为是直觉的、投入的、超越
规则的主体，"专家的技能在很大程度上已经成为他的一个组成部分，以至于他应当
了解技能与了解自己的身体一样"，这个基本定义适用于这两种情况。在现象学的层
面，专家级的象棋选手和小车司机在功能上相当于专家的身份④。

　　最后，德雷福斯的解释提供了一种实践的专家的观点。以他的模型的普遍范

---

① H. Dreyfus, "The Primacy of Phenomenology over Logical Analysis", *Philosophical Topics*, Vol. 27. No. 2,
1999b, pp. 4-5. 就德雷福斯对所谓的"专家"计算机系统的批评而言，如此强调情感是重要的，他认为，
"专家"计算机系统能够近似计算能胜任的人类的行为表现。对于德雷福斯来说，技能获得不仅包括主体的
认知获得，而且包括计算机不能体验到的一种情感的转变。在用来突出德雷福斯的现象学背景时，这一点是
特别重要的。现象学家长期以来一直认为，成为一个主体就是意味着拥有与世界的意向关系，这样，主体的
改变与主体世界的改变是相互关联的。通过专业学徒训练的发展过程的主体与刚开始进入这个过程的主体不
一样，对于后者而言，世界同样是没有意义的；专家和非专家确实是不同的主体。他们是不同类型的人，他
们的考虑和感觉都不一样，而且，对世界的回应也不同。专家不仅比初学者做的事多，而且，他们的整个情
感态度发生了变化。专家对他们的实践活动的关注方式是从他们刚入门时变化而来的，从相对分离进步到积
极介入。这就是为什么德雷福斯把五个发展阶段描述为对什么是一项任务和用什么样的决策模式完成这项任
务的"在定性意义上不同的感性认识"的原因所在（Dreyfus, H. and Dreyfus, S., *Mind Over Machine*: The
Power of Human Intuition and Expertise in the Era of the Computer, New York: Free Press, 1986, p. 19）。

② Dreyfus, H. and Dreyfus, S., *Mind Over Machine*: The Power of Human Intuition and Expertise in the Era
of the Computer, New York: Free Press, 1986, p. 33.

③ 尽管并不是每一种实践都能使每一位初学者有机会达到专家级水平的驾驭。但许多实践是如此：
"不是所有的人在技术上都能达到专家级的水平。某些领域的技能的特点是，只有极少部分的初学者能够
精通这个领域。"（Dreyfus, H. and Dreyfus, S., *Mind Over Machine*: The Power of Human Intuition and
Expertise in the Era of the Computer, New York: Free Press, 1986, p. 21）

④ Dreyfus, H. and Dreyfus, S., *Mind Over Machine*: The Power of Human Intuition and Expertise in the Era of
the Computer, New York: Free Press, 1986, pp. 21-35.

围为基础，他试图描述所有专家都共享的与世界的共同认知关系和情感关系，他之所以这么做，是为辩护专家的诺言提供基础。当其他理论家试图勾勒这种"专家的观点"时，他们大多数人都是从理论整体论的视角进行，指出专家对一个领域的概要式的展望。例如，斯科特·布鲁尔（Scott Brewer）认为，非专家只能知道与一个领域相关的特殊的、真实的事实和方法，只有专家才能知道，一个领域的相关特征（比如，"事业心"和"价值论"的特性）如何关系到为下列问题提供共享的意义：这个领域内的实践者把什么断言为是真的，如何断言为是真的，为什么断言为是真的。① 布鲁尔注意到，他对理论知识的系统性的关注是有点柏拉图式的②。相比之下，德雷福斯经常把他的技能模型看成是反柏拉图式的。他

---

① S. Brewer, "Scientific Expert Testimony and Intellectual Due Process", *The Yale Law Review*, Vol. 107, No. 4, 1998, pp. 1568-1593. 布鲁尔（Brewer）像其他许多哲学家和法学理论家一样，要求把"观点"作为表达视角与辩护如何相关的一种分析手段（S. Brewer, "Scientific Expert Testimony and Intellectual Due Process", *The Yale Law Review*, Vol. 107, No. 4, 1998, pp. 1568-1570）。他利用观点来阐述适用于所有科学专家的共同的理论视角，可以认为，对于所有的专长领域内的所有专家来说，这都是普遍的，不管实践者是哪个特殊的科学领域内的专家。他写道，"人们用一种观点来辩护某种断言。为了服务于这种辩护功能，假定观点是达到某一理性事业的（显在的或潜在的）目标的可靠方法"（S. Brewer, "Scientific Expert Testimony and Intellectual Due Process", *The Yale Law Review*, Vol. 107, No. 4, 1998, p. 1575）。布鲁尔的推理思路是，人们利用观点，为我们应该相信什么的理论主张或应该如何行动的实践主张，作出合理的辩护。这类证实的独特之处是，它把对一种主张的辩护与一种不同的然而是"可靠的"方法相联系，选择这种方法来到达特殊的认知目标。所有"合理的"观点共有的"可靠"方法是根据两个特征来定义的，即"规划"的理念和"价值论"的设想。一种"规划"被定义为为了服务于某些特殊的认知目标，选择和运用一种特殊的分析方法。他承认，即使在"相同的通用规划"中，实践者也会对"完全明确的规划目标"持有异议（S. Brewer, "Scientific Expert Testimony and Intellectual Due Process", *The Yale Law Review*, Vol. 107, No. 4, 1998, p. 1571）。但这样一种歧义将会在涉及"价值论"分量的"整体"网络中发生。当布鲁尔讨论辩护的"价值论"维度时，他是在劳丹（Larry Laudan）用语的意义上这么做的。劳丹在分析推理情况时，在"事实的"、"方法论的"分析层次和"价值论"分析层次之间作出了区分。对于科学家来说，事实分析关注世界上存在着什么，包括理论的和不可观察的实体在内。在方法论层次上，特定领域内的实践者共享着既精确又模糊的规则。就科学家而言，相对于精确规则（比如，根据标准"y"来校准仪器"x"）来说，这可能包括模糊规则，比如，回避特设性说明。通常以规则形式加以说明的价值论层次指明了认知目标。布鲁尔像劳丹一样认为，事实、方法和价值论目标之间的关系不应该被理解为"简单的线性层次关系"，而是被理解为在目标、方法和信念的整体论的网络中是多方向的"错综复杂"的约束（S. Brewer, "Scientific Expert Testimony and Intellectual Due Process", *The Yale Law Review*, Vol. 107, No. 4, 1998, p. 1575）。事实、方法和价值论的目标在"多方向的"意义上是相关的，因为它们三者之间彼此"约束"具有同样的价值。事实、方法和价值论的目标在"整体论的"意义上是相关的，因为它们共同构成的观点把三种特征联系在一起。换言之，一种观点就是一个完备的和系统的视角，不可还原为孤立的观察。最后，由于观点的整体论本性，布鲁尔把观点的认知地位描述为"理解"而不是"知识"。本尔也特（Miles Burnyeat）断言，布鲁尔也重申，"知识与理解之间的重要差别是，知识可能是零碎的，可能是一个接一个地掌握单独的真理，相反，理解总是包含有看到已知项之间的连接和关系"（S. Brewer, "Scientific Expert Testimony and Intellectual Due Process", *The Yale Law Review*, Vol. 107, No. 4, 1998, p. 1591）。

② S. Brewer, "Scientific Expert Testimony and Intellectual Due Process", *The Yale Law Review*, Vol. 107, No. 4, 1998, p. 1591.

对实践理解的共同特征提出了一种描述，而不是从理论上对"专家的观点"提出一种整体论的解释。专家的实践理解不是来自信念或理论承诺，而是来自获得的体知型技能（embodied skill）。因此，德雷福斯写道，"五阶段模型的教益是，智能行为不只是计算的合理性"①。

四、规范的意义

德雷福斯的解释描述了对专家的服务抱有哪些期望是正当的，提出哪些要求是合法的，哪些要求是不合法的。因此，德雷福斯的解释具有规范意义，他以弹道学主考人、雏鸡雌雄鉴别师、市民、法官、护士、医务人员、医师、科学顾问和教师等人为例，从经验上讨论了这些规范的意义。② 德雷福斯建议，我们可以通过某些手段来实现牵涉到专家的某些社会目标和政治目标，比如，由于个人、制度和法律的原因咨询专家，对专家决策本性的现象学理解，有必要成为辨别这些手段的基础。尽管从描述性的基础上推断规范责任，可能意味着，德雷福斯犯了自然主义的错误推断：从只陈述实际情况的前提，推断出应该做什么的责任。他主张，现象学与规范性之间的关系是一个"优先权"的论题。③

这些规范的意义之所以重要，有如下几种理由。第一种理由是，专家很难与他人沟通，这是出了名的。尽管有时把此归因于技术语言与日常语言之间的差异，有时归因于像傲气之类的心理因素，但德雷福斯的解释提出了更深刻的原因，我们后面会返回这个论题。他的解释的规范意义之所以重要的第二种理由是，当专家充当个人顾问、制度顾问和法律顾问时，他们的动机与偏向不断地受到挑战，这可能关系到：①他们的职业训练，这会影响到专家如何概念化和约束

---

① Dreyfus, H. and Dreyfus, S., *Mind Over Machine: The Power of Human Intuition and Expertise in the Era of the Computer*, New York: Free Press, 1986, p. 36.

② Dreyfus, H. and Dreyfus, S., *Mind Over Machine: The Power of Human Intuition and Expertise in the Era of the Computer*, New York: Free Press, 1986, pp. 196-201; Dreyfus, H., Spinosa, C., and Flores, F., *Disclosing Worlds: Entrepreneurship, Democratic Action, and the Cultivation of Solidarity*, Cambridge: MIT Press, 1997, p. 106.

③ 德雷福斯认为，"优先考虑"能动者实际上如何回应具体情况的现象学家弄清了规范的责任，在这种意义上，现象与规范性之间的关系涉及"优先考虑"问题。根据德雷福斯的观点，黑格尔的伦理学传统中的那些人也有这种"优先考虑"的问题，比如，威廉姆斯（Bernard Willams）、泰勒（Charles Taylor）、吉利根（Carol Gilligan）。相比之下，德雷福斯认为，康德的道德传统中的那些理论家，比如，哈贝马斯、罗尔斯（John Rawls）和柯尔伯格（Lawrence Kohlberg），"优先考虑"分离的原理：详述如何做正确的事情，超过了对允许做出某些决定的经验条件的理解 [Dreyfus, H. and Dreyfus, S., "What is Morality? A phenomenological Account of the Development f Ethical Expertise", In D. Rasmussen (ed.), *Universalism vs. Communitarianism: Contemporary Debates in Ethics*, Cambridge: MIT Press, 1990, p. 237].

他们处理的问题；②他们的老板，这会在政治上影响到专家得出的结论；③经济利益，这会使专家变成职业杀手，他们的推荐被以适当的价格购买；④渴望得到认可和美名——我们下面将要返回的另一个论题。相比之下，德雷福斯建议，有些社会势力有时影响了专家所作出的断言，在不考虑这些社会势力的前提下，也能够确定围绕专家级的行为表现的基本的规范限制——在我们的批评中，将要返回的又一个论题。

这些规范意义源于这样的断言：不管社会环境是多么可怕或紧迫，在任何一个领域内，新手都不能超越规则来达到专家能够练出来的直觉。①按照德雷福斯的观点，期望他们以命题陈述②的形式描述他们的决策过程，是不合法的，因为他们的决策是在默默地运用直觉的基础上作出的。不仅象棋大师凭直觉采取行动，而且弹道学专家也是如此，他不可能以真实的方式用命题来表达，在一次特殊的枪击中他或她如何确定一发特殊的子弹是否能发射出来。③ 然而，在假定专家级的法定证人的角色时，弹道学专家将被期望这么做，即把结论与规则联系起来，并且只有在这么做时，他或她才有可能相信评审委员会。但德雷福斯认为，这种说服力是以"形式"优先于说明的"内容"为代价得出的。④ 的确，专家在基于规则说明他或她的决策过程时，甚至"丧失了"专长，因为那会牵涉到放弃引导决策过程的直觉体验。⑤ 德雷福斯认为，专家决策的理性重建不能准确地代表原则上不可描述的过程。当非专家要求专家领着他们一步一步地走向他们的决策过程，以便他们能够遵守专家的一连串演绎和推理（也许希望为他们自己提供这些一连串的演绎和推理）时，按照德雷福斯的观点，他们不再允许专家起到专家的作用，而是相反，使专家对他或她的专长产生了派生的极其错误的表达。因此，德雷福斯认为，对于非专家来说，不应该给专家施加太多的压力，以使他

① 例如，这对费耶阿本德来说是一个重要的修正，他没有看到这一点。他认为，当科学权威不需要过分复杂时，社会情况是，不能过分地尊重专家，比如，在战争年代，人们能够有效地简化医药程序。他断言，就此而言，在那些新兵身上就能找到证据，历史已经证明，这些新兵只接受了半年的训练，就能被委任为是医生（P. Feyerabend, *Science in a Free Society*, London: Verso, 1987, p. 307）。费耶阿本德错过的观点是，新兵可能很快就被指定为胜任不同方面的医药工作，比如，验伤，但这种训练不会产生出专家级的医生，或者削弱了规范训练方式的基础，规范训练方式使专家级的医生比集训医生能做更多的事情。

② 原文"prepositional statements"有误，与作者沟通后，应改为"propositional statements"。——译者注

③ Dreyfus, H. and Dreyfus, S., *Mind Over Machine: The Power of Human Intuition and Expertise in the Era of the Computer*, New York: Free Press, 1986, p. 199.

④ Dreyfus, H. and Dreyfus, S., *Mind Over Machine: The Power of Human Intuition and Expertise in the Era of the Computer*, New York: Free Press, 1986, p. 199.

⑤ Dreyfus, H. and Dreyfus, S., *Mind Over Machine: The Power of Human Intuition and Expertise in the Era of the Computer*, New York: Free Press, 1986, p. 196.

们的"直觉"的决策过程"合理化"①。

这样，德雷福斯的规范立场相当于是道格拉斯·沃尔顿（Douglas Walton）所说的"不可接近性论点（inaccessibility thesis）的强形式"（以下简称 IT）："专家的结论不可能被追溯到为专家判断提供基础的一组前提和推理规则（已知的事实与规则）。"② 根据 IT，当专家根据他们的专长给出一种裁定时，专家不可能以命题的形式表达，"通过逻辑推理的演绎（乃至归纳）步骤，将会显示结论（即专家所得出的结论）的意义的一组定律和初始条件（原理和事实）"，在这种意义上，对非专家来说，他们的判断是不可接近的。③ 由于专家依赖于难以表达的直觉维度，IT 建议，"一个领域内的顶尖专家，即出色地掌握了所属领域的技能之人，最不能传播他的知识"④。这不仅仅是专家不能向非专家传播他们如何获得知识；对于他们自己和其他专家说来说，他们的知识同样也是难以理解的。

因此，德雷福斯的主要的规范性问题，牵涉到非专家可能危害到专家的某些方面，而不是专家如何能危害到非专家。"专家是一个濒危物种（endangered species）。"⑤ 他把这刻画为是美国面临的一个问题，特别是："要求其专家能够说明，他们做的工作如何很不利于像我们的文化那样的理性文化，这种文化与像日本文化那样的直觉文化形成对比。"⑥ 此外，德雷福斯认为，历史变化加速了专家当下的危机。他主张，近来计算机技术的进步和官僚主义的社会组织的发展，使这种文化的问题更加恶化："如果不是发明了现代数字计算机的话，使社会合理化的欲望依然只是一个梦想。未来，由于对合理性的过分信赖，将会丧失技能和专长，社会的官僚主义本性的日益增加，正在强化着这种危险。"⑦ 为了

---

① 通常的愿望是，专家捍卫他们的提议，反对其他的专家，或者反对某一方面的盘问，结果，这些情感会质疑他们的预设。如果接受这一点意味着，专家必须阐述他的价值、规则和事实性假设，那么，考察就成为在抛弃专长和浪费时间的理性化中的一种无价值的练习（Dreyfus, H. and Dreyfus, S., *Mind Over Machine*：*The Power of Human Intuition and Expertise in the Era of the Computer*, New York：Free Press, 1986, p. 196）。

② D. Walton, *Appeal to Expert Opinion*：*Arguments from Authority*, University Park：Pennsylvania State University Press, 1997, p. 110.

③ D. Walton, *Appeal to Expert Opinion*：*Arguments from Authority*, University Park：Pennsylvania State University Press, 1997, p. 109.

④ D. Walton, *Appeal to Expert Opinion*：*Arguments from Authority*, University Park：Pennsylvania State University Press, 1997, p. 113.

⑤ Dreyfus, H. and Dreyfus, S., *Mind Over Machine*：*The Power of Human Intuition and Expertise in the Era of the Computer*, New York：Free Press, 1986, p. 206.

⑥ Dreyfus, H. and Dreyfus, S., *Mind Over Machine*：*The Power of Human Intuition and Expertise in the Era of the Computer*, New York：Free Press, 1986, p. 196.

⑦ Dreyfus, H. and Dreyfus, S., *Mind Over Machine*：*The Power of Human Intuition and Expertise in the Era of the Computer*, New York：Free Press, 1986, p. 195.

解决专长消失的问题，德雷福斯试图再次教导人们关注"知道如何去做"和"知道是什么"之间的差别。德雷福斯所说的未来并不遥远。他建议，"在一代人当中"，我们就可能失去"我们的职业专家"①。目标是动员人们去鉴赏直觉和理性考虑的极限的价值。②

德雷福斯既从反社会的视角分析专长，也从社会的视角分析专长。在反社会的层面，他在现象学意义上研究了全人类是如何获得技能的，不管他们是谁，不管他们是哪个领域内的学徒工，也不管他们何时和在哪里学习他们的技能。在社会的层面，他提出了一种规范的立场来决定，当请专家担任顾问时，应该如何对待专家。社会的论证与反社会的论证的这种特殊结合的安排，是基于德雷福斯的这个基本假设：专家应该做什么的社会预期应该与限定他们事实上能做什么所体现出的限制（embodied limitations）相一致。这个假设是支持他的下列观点的辩护机制：拥有专家级水平的技能，是成为一名专家的充分必要条件；而且，把这种能力合理化为专长的标志是错误的。

五、德雷福斯的描述性解释的问题

无论如何，德雷福斯的解释允许使不属于专家的几种人作为专家，并且，遗漏了属于专家的几种人。比如，德雷福斯断言，在走路和说话时，成年人是"专家"——正如我们所看到的那样，这对他的解释来说是基本的，因为他的解释把成年人放在与传统的专家行为一样的生活世界中——这与日常用法相矛盾。我们不是把只会走动或言语的人称为"专家级的"步行者或说话者，而是为经历了特殊训练的那些人保留这个形容词，提出职业忠告等。我们不是把有驾照的司机都称为"专家"，甚至也不把爱好者或有竞争力的业余爱好者称为"专家"，即使他们对自己开的车很有直觉。更确切地说，我们为属于职业驾校、参加某类竞赛等的司机保留这个词。

同时，德雷福斯的解释也从属于专家的人中排除了几种人。比如，我们之一已经把"专家级的 X"和"在 X 方面的专家"区分开来，"专家级的 X"是把

---

① Dreyfus, H. and Dreyfus, S., *Mind Over Machine：The Power of Human Intuition and Expertise in the Era of the Computer*, New York：Free Press, 1986, p. 206.

② 德雷福斯写道："社会……必须鼓励其孩子培育他们的直觉能力，以便他们可以获得专长，不是鼓励他们成为人类逻辑机器。而且，专长一旦获得，就必须得到承认，并认识其价值。把专家的常识、智慧、成熟的判断与今天的人工智能相混淆，或者，降低它们的价值，真的很愚蠢。"（Dreyfus, H. and Dreyfus, S., *Mind Over Machine：The Power of Human Intuition and Expertise in the Era of the Computer*, New York：Free Press, 1986, p. 201）

"专家"作为形容词来使用的，源于拉丁语"expertus"（内行的）；"在 X 方面的专家"实质上是把"专家"当做名词①。用德雷福斯的术语来说，"专家级的 X"对应于"知道如何去做"，而"在 X 方面的专家"对应于"知道是什么"。一位专家级的 X 可能会是"专家级的农民"，而一位"在 X 方面的专家"则可能会是"在种田方面"的专家。"在种田方面"的专家能有效地传递、协调和综合关于种田的准确的建议性信息——能够成为农业部长——即使非常害怕犁耕和开拖拉机。把运动员的过去行为与当前的情况关联起来的一名"运动方面的专家"可能是跛子，并丧失了运动能力；一位"音乐方面的专家"可能是一位极差的音乐家。尽管如此，德雷福斯否认，"在 X 方面的专家"定义的命题性知识是专业知识（expert knowledge）。

> 倾听……至少把一半时间花在博学的谈话节目中的评论员，与倾听表达力强的象棋观众一样，这位观众对每步棋都有看法，并诉诸大量的原则，但他没有致力于承担象棋锦标赛的压力与风险，因而没有专家的资格。②

德雷福斯把"在 X 方面的专家"（"评论员"或"乱出点子的人"）的知识与"削弱了源于实践合理性的承诺基础"的"公共领域"的定论的"闲聊"联系起来。③"在 X 方面的专家"的推理"不是基于局部的实践"，而是基于"抽象的解答"和不能体现"智慧的""千篇一律的原则"④。此外，德雷福斯认为，一位"在 X 方面的专家"缺乏真正的"专家级的 X"拥有的肢体投入；只有一位"专家级的 X"，才能从情感上关心一种情况的后果，并体验"冒险的"表现。但这种描述并不准确。可以推测，一位"艺术史方面的专家"不是滔滔不绝地说出抽象概念和概括性的话，而是有助于指导他人，有时甚至本身就是艺术家，能鉴赏作品的意义和艺术过程本身的意义。乔治·斯坦纳（George Steiner）曾经提出了相当于是一种有效的反驳，降低了德雷福斯在这里的地位：他试图设想一个没有批评家和评论员的社会，在这个社会里，"禁止所有关于艺术、音乐和文学的讨论……认为这种讨论是一种不正当的空谈"⑤。当斯坦纳发现他自己渴望这样一个"反柏拉图式的共和国"时，像柏拉图一样，他也承认，他的思

---

① 接受作者的建议，这里去掉了原文中附加的参考文献。——译者注

② Dreyfus, H., Spinosa, C., and Flores, F., *Disclosing Worlds*: *Entrepreneurship*, *Democratic Action*, *and the Cultivation of Solidarity*, Cambridge: MIT Press, 1997, p. 87.

③ Dreyfus, H., Spinosa, C., and Flores, F., *Disclosing Worlds*: *Entrepreneurship*, *Democratic Action*, *and the Cultivation of Solidarity*, Cambridge: MIT Press, 1997, p. 86.

④ Dreyfus, H., Spinosa, C., and Flores, F., *Disclosing Worlds*: *Entrepreneurship*, *Democratic Action*, *and the Cultivation of Solidarity*, Cambridge: MIT Press, 1997, p. 87.

⑤ E. Steiner, The Paradox of Expertise, *Ph. D. dissertation*, Stony Brook University, 1989, pp. 4-5.

想实验是不可能的。最终，斯坦纳意识到，这种结果不仅会扼杀艺术和人的富有创造的想象力，而且会扼杀生活世界本身。可以推测，德雷福斯可能同意，他的头三个阶段需要达到言语表达的程度，因此，专长不可能是在这样一个世界里获得的——但斯坦纳的思想实验的精神是，即使艺术家、音乐家和作家，也只能在对他们付诸的实践作出高层次讨论的海洋里兴旺发达。

德雷福斯对技能获得的解释所忽视的另一个专长范畴是"教练"。德雷福斯的解释把指导教师设想为是以标准化的方式对许多人同时宣布规则和命令——在经过学习的头三个阶段之后，指导教师不见了。即使德雷福斯稍微作了修改的七阶段模型，包括模仿"师傅"在内，但师傅不是教练，只是以某种无法详述的方式，列举一个形成"风格"的例子，帮助学习者发展他或她自己。① 但即使是从语言上仲裁，也是错误地把训练刻画为以标准方式分配规则；更确切地说，这关系到模仿和示范，并关注体知型学习者的特殊表现。既然这些表现因身体而异，所以，不能把训练中的体知型知识降低为经验规则。在头三个阶段之后，它也不会消失；总之，它变得越来越重要。许多职业运动员、音乐家、演员和各种类型的表演者——包括德雷福斯的典范性专家鲍比·菲舍尔在内——都会发现，没有教练帮助评价和强化他们的演技就去表演，是不可想象的。因为教练不止是纠正不好的表演；教练也促使好的表演变得更好。

诚然，专家级的表演者与教练或"在 X 方面的专家"发挥着不同的作用。然而，他们是密切相关的，足以意味着，德雷福斯依赖的二分法是错误的，即在"做"的那些人与评论员（那些乱出点子的人）或最好作为预备教育指导"做"的那些人之间的二分法（有人故意作对地想知道，在这里，对明白地讲授抱有偏见，是否并不是学者的职业病）。一位有生活阅历的体知型表演者，对自己的表演过程，缺乏客观的评价。虽然德雷福斯显然不会同意，但那位表演者的演出因而总是在原则上更好地对外行公开，外行的仲裁就有助于表演者演得更好。当然，这不是说，由于一种客观的态度，成就了外行的见识；更确切地说，这是因为，对外行有不同的定位。此外，训练不仅要理解表演者在做什么、什么是好的表演，而且为了促进学生的表演，还理解如何调整到最好。因此，训练是第一人称过程，需要不同类型的语境敏感性；它有自己的技巧、自己的意向弧以及自己对最佳掌握的追求（在这种语境中，这意味着，有能力使条件最优化，允许学生超越他或她当前的表演能力，教练也是如此）。简而言之，训练不仅是使人能成为专家的一个实践过程，而且是专长本身的一种形式——用德雷福斯自己的术语来说。即使专长不等同于专家级的表演者的专长，但两者密切相关，共享着共同

---

① H. Dreyfus, *On the Internet*, New York: Routledge, 2001, pp. 43-46.

的目标：达到好的表演效果。

德雷福斯的解释求助于"知道如何去做"和"知道是什么"之间的不同，这种错误的二分法不仅暴露了这种求助带来的一个问题，而且也暴露了他的解释中的更深层次的缺陷。德雷福斯假定，获得技能的身体与人生经历、基因、种族或年龄无关。① 他确实承认，"文化风格"影响了如何学习技能，例如，注意到在美国人和日本人之间，他们"对待"婴儿的方法是不同的。② 然而，超越未经证实的普遍性，不可能提出"文化风格"这个概念，而且，"文化风格"概念也假定了，共享一种文化的个体会有文化经历的差异，这是无意义的。从德雷福斯的视角来看，人们只要沉浸在实践中，就能开发出专家的情感态度和直觉能力；假定了技能的身体至少原则上能成为直觉的中心所在，处于学徒期的人不受外在实践力量的影响。

席茨-约翰斯通（Sheets-Johnstone）对这种"成年人行为歧视"（adulist）的态度提出了尖锐的批评。她认为，把成年人的身体看成是现象学的起点是一种方法论的错误，因为它忽视了"我们的知识、我们的能力、我们的存在的发源地"③。成年人身体的行为不是简单地发生，而是幼年时"学徒训练"的结果，学徒训练使得成年人的身体充满了意义和体验，对他们所有的未来行为（甚至包括他们继续学习和应用的技能在内）来说，这些体验都是基本的。"不管成年人学习的技能有多么特殊，弹钢琴、开车、下棋、做裤子——这都是由沉淀技能的体验和我们的历史自然形成的概念构成的。"④

诚然，席茨-约翰斯通所意指的学徒训练不同于德雷福斯的意思。对于她来说，学徒训练涉及全然不知的持久体验和起因于它的规则——即使是经验规则——而不是先于它的规则。另一方面，对于德雷福斯来说，学徒训练是一个明确的经过慎重考虑的过程，而且，遵守形成的规则。尽管如此，她指出了德雷福斯解释中所忽视的某种东西：学习的身体总是已经被嵌入（embedded），而不是使身体脱离技能获得的过程。比如，为了学习像舞蹈之类的技能，我必须已经能够活动我的身体。舞蹈的每个身体已经嵌入到活动的方式之中，学习舞蹈不是学

① I. Young, "Throwing Like a Girl", In D. Welton（ed.）, *Body and Flesh: A Philosophical Reader*, Oxford and Maiden, MA: Blackwell Publishers, 1998; M. Sheets-Johnston, "Kinetic Tactile-Kinesthetic Bodies: Ontogenetical Foundations of Apprenticeship Learning", *Human Studies*, No. 23, 2000.

② H, Dreyfus, "Could Anything be More Intelligible than Everyday Intelligibility?" Reinterpreting Diviion I of *Being and Time* in the Light of Division II", In J. Faulconer and M. Wrathall（eds.）, Appropriating Heidegger, Cambridge: Cambridge University Press, 2000, pp. 46-47.

③ Sheets-Johnston, M., *The Primacy of Movement*, Philadelphia: John Benjamins, 1999, p. 232.

④ Sheets-Johnston, M., "Kinetic Tactile-Kinesthetic Bodies: Ontogenetical Foundations of Apprenticeship Learning", *Human Studies*, No. 23, 2000, p. 359.

习简单活动的问题——正如学习一种语言在很大程度上不同于学习说话一样。成年人的身体绝不是以一般的方式活动，而总是以这种方式或那种以反映如何教育人和培养人的方式来活动。

因此，不能说学习舞蹈举例说明了技能获得的德雷福斯式的进展，其中，学习者从遵守规则开始，在高级的初学者阶段了解到了破例的规则，在更高级的阶段学会自发地和流利地制定直到"称之为"情境的规则，等等。学习舞蹈作为一项技能几乎总是需要接受某人的训练，这位教练对个人身体的演技之间的差别和通过模仿与示范调整这些演技的方式是很敏感的。学习者没有丢弃这种嵌入性，他们从人们已有的活动方式和形成新的活动方式出发，而不是进一步发展和改变这种嵌入性。[①] 因此，即使实践的情感态度和直觉能力，也与技能实践最没有直接关系的更广泛的社会影响分不开。这些主张有时是从鲍比·菲舍尔的人生经历中提出的，比如，他独特的咄咄逼人的下棋风格并不只是通过玩许多美国式的象棋游戏养成的，而在某种程度上是通过童年时代的亲身体验形成的。这一点是有道理的，更有说服力；如果真是这样，他的"专家级的回应"，即对特殊行动的选择，不完全是通过语境敏感性和体验来锻造的，而且，下棋比德雷福斯的推测更有生命力。

奇怪的是，在已知德雷福斯的哲学评论中，他的解释缺乏解释学的敏感性。他假设，有技能的行为明确地超越了语境敏感性和体验，个人或文化的经历并不起作用，他的这个假设中的缺陷能被追溯到无法说明这样的事实：体知型主体，即使行为表现得很老练，也面临着从历史上和文化上传承这种情况，甚至主体不可能马上在认知意义上掌握一切。结果，专家级的个体表演者没有完全获得他或她自己的专业行为。因此，与德雷福斯相反，对于一位专家级的表演者来说，从另一个语境敏感的人获得自己的演技原则上将总是可能的——再一次，虽然这不是因为他人设法获得了一种客观的立场，而是因为他人的处境不同。

六、德雷福斯的规范性解释的问题

缺乏解释学的敏感性也影响了德雷福斯的规范解释，因为他的解释假设了专家训练的自主性，这个假设建议，告诫新手要"信任专家"。当一位初学者进入到在文化或情境意义上嵌入了偏见、意识形态或隐藏动机的一种训练程序时，在他达到专家阶段的时候，这些都会被丢弃；正如我们看到的那样，对于德雷福斯来说，一位专家的知识明确地超越了语境敏感性和体验。但如果人们不能拒绝解

---

① R. P. Crease, "Compromising Peer Review", *Physics World*, January, 2002.

释学的循环，那么，他们最多只是转变自己的嵌入性，而不是摆脱它。专长的获得不是对嵌入性和语境的一种超越，而是深化和扩展人们与嵌入性和语境的关系。① 这也意味着，专家将不能使他们自己先验地摆脱这样的怀疑：偏见、意识形态或隐藏的动机潜伏在描述专家评价的前反思性关系中。② 这种怀疑不仅是可以料到的；它的缺乏将造成社会危险。这导致了认可问题（recognition problem），在涉及需要专家建议的许多与技术相关的实际争论中，认可问题扮演着这样一种重要的角色。但在德雷福斯的解释中，这个认可问题无从谈起。他没有给出理由来理解，专家如何合法地受到挑战（或接受教育，就此而言，像在敏感性训练、非专家评审组等事例中那样）。根据德雷福斯的解释，人们根本想象不到社会有可能受到专家的危害，只能想到社会的预期和行动如何能危害专家。我们将会认为，实际争论的故事不仅表明，事情并不像德雷福斯所说的那样发展，而且，如果真是这样，也不太有益。我们断言，这些故事，对于德雷福斯的规范主张来说，相当于是反例，并指出了他的论证中的严重缺陷。

让我们回忆一下，德雷福斯从一开始就假定，拥有专家级技能的人同样也应该在社会上被公认为是专家的人。对于他来说，在特定的社会情境中，谁"算作"是专家，是不成问题的；他假定，不存在认可问题。从他举的例子来看，这是显而易见的。一方面，他提到的人是得到社会公认的专家，比如，飞行员、外

---

① R. P. Crease, "Hermeneutics and the Natural Sciences: Introduction", In R. Crease (ed.), *Hermeneutics and Natural Sciences*, Dordrecht: Kluwer, 1997.

② 有一次，德雷福斯建议说，如果专家提供某类叙述，就能合法地得到信任。当他提出这个建议时，德雷福斯似乎缓解他的 IT 论题。他建议，只要专家不需要对他们制定决策时遵守的规则作出演绎说明，他们就可能是有效的交流者："在直觉文化中对职称专家的盘问可能会出现相互冲突的解释形式，其中，要求每一位专家产生和捍卫一种一致性的叙述，这种叙述致使自然地接受他的观点。"（Dreyfus, H. and Dreyfus, S., *Mind Over Machine: The Power of Human Intuition and Expertise in the Era of the Computer*, New York: Free Press, 1986, p.196）这一段是有趣的，因为它意味着，专家不仅能产生而且能"捍卫"所谓"一致性叙述"的东西，可以在直觉文化中评价这种叙述，没有危及专长。这一段是成问题的，因为德雷福斯没有说明，什么是"一致性的叙述"，为什么它是如此的有效，以至于"致使自然地"（另一个未定义的短语）接受一种观点。这种含义是，在某种程度上，专家能够回避 IT 问题，这与我所说的"实践的观点"相关，由于在他们的"一致性的叙述"中，提出了另一种观点。"一致性的"这个词可能被认为意味着整体论，德雷福斯想到的观点，就像布鲁尔的"理论整体论的"专家观点一样。但如果德雷福斯正在唤醒理论整体论，那么，他并没有说明，他如何能指望专家有能力提供这样一种作为专家的叙述，根据 IT 观点，专家在产生命题内容（原文"prepositional content"有误，这里应该是"propositional content"——译者注）时"放弃了"他们的专长。这种引导性预设是，这个"一致性的叙述"无论是什么，专家无疑只是在像日本文化之类的"直觉文化"中生产。这种预设不仅以不受支持的文化本质主义为依据，而且这对 RT（recognition problem）产生了质疑。在演绎程序步骤的基础上，非专家可能无法认可专家，但德雷福斯表明，认可可能在对"一致性的叙述"内容的自然回应之基础上产生。没有说明存在着什么，至于为什么非专家应该信任专家的直觉，德雷福斯回避了问题的实质。

科医生、象棋大师，目标是举例说明体知型的专家级的行为表现是如何起作用的。① 他举的这些事例从叙述上看是有负载的，因为它们用社会上承认的专家作为对专长的反社会的现象学描述的证据。另一方面，他把世俗的日常活动的例子，比如，开车、走路、说话和交谈，描绘为专家如何表现的范例，即使在社会意义上，也不能规范地认为这些活动是由专家完成的。② 社会上公认的是，在专家层次上，只出现高超的技能表现，不会出现世俗的技能表现，但在这两种技能表现中，有相同类型的专家资格。通过这种论证，德雷福斯促进了他实现去除表面上有迷人力量的专家评价的神秘性的目标，确立了专家和每天的生活世界活动之间的连续性。对于他而言，在这两类事例中，专长是技能与环境之间的一种前反思关系："专家的技能已经成为他的很大一部分，以至于他需要对技能的觉察就像对自己身体的觉察一样。"③。

从德雷福斯的视角来看，当社会的能动者无法承认由现象学的研究者明确地提供的专长的基本品质时，就会产生社会问题。这些社会问题的来源与专长的来源不同；它们是由政治、社会和文化参与的干预导致的。但这种解释学循环将会把这一点抛在脑后；这些参与首先致使承认和信任专家。即使人们听从传统的"专家"，那种听从也是人们的特殊背景和生活世界参与的结果。因为专长是一种双向关系：对专长本身的认定涉及社会需求；这种认定不只是中立的识别标志，而且是他人应该听从专家判断的一个声明。因此，专长现象最终与其社会效用有着千丝万缕的联系；专家不仅"在"一个领域内，而且"代表了"群众。

一个显而易见的例子是专家的"选择"问题。在许多事例中，支持同一领域内的不同结论的专家可能与另一些被认为是"反对的专家"相抗衡。"反对的专家"的共同策略在于要求主张，专家提供的判断没有信任度，归因于存在偏见、意识形态或隐藏的动机。"反对的专家"在司法界是特别普遍的，在那里，

① Dreyfus, H. and Dreyfus, S., *Mind Over Machine*: *The Power of Human Intuition and Expertise in the Era of the Computer*, New York: Free Press, 1986, pp. 30-35.

② Dreyfus, H. and Dreyfus, S., *Mind Over Machine*: *The Power of Human Intuition and Expertise in the Era of the Computer*, New York: Free Press, 1986, p. 30.

③ Dreyfus, H. and Dreyfus, S., *Mind Over Machine*: *The Power of Human Intuition and Expertise in the Era of the Computer*, New York: Free Press, 1986, p. 30.

专家鉴定人发挥的作用符合为原告和被告提供证词的抗辩系统的逻辑。①

斯蒂芬·特纳（Stephen Turner）认为，某人为了成为一名专家，他或她不仅需要有技能，而且需要有群众，群众在社会意义上承认他或她的技能类型为有技能的专长，他的专长解释由于指出这一点提出了认可问题。②特纳主张，尽管默顿所说的"认知权威"既不是一个"能被分配的对象"，也不是能被"简单转让的某种东西"，然而，它基于不同群众的评价"通向反抗和屈服"。③相比之下，德雷福斯只在技能获得与运用的基础上定义专家，在方法论上，他把群众排除在专家与专长的描述之外，把专家定义为对情境具有正确情感态度和直觉回应的人。

这种方法论的策略比德雷福斯的现象学有许多优势。它说明了为什么：①专长不是一个稳定的特性，但却能被获得和失去；②专长的讨论为什么往往不仅集中于认识论问题，而且集中于政治问题；③专长的感知为什么能被建立在历史转型的基础上。在把专长与这些联系起来时，特纳放弃了揭示普遍结构和超验的特殊性的哲学目标（即只能讨论与局部特征相关的专长）。例如，他提供了说明大致把专家区分为五种类型的下列"分类"：

他们的专长通常会得到承认的群体成员的专家，比如，物理学家；

---

① 正如希拉·加萨诺夫（Shelia Jasanoff）指出的那样，选择真正专家的问题在法庭上是如此之困难，以至于最初1923年在美国的弗赖伊案件中不得不提炼"普遍接受"的使用标准，因为它"关于在多大程度上或在多少人中间达成共识是足够的，并没有提供指导"（Jasanoff, S., *Science at the Bar*: *Law*, *Science*, *and Technology in American*, Cambridge: Harvard University Press, 1995, p. 62）。普遍接受是RT的一个潜在预设，没有起到合法的作用，因为它不能：a. 澄清在多大程度上需要建立普遍接受的共识；b. 为应该分析边界工作的变化为何会产生矛盾结果确立指导原则；c. 决定如何权衡相对于现有科学的边界所产生的结果（S. Jasanoff, *Science at the Bar*: *Law*, *Science*, *and Technology in America*, Cambridge: Harvard University Press, 1995, p. 62）。既然在司法系统中，把普遍接受作为判断专家共识的标准证明是模糊的和无效的，因此，德雷福斯把它延伸到使IT与RT潜在地关联起来是成问题的。通过表明，一个问题是如此之复杂，以至于专家无法对如何解决问题达成共识，就产生了感觉主义，在这种情况下，新闻界也惯例性地使用普遍接受的标准。但对于德雷福斯来说，这并不是问题。德雷福斯总的来说高估了专家的可信赖性，因为他把专家解释为普遍范畴，在这方面，无法承认专家的共识是特定领域。既然专家共识的层次是特定领域的，由此推出，在德雷福斯的一般术语中，专家直觉的可信赖性是不能被阐述的东西。

② S. Turner, "What Is the Problem With Experts?" *Social Studies of Science*, No. 31, 2001, p. 138. 这并不意味着，比较群众规模的大小认可每个专长领域。例如，物理学中的专家具有的专业技能比神学家拥有的专业技能得到了更广泛的认可，神学家只被某一特殊教派所认可，特纳称之为"有限制的群众"（S. Turner, "What Is the Problem With Experts?" *Social Studies of Science*No. 31, 2001, p. 131）。

③ S. Turner, "What Is the Problem With Experts?" *Social Studies of Science*, No. 31, 2001, p. 128. 例如，在讨论按摩师与认可之间的关系时，特纳注意到，有些人觉得，他们从按摩疗法中获益，而另外一些人并不觉得履行了按摩疗法的承诺。因此，只有前者，才会把按摩师看成是专家："这样，按摩师已经……创建了群众、一群追随者，对于这些人来说，他们是专家，因为他们通过自己的行动向这群人证明了他们自己。"（S. Turner, "What Is the Problem With Experts?" *Social Studies of Science*, No. 31, 2001, p. 131）

他们的个人专长经受了考验并被人们所接受的专家，比如，自助书（self-help book）的作者；他们的专长只被特殊群体所接受的专家，比如，神学家，他们的权威只被他们的宗教人士所接受；他们的群众是公众但从有兴趣把他们的意见接受为是权威性的政党获得资金支持的专家；以及他们的群众是有自由决定权的官僚主义者的专家，比如，公共管理方面的专家，他们的观点被公共管理者接受为是权威性的。①

这样，特纳的解释就使专长现象关系到变化的历史感知和社会感知。德雷福斯很可能反驳道，没有理由把他的现象学模型理解为与特纳的模型不一致——特纳只是扩展了德雷福斯的模型，因为德雷福斯描述了什么是专家级的技能，特纳描述了不同的群众如何最终辨认出这些技能是专家水平的技能。但批评的方法论的不同之处是，特纳没有把技能的拥有看成是专长的充分必要条件，遗漏了提到的群众，而德雷福斯正相反。②

下列悖论突出了德雷福斯为什么没有提出认可问题的原因。一方面，他认为，只有专家，才能把另一个人认可为是真正的专家，因为非专家在评价某个人是否有技能时，不知道寻找什么。③只有专家，才能懂得这个人。这被称为差异性主张（difference claim，DC）：专家"与我们不一样"。DC完全不会招致反对，乃至特纳坚持一种看法："只有论证了专家断言的真理性，才能说服其他专家，这是专长的特征；我们大家必须把专家的断言接受为是真的根据，与其他专家把专家的断言接受为是真的根据，是不同的。"④另一方面，德雷福斯也提出了相似性主张（similarity claim，SC），据此，"他们"——专家——"与我们很像"。他们的行为方式与我们大家在每天活动中的行为方式相类似。在每天处理日常事

---

① S. Turner, "What Is the Problem With Experts?" *Social Studies of Science*, No. 31, 2001, p. 140.

② 然而，人们可能担心，特纳通过坚持认可专长的本质主义维度，削弱了专长的客观性基础。历史上充满了在一个瞬间被公认为专家在另一个瞬间被声讨为江湖骗子的人的例子。特纳承认，"所谓的'专家'是约定的、易变的和多变的，通过可变的、多变的约定，说服人们相信专家的断言"（S. Turner, "What Is the Problem With Experts?", *Social Studies of Science* No. 31, 2001, p. 145）。尽管如此，他声称，纠正错误判断的特权是民主生活的一个关键部分，而且，坚持一个高标准是"不切实际的"（S. Turner, "What Is the Problem With Experts?" *Social Studies of Science* No. 31, 2001, p. 146）。

③ 再一次，考虑德雷福斯的例子是有益的。在德雷福斯描述的一个实验中，能够找到DC的一个例子，在这个实验中，学生、有经验的护士和CPR（即Cardiopulmonary Resuscitation，指心肺复苏术——译者注）教师观看为病人提供的六名模范人员（五名学生和一名有经验的护士）的录像带，这样，向目标群提出的问题是，他或她将选择哪一名模范人员来拯救他自己的生命。德雷福斯写道："结果是有启发意义的。在护士人群中，十个人中有九个人选择有经验的护士。十个学生中有五个人选择学生。教师试图寻找一名护士，方法是，看一下知道规则的哪一名护士严格遵循规则，结果，他们没有能找到这名专家，因为一名有经验的护士已经超越了遵循规则的阶段！"（Dreyfus, H. and Dreyfus, S., *Mind Over Machine: The Power of Human Intuition and Expertise in the Era of the Computer*, New York: Free Press, 1986, p. 201）

④ Dreyfus, H., *On the Internet*, New York: Routledge, 2001, p. 129.

务的最基本的层次上，每一个人都值得被刻画为是一名专家："公民都有讲话的专长，不管他们是善于根据外交文化与政府做交易的大学教授，还是议论需要立法来解决具体问题的农民或小店老板。"① 这样，我们能在影射意义上认同专家，理解在他们的判断中所运用的那种知识。问题不仅仅是，德雷福斯需要促进这两者，到最后，他需要 SC 胜过 DC。否则，非专家将没有任何根据去承认、接受和信任专家拥有的那种知识。

这一点能够参照易卜生的戏剧《人民的敌人》中描绘的情境得到最直接的揭示。核心角色托马斯·斯托克曼（Thomas Stockmann）是一位温泉疗养医生，他家乡的人以温泉为生。他认为，一种无形的毒物正在污染泉水，他把水样送给大学里的一位专家化验后，得到了确证。他企图通知乡镇人员，提醒他们注意危险，认为他们会因此而感激他。但一位朋友警告他说，并不能这么肯定他们会作出怎样的回应："你是一名医生，一名科学家，对你来说，这种水的生意是可以单独考虑的事。我认为，你也许还没有意识到，这与其他事情联系在一起。"斯托克曼引证了专家的化验结果，坚持认为，"机灵聪明"之人将会"被迫"接受这个消息。镇长却指出，对于市民来说，事情从经济角度考虑比从科学角度考虑更重要；的确，市民们把斯托克曼谴责为是一位"人民的敌人。"这种基本冲突因此涉及科学家和市民，科学家把专家的技术忠告接受为是权威的，而市民没有发现专家忠告的权威性，却发现它威胁到了他们的世界，并寻找其他人的指导。

易卜生虚拟的情境像是这样一个范例，它剥去非基本的细节来澄清一种情境的基本势力。这种情境涉及对有关科学技术方面的一种反复无常的争论，在这方面，人们已经对传统的"专家"失去信心。站在群众的立场上，我们看到，这两个故事能从两种不同的视角来谈论这样一种情境。从专家的视角来看，即从斯托克曼和大学化学家的视角来看，关于下列问题的争论根本就是不确定的或没有根据的：谁是专家、人们需要什么类型的训练，以及什么类型的信息（水质的技术问题）与那种特殊社会情境的利益的最大化相关。专家像市民一样希望使乡镇利益最大化；他们根据他们的特殊知识这么做。市民不认可专家，也不听从专家的忠告，是由于受到了外围的乃至有害的和腐蚀的经济与政治动机的影响，以及市民与政治家、媒体和其他非专家权威的参与的影响。另一方面，从市民的视角来看，事情更混乱。市民的背景和生活世界的卷入意味着，经济和政治动机的作用似乎比斯托克曼和化学家动机的作用大得多，而且建议说，他们为了寻找促进他们的福利，应该听从不同的人的忠告。从这个视角来看，所谓的"专

---

① Dreyfus, H., Spinosa, C., and Flores, F., *Disclosing Worlds*: *Entrepreneurship*, *Democratic Action*, *and the Cultivation of Solidarity*, Cambridge：MIT Press, 1997, p. 107.

家"——斯托克曼和化学家——确实与其他市民不一样，因为市民有不同的动机（而且，这位化学家真的是一名外行）。

我们从第三人称视角，即代表群众的视角，看到了这两种定位视角之间的差别。这种第三人称视角也不是中立的，其效果是突出了相似性主张；我们承认，斯托克曼和市民有共同的人性，他的知识与他们的福利相关。当人们完全重视这种冲突的严重性及其解决方案的不可能性的时候，人们才能意识到，市民的福利要求他们听从专家。但我们在群众中没有怀疑谁是真正的专家和在基本的外围参与之间的差异。群众没有站在某种立场上，即没有站在与这方面相关的立场上；简而言之，这种第三人称的立场在解释学意义不是一种敏感的立场。

然而，在任何一种真正的争论中，没有人站在群众的立场上；每个人事实上都是"在舞台上"占据某一位置，即根据对剧情提供一种特殊理解的特殊的参与采取某种观点，并据此倾向于把某人而不是其他人接受为权威。哪些参与是基本的，哪些参与是特殊的；谁的行动抓住了偏见、意识形态和隐藏的动机，谁的行动对社会最有利，这些事先都不清楚。与此同时，专家也持有一种特殊的立场，即站在某种立场上，有某种特殊的参与，而且，专家评价的要求是应该听从某人的一个命令、一种心理效价、一种需求。在真实生活中，涉及专家评价的每一种争论，都采取了在下列两类人之间作出调节的形式：一些人是提高对专家评价的要求，来提高他们的权威；另一些人是寻找要听从的正确权威。这是解释学的困境：没有从对一种特定情境的特殊参与中解脱出来。专家评价没有，也不可能有护身符。

这个问题在关于涉及技术维度——特别是涉及公共安全——的反复无常的公共争论中是最明显的，在那里，传统的权威来源已经失去信任。在这些情况下，谁是权威的问题，即谁是专家的问题，是有争论的。每一位市民，以及被提升为专家的每一个人，都有一个特殊的介入集合（set of involvements），而且，根本没有安全的听众立场，能用来在一名专家的判断中挑选出本质的介入和非本质的介入。

比如，考虑关闭国家氚标记设施（NTLF）的案例——碰巧，德雷福斯的妻子在那里工作。这种设施通过把氚放入特殊分子的适当位置，独特地创造了加标记的分子，即生成一种在基础研究和应用研究中用来研究生化转化机制的追踪剂。但反核积极分子反对这一事实，因为此过程会把一些氚释放到环境中。科学家、当地州政府和联邦的公共卫生专家，在认真考察了这种情况之后，声称排放物是安全的；环保局的一部分人建议限量排放，限量排放反过来是一部分本底水平的辐射。但反核积极分子企图怀疑那些主张，说它们要么是由那些一线工作人员或在某种方面与这个设施相关的机构提出的，要么是由太了解氚以至于是有私心的科学家提出的。这些反核积极分子对正常的社会谈判程序的干扰是有效的，因为他们权威性地讲到了安全问题，最后，这种设施被关闭了。[①]

或者再一次考虑围绕 1976 年从布鲁克海文国家实验室的高能反应堆运输用过的燃料棒的争论。[②] 这场争论是反核积极分子与科学家之间的对立，反核积极分子把研究反应堆与动力反应堆/核武器/军队联系起来；对于科学家来说，这样的联系根本没有意义。这种情境产生了特纳的分类所描述的那类一连串的专家。在牵涉到布鲁克海文实验室的另一个争论中，下列项目也遭到了攻击：这个项目在追查研究马歇尔岛民的健康状况时发现，岛民偶然受到核武器试验的辐射微尘的辐射，同时，一个复杂的因素是，这个项目涉及工作在第三世界国家并不熟悉其语言和风俗的美国科学家的典型的殖民情况。[③]

这些争论中的每一场争论都是复杂的，并牵涉到很大的利害关系。每一场争论都与科学技术问题相关，因此，有必要求助于专家意见。然而，谁是"真正的"专家，被视为是人们依赖于谁，站在谁的立场上。要理解关于专家和专家意见的这些争论，要求超越实践的专家的观点。

七、结　　论

德雷福斯的专家级的技能获得模型在哲学上是重要的，因为对于体知型的认知和情感来说，它把焦点从它在 STS 中的社会的外在化和技术的外在化，以及在经典科学哲学中排除发现的历史语境和心理学语境，转向了专长。在这么做时，他说明了为什么把专家描述为是意识形态的拥护者不是最恰如其分的，以及为什么他们的权威性不是一概建立在社会网络基础之上的。此外，他通过从第一人称视角对专长的现象学分析，揭示了从第三人称视角研究专长的局限性和有时造成的表面论述。因此，他表明，基本的科学问题，只有借助于现象学的工具才能得到全面阐述，专长是这种问题的最典型的一个事例。

然而，德雷福斯的描述性模型和他的规范性要求，由于缺乏解释学的敏感性，因而是有缺陷的。也就是说，他假定，专家的知识在语境敏感性和体验范围之外是明确的，而且，专家在训练过程中已经摆脱了人们开始时拥有的任何偏

---

① R. P. Crease, "Compromising Peer Review", *Physics World*, January, 2002.

② R. P. Crease, "Conflicting Interpretations of Risk: The Case of Brookhaven's Spent Fuel Rods", *Technology*, No. 6, 1999.

③ R. P. Crease, "Fallout: Issues in the Study, Treatment, and Reparations of Exposed Marshall Islanders", In R. Figueroa and S. Harding (eds.), *Exploring Diversity in the Philosophy of Science and Technology*, Routledge, 2002.

见、意识形态或隐藏的动机。这个假设不仅在德雷福斯的描述性解释中是有缺陷的，而且在他的规范性解释中也是有缺陷的。

现象学的目标是揭露潜藏在自由态度中没有被捕获到的预设。现象学的体验揭示了，要揭露和控制的最困难的预设是最接近事实的那些预设。本着这种精神，虽然可能带有刚愎自用的痕迹，但人们可能通过提出下列问题揭露这种局限性："为什么吉纳维夫·德雷福斯不再在国家定氚标记设施处工作呢？"

从德雷福斯自己的解释来看，这是因为损坏和腐蚀了一种合法的和在现象学意义上可辩护的权威。一个反核改革小组设法操纵社会谈判过程，说服管理部门监督强健的科学和专家的忠告，并作出一种纯粹的政治决定。但从另一个很普通、很重要且很可能具有强有力的解释学敏感性的观点来看，人们可能会讲述一个相当不同的故事。个体不是赤裸裸地接受信息项，即使是科学信息。重要的是谁传递信息以及在什么语境中传递信息。伯克利反核积极分子与科学家相比，生活在一个不同解释的世界里，支持一套不同的行为、价值和风俗习惯；他们用来解释事实、原理及其应用的产生意义的过程是很不同的。① 因此，不能把科学家与反核积极分子之间的差别看成是认知者与破坏或出卖那种知识的那些人之间的差别，而这更像是一种文化的成员与另一个文化的成员之间的关系。无疑，对这种解释学困境的描述，适合于对专长的任何一种解释，德雷福斯自己的解释中没有这种困境。

最后，我们相信，当从内在的视角评价时，德雷福斯的专长解释，按照他自己的哲学标准来说是失败的。在许多出版物中，德雷福斯更加捍卫海德格尔的现象学进路，而不是胡塞尔的现象学进路。他甚至认为，海德格尔在《存在与时间》第二部分中对可靠的 Dasein（此在）的解释使 Dasein 成为一名"专家"②。然而，由于作为一种静态的方案来接近信任（其中，非专家恳求专家的忠告），所以，德雷福斯诉诸隐含的胡塞尔的主体间性的框架。SC 的本质是类推：正如我不指望自己能明确阐述如何开车或骑车的规则那样，我不应该指望诸如弹道学的主考官、护士和生态学家之类的专家，来辩护他们参照规则作出的决定。德雷福斯潜在地认为，即使信任是以专家的业绩为基础而在社会层面确立的，但在现象学层面，信任是通过胡塞尔所说的"主体间的配对"（intersubjective pairing）确立的。通过从我的行为到专家的行为进行前反思的类推，我承认，专家的行为

---

① R. P. Crease, "Conflicting Interpretations of Risk: The Case of Brookhaven's Spent Fuel Rods", *Technology*, No. 6, 1999, p. 498.

② H. Dreyfus, "Could Anything be More Intelligible than Everyday Intelligibility?", Reinterpreting Division I of *Being and Time* in the Light of Division II", In J. Faulconer and M. Wrathall（eds.）, *Appropriating Heidegger*, Cambridge: Cambridge University Press, 2000.

基本上类似于我自己的行为。这名专家是一名专家的"另一个我"（alter ego）。在德雷福斯的框架中，我应该信任专家，因为：①我相信我自己以类似于直觉的方式作决定；②我相信我的决定是好的决定，即使我像专家一样，以命题的形式根据规则辩护这些决定。即使专家比我有更多的训练，我们在认知上的相似性超过了我们在技术上的差异性。

德雷福斯在这里对主体间性的解释，由于缺乏解释的敏感性，使人更多地想到的是胡塞尔，而不是海德格尔。的确，他描绘的专家画像，使人想起海德格尔的漫画，据此，可靠的 Dasein 被赋予了太崇高的自由。在这幅专家画像中，专家在通过所有的发展阶段之后，掌握了他或她自己的相关专长，并且，专家觉得不需要寻找对他的能力的外在认可。但海德格尔对世界性（worldhood）的解释学维度的敏感性，不得不把专家描绘为卷入了一个更加脆弱的易受攻击的过程，在这个过程中，专长似乎不像是最终目标；最终目标是，个人在那个世界中超越了人的嵌入性。专长总是一个逐渐形成（becoming）的过程，至少在原则上，对于教练员、评论员和自己的专长与履行者逃避掌握的专家的行为表现部分一致的那些人来说，这个过程总是可能的。

（成素梅 译）

Ultimate Justification in Husserl and Wittgenstein

Dagfinn Føllesdal

The aim of the 27th International Wittgenstein Symposium was to investigate the relations between phenomenology and analytic philosophy. It is gratifying that these various traditions in contemporary philosophy are now communicating and concentrating on the philosophical issues. Some decades ago the labels were mostly used polemically. One criticized various "-isms" or traditions: positivism, existentialism, phenomenology, analytic philosophy, etc. These "-isms" were normally not defined and the "arguments" directed against them were correspondingly loose. As could be expected, these arguments had no effect on those who were criticized—they had difficulties recognizing themselves in the views criticized. The so-called "Positivismus-Streit" left no traces, except in the minds of some students who abstained from reading "positivist" authors. Fortunately, such label-throwing has gone out of fashion; it is rare that one hears somebody write off whole groups of philosophers without having read them. To take phenomenology as an example, it is no longer rejected off-hand without examination. During the last decades numerous so-called "analytic" philosophers have discovered that Husserl and other phenomenologists had a lot to say on topics that interest them, such as perception, intersubjectivity, objectivity, action, and above all intentionality, where Husserl more than a hundred years ago set forth ideas and analyses that are strikingly similar and in some cases better than those that have emerged during the last decades.

In this paper I will discuss a much neglected topic in Husserl, that of ultimate justification, and compare and contrast his view with that of Wittgenstein, mainly in *On Certainty*.

Two elements are involved in justification: perception and transfer of evidence. On both of them Husserl had interesting things to say, and on both he anticipated Wittgenstein.

1. Husserl on Perception[1]

Already in his *Philosophy of Arithmetic* (1891) Husserl anticipated Gestalt psychology, and the ambiguous pictures that were so popular in Gestalt psychology can be used to illustrate Husserl's ideas on perception and also to bring out the similarities and differences between him and Wittgenstein on this point. Jastrow's duck/rabbit picture that became famous through Wittgenstein is a good starting-point:

Husserl maintained that perception is underdetermined: what reaches our senses is never sufficient to uniquely determine what we experience. According to Husserl, our consciousness structures what we experience, and our experience in a given situation can always in principle be structured in different ways. How it is structured, depends on our previous experiences, the whole setting of our present experience and a number of other factors. Thus, if we had grown up surrounded by ducks, but had never even heard of rabbits, we would have been more likely to see a duck than a rabbit when confronted with the duck/rabbit picture; the idea of a rabbit would probably not even have occurred to us. In a few rare cases, such as in the duck/rabbit example, we can go back and forth at will between different ways of structuring our experience. Usually we are not even aware of any structuring going on; objects are simply experienced by us as having a structure.

The structuring always take place in such a way that the many different features of the object are experienced as connected with one another, as features of one and the

---

[1]   A fuller discussion of Husserl's theory of perception may be found in Føllesdal 1974 and 1978. A clear and pedagogical presentation and discussion of Husserl's theory of perception is given in Miller 1984. See also the excellent Mulligan 1995.

same object. When, for example, we see a rabbit, we do not merely see a collection of colored patches, various shades of brown spread out over our field of vision (incidentally, even seeing colored patches involves intentionality, a patch is also a kind of object, but a different kind of object than a rabbit). We see a rabbit, with a determinate shape and a determinate color, with the ability to eat, jump, etc. It has a side that is turned toward us and one that is turned away from us. We do not see the other side from where we are, but we see something that has another side.

It is this peculiarity of our consciousness that Husserl labels *intentionality*, or directedness. That seeing is intentional, or object-directed, means just this, that the near side of the object we have in front of us is regarded as a side of a thing, and that the thing we see has other sides and features that are cointended, in the sense that the thing is regarded as more than just this one side. This structure that makes up the directedness of consciousness, Husserl called the *noema*. The noema is the comprehensive system of determinations that gives unity to this manifold of features and makes them aspects of one and the same object.

It is important at this point to note that the various sides, appearances or perspectives of the object are constituted together with the object. There are no sides and perspectives floating around before we start perceiving, which are then synthesized into objects when intentionality sets in. There are no objects of any kind, whether they be physical objects, sides of objects, appearances of objects or perspectives of objects without intentionality. And intentionality does not work in steps. We do not start by constituting six sides and then synthesize these into a die; we constitute the die and the six sides of it in one step.

The word "object" must, as we have noted, be taken in a very broad sense. It comprises not only physical things, but also, as we have seen, animals, and likewise persons, events, actions and processes, and sides, aspects and appearances of such entities.

We should also note that when we experience a person, we do not experience a physical object, a body, and then infer that a person is there. We experience a full-fledged person, we are encountering somebody who structures the world, experiences it from his or her own perspective. Our noema is a noema of a person, no inference is involved. Seeing persons is no more mysterious than seeing physical objects, no inference is involved in either case. When we see a physical object we do not see sense data or the like and then infer that there is a physical object there, our noema is the noema of a physical object. Similarly, when we see an action, what we see is a full-fledged action, not a bodily movement from which we infer that there is an action.

2. Filling, the Hyle

In the case of an act of perception, its noema can also be characterized as a very complex set of expectations or anticipations concerning what kind of experiences we will have when we move around the object and perceive it, using our various senses. We anticipate different further experiences when we see a duck and when we see a rabbit. In the first case we anticipate, for example, that we will feel feathers when we touch the object, in the latter case we expect to find fur. When we get the experiences we anticipate, the corresponding component of the noema is said to be *filled*. In all perception there will be some filling: the components of the noema that correspond to what presently "meets the eye" are filled, similarly for the other senses.

Such anticipation and filling is what distinguishes perception from other modes of consciousness, for example imagination or remembering. If we merely imagine things, our noema can be of anything whatsoever, an elephant or a locomotive standing here beside me. In perception, however, my sensory experiences are involved; the noema has to fit in with my sensory experiences. This eliminates a number of noemata which I could have had if I were just imagining. In my present situation I cannot have a noema corresponding to the perception of an elephant. This does not reduce the number of perceptual noemata I can have just now to one, for example, of you if you were sitting in front of me.

As we noted earlier, a central point in Husserl's phenomenology is that I can have a variety of different perceptual noemata that are compatible with the present impingements upon my sensory surfaces. In the duck/rabbit case this was obvious; we could go back and forth at will between having the noema of a duck and having the noema of a rabbit. In most cases, however, we are not aware of this possibility. Only when something untoward happens, when I encounter "recalcitrant" experience that does not fit in with the anticipations in my noema, do I start seeing a different object from the one I thought I saw earlier. My noema "explodes", to use Husserl's phrase, and I come to have a noema quite different from the previous one, with new anticipations. This is always possible, says Husserl. Perception always involves anticipations that go beyond what presently "meets the eye", and there is always a risk that we may go wrong, regardless of how confident and certain we might feel. Misperception is always possible.

The experiences I typically have when my sensory organs are affected, and which constrain my acts of perception, Husserl calls "hyle", using the Greek word for matter. He regarded his view on perception as a variant of hylomorphism; the hyle, my sensory experiences, constrain and are informed by the noema, the "form". Note that the hyle, for Husserl, are experiences, they are not objects of acts, they are components of acts. We can turn them into objects by reflecting on them, but in the normal acts of perception they are not perceived, but constrain our acts of perception. For Husserl there is not something unstructured "given" in perception, no intermediaries of the kind appealed to by sense datum theorists.

3. The World and the Past

We structure, or to use Husserl's word, "constitute" the thing not only so that it has various properties, but also so that it stands in relations to other objects. If, for example, I see a tree, the tree is conceived of as something which is in front of me, as perhaps situated among other trees, as seen by other people than myself, etc. It is also conceived of as something which has a history: it was there before I saw it, it will remain after I have left, perhaps it will eventually be cut and transported to some other place. However, like all material things, it does not simply disappear from the world.

My consciousness of the tree is in this way also a consciousness of the world in space and time in which the tree is located. My consciousness constitutes the tree, but at the same time it constitutes the world in which the tree and I are living. If my further experience makes me give up the belief that I have a tree ahead of me because, for example, I do not find a treelike far side or because some of my other expectations prove false, this affects not only my conception of what there is, but also my conception of what has been and what will be. Thus in this case, not just the present, but also the past and the future are reconstituted by me. To illustrate how changes in my present perception lead me to reconstitute not just the present, but also the past, Husserl uses an example of a ball which I initially take to be red all over and spherical. As it turns, I discover that it is green on the other side and has a dent:

... the sense of the perception is not only changed in the momentary new stretch of perception; the noematic modification streams back in the form of a retroactive cancellation in the retentional sphere and modifies the production of sense stemming from earlier phases of the perception. The earlier apperception, which was attuned to the harmonious development of the "red and uniformly round", is implicitly "reinterpreted" to "green on one side and dented". [1]

4. Values, Practical Function

So far I have mentioned only the factual properties of things. However, things also have value properties, and these properties are constituted in a corresponding manner, Husserl says. The world within which we live is experienced as a world in which certain things and actions have a positive value, others a negative. Our norms and values, too, are subject to change. Changes in our views on matters of fact are often accompanied by changes in our evaluations.

Husserl emphasizes that our perspectives and anticipations are not predominantly factual. We are not living a purely theoretical life. According to Husserl, we encounter the world around us primarily "in the attitude of the natural pursuit of life", as "living functioning subjects involved in the circle of other functioning subjects". Husserl says this in a manuscript from 1917, but he has similar ideas about the practical both earlier and later. Thus in the *Ideas* (1913) he says:

> ... this world is there for me not only as a world of mere things, but also with the same immediacy as a world of values, a world of goods, a practical world. [2]

Just as Husserl never held that we first perceive bodies and bodily movements and then infer that there are persons and actions, or that we first perceive sense data, or perspectives or appearances, which are then synthesized into physical objects, so it would be a grave misunderstanding of Husserl to attribute to him the view that we first perceive objects that have merely physical properties and then assign to them a value or a practical function. Things are directly experienced by us as having the features, functional and evaluational as well as factual, that are of concern for us in our natural pursuit of life.

---

[1] Husserl 1939, § 21a, 96 = p. 89 of Churchill and Ameriks' English translation.

[2] Husserl 1913, § 27, Husserliana III, 1, 58. 13-19 = Kersten's translation, 53. I have changed his translation slightly.

5. Seeing as: Husserl versus Wittgenstein

These basics of Husserl's theory of perception go far beyond Wittgenstein. However, there is one important difference between Husserl and Wittgenstein on perception that should be mentioned. Wittgenstein is often praised for having observed that all seeing is "seeing as". Husserl might seem to maintain the same. However, Husserl's view is different. In normal perception there is no object that is seen as one thing, for example a duck, on one occasion and as another, say a rabbit, on another occasion. There is no underlying more basic object that is seen as one object or another. There is hyle, but the hyle are not objects seen; they are experiences that constrain our structuring, they are not objects experienced. Wittgenstein, unfortunately, started out with pictures, where there is an object, viz. the picture, that can be seen one way or the other, and he modeled his view on perception on this.

If we shall get a proper understanding of Husserl's theory of perception, we should therefore not use Jastrow's picture as our example, but rather look at a case of normal perception, where we are out one evening and see a silhouette like the Jastrow picture against the horizon. In such a case I can see a duck or a rabbit, but there is no picture or other object that is seen as a duck or a rabbit. One might think that the silhouette is such an object, just like the picture in Jastrow's and Wittgenstein's example. However, one would then miss a fundamental point in Husserl's theory of perception, a point that saves his theory from problems that have become stumbling blocks for most other theories of perception: there are no intermediaries in perception, neither sense data, pictures nor silhouettes.

There are passages in Wittgenstein that indicate awareness that perception should not be modeled on seeing pictures. However, the enthusiasm about "seeing as" has tended to overshadow the fundamental difficulties of basing a theory of perception on picture-seeing.

6. Reflective Equilibrium

As will be obvious from the preceding, Husserl did not regard perception as a firm, unshakeable foundation for science and knowledge. There is interplay between perception and our various past experiences and theories of the world that make perception malleable. This is one feature of Husserl's general view on justification that it shares with more recent views to which we will now turn.

These modern views on justification, which are very similar to that of Husserl, are often called "reflective equilibrium" views, a label introduced by Rawls. (Rawls 1971) One of the clearest statements of the view is given by Nelson Goodman, in *Fact, Fiction and Forecast*, who here discusses the problem of justification of *deduction*, which many philosophers have taken to be obvious, not in need of justification:

> How do we justify a deduction? Plainly by showing that it conforms to the general rules of deductive inference.
>
> . . . . . .
>
> But how is the validity of rules to be determined? Here again we encounter philosophers who insist that these rules follow from some self-evident axiom, and others who try to show that the rules are grounded in the very nature of the human mind. I think the answer lies much nearer the surface. Principles of deductive inference are justified by their conformity with accepted deductive practice. Their validity depends upon accordance with the particular deductive inferences we actually make and sanction. If a rule yields inacceptable inferences, we drop it as invalid. Justification of general rules thus derives from judgments rejecting or accepting particular deductive inferences.
>
> This looks flagrantly circular. I have said that deductive inferences are justified by their conformity to valid general rules, and that the general rules are justified by their conformity to valid inferences. But the circle is a virtuous one. The point is that rules and particular inferences alike are justified by being brought into agreement with each other. *A rule is amended if it yields an inference we are unwilling to accept; an inference is rejected if it violates a rule we are unwilling to amend.* The process of justification is the delicate one of making mutual adjustments between rules and accepted inferences; and in the agreement achieved lies the only justification needed for either. [1]

I will now state five main features of this approach to justification, all of them found also in Husserl.

---

[1] Goodman 1955, 66-67. The italics are Goodman's.

（i） Coherence

The method emphasizes the *coherence* of one's views. The coherence is of the kind that we typically strive for in scientific theories, deductive logical inference plays an important role (but deduction itself has to be justified, as pointed out by Goodman), so do simplicity and other considerations; some might, for example, want to make use of what is often called "inference to the best explanation". Typically, general statements are justified by the fact that the desired particular statements follow from them; but on the other hand the particular statements in their turn are to a certain extent justified by being deduced from more general statements.

This latter point, that particular statements are to some extent justified by being deduced from more general statements, is what distinguishes the method of reflective equilibrium from the traditional hypothetico-deductive method, where the particular statements, usually statements about observation, are looked upon as incorrigible, or at least not influenced by theory. We will now turn to this feature.

（ii） Total corrigibility

No statement in one's "theory" is immune to revision (I use the word "theory" in a broad sense, which does not require a theory to be fully worked out into a deductive system, but only requires the statements to be sufficiently related to permit transfer of evidence between them). Any statement may be given up when we find that giving it up brings about simplifications and greater coherence in our overall theory. The views of some of the logical empiricists on "protocol statements" as nonrevisable and unaffected by theory, are incompatible with this, and their methods are therefore not examples of reflective equilibrium, although they are examples of the hypothetico-deductive method. The method of reflective equilibrium is also incompatible with a theory of "sense data" where statements about sense data are supposed to be incorrigible. Adherents of reflective equilibrium are fallibilists not only with respect to some or most statements, such as the hypotheses of the theory in the case of the hypothetico-deductive method, but with respect to all, including reports of observations and other "data". Husserl's theory of perception fits perfectly into such a "holistic" view on justification.

（iii） Different fields of application

The method of reflective equilibrium can be applied in a number of different *fields*, four prominent ones being empirical science, mathematics, logic and ethics. Philosophers can regard the method as appropriate for one, two, three, or all four of these fields. Philosophers can also be distinguished according to whether they regard these four fields as separate fields, where evidence from one field does not transfer to the other fields, or whether they are what we could call "unbounded" holists, and regard all four fields as part of one whole, where coherence considerations involve all four of them and where evidence accordingly is transferred from one area to the other. Evidence from the empirical sciences will thus be relevant for questions of values and norms, and more remarkable, evidence from ethics may have a bearing on questions in mathematics, logic or empirical science.

The foremost representative of such unbounded holism is Morton White, in *Toward Reunion in Philosophy* (1956, 254-256 and 263), and in *What Is and What Ought To Be Done* (1981). White here argues that all the four areas mentioned are interrelated in such a way that statements from all four areas can be put to test together, and that sometimes "we may reject or revise a *descriptive* statement in response to a recalcitrant moral feeling" (White 1981, 122). By including statements from all four areas in the body of statements to be tested White includes more than Quine, who did not discuss ethics, and much more than Duhem, in *The Aim and Structure of Physical Theory*, who was an important "holist" but did not include mathematics and logic and did not discuss ethics.

However, on the other hand White includes less than Quine in each separate test. While Quine holds that "*every* one of our beliefs is on trial in *any* experiment or test" (White 1981, 22-23)[1], White, like Duhem, thinks that only part of our "web of belief" is involved in each test. To distinguish his view from Quine's White therefore calls his view "limited corporatism". He uses "corporatism" the way I am using the word "holism", for the view that "we do not test isolated individual statements but bodies, or conjunctions, of statements" (White 1981, 15). The qualification "limited" indicates that White, like Duhem and unlike Quine, holds that the bodies of statements that are tested in any one test are less comprehensive than they are thought to be by Quine. Avoiding the word "corporatism" that is also used in political theory, I will

---

[1] See, however, Quine's more detailed statement of his doctrine in § 3 of *Word and Object*, where he writes that "some middle-sized scrap of theory usually will embody all the connections that are likely to affect our adjudication of a given sentence" (Quine 1960, 13).

continue to use "holism" for what White calls "corporatism". One might, if one wanted to, introduce different labels for the different variants of holism, for example "piecemeal holism" for Duhem and White's view, where the bodies that are tested in any single test make up only limited pieces of our whole web of belief, and "bounded holism" for a view like Quine's, where not all four fields are included. Duhem's view would hence be a "bounded piecemeal holism", while White's would be an "unbounded piecemeal holism". However, rather than burdening my reader's memory, I will spell out what kind of holism I am discussing whenever that is pertinent.

In *Science and Sentiment in America*, White shows that William James vacillated between a "trialistic" view and a holistic view. The former predominates in the *Psychology* and in *The Will to Believe*, where natural science, logic/mathematics, and ethics are three separate fields, each subject to a method like that of a reflective equilibrium, but with no transfer of evidence between the fields. The latter, holistic view, White finds advanced in some parts of *Pragmatism* and in *A Pluralistic Universe*, where all three fields are regarded as part of one unified whole, one stock of beliefs, in a broad sense, with evidence being transferred between the fields: strains in one field may be increased or reduced by what is happening in the other fields. According to White, "an unsatisfied desire may challenge the stock as much as the discovery of a logical contradiction or a recalcitrant fact, and it is James' belief in the parity of unsatisfied desire with the two other creators of strain that distinguishes his later position" (White 1972, 205). Husserl regarded James very highly and it is not implausible that he may have received some impulses from James in his view on justification, as he definitely did on some other points. Also Wittgenstein was strongly engaged by James and even recognized that his own view amounts to "something that sounds like pragmatism" (See, for example, Goodman 2002).

White's latest book, *A Philosophy of Culture: The Scope of Holistic Pragmatism* gives a thorough historical and systematic examination of the development of this holistic view. One important theme that is more thoroughly discussed in *A Philosophy of Culture* than in the earlier works is the status of the holism itself: "I have maintained that thinkers who seek knowledge do and should use the method of holistic pragmatism in testing their views." (184-185) Can pragmatic holism itself be tested and given up? White responds that we should regard it as a rule of good scientific methodology. It reflects the idea that epistemology is a normative discipline, but it should not be regarded as *a priori*, necessary or immutable. Like some other rules in ethics and science "they are entrenched, but they may be removed from their trench for good reason" (182) .

Note here the three features of the method of reflective equilibrium that we have discussed so far: justification, coherence and total corrigibility. The passage I quoted from Goodman brings into focus a further, highly important feature, to which we shall now turn: pre-reflective, intuitive acceptance as the basic source of evidence.

(iv) Pre-reflective, intuitive acceptance

The method of reflective equilibrium makes crucial use of our *pre-reflective, intuitive acceptance* of various statements. Through reflection, systematization and observation it seeks to gradually modify our acceptances, strengthen some of them and weaken others, but it does not attempt a wholesale rejection of all of them in order to replace them with something radically new. There is no source of evidence upon which such a new edifice could be built, all the evidence there is, is imparted through these intuitive acceptances.

(v) Perception and other sources of evidence

Our intuitive acceptances come in various strengths and are influenced by various factors, some of which we consider more reliable than others. Perception influences many of our acceptances of what the world that affects our senses is like. Perception would, at least by empiricists, be looked upon as one privileged source of evidence, which, although not infallible, provides whatever evidence there is, in addition to the coherence considerations.

While most philosophers would attribute to perception and observation such a privileged role in the sciences, the situation is not so clear in ethics. Rawls, in his early article "Outline of a Decision Procedure for Ethics", seems to hold that some *particular* moral judgments have such a privileged status, and that the *general* ethical principles have whatever acceptability they have in virtue of how well they systematize our particular moral judgments. Our acceptance of some of the particular judgments may be modified through this systematization, but the particular judgments remain the ultimate source of evidence for the ethical principles, much as in science the particular observation statements are the ultimate source of evidence for the general hypotheses of the theory.

However, Rawls in his later writings no longer gives particular moral judgments a privileged status when compared with the general judgments, but holds that judgments of both kinds, particular and general, serve as evidence and that both kinds of judgments become modified through our attempts to create a coherent theory of ethics, where our general principles and our judgment in the individual cases are in equilibrium with one another.

Our intuitive acceptances of ethical statements are often unreliable, depending on egoistic considerations, cultural influences, etc. Reflecting upon them we come to regard some as less reliable than others. These considerations of reliability are part of the reflections we perform in order to arrive at a reflective equilibrium and these considerations themselves have to fit into the reflective equilibrium. Only a careful study of how various observations, experiences, and changes in our system affect our acceptance can tell us whether, in addition to coherence considerations, which are crucial to the method of reflective equilibrium, there is any source of evidence that is of particularly great importance.

7. Reflective Equilibrium in Husserl

Let us now turn to Husserl and his view on justification. Husserl held a reflective equilibrium view on evidence. I have argued for this in an earlier article and shall not repeat all the evidence here. Husserl accepted all the features of the view that we have discussed: coherence, total corrigibility, pre-reflective acceptances, etc. He does so separately within all the four areas that we have described, natural science, mathematics, logic and ethics. In particular, his view on ethics differs from that of Wittgenstein. Although Wittgenstein's view on ethics did not remain constant throughout his philosophical development, I have found no evidence that he ever held that there could be anything like justification of ethical statements.

Husserl argued that ethical statements can be justified. He asks:

And thus also in ethics we have to ask: where is the source of the primitive ethical concepts, where are the experiences [*Erlebnisse*], on the basis of which I can grant these concepts the evidence of conceptual validity?[1]

He answers, like the English "moral sense" philosophers, that sentiment [*Gemüt*]

---

[1]   *Husserl-manuscript F* I 20, 106. Quoted in Diemer [2] 1965, 316.

is the basic source of evidence for moral philosophy:

> The English moral sense philosophy [*Gefühlsmoral*] has after all established beyond doubt: If we imagine a being, who is *sentiment-blind* in the same way as we know beings who are *color-blind*, then everything moral loses its content, the moral concepts become words without sense. [1]

From this basis, Husserl builds up an ethics. He also develops the method of reflective equilibrium as I have described it above. However, he gives a special twist to the method, based on how what he calls the "life-world" plays a crucial role in justification. And now we come to a main point of similarity between Husserl and Wittgenstein in their view on justification, although here, too, there are interesting differences.

8. The Life-World and Its Role in Justification

The life-world is for Husserl the world as we experience it, each of us from our own point of view, shaped by our culture, our education, our previous experiences and our reflection. What is particularly important about the life-world is that most of it we are not consciously aware of. All that we experience and all that we do in our lives leave sediments in the lifeworld, but only seldom are we reflecting on this and taking a conscious stand, judging that the world is so-and-so. We could simply not go on living if we should constantly bother to pay attention to all that is constantly coming in. However, we have to reflect on it in order to understand how justification works:

> There has never been a scientific inquiry into the way in which the life-world constantly functions as a subsoil, into how *its manifold prelogical validities act as ground for the logical ones*, for theoretical truths. And perhaps the scientific discipline which this life-world as such, in its universality, requires is a peculiar one, one which is precisely not objective and logical but which, as the ultimate grounding one, is not inferior but superior in value. [2]

Note how Husserl here expresses a view very similar to that of Goodman: the pre-logical validities act as ground for the logical ones, the life-world functions as a

---

①   *Husserl-manuscript F* I 20, 227 (Diemer [2]1965, 317 and also 48, n. 106) .

②   Husserl 1923-24, § 34, Husserliana VI, 127. 13-20 = p. 124 of David Carr's translation. The emphasis is mine.

subsoil. Remember how, according to Goodman, "principles of deductive inference are justified by their conformity with accepted deductive practice. Their validity depends upon accordance with the particular deductive inferences we actually make and sanction" (Goodman 1955, 67) . Similarly, Husserl says:

every objective logic, every a priori science in the usual sense. . . is to be grounded. . . no longer "logically" but by being traced back to the universal prelogical apriori [i. e. the life-world] through which everything logical, the total edifice of objective theory in all its meth-odological forms, demonstrates its legitimate sense and from which, then, all logic itself must receive its norms. ①

There are a number of similar passages in Husserl's *Crisis* and in his *Experience and Judgment.* Many of them are quoted in my article " Husserl on Evidence and Justification" (Føllesdal 1988) . Here I will only quote a few passages that are particularly pertinent to a comparison with Wittgenstein.

First, Husserl does not regard the life-world as fixed and unchanging. It changes with our new experiences and is also influenced by science and developments in our scientific theories. The latter goes against the expectations of those who think that the life-world is something like the opposite of the world of science. However, the textual evidence is clear. The life-world is not a realm separate from that of the sciences. As we have seen, the sciences have the life-world as their evidential basis. The scientific theories also get their meaning through their connection with the life-world. The sciences in turn gradually change the life-world. As Husserl puts it in *Experience and Judgment*:

. . . everything which contemporary natural science has furnished as determinations of what exists also belong to us, to the world, as this world is pregiven to the adults of our time. And even if we are not personally interested in natural science, and even if we know nothing of its results, still, what exists is pregiven to us in advance as determined in such a way that we at least grasp it as being in principle scientifically determinable. ②

The reason why science belongs to the life-world is, according to Husserl, that it is conceived of as being valid, as making a claim to truth:

Though the peculiar accomplishment of our modern objective science may

---

① Husserl 1923-1924, § 36, Husserliana VI, 144. 27-34 = Carr, 141.

② Husserl 1939, § 10, p. 39 = Churchill and Ameriks, 42.

still not be understood, nothing changes the fact that it is a validity for the life-world, arising out of particular activities, and that it belongs itself to the concreteness of the life-world. ①

Finally, I come to a most important point: the life-world for Husserl is an ultimate court of appeal, behind which there is no point in asking for further justification. The main reason Husserl gives for this, is that most of the life-world consists of acceptances that we have never made thematic to ourselves and which have therefore *never been the subject of any explicit judicative decision*:

> ... where such completely self-giving intuition of the judicative substrates takes place, there is absolutely no possible doubt with regard to the " so " or " otherwise " and hence no occasion for an explicit judicative decision. ②

9. Wittgenstein on Ultimate Justification

After having examined Husserl's view on ultimate justification, let us now turn to Wittgenstein. I will concentrate on *On Certainty*③, for two reasons: it is the work where Wittgenstein most explicitly addresses this issue; and it gives his final view: the book was his last work, the last entry is two days before his death on April 29th 1951.

First, Wittgenstein's view on justification clearly shares some of the characteristic features of Husserl's reflective equilibrium view. The notion of a **system** is central and comes up in several entries:

> #105 All testing, all confirmation and disconfirmation of a hypothesis takes place already within a **system**. And this system is not a more or less arbitrary and doubtful point of departure for all our arguments: no, it belongs to the essence of what we call an **argument**. The system is not so much the point of departure, as the element in which arguments have their life.

> #144 The child learns to believe a host of things. I. e. it learns to act according to these beliefs. Bit by bit there forms a **system** of what is believed, and in that system some things stand.

> #225 What I hold fast to, is not one proposition, but a **nest** of

---

① Husserl 1923-24, § 34f, Husserliana VI, 136. 18-22 = Carr, 133.

② Husserl 1939, § 67, 330 = Churchill and Ameriks, 275.

③ Wittgenstein 1977. In the passages I quote from this book, the *italics* are Wittgenstein's while the **bold** types are my emphasis.

propositions.

#410 Our knowledge forms an enormous **system**. And only within this system has a particular bit the value [Wert] we give it.

Particularly important is the following entry, which states a core idea in Husserl's view:

142 It is not the single axioms that strike me as obvious, it is a system in which consequences and premises give one another *mutual* support.

However, there are also passages that do not quite fit with Husserl's view, such as the following:

185 It would strike me as ridiculous to want to doubt the existence of Napoleon, but if someone doubted the existence of the earth 150 years ago, perhaps I should be more willing to listen, for now he is doubting our whole system of evidence. It does not strike me as if this system were more certain than a certainty within it.

When the entry is read in context, however, one sees that Wittgenstein's point is not that he considers the existence of Napoleon as more credible than the existence of the earth 150 years ago, but that he is interested in the notion of evidence; this is why he, the philosopher, is more eager to listen to the one who doubts our whole system of evidence.

A second main point in Husserl's view on justification is that simplicity and coherence makes a view more justified and also makes it more persuasive. Wittgenstein makes a similar observation:

92, end: Remember that one is sometimes convinced of the *correctness* of a view by its *simplicity* or *symmetry*, i. e. , these are what induce one to go over to this point of view. One then simply says something like: "That's how it must be. "

Husserl held that the life-world is taken over from the culture in which I live. It is not a result of conscious exploration and choice, but is mostly taken over without our noticing it. The life-world is unthematized and consists mostly of tacit knowledge. Wittgenstein expresses a similar view:

#159 As children we learn facts; e. g. , that every human being has a brain, and we take them on trust. I believe that there is an island, Australia, of such-and-such a shape, and so on and so on; I believe that I had greatgrandparents, that the people who gave themselves out as my parents really were my parents, etc. **This belief may never have been expressed; even the thought that it was so, never thought.**

#398 And don't I know that there is no stairway in this house going six floors deep into the earth, even though I have **never thought** about it?

The following statement seems to express the same view, although it is here related to the scientist's use of auxiliary hypotheses:

165 end: I say world-picture and not hypothesis, because it is the matter-of-course foundation for his **research** and as such also goes **unmentioned.**

The beginning of the following entry makes the same point, that what Wittgenstein calls the world-picture and Husserl calls the life-world, is not a result of conscious exploration and choice, but is taken over without our noticing it:

94 But I did not get my picture of the world by satisfying myself of its correctness; nor do I have it because I am satisfied of its correctness. No: it is the inherited [*überkommene*] background **against which I distinguish between true and false.**

I have emphasized the last words of the entry, which make a new point, also made by Husserl: The life-world makes sense of our distinction between true and false. It also gives sense to our notion of reality. Husserl's answer to the skeptic and also to the solipsist is that in trying to express their position they are undercutting it, they are sawing off the branch they are sitting on. A similar thought comes to expression in the following entry:

450 end: A doubt that doubted everything would not be a doubt.

So far, the similarities between Husserl and Wittgenstein in their view on ultimate justification are striking. However, there are two important points where their ways part and where I think we should rather follow Husserl.

First, Husserl, as we saw, held that the life-world as a whole provides the background against which we distinguish between true and false, and gives meaning to our scientific theories and is their ultimate testing ground. However, we also noted that the life-world is changing. Husserl held that **there is no individual statement that is held fixed and is the basis for certainty.** Wittgenstein, however, often appears to think that there are such fixpoints. Consider the following passages.

88 It may be for example that all *inquiry on our part* is set so as to exempt certain propositions from doubt, if they are ever formulated. They lie

apart from the route traveled by enquiry.

Are these propositions that are exempted from doubt always the same? Or are they some at one time, others at other times? Husserl goes for the latter.

341 That is to say, the *questions* that we raise and our *doubts depend* on the fact that some propositions are exempt from doubt, as if they were like hinges on which they turn.

Are the hinges always the same, or are the hinges in one case doubted in another?

342 That is to say, it belongs to the logic of our scientific investigations that certain things are *in deed* not doubted.

Again, are there some things that are never doubted or is Wittgenstein just making the point that at any time some thing or other is not doubted? The order of quantifiers is important.

The second point where Husserl and Wittgenstein disagree, is with regard to the ultimate foundations for justification. Are the foundations subjects of **decision**? We noticed above that for Husserl no decision is involved:

... where such completely self-giving intuition of the judicative substrates takes place, there is absolutely no possible doubt with regard to the "so" or "otherwise" and hence *no occasion for an explicit judicative decision.* ①

This is a very important point for Husserl, at the core of his view on justification. I have found one passage where Wittgenstein seems to be moving in this direction:

359 But that means I want to conceive it as something that lies beyond being justified or unjustified; as it were, as something **animal**.

However, there are many entries where Wittgenstein states that there is a decision involved, such as the following:

#362 But doesn't it come out here that knowledge is related to a decision?

#146 end: —but somewhere I must begin with an assumption or a **decision**.

10. Conclusion

There are many similarities between Husserl and Wittgenstein in their views on justification, and between their views and the reflective equilibrium view. However,

---

① Husserl 1939, § 67, 330 = Churchill and Ameriks, 275. My italics.

there are also differences between them, with regard to whether some specific propositions are exempted from doubt and, most importantly, with regard to the role of decision in ultimate justification. Here I find Husserl's position particularly interesting. His view, that at this ultimate level there is no occasion for an explicit judicative decision, plays a crucial role in his theory of justification. It throws light on why the method of reflective equilibrium actually provides justification and is not just a means for settling disagreements. However, this is a subject for a treatise and goes far beyond a comparison between Husserl and Wittgenstein.

References

Duhem, Pierre 1906 *La Théorie Physique: Son Objet, Sa Structure*, Paris: Rivière. [English translation: *The Aim and Structure of Physical Theory*, Princeton: Princeton University Press, 1954.

Dreyfus Hubert L. (ed.) 1982 *Husserl, Intentionality and Cognitive Science*, Cambridge, Mass. : MIT Press (Bradford Books) .

Føllesdal, Dagfinn 1974 "Phenomenology", in: Edward C. Carterette and Morton P. Friedman (eds.), *Handbook of Perception*, Vol. I, New York: Academic Press, 377-386 (Chapter 19) . [Reprinted in Dreyfus 1982.]

—1988 "Husserl on Evidence and Justification", in: Robert Sokolowski (ed.), *Edmund Husserl and the Phenomenological Tradition: Essays in Phenomenology* (Proceedings of a lecture series in the Fall of 1985. *Studies in Philosophy and the History of Philosophy*, Vol. 18), Washington: The Catholic University of America Press, 107-129.

—1978 " Brentano and Husserl on Intentional Objects and Perception ", in: Goodman, Nelson 1955 *Fact, Fiction and Forecast*, Cambridge, Mass. : Harvard University Press. Roderick M. Chisholm and Rudolf Haller (eds.), *Die Philosophie Franz Brentanos: Beiträge zur Brentano-Konferenz, Graz, 4. -8. September* 1977 (*Grazer Philosophische Studien* 5) , 83-94. [Reprinted in Dreyfus 1982.]

Goodman, Russell B. 2002 *Wittgenstein and William James*, Cambridge: Cambridge University Press.

Husserl, Edmund 1891 *Philosophie der Arithmetik*, Husserliana XII, The Hague: M. Nijhoff 1970.

—1913 *Ideen zu einer reinen Phänomenologie und phänomenologischen Philosophie*, Husserliana III, The Hague: M. Nijhoff 1950. [Translated by Fred Kersten, The

Hague：Nijhoff，1982.］

—1939 *Erfahrung und Urteil*，Prag：Akademia Verlagsbuchhandlung.［English translation by James S. Churchill and Karl Ameriks，Evanston，Ill.：Northwestern University Press，1973.］

—1923-24 *Die Krisis der europäischen Wissenschaften und die transcendentale Phänomenologie*，Husserliana VI，The Hague：M. Nijhoff 1976.［Translated by David Carr，Evanston，Ill：Northwestern University Press，1970.］

Miller，Izchak 1984 *Husserl，Perception，and Temporal Awareness*，Cambridge，Mass：MIT Press（Bradford Books）.

Mulligan，Kevin 1995 "Perception"，in：Barry Smith and David Woodruff Smith （eds.），*The Cambridge Companion to Husserl*，Cambridge：Cambridge University Press，168-238.

Rawls，John 1951 "Outline of a Decision Procedure for Ethics"，*Philosophical Review* 60，177-191.

—1971 *A Theory of Justice*，Cambridge，Mass.：Harvard University Press. Quine，Willard van Orman 1960 *Word and Object*，Cambridge，Mass.：MIT Press.

White，Morton 1956 *Toward Reunion in Philosophy*，Cambridge，Mass.：Harvard University Press.

—1972 *Science and Sentiment in America*：*Philosophical Thought from Jonathan Edwards to John Dewey*，New York：Oxford University Press.

—1981 *What Is and What Ought to Be Done*，New York：Oxford University Press.

—2002 *A Philosophy of Culture*：*The Scope of Holistic Pragmatism*. Princeton：Princeton University Press.

Wittgenstein，Ludwig 1977 *On Certainty*；*Über Gewissheit*，translated by Denis Paul and G. E. M. Anscombe，Oxford：Blackwell.

胡塞尔和维特根斯坦论终极辩护*

达格芬·弗罗斯达尔

第27届国际维特根斯坦研讨会的目标是探讨现象学和分析哲学之间的关系。令人高兴的是，当代哲学中的这些不同传统目前正在共同关注一些哲学论题并且进行沟通。几十年前，这些哲学标签的使用通常是很有争议的。人们批评各种"主义"或传统：实证主义、存在主义、现象学、分析哲学等。这些"主义"通常是不明确的，反对它们的"论证"相应地也不太严密。正如所料，这些论证对受到批评的那些人没有产生影响——他们难以承认他们自己持有被批评的观点。除了从来不读"实证主义者"的著作的某些学生的想法之外，所谓的"实证主义的争论"没有留下任何痕迹。幸运的是，这些乱贴标签的做法已经不再流行了；某人在没有阅读过哲学家著作的前提下，就一口气全部写出各种不同群体的哲学家的名字的事已经很少。以现象学为例，如果不进行考察，人们就不再立刻拒绝它。在过去的几十年里，许多所谓的"分析"哲学家发现，胡塞尔等现象学家也致力于他们所感兴趣的话题，比如，知觉、主体间性、客观性、行动，而最重要的是意向性。在这里，胡塞尔在100多年前提出了很类似于意向性的一些观点和分析，而且，在某些事例中，比近几十年内出现的观点与分析更好。

在本文中我将讨论在胡塞尔哲学中被严重忽视的一个话题，即终极辩护的话题，并把他的观点与维特根斯坦主要在《论确定性》一书中的观点进行比较和对比。

辩护涉及两个要素：知觉和证据转移。胡塞尔先于维特根斯坦对两者进行了有趣的论述。

一、胡塞尔论知觉①

胡塞尔在他的《算术哲学》一书中已经预先考虑过格式塔心理学和格式塔心

---

* 达格芬·弗罗斯达尔（Dagfinn Føuesdal），挪威奥斯陆大学与美国斯坦福大学教授。译自 M. E. Reicher, J. C. Marek（Eds.），*Experience and Analysis.* Vienna：ÖBT & HPT, 2005, 127-142. ——译者注

① 对胡塞尔知觉理论的一个更充分的讨论可以在 Føllesdal 的作品中（1974年和1978年）找到。关于胡塞尔知觉理论的一个清晰的教学式的陈述和讨论可以参见 Miller（1984）；同样也可参见 Mulligan（1995）中精彩的论述。

理学中非常流行的那些模棱两可的图像，这些图像不仅能被用来举例说明胡塞尔关于知觉的观念，也揭示了他和维特根斯坦对这个问题看法上的异同之处。查斯特罗（Jastrow）的鸭兔图（它通过维特根斯坦变得著名起来）是一个很好的切入点：

胡塞尔坚持认为，知觉是不能充分确定的：我们的感官感知到的东西不能充分唯一地确定我们所体验到的东西。根据胡塞尔的观点，我们的意识构造了我们所体验到的东西，而且，我们在特定情境中的经验，原则上，总能以不同的方式来构造。如何构造经验，依赖于我们先前的经验，即我们当前经验的整个背景和许多其他因素。因此，如果我们是在鸭子的环境中长大，甚至从未听说过兔子，那么，当我们面对鸭兔图时，我们看出是鸭子的可能性比看出是兔子的可能性要大；我们甚至不会产生兔子的观念。在少数罕见的情况下，比如，在鸭兔图的例子中，我们能在构造我们经验的不同方式之间随意转换。通常，我们甚至意识不到构造的进行；对象因为有结构，才被我们简单地体验到。

构造总是以这样一种方式发生：对象的许多不同特征被体验为是彼此关联的，即体验为是同一个对象的特征。例如，当我们看到一只兔子时，我们不只是看到了一系列色块的集合，即在我们的视域中，分散着各种不同的棕色色阶（顺便提一句，即使看色块，也含有意向性，一个色块也是一类对象，不过是一个不同类型的对象，而不是一只兔子）。我们看见一只兔子，有确定的形状和确定的颜色，能吃、能跳等。它一面朝向我们，另一面背离我们。从我们所在的位置，我们看不见另一面，但我们看出大致有另一面。

胡塞尔正是把我们意识的这种特性贴上了意向性或指向性的标签。看的过程是意向的，或者是指向对象的，这恰好意味着是这样的：在一个东西被看成是有多个侧面的意义上，我们把对象面向我们的那一面看成是这个东西的一个侧面，而且，我们看见的这个东西还有共同预期的其他不同侧面和特征。这种结构构成了意识的指向性，胡塞尔把它叫做意向对象。意向对象是一个综合确定的系统，这个系统使这些多方面的特征成为统一体，并使他们成为一个对象和相同对象的不同方面。

在这一点上，重要的是注意到，对象的不同侧面、外表或透视图共同构成了这个对象。在我们开始感知之前，根本没有浮现出任何一个侧面和透视图，当确立了意向性之后，才能把它们综合到不同对象中去。如果没有意向性，就没有任何类型的对象，无论它们是物理对象，还是不同对象的不同侧面、不同对象的不同外表，或者，不同对象的不同透视图。而且，意向性并不是逐步起作用的。我们不是先指定六个面，然后把它们综合成为一个六面体；我们是一步到位地构成这个六面体及其六个面。

正像我们已经提到的那样，我们必须在一个相当宽泛的意义上采用"对象"这个词。它不仅包括物理的东西，而且正如我们看到的那样，还包括动物，同样也包括人、事件、行动和过程，还有这些实体的不同侧面、不同方面和不同外表。

我们也应该注意到，当我们发现了一个人时，我们并不是发现了一个物理对象、一个身体，然后推断出这里有一个人。我们发现了一个合格的人，即我们遇到了能够构造世界、从他或者她自身的视角体验世界的某个人。我们的意向对象是一个人的意向对象，不涉及任何推理。看人不比看物体更神秘，在这两种情况中，都不涉及任何推理。当我们看到一个物体时，我们不是先看到感觉材料或者类似的东西，然后，推出这里有一个物体，我们的意向对象是一个物体的意向对象。同样，当我们看到一种行动时，我们看到的是一个完整的行动，而不是一个身体的移动，然后，据此推出这是一种行动。

二、充实：感觉材料

在一种感知行为的情况下，也能把它的意向对象描述为，当我们绕着这个对象转并用我们的各种感官感知它时，关于我们所拥有的那类经验的不同期望或预期的一个很复杂的集合。当我们看见一只鸭子和看见一只兔子时，我们进一步的经验预期是不同的。例如，在前一种情况下，我们预期，当我们摸这个对象时，会感到有羽毛；在后一种情况下，我们期望发现皮毛。当我们获得了我们所预期的经验时，就可以说，充实了意向对象的相应成分。在所有的知觉中，都存在某个充实的过程：与当前"看到的东西"相对应的意向对象的成分得到了充实，这对于其他感官也是类似的。

这样的预期和充实是把知觉与其他意识模式（例如，想象和记忆）区分开来的东西。如果我们只是想象不同的东西，我们的意向对象可能是任何东西，在这里，可能是站在我旁边的一头大象，或者，停放在我旁边的一辆机车。然而，在知觉中包含了我的感觉经验；意向对象必须由我的感觉经验来填充。这排除了许多当我只是想象时可能具有的意向相关项。在我当前的情境中，我不可能拥有与一头大象的知觉相对应的一个意向对象。这并没有把我刚才可能具有的知觉意向相关项数量减少到（比方说）你的意向相关项，假如你坐在我前面的话。

正如我们前面所注意到的那样，胡塞尔现象学的一个核心观点是，我会有各种不同的知觉意向相关项，它们与对我的感知面产生的当前影响相一致。在鸭兔图的例子中，这是显而易见的；我们能在拥有一只鸭子的意向对象和拥有一只兔子的意向对象之间随意地转换。然而，在大多数情况下，我们没有意识到这种可能性。只有当发生某种意外时，即当我遇到不能用我的意向对象中的预期填充的"顽固的"经验时，我才确实开始看到一个与我认为前面看到的对象不同的对象。用胡塞尔的术语来说，我的意向对象"突变了"，而且，根据新的预期，我获得了一个与前面的意向对象相当不同的意向对象。胡塞尔说，这总是可能的。知觉总是包含了超越当前所"看到的"东西的那些预期，并且，不管我们可能觉得多么自信和确定，总是存在着我们有可能出错的风险。感知错误总是可能的。

胡塞尔运用了希腊语中的质料一词，把当我的感知器官受到影响时典型地具有的那些经验，以及约束了我的行动的那些经验，称为"感觉材料"（hyle）。他把他的知觉观看成是亚里士多德的"形式质料说"的变种；感觉材料，即我的感知经验，约束了这种"形式"，而且，通过这种意向对象来告知。请注意，对于胡塞尔来说，感觉材料就是经验，它们不是行动的对象，它们是行动的组成部分。我们能通过对它们的反思把它们转变为对象，但在正常的知觉行动中它们不能被感知到，却约束了我们的感知行动。对于胡塞尔来说，在知觉中，没有非结构的"给定的"东西，这类中介没有一个对感知材料的理论家有吸引力。

三、世界和过去

我们构造，或用胡塞尔的术语来说，"构成"这个东西，不仅为了使它有各种不同的特性，而且为了使它与其他对象相关。例如，如果我看一棵树，这棵树被设想为是在我面前的某个东西，也许设想为在其他树当中，设想为被其他人而不是我自己所看到，等等。也能把它设想为是有历史的某个东西：在我看见它之前，它就在那里，当我离开后，它依旧在那里，也许，它最终会被砍掉，并被运送到某个别的地方。然而，像所有的物质一样，它不会从世界中简单地消失。

在这方面，我对这棵树的意识，也是这棵树所在时空中的世界的意识。我的意识构成了这棵树，但同时，也构成了这棵树和我生活于其中的世界。如果我进一步的经验使我放弃在我面前有一棵树的信念，因为，例如，我没有发现远处有一个类似树的东西，或者，因为我的某些其他期望证明这是错误的，那么，这不仅影响了我关于这里是什么的概念，而且影响了我关于曾经是什么和未来是什么的概念。因此，在这个例子中，我们不仅重构了现在，而且也重构了过去和未来。为了说明在我当前感知中的变化，如何致使我不仅重构现在，而且重构过去，胡塞尔举了一个球的例子，我最初把这个球看成完全是红的和球形的。当它转过来后，我发现它的另一侧是绿的，还有一个凹坑：

> ……这种知觉感不是只在产生新感知的那一瞬间发生变化；意向相关项的修改以追溯力终止的形式不断地进入记忆阈，并且修改了从较早的知觉阶段产生的感知。较早的统觉，即与"红色和很圆"的协调发展相适应的统觉，相对于"绿的和凹下去的一面"，潜在地得到了"重新解释"①。

四、价值、实践功能

到目前为止，我只提到了事物的事实属性。然而，胡塞尔说，事物也有价值属性，而且，我们用相应的方式构成这些属性。我们生活的世界被体验为这样一个世界：在这个世界中，有些事物和行动有正面价值，另一些事物和行动有负面价值。我们的规范和价值也会发生变化。我们对事实问题的观点发生了变化，随之我们对价值的评价也会发生变化。

胡塞尔强调说，我们的视角和预期主要不是事实的。我们不是过着纯理论的生活。根据胡塞尔的观点，我们首先以"一种自然的追求生活的态度"遇到我们周围的世界，因为"生活的功能主体卷入了其他主体的圈子内"。胡塞尔在1917年的一篇手稿中指出这一点，但他在早期和晚期对实践的世界持有相似的观点。因此，他在《观念》一书中写道：

> ……对于我来说，这个世界就在这里，不仅作为只有事物的世界，而且同样直接地作为一个价值的世界、一个善的世界、一个实践的世界。②

胡塞尔从来不认为，我们首先感觉身体和身体的移动，然后推断出有人和行动，或者，我们首先感觉到感知材料、透视图或外表，然后，把它们综合到物体上。因此，归因于胡塞尔的下列观点是对胡塞尔的最大误解：我们首先感觉到只有物理属性的对象，然后，再赋予它们价值或实践的功能。事物直接被我们体验为具有这些特征：功能的和评价的，还有事实的，在我们对生活的自然追求中，这对于我们来说是有意义的。

---

① Husserl 1939，§ 21a，96 = p. 89，Churchill 与 Ameriks 的英译本。

② Husserl 1913，§ 27，Husserliana III，1，58. 13-19 = Kersten 的译本，53. 我稍微改变了一下他的翻译。

五、看做：胡塞尔和维特根斯坦

胡塞尔知觉理论的这些基础极大地超越了维特根斯坦。不过，应该说，在胡塞尔和维特根斯坦的知觉理论之间，有一个重要的不同。维特根斯坦经常受人赞赏的是，他注意到，所有的看见（seeing）都是"看做"（seeing as）。胡塞尔似乎也持有同样的观点。然而，胡塞尔的观点是不同的。在通常的知觉中，根本没有一个对象，在一种场合，被看做是一种东西，例如，一只鸭子，在另一种场合被看做是另一种东西，比如一只兔子。根本没有任何潜在的更基本的对象，被看做是这个对象或那个对象。存在有感觉材料，但这些感觉材料不是所看见的对象；它们是约束我们进行构造的经验，它们不是所体验的对象。不幸的是，维特根斯坦以图像为出发点，也就是说，在有一个对象的情况下，无论怎么看，都能看到这个图像，而且，他效仿这一点形成了他的知觉观。

如果我们希望对胡塞尔的知觉理论作出正确的理解，那么我们就不应该用查斯特罗的鸭兔图作为我们的例子，而是考虑一个正常知觉的例子，例如，一天夜晚我们在外面的一个地方，看到在地平线上有一个像查斯特罗的鸭兔图的一个影子。在这种情况下，我能看出一只鸭子或一只兔子，但根本没有看做是一只鸭子或一只兔子的图像或别的对象。人们可能认为，这个影子是这样一个对象，它恰好像查斯特罗和维特根斯坦的例子中的图像一样。然而，人们错过了胡塞尔知觉理论中的一个基本观点，即把胡塞尔的知觉理论从成为大多数其他知觉理论的绊脚石的问题中拯救出来的一个观点：在知觉中，根本没有中介，既没有感觉材料、图像，也没有影子。

在维特根斯坦的作品中有几段表明他已经意识到，知觉不应该效仿被看见的图像。然而，关于"看做"的热情，往往遮蔽了把知觉理论建立在观看图像之基础上的基本困难。

六、反　思　平　衡

从上面显然能看出，胡塞尔没有把知觉看成是科学和知识的一种牢固的不可动摇的基础。在知觉与使知觉易受影响的我们关于世界的理论和过去的不同经验之间，存在着相互作用。这是胡塞尔的总的辩护观的一个特征，我现在将要转而探讨的更新近的观点也分享了这种辩护观。

这些现代的辩护观，与胡塞尔的观点很类似，通常被称为"反思平衡"的观点，这是由罗尔斯（Rawls）提出引入的一个称号（Rawls, 1971）。这种观点的最明确的陈述之一是由尼尔森·古德曼（Nelson Goodman）在《事实、虚构与预测》一书中提供的，在这本书中，他探讨了演绎辩护问题，许多哲学家把演绎辩护问题看成是显而易见的，没有辩护的必要：

> 我们如何证明一种演绎是正当的呢？很简单，通过表明，它合乎演绎推理的一般规则。
>
> ……
>
> 但如何确定规则的有效性呢？在这里，我们再次遇到了两类哲学家：一类哲学家坚持认为，这些规则是根据某个自明的公理得出的；另一类哲学家试图表明，这些规则建立在人类心灵的本性之基础上。我认为，这样的回答是很肤浅的。演绎推理原理通过它们合乎公认的演绎实践被证明是合理的。它们的有效性是根据我们实际上制定和认可的特殊的演绎推理而定的。如果一条规则产生了不可接受的推理，我们就视其为无效而将其丢弃。一般规则的辩护因而源于对拒斥或者接受特殊的演绎推理的判断。
>
> 这看起来显然是循环论证。我曾说过，演绎推理通过它们合乎一般规则被证明是合理的，一般规则又通过它们合乎有效的推理被证明是合理的。但这种循环是一个良性循环。关键在于，规则和特殊推理两者都是通过达到互相一致被证明是合理的。如果一条规则产生了我们不愿意接受的推理，那么，它就会被修改；如果一种推理违背了我们不愿意修改的一条规则，那么，它就会被拒绝。辩护的过程是，在规则与公认的推理之间作出一种精致的相互调整；并且，各自需要的唯一辩护在于所达成的共识。[①]

现在，我陈述这条辩护进路的五个主要特征，都能在胡塞尔的著作中找到。

1. 一致性

这种方法强调了人们的诸多观点的一致性。这种一致性是我们在科学理论中典型地去力求的那种一致性，演绎逻辑推理起到重要的作用（但正如古德曼指出的那样，演绎本身必须被证明是合理的），简单性和其他考虑也是如此，例如，一些人可能希望利用通常所谓的"最佳说明推理"。在有代表性的意义上，一般陈述通过下列事实被证明是合理的：渴望得到的特殊陈述是从一般陈述中得出的。但另一方面，这些特殊陈述依次在某种程度上通过从更一般的陈述中推出被证明是合理的。

最后这一点，即特殊陈述在某种程度上通过从更一般的陈述中推出被证明是合理的，把反思平衡方法与传统的假说-演绎方法区分开来，在这里，特殊陈述

---

① Goodman 1955，66-67.

（通常是关于观察的陈述）是不可修改的，或者，至少是不受理论影响的。我们现在转向这个特征。

2. 整体的可修改性（total corrigibility）

在人们的"理论"中没有一种陈述是免于修改的（我在一个宽泛的意义上用"理论"这个词，不要求一个理论完全发展成为一个演绎体系，只要求这些陈述与它们之间允许的证据转移密切相关）。任何一个陈述，当我们发现，放弃它会使我们的全部理论简单化和具有更大的一致性时，它就可能被放弃。某些逻辑经验主义者关于作为不可修改的和不受理论影响的"经验陈述"（protocol statements）的观点，与此相矛盾。他们的方法因此而不是反思平衡的例子，尽管它们是假说演绎方法的例子。反思平衡方法也与"感觉材料"的理论不相容，在"感觉材料"的理论中，关于感觉材料的陈述应该是不可修改的。反思平衡的支持者是证伪主义者，不仅是关于某些或大多数陈述的证伪主义者，比如，在假说演绎方法情况下的理论假说，而且是关于包括观察报告和其他"材料"在内的所有陈述的证伪主义者。胡塞尔的知觉理论完全符合这样一种"整体论的"辩护观。

3. 不同领域的应用

反思平衡方法能应用于许多不同的领域，四个最突出的应用领域是经验科学、数学、逻辑学和伦理学。哲学家会把这个方法看成是适用于一个、两个、三个或者所有这四个领域的。区分哲学家的根据是，他们是否把这四个领域看成是分离的，在这里，一个领域的证据不能转移为另一个领域的证据；或者，他们是否是我们所能称之为的"无约束的"整体论者，并把所有的四个领域都看成是一个整体的部分，在这里，一致性考虑包括所有这四个领域，相应的证据可以从一个领域转移到另一个领域。经验科学的证据就与价值和规范的问题相关，更为引人注目的是，伦理学的证据可能关系到数学、逻辑学或经验科学中的问题。

这种无约束的整体论的最重要的代表作是默顿·怀特（Morton White）的《走向哲学中的再联合》（1956，254-256 and 263）与《做了什么和应该做什么》（1981）。怀特在这里认为，所提到的这四个领域以这样一种方式相互关联：这四个领域的陈述能被一起检验，有时，"我们为了回应一种顽固的道德情感，可以拒斥或修正一个描述的陈述"（White 1981，122）。由于在被检验的陈述中包括了所有这四个领域的陈述，所以，怀特超越了不讨论伦理学的奎因，更超越了迪昂的《物理学理论的目的和结构》，迪昂是一位重要的"整体论者"，但不包括数学和逻辑学，也不讨论伦理学。

然而，另一方面，在每一次单独检验中，怀特没有超越奎因。奎因相信"在任何一次实验或检验中，我们的每个信念都是未经证实的"① (White 1981, 22-23)，而怀特像迪昂一样，认为在每次检验时，只包含我们的部分"信念之网"。怀特为了把他的观点与奎因的观点区分开来，而把他的观点称为"有限的社团主义"（limit corporatism）。他以我使用"整体论"这个词的同样方式使用"社团主义"，这是由于"我们不是检验孤立的个别陈述，而是检验陈述的体系或联合"这一观点（White 1981）。"有限的"限定性条件表明，怀特与迪昂一样，但与奎因不同，他认为，在任何一次检验中检验的陈述体系，都不如奎因所认为的陈述体系更全面。为了避免也会用在政治理论中的"社团主义"这个术语，我将继续用"整体论"来取代怀特所称的"社团主义"。人们可以（如果愿意的话）为整体论不同变种赋予不同的称号，例如，把迪昂和怀特的观点称为"部分整体论"，在这里，任何一个单独的检验陈述体系，只是由我们的信念之网中的有限部分构成的，而且，用"受约束的整体论"替代了像奎因那样的整体论的观点，在这里没有全部包括这四个领域。迪昂的观点因此是"受约束的部分整体论"，而怀特的观点是"不受约束的部分整体论"。然而，为了不增加我的读者的记忆负担，我将阐明我所讨论的所有相关的整体论的类型。

怀特在《美国的科学与情操》一书中表明，威廉·詹姆斯（William James）在"三元论者的"观点和整体论者的观点之间摇摆不定。前者在《心理学》和《信念的意志》中占有统治地位，在这里，自然科学、逻辑学/数学以及伦理学是三个独立的领域，每一门学科都服从一种类似反思平衡的方法，但在这三个领域之间没有证据转移。怀特发现，后者，即整体论的观点，是在《实用主义》和《多元的宇宙》的部分章节中提出的，在这里，所有的三个领域都被看成是一个统一整体（即一系列信念）的组成部分，在广泛意义上，证据在三个领域之间发生了转移：在一个领域中的紧张局势可以通过其他两个领域内所发生的情况来减轻或加强。根据怀特的观点，"一种感到不满的愿望可能对这个信念系列提出的挑战，与发现一个逻辑矛盾或一个顽固的事实提出的挑战一样大，正是詹姆斯相信感到不满的愿望等同于其他两个领域带来的紧张局势，区分出他后面的立场"（White 1972, 205）。胡塞尔很尊敬詹姆斯，而且，令人信服的是，在他的辩护观中，他可能受到了詹姆斯的激励，正如他在其他某些方面明确地研究的那样。同样，维特根斯坦也被詹姆斯极大地吸引住了，甚至承认，他自己的观点相当于"听起来像实用主义的某种观点"（参见，例如，Goodman 2002）。

---

① 不过，在《词语与对象》的第三章可以看到奎因关于他的观点的更详细的陈述，在那里他写道，"理论的某个中等大小的片段通常将体现出很可能影响我们对一个给定语句作出裁定的所有联系"（Quine 1960, 13）。

怀特最近的一本著作《文化哲学：整体论实用主义的视野》对这种整体论观点的发展进行了历史的和系统的全面考察。《文化哲学：整体论实用主义的视野》比他早期的著作更全面地探讨的一个重要论题是，整体论本身的地位问题："我一直坚持认为，探索知识的思想家，在检验他们的观点时，确实使用和应该使用整体论的实用主义的方法。"（184-185）实用主义的整体论本身能够被检验和被抛弃吗？怀特回答说，我们应该把它看成是一个好的科学方法论的法则。它反映了这样的观点：认识论是一门规范的学科，但不应该被看成是先天的、必然的或不可改变的。像伦理学和科学中的某些其他规则一样，"它们是根深蒂固的，但把它们从它们的战壕中排除出去，是有道理的"（182）。

这里请注意，到目前为止，我们已经讨论了反思平衡方法的三个特征：辩护、一致性和整体的可修改性。我所引用的古德曼的这段话，开始进一步关注我们现在将转向讨论的更重要的特征：前反思的直觉接受被看成是证据的基本来源。

4. 前反思的直觉接受

反思平衡的方法的关键是利用了我们对不同陈述的前反思的直觉接受。通过反思，设法逐渐地修改我们接受的系统化和观察，强化了其中的某些陈述，弱化了另一些陈述，但不是试图全部拒绝所有的陈述，以便用一些全新的陈述取而代之。根本没有能够建立这样一个"新大厦"的证据来源，现有的所有证据都是通过这些直觉接受赋予的。

5. 知觉和其他的证据来源

我们的直觉接受发挥着各种不同的力量，也受到了各种不同因素的影响，我们认为，其中的某些因素比另一些因素更加可靠。就像世界影响了我们的感觉一样，知觉影响了我们的许多接受。知觉至少被经验主义者看成是一种享有特权的证据来源，除了一致性考虑之外，这种证据来源尽管不是绝对可靠的，但提供了现存的任何证据。

当大多数哲学家把在各门学科中这种享有特权的作用归因于知觉和观察时，这种情境在伦理学中还不是如此明确。罗尔斯在他早期的一篇文章《伦理学的一种判定程序的纲要》中似乎认为，某些特殊的道德判断具有这种享有特权的地位，普遍的伦理学原理具有的任何可接受性，都是借助于它们多么好地使我们的特殊的道德判断系统化。我们接受的某些特殊判断，可以通过这种系统化加以修改，但这些特殊判断仍然是这些伦理学原则的最终证据来源，非常像在科学中，特殊的观察陈述是普遍的理论假设的最终证据来源。

然而，罗尔斯在他晚期的作品中，当与普遍判断相比时，不再赋予特殊的道德判断特权地位，反而认为，这两种类型的判断，普遍判断和特殊判断，都作为证据使用；而且，这两种类型的判断，通过我们创造一个一致的伦理学理论的企图，变成可以修改的，在这里，我们在个别情况下的普遍原理和判断，与另一些情况下的普遍原理和判断相平衡。

我们对伦理学陈述的直觉接受通常是不可靠的，依赖于利己主义的考虑、文化影响等。对它们进行反思之后，我们得出的结论是，一些特征不如另一些特征可靠。对可靠性的这些考虑是我们为了达到反思平衡进行反思的一个组成部分，并且，这些考虑本身必须与这种反思平衡相符合。只有认真地研究在我们的系统中的观察、实验和变化对我们的接受产生的影响是多么的不同，我们才能知道，除了一致性的考虑（这是反思平衡方法的关键）之外，是否存在特别具有重要性的证据的任何来源。

七、胡塞尔的反思平衡

现在让我们转向胡塞尔和他的辩护观。胡塞尔坚持关于证据的反思平衡观点。我在较早的一篇文章中对此作了论证，这里，不再重复所有的证据。胡塞尔接受我们所讨论的观点的所有特征：一致性、整体可修改性、前反思的直觉接受等。在我们所谈论的这四个领域（即自然科学、数学、逻辑学及伦理学）中，胡塞尔分别都是这么认可的。特别是，他关于伦理学的观点与维特根斯坦的观点不同。尽管维特根斯坦关于伦理学的观点在他的整个哲学发展中不是永久不变的，但我没有找到证据证明，他曾经认为对伦理学陈述的辩护是可能存在的。

胡塞尔论证说，伦理学陈述能够被证明是合理的。他问道：

于是，在伦理学中，我们也必须要问：原始的伦理学概念的来源在哪里？我们能据以赋予这些原始概念为概念有效性之证据的那些经验，它们在哪里？[①]

像英国的"道德感"哲学家一样，他回答说，情感是道德哲学证据的基本来源：

英国的道德感哲学家无疑终究确定了：如果我们设想一个人是情感盲，与我们知道的色盲的人一样，那么，一切道德规范都会失去它的内容，道德概念将成为无意义的词语。[②]

根据这个基础，胡塞尔建立了一种伦理学。他也发展了反思平衡方法，正如我

---

① Husserl-manuscript F I20, 106. Quoted in Diemer [2] 1965, 316.

② Husserl-manuscript F I20, 227 (Diemer[2]1965, 317 and also 48, n. 106).

上面描述的那样。然而，他以他所谓的"生活世界"如何在辩护中发挥重要作用为基础，对这种方法赋予了一种特殊的曲解。于是，现在尽管在胡塞尔和维特根斯坦的辩护观中也有许多有趣的差异，但我们得到了他们之间的主要的相似点。

八、生活世界及其在辩护中的作用

对于胡塞尔来说，生活世界是我们每个人从自己的视角来体验的，由我们的文化、教育、先前的经验和反思所塑造的世界。关于生活世界特别重要的是，它的绝大部分我们并没有自觉地意识到。我们的所有体验和所作所为，都沉积在生活世界中，只是我们很少反思这一点，很少采纳一种有意识的立场来判断世界是怎么样的。如果我们不断费心地注意经常发生的一切，我们就不能简单地保持生活。无论如何，我们必须对此进行反思，以便理解辩护是如何起作用的：

> 生活世界经常起到基础的作用，在这方面，人们从来没有进行过科学探索，对于理论真理而言，它的多种多样的前逻辑的有效性问题也没有进行过科学探索。也许这个生活世界本身在其普遍性方面所要求的科学的学科是一门独特的学科，恰好不是客观的和有逻辑的一门学科，而是作为终极根据的一门学科，这门学科在价值上不是低级的，而是高级的。①

注意，胡塞尔在这里如何表达了一种很类似于古德曼的观点：前逻辑的有效性充当了逻辑有效性的根据，生活世界起着基础性作用。请记住，根据古德曼的观点，"演绎推理原理如何通过合乎公认的演绎实践被证明是合理的。它们的有效性是根据我们实际所做出的和认可的特殊的演绎推理来决定的"（Goodman 1955, 67）。类似地，胡塞尔说：

> 在通常意义上的每一门客观的逻辑学，每一门先验的科学……都不再是建立在"逻辑的"基础上，而是建立在追溯到普遍的前逻辑的先验（即生活世界）的基础上，通过生活世界，客观理论的一切逻辑的整体大厦，在它的所有的方法论形式中，证实了它的合法意义，然后，整个逻辑学本身从生活世界中获得它的规范。②

在胡塞尔的《危机》和《经验与判断》中还有许多类似的段落。我在《胡塞尔论证据与辩护》（Føllesdal, 1988）一文中引用了其中的许多段落。这里我只引用了与维特根斯坦的比较特别相关的一小段。

---

① Husserl 1923-24，§ 34，Husserliana VI, 127. 13-20 = p. 124，David Carr 的译本，强调的部分是我做出的。

② Husserl 1923-24，§ 36，Husserliana VI, 144. 27-34 = Carr, 141.

首先，胡塞尔没有把生活世界看成是固定的和不变的。它随我们的新的经验变化而变化，而且，也受到了科学和发展我们的科学理论的影响。后者出乎这些人的预料：他们认为，生活世界与科学世界有点像是对立的。然而，文本的证据是明确的。生活世界并不是与科学世界相分离的一个领域。正像我们已经看到的那样，科学把生活世界作为它们的证据基础。科学理论也通过它们与生活世界的关联获得它们的意义。科学依次逐渐地改变了生活世界。正如胡塞尔在《经验与判断》中指出的：

> ……当代自然科学为所存在的东西提供证实，对我们时代的成年人来说，当这个世界是预先给定时，这一切属于我们，也属于世界。即使我们就个人而言对自然科学没有兴趣，即使我们对自然科学的成果一无所知，对我们来说，所存在的东西仍然是预先给定的，事先被看成是以这样一种方式来确定的：我们至少把它看成是原则上可科学地确定的。[1]

根据胡塞尔的观点，科学为什么属于生活世界的理由是，我们以为科学是有效的，以为作出了真理的主张：

> 尽管我们近代的客观科学的特殊成就可能仍然没有得到理解，但下列事实是无法改变的：它是由于特殊活动而产生的一种生活世界的有效性，而且，它本身属于生活世界的具体化。[2]

最后，我们得出最重要的观点：胡塞尔的生活世界是最终的上诉法庭，要求进一步作出支持它的辩护，这是毫无意义的。胡塞尔为此提供的主要理由是，我们的大部分生活世界是由各种接受组成的，我们从来没有提出我们自己的主题，因此也绝不是任何明确的判决结果的主题：

> ……在这种完全自然发生的直觉成为判决基底的地方，关于"这样"或"那样"的怀疑，是绝对不可能的，因此，没有必要作出一个明确的判决结果。[3]

九、维特根斯坦论终极辩护

在考察了胡塞尔的终极辩护观之后，我们转向维特根斯坦。出于两个原因，我将把注意力集中在《论确定性》[4]上，正是在这部著作中，维特根斯坦最明确

---

① Husserl 1939, § 10, p. 39 = Churchill and Ameriks, 42.
② Husserl 1923-24, § 34f, Husserliana VI, 136. 18-22 = Carr, 133.
③ Husserl 1939, § 67, 330 = Churchill and Ameriks, 275.
④ Wittgenstein 1977. 在我从这本书中引用的段落中，楷体字是维特根斯坦的原文，而**黑体字**是我所强调的。

地提出了这个问题；它给出了他的最终观点：这本书是他的最后一部著作，最后的条目是他在 1951 年 4 月 29 日，即逝世的前两天写下的。

首先，维特根斯坦的辩护观显然分享了胡塞尔的反思平衡观点的某些典型特征。一个**体系**的概念是核心概念，它出现在几个条目中：

#105 对一个假设的一切检验、一切证实或否证，都已经在一个**体系**之中发生。而且，这个体系不是我们所有**论证**的几乎是任意的和可疑的出发点：不，它属于我们称之为论证的本质。这个体系与其说是论证的出发点，不如说是论证有其生命力的一个要素。

#144 儿童学会了相信许多东西，即学会依照这些信念行动，逐渐地形成了所相信的一个**体系**，在这个体系中，某些东西是保持不变的。

#225 我很快所要坚持的，不是一个命题，而是一**组**命题。

#410 我们的知识形成了一个庞大的**体系**。而且，只有在这个体系内，一个特殊的部分，才有我们赋予它的价值。

特别重要的是下面这个条目，这个条目陈述了胡塞尔观点的一个核心观念：

#142 给我留下明显印象的不是单独的公理，而是一个体系，在这个体系中，前提与结论提供了相互支持。

然而，也有一些段落完全不符合胡塞尔的观点，比如下面这一段：

#185 给我留下荒谬印象的是，希望怀疑拿破仑的存在；但如果有人怀疑在 150 年前地球的存在，也许我应该更愿意洗耳恭听，因为现在他正在怀疑我们的整个证据体系。没有给我留下印象的是，好像这个体系比其中的一种确定性更加肯定。

然而，当在语境中读这个条目时，人们看到，维特根斯坦的观点并不是这样，他把拿破仑的存在看成是比 150 年前地球的存在更可信，而是他对证据概念感兴趣；这就是为什么，他这位哲学家更渴望倾听怀疑我们的整个证据体系的这个人的观点的原因。

胡塞尔辩护观中的第二个要点是，简单性和一致性使得一种观点被证明是更合理的，也使得它更具有说服力。维特根斯坦作出了类似的观察：

#92 结论：请记住，人们有时根据一种观点的简单性或对称性而确信它的正确性，也就是说，这些致使人们转向这种观点。人们这时只说些类似于这样的话："它必须是这样。"

胡塞尔认为，生活世界被我们生活于其中的文化取而代之。它不是有意探索和选择的结果，而主要是被无意地接受。生活世界没有被主题化，主要是由意会知识构成的。维特根斯坦表达了类似的观点：

#159 作为孩子，我们了解事实；每个人都有头脑，而且，我们相

信事实。我相信，有一个岛，叫澳大利亚，它有着如此这般的形状，等等；我相信，我有曾祖父和曾祖母，他们生育后代，因为我的父母实际上就是我的父母，等等。**这种信念可能从来没有得到表达；甚至事实如此的思想，从未得到思考。**

#398 于是，即使我**从未考虑过**这个问题，难道我不知道，在这所房子里没有直通地下室六层的楼梯吗？

下面的陈述似乎表达了同样的观点，尽管它在这里与科学家运用的辅助假设相关：

#165 结论：我说世界图像，而不是假设，因为这是他的**研究**的必然基础，而且，就其本身而言，**也未提到过**。

下面这个条目的开头提出了同样的观点，维特根斯坦所谓的世界图像和胡塞尔所谓的生活世界，并非有意探索和选择的结果，而是我们在无意中接受的：

#94 但是，我通过使我自己相信它的正确性，得不到我的世界图像；我也不会拥有它，因为我确信它的正确性。不：它是我**用来分辨真假的**可继承的背景。

我强调了这个条目中的最后几个词，它们提出了一个新的观点，也是胡塞尔也提出的观点：生活世界使我们对真假的区分具有意义。它也赋予我们实在概念的意义。胡塞尔对怀疑论者和唯我论者的问题的回答是，在试图表达他们的立场时，他们削弱了这个立场，他们正在砍掉他们就座的树枝。下面这个条目表达了类似的思想：

#450 结论：对所怀疑的一切的怀疑，不是一种怀疑。

到目前为止，胡塞尔和维特根斯坦关于终极辩护的观点之间的相似性是令人吃惊的。然而，由于两个重要的观点，他们分道扬镳了，而我认为，我们应该宁愿追随胡塞尔。

首先，正像我们所看到的那样，胡塞尔认为，生活世界作为一个整体，提供了我们据此区分真假的背景，赋予我们的科学理论的意义，而且是最终检验它们的根据。然而，我们也注意到，生活世界正在发生变化。胡塞尔认为，**根本没有这样的个别陈述：它被认为是固定不变的，而且，是确定性的基础**。然而，维特根斯坦通常似乎认为，存在这样的固定点。考虑下面这一段：

#88 例如，可能是，对我们来说，进行探索，是为了使确定的命题免遭怀疑，如果这些命题曾经被阐述过的话。它们脱离了探索的路线。

免遭怀疑的这些命题总是相同的吗？或者，它们在一段时间是这样的，在另一段时间是那样的呢？胡塞尔支持后者。

#341 这就是说，我们提出的问题和我们的怀疑依赖于这样的事实：

有些命题是免遭怀疑的，好像它们像它们依赖的枢纽一样。

这些枢纽总是相同的吗？或者，这些枢纽在一种情况下是可疑的，在另一种情况下并非如此吗？

#342 这就是说，它属于我们科学研究的逻辑：有些东西确实是不受怀疑的。

再说一遍，存在着从来都不受怀疑的某些东西吗？或者，维特根斯坦恰好提出了某个东西或另一个东西无论何时都不受怀疑的观点吗？量词的顺序是很重要的。

胡塞尔和维特根斯坦有分歧的第二个观点，是关于辩护的最终基础。这些基础是**决定**的理由吗？我们前面注意到，对于胡塞尔来说，完全与决定无关：

  ……在这种完全自然发生的直觉成为判决基底的地方，关于"这样"或"那样"的怀疑，是绝对不可能的，因此，没有必要作出一个明确的判决结果。①

对于胡塞尔来说，这是很重要的观点，是他的辩护观的核心。我找到了维特根斯坦似乎向这个方向发展的一段：

#359 但这意味着，我希望把它设想成超越了是合理的或不合理的某个东西；可以说，设想成是某种**动物**。

然而，有许多条目中，维特根斯坦主张，确实有一个相关的决定，比如下面这些：

#362 但这不是揭示了，在这里，知识与决定相关吗？

#146 结论：——但是在有些地方，我必须从一个假设或一个**决定**开始。

十、结　　论

胡塞尔和维特根斯坦，在他们的辩护观之间和他们的观点与反思平衡观点之间有许多相似性。然而，在他们之间，关于某些特殊命题是否是免遭怀疑的观点，最重要的是，关于决定在终极辩护中的作用的观点，也有差异。在这里，我发现，胡塞尔的立场是特别有趣的。他的观点，即在最根本的层次上没有必要作出一个明确的判决结果，在他的辩护理论中起到了关键性的作用。这阐明了为什么反思平衡方法实际上提供了辩护，而不只是解决分歧的一种手段。然而，这是一篇论文的主题，远远超出了比较胡塞尔和维特根斯坦的范围。

（戴潘 译，郑晓松　成素梅 校）

---

① Husserl 1939, § 67, 330 = Churchill and Ameriks, 275. My italics.

Function and Meaning: The Double Aspects of Technology

Andrew Feenberg

1

Every new automobile and computer is equipped with an operating manual. These manuals explain the functions of the device they describe and how to use it. They are apparently exhaustive, that is, once you have understood the device on their terms, it seems you understand it fully. But of course there is more to say, lots more. Automobiles and computers belong to a social world in which they play a complicated part. They are linked to so many other features of that world it is impossible to explain them all, to grasp the totality of their involvements, some symbolic, others causal. A few examples illustrate this complexity. Automobiles shape urban designs, they signify the status of their owners, they are polluters, and so on. Computers similarly transform intellectual property regimes, alter the relations between individual self-expression and the mass media, overcome various types of social isolation, and so on. One could continue these lists ad infinitum. We need a term for signifying this wider range of involvements. I will call it the "meaning" or "significance" of technology.

Philosophers have shown an interest in the relation between the socially determined functions of devices and the natural causality that enables them to "work". They have discovered that this is a more complicated relation than it at first appears. My subject is complementary to that one and still more complicated. I want to understand the connection between the strictly functional dimension of a device such as one finds in an operating manual and its social meaning. This is the question to which this paper is addressed, however, the path I will take in answering it is quite different from the analytic approaches familiar from the established study of function. There is a good reason for this difference: the relation between function and meaning is not primarily a conceptual problem but strikes at the heart of modernity as a unique social formation. It raises the question of how the dominant paradigm of knowledge in modern societies relates to the dimension of meaning in the sense I have given the term. I will therefore begin with some considerations on this aspect of the problem.

2

Modern societies understand themselves and the world in terms of a stripped down functional logic, freed from sentimental and teleological accretions. This has enabled them to create effective technical means of controlling the natural world and markets which support unprecedented economic growth. The human personality itself is a collection of functions that are maintained when they break down by more or less efficient medical, psychological, and social interventions.

The attempt to generalize functionality as a culture, to found a civilization on it, is so bizarre that it commands our attention once it is noticed. This is the dystopian paradigm of modernity in the 20th century. The critique of this astounding project in thinkers as diverse as Weber, Lukács and Heidegger, Marcuse, Habermas and Foucault, should also command our attention. It points the way to a new current of social theory, a critical theory of science and technology grasped not as a specialized activity but as central to social life.

I would summarize the central insight of this current as the belief that functional understandings replace more complex meanings in the dominant culture. But before elaborating on this proposition, I must deal with an objection. Recent philosophical writing on function points out that it has a hermeneutic dimension. To recognize a function is already an interpretive act. A hammer is useful only insofar as it is recognized as such. This seems to confute the claim that modern societies are hostile to meaning. But this is to confuse two different definitions of the word "meaning". Anticipating my conclusion, I will argue that a bare minimum of meaning necessary to use a tool is always an abstraction from a broader range of connotations and connections possessed by any object in its social context. The abstraction is of course a useful one, but it is not the whole story. A society which attempts to restrict the understanding of meaning to the bare minimum is different from one that admits the relevance of its whole range.

The influential Weberian analysis of this distinction is based on what is called the "differentiation of cultural spheres". Applied to science, technology and management, this is the social ontological equivalent of the epistemological notion of pure rationality. It asserts the effective institutional separation of functional aspects of objects from their broader significance in their social context. Means and ends are no longer united conceptually and practically but have been broken apart. The premodern notion of "essence", with its teleological conception of meaning that embraced a wide range of connections, gives way to a narrow rationality organized around a modern notion of causality.

The methodological basis of this Weberian view originated ultimately in Marx who discovered that the market has a unique rational form imposed by sundering capitalist economic exchange from tradition, religion, and politics. The larger context of use value, which situates objects within the way of life to which they belong, is replaced as an effective basis of economic action by the narrow concept of exchange value. Marx also showed that apparently "differentiated" and autonomous market rationality is tied to the rise of a specific class and creates a class biased society. Neutral rationality and class bias are conjoined in the market.

Although Weber was Marx's most influential if by no means orthodox follower, he focused on the notion of autonomization and ignored the critique. Only later, in the 1920s did Lukács recover Marx's critical theory of rationality. Following Weber, Lukács generalized the Marxian critique of market rationality to cover the whole surface of modern capitalist society, technology, administration, the media, and so on. This is the famous theory of reification which revolutionized Marxism, if not capitalism, and bore fruit in the Frankfurt School, influencing also directly or indirectly Heidegger and many other critics of modernity. Lukács offered the first version of an argument according to which behind modernity's apparently autonomous, value neutral rational systems lie power relations of a new type. The differentiation of rationality from other cultural spheres is simultaneously the subjection of society as a whole to capitalism.

Lukács notes the similarity between scientific knowledge and the laws of the market Marx criticized. The market is a "second nature" with laws as pitiless and mathematically precise as those of the cosmos. Like the worker confronted by the machine, the agent in a market society can only manipulate these laws to advantage, not change them. Lukács takes over Weber's analysis of bureaucratic and legal systems which reveals a related formalistic paradigm at work. He argues that capitalism reorganizes society around the kinds of abstractions characteristic of modern science and technology.

His position depends on his critique of the paradigmatic role of mathematical physics in the structure of modern knowledge and social practice. Since the 17th Century, physical law has been the model for all true knowledge and effective rational action has been identified with the kind of technical manipulation that can be based on such law. As Lukács writes,

> What is important is to recognize clearly that all human relations (viewed as the objects of social activity) assume increasingly the objective forms of the abstract elements of the conceptual systems of natural science and of the abstract substrata of the laws of nature. And also, the subject of this action likewise assumes increasingly the attitude of the pure observer of these-artificially abstract-processes, the attitude of the experimenter. [1]

These laws are formal universals, abstracted from all specific time-space coordinates and from the developmental process of their objects. They isolate the functional aspect of social objects through which they can be controlled technically. Their cognitive universality promises equally universal technical control of all aspects of nature and society. But insofar as they are purely formal, they are incapable of comprehending social practice and the ever new historical contents it produces. Resistances testify to the living human content that cannot be fully accommodated to the reified forms. Lukács found in class struggle the exemplary instance of this dialectic of reification and the life process.

Lukács analyzed this dialectic in terms of Hegel's critique of Kant's notion of formal-analytic rationality, also based on the model of physics. He transposed the Hegelian critique into the social realm, and identified correspondences between it and the Marxian critique of capitalist market rationality. The Marxian critique was thus raised to the highest level of abstraction and became the basis for an alternative cognitive paradigm.

This is the background to Marcuse's critique of one-dimensional society. The fact that Marcuse did not cite Lukács in his later work is, I believe, due to the uptake of these ideas by the Frankfurt School. By the post-World War II phase of his career,

---

[1] The passivity of the experimenter to which Lukács refers is only apparent: The experimenter actively constructs the observed object but, at least in Lukács's view, is not aware of having done so and interprets the experiment as the voice of nature. While Lukács does not criticize the epistemological consequences of this illusion in natural science, in the social arena it is defining for reification. Georg Lukács, *History and Class Consciousness*, trans. R. Livingstone (Cambridge, MA: MIT Press, 1971), 131.

Marcuse could take for granted the basic Lukácsian approach to the understanding of the relationship between capitalism and science and technology. However, there is a second partially hidden source of Marcuse's thought in this period: Heidegger's late critique of technology.

Heidegger argues that the modern world is a sum of resources, raw materials and system components. Nothing any longer has its own inner principle of movement, its own essential core of being, but rather everything is exposed to transformation to serve a role in the technical system. Objects are ripped from their contexts and reduced to their useful properties. These decontextualizations and reductions are inherently one-sided and violent. In this respect modern technology differs from craft work in which a pre-existing essential form embracing a wide range of values and meanings is realized by the craftsperson in materials conceived as predestined for the work. Instead, modern societies impose plans on passive materials. Marcuse's approach is shaped by this Heideggerian theory of the enframing, and the contrast between ancient technē and modern technology on which it is based.

Interestingly, the two critiques, that of Lukács and that of Heidegger, converge around several basic themes that reappear in Marcuse. These are:

1. The emergence of scientific-technical rationality as a dominant cultural framework;

2. The neutrality of this formalistic paradigm of rationality, i. e. its differentiation from meanings and values circulating in the lifeworld;

3. The predominance of technology over every other relation to reality;

4. The consequent loss of an authoritative cognitive grasp of significant aspects of the world;

5. The potential for catastrophe implicit in this limitation of the dominant culture to technical manipulation.

Naturally, the way in which Lukács and Heidegger develop these themes is quite different, but in Marcuse's appropriation a kind of synthesis is achieved. The central notion of this synthesis is the paradox of value neutrality which appears to isolate science and technology from the social while in reality it integrates them to it in a new way. This is the basis of Marcuse's critique of what he calls "technological rationality", a form of rationality that grasps its objects on purely functional terms without presupposing any goal except its own application and extension.

In chapter six of *One-Dimensional Man*, Marcuse writes, "This interpretation would tie the scientific project（method and theory）, *prior* to all application and utilization, to a specific societal project, and would see the tie precisely in the inner form of scientific rationality... It is precisely its neutral character which relates objectivity to a specific historical Subject-namely, the consciousness that prevails in the society... "[1] Marcuse's approach, as exemplified in these passages, is based on the notion that the differentiation of modern scientific-technical rationality is linked to domination. Neutrality is just the reverse side of the insistence on quantifying and controlling all objects indifferent to their own inherent potentialities. Other forms of action associated with artistic production, craft, the care of human beings and the cultivation of nature, which are based on a relation to the potentialities of their objects, do not offer the prospect of full control and so are dismissed as prescientific or irrational.

Why is neutrality specifically linked to the project of domination of capitalism？ Technological production breaks with the past and all the restraints it placed on the pursuit of productivity and profit. Traditional forms of knowledge are too closely integrated to the very lifeworld which capitalism must destroy in the course of its advance. They condense cognitive and valuative dimensions in ways that block technological rationalization, for example, by limiting the exploitation of labor or the natural environment, or preventing the optimization of resources and land.

Scientific-technological knowledge is adapted to the pursuit of power by its selective focus on quantitative aspects of its objects through which they can be broken down and transformed. Organic and essentialist paradigms of knowledge that presuppose some sort of teleology have no place here and give way to a mechanistic approach based on the measurable attributes of things. The neutrality of modern knowledge is thus both real and unreal, breaking the chains of tradition only to enter the prison house of power.

Marcuse's application of the reification/enframing thesis he derived from Lukács and Heidegger leads to the demand for a restoration of meaning through a transformation of the paradigm of knowledge and the technology that depends on it. Marcuse calls for radical technological transformation：

> Only if the vast capabilities of science and technology, of the scientific
> and artistic imagination direct the construction of a sensuous environment,
> only if the work world loses its alienating features and becomes a world of
> human relationships, only if productivity becomes creativity, are the roots of

---

[1] Herbert Marcuse, *One-Dimensional Man*（Boston：Beacon Press, 1964）, 156-159.

domination dried up in the individuals. No return to precapitalist, pre-industrial artisanship, but on the contrary, perfection of the new mutilated and distorted science and technology in the formation of the object world in accordance with "the laws of beauty". And "beauty" here defines an ontological condition—not of an oeuvre d'art isolated from real existence... but that harmony between man and his world which would shape the form of society. ①

Technologically instituted meanings are required both for human life to make sense once again and as cognitively valid guides to improving technological processes that threaten human wellbeing and survival. But Marcuse insists that this restoration must not take the form of a return to premodern modes of thought. A "qualitative physics" is excluded at the outset. Instead, he promises a kind of synthesis of art and science, an aestheticization of technology that would bring values into the design process as quantifiable parameters.

This development would depend on the emergence of an "aesthetic *Lebenswelt*", a new structure of experience encompassing aesthetic criteria. Aestheticized perception would embrace functional aspects of objects in the larger framework of their relation to life as a value. Instead of a purely empirical understanding of objects based on modern scientific-technical rationality, or a teleological notion of essence, articulating the place of objects in a traditional form of social life, an imaginative grasp of objects would locate them in a freely chosen way of life oriented toward peace and fulfilment.

While provocative the rather vague positive outcome of Marcuse's persuasive critique is bound to be disappointing. He never worked out the alternative satisfactorily although he developed a convincing diagnosis of the problem. The impression of pessimism and indeed dystopian despair left by his contribution is due in large part to the disproportionate effectiveness of the critique relative to the rather weak positive perspective on the future.

3

The philosophy of technology has encountered the very same question that

---

① Herbert Marcuse, *Towards a Critical Theory of Society*, ed. Douglas Kellner (New York: Routledge, 2001), 138-139.

preoccupied Marcuse, namely, how can meaning be restored in the context of a civilization based on a paradigm of rationality for which only causes and functions are real.

Albert Borgmann and Lorenzo Simpson, two philosophers influenced by the later Heidegger, have addressed this question in their writings. [1]In his later work Heidegger called for a "free relation" to technology. This alternative to a technological civilization would not require Marcuse's transformed technology but rather a change in attitude toward technology as we know it. Presumably, if we could use technology without interpreting all of reality in technological terms, we could enjoy the best of both worlds, a world of functional efficiencies and a world rich in meaning.

Simpson and Borgmann appear to be working with Heidegger's program. They hope to restore and validate the concept of meaning through a loosely phenomenological strategy. They address two problems associated with this project: first, the lack of rational grounds for culturally general meanings, and second, the problem of consensus in a postconventional society. Both Simpson and Borgmann reject the relativistic refusal of local meaning in terms of some absolute standard of rationality they deem inappropriately applied to culture. They also reject the notion that cultural differences in modern societies can be adequately represented as differing concepts of the good. This supposed "good" is a matter of opinion rather than the articulation of a way of life and as such it is subject to infinite and arbitrary variation. They agree with Lukács, Heidegger and Marcuse that there *is* a very definite consensus around a way of life in technologically advanced societies. This consensus is not located at the level of opinion but at the practical level, where technology shapes a common framework of experience and action. Simpson and Borgmann argue that a desirable way of life will sustain human relations and community independent of the technological fixation of modern societies.

Simpson's and Borgmann's critiques are based on the distinction between function and meaning. They argue against the overestimation of the former at the expense of the latter. Today, individuals commonly talk as though the objects by which they are surrounded were essentially functional. Indeed, they even extend a functional understanding to themselves and their human relations. This is the sociological crux of

---

① Albert Borgmann, *Technology and the Character of Contemporary Life* (Chicago: University of Chicago Press, 1984); Lorenzo Simpson, *Technology, Time, and the Conversations of Modernity* (New York: Routledge, 1995).

the one-dimensionality thesis. Such an understanding cannot be said to be false, but it is certainly partial.

A purely functional understanding is encouraged by the existence and prestige of technical disciplines in modern societies. The transportation function of the automobile is of special concern to automotive engineers. They are obviously justified in narrowing their focus in their work. But there is a sense in which an ordinary non-professional interpreter who understands the automobile exclusively in terms of its function adopts the engineers point of view in an inappropriate context. There is a risk that the legitimate limits of the engineer's standpoint, which open up a realm of technical knowledge, may become misleading obstacles to broader understanding for the theorist, the user and the citizen. Yet something like this confusion is implicit in much discussion of technology, and indeed in everyday attitudes as well.

The intellectual heritage of this type of argument should be clear by now. The differentiation of society has enabled function to be distinguished from the concrete relationship of artifacts to social life. One-dimensionality results from the attempt to totalize a functional view of the world, denying the residue of meaning excluded by the differentiation of function. The effectiveness of this totalization is not a theoretical but an empirical question. In Marcuse's view as in the views of Simpson and Borgmann, it is sufficiently effective to suppress awareness of the most important "dimension" of social life, the meanings or potentialities that enable humane understanding and self-understanding, and, Marcuse would emphasize, social progress.

Lorenzo Simpson addresses this problem as the reduction of "meaning" to "value". Values as he defines them are simple ends or goals. They can be abstracted from the complex web of meanings in which they emerge in actual life and represented independent of that context. Such abstractions have their uses, but when they are substituted for the larger framework of meaning, the results are dispiriting.

Meanings are built out of myriad connections between experiences and spheres of life. They are not definite, bounded things we have at our disposal, but structures or frameworks which we inhabit and which contribute to making us what and who we are. Meanings are enacted in our perceptions and practices. They are not chosen but rather they "claim us". Purpose is only one aspect of the phenomenon of meaning, but it can be isolated and privileged as the significance of the whole. The technologies that mediate the achievement of purpose then appear peculiarly central. The pursuit of ends with means, preferably technically efficient means, replaces an understanding of meaning. The focus on the means leads to a forgetfulness of the complexity of the structure of meaning and eventually to the lopping off of whole dimensions of the original experience as they appear irrelevant to maximum efficiency.

Simpson points out a second consequence of the reduction of meaning to value. Structures of meaning belong to a way of life. They can be justified only within that framework, by reference to each other and to the general virtues of the way of life in question. Values, on the other hand, seem arbitrary unless justified by arguments with rational appeal under any and all conditions. But such arguments invariably fail and so the values-perspective leads directly to a relativism that devalues the idea of the good life generally.

Simpson argues that the values perspective presupposes an absolutely detached observer. But the "death of God" is also the death of an absolute knower, uninvolved in any social world and tradition. Simpson writes,

> What happens in such a transformation of meaning into value? As meaning becomes thematized as value, the manifold connections which operate in part "behind our backs" and which, through informing and shaping our experience, predispose us to experience in a characteristic way, are transformed into premises. The validity of these value-premises, apart from the referential anchoring in the meaning which gave rise to the value, stands or falls with the rational evaluation of those premises. Our inability to provide *purely rational* foundations for such premises, in abstraction from the meaning that gives them point, results in our inability to experience as binding in a nonarbitrary way. That is, such values *qua* values, that is, in isolation from meaningful practices, cannot claim us. ①

If the position of the participant is privileged rather than that of the outside observer, relativism is avoided by reference to the internal significance of the meanings that circulate in a way of life. Those meanings do not have a compulsory evidence; they can be thematized and criticized. But the exercise of critical intelligence is a moment within the way of life, not escape beyond any and all involvement. Criticism does not automatically devalue meaning in general, but enables a more refined and appropriate relation to meaning in the particular situation of the participating individual. In sum, experience is not transcended in knowledge but forms its horizon. Simpson calls this a

---

① Simpson, *Technology, Time, and the Conversations of Modernity*, 47.

"*sittliche* account of rationality", referring to the Hegelian notion of value as immanent in the way of life of the community rather than speculatively constructed in abstraction from any involvement. ①

Unfortunately, Simpson does not see that the very reasons he adduces for insisting that values not be disengaged from their background in meanings apply to technologies as well. Technologies, considered apart from their context, are just as abstract as goals artificially isolated from the framework within which they are pursued. As a result, Simpson's account is vitiated by an unconvincing opposition between technology and meaning. Simpson distinguishes on several occasions between a technological mentality and actual technologies, but he fails to locate his critique clearly on one or the other side of the line between them. ② Thus he recognizes that a different cultural environment would generate different technologies and yet he also wants to insist that the properties of technology he criticizes constitute a "residue" characteristic of technology in general. ③Unfortunately the residue seems so loaded with undesirable features it is incompatible with benign alternatives. Meaning then shrinks to the margins of his conception of modern life, repelled by the very technical means on which it depends today for its context and realization.

Borgmann's argument is similar. He contrasts a consumer oriented way of life with an alternative organized around what he calls "focal things" in something like the Heideggerian sense. For Heidegger the concept of the thing refers not merely to an existing entity but to the gathering power of the objects around which the rituals of everyday life are organized. Things "thing" according to Heidegger in the sense that they lay out the framework of a local world within which relationships and identities are formed. In that world individuals are active participants rather than passive consumers, although Borgmann insists that their action is not arbitrary but is shaped by the possibilities opened by the things around which it is organized.

Borgmann believes that we have become so obsessed with the acquisition of commodities that we have lost touch with things in this Heideggerian sense. Technology teaches the sharp difference between means and ends where formerly each implicated the other. The complex involvements of individuals with others and with nature of an earlier

---

① Ibid. , 131.
② Ibid. , 8.
③ Ibid. , 174-175, 182.

time, when activities were less effectively mediated by technology, give way to hollow technical control. "Devices... dissolve the coherent and engaging character of the pretechnological world of things. In a device, the relatedness of the world is replaced by a machinery, but the machinery is concealed, and the commodities, which are made available by a device, are enjoyed without the encumbrance of or the engagement with a context."①

Consumer society is made possible by a technology sufficiently advanced to provide abundance. But the role of technology is not innocent. It is not merely a means to extrinsic ends. The ready availability of technological means to specific types of satisfactions tends to bias socially sanctioned desires toward just those satisfactions. Facility and convenience exercise a hidden tyranny which Borgmann calls the "device paradigm". A whole way of life is thus implied in technology and the consensus it organizes practically is difficult to criticize much less to challenge and overcome. "Technology", he claims, is "the new orthodoxy, the dominant character of reality".②

Meaning arises from engagement with focal things, things that exercise the gathering power to constitute worlds. Such things may be celebrations or occasions as well as objects. They require effort and engagement, a practice "that can center and illuminate our lives"③. They develop the relationships and skills of those who engage with them. They provide a focus from out of which to experience a context rather than supplying a commodity with efficiency and ease. Borgmann readily admits that focal things are not subject to proof or justification in any scientific sense. A "deictic" discourse can however point to the features of the world that engage our focus. In deictic discourse we can testify to the importance of the focal things in our lives and bring them to the attention of others with the hope of engaging them too in their gathering power. This notion resembles the Kantian "reflective judgment" which is also a testimony and an appeal based on an implicit concept of human nature rather than an absolute ground.

Like Simpson, Borgmann wants to withdraw from technology into activities on the margins, but he too formulates his program in ambiguous terms. Borgmann rejects

---

① Borgmann, *Technology and the Character of Contemporary Life*, 47.
② Ibid. , 189.
③ Ibid. , 4.

regression and claims that a renewed emphasis on focal things will enable us to rectify our relationship to technology. Thus technology as such is not the problem but rather the device paradigm which frames life as the application of efficient means to the pursuit of goals abstracted from a context of meaning. [1] Yet at the end of his book, Borgmann calls for a "two sector economy" in which technology as device will coexist with craft production. This seems to imply that technology is after all the problem and that bounding it is the solution. Once again the argument wavers between condemning the technological mentality and condemning technology itself. [2]

4

Marcuse argued that fundamental change in technology must be an aspect of fundamental social change. On his terms a return to focal things would imply not only a new attitude toward technology but also new technology. Although his reflections on this prospect were too vague to carry conviction, there is evidence that something like this process of change has already begun in small ways. This is especially clear in the case of the Internet. Not only has the Internet served as a scene on which new forms of sociability have been created, but users have played an unprecedented role in shaping and reshaping the technology. This example is important for revealing new forms of dialectical interaction between technology and the underlying population. We do not return to Lukács's vision of class resistance to reification with this example, but nor are we locked into a Heideggerian enframing. Marcuse's hopeful speculation appears relevant in a general way to this case.

Sociability was not in the original plans of the Internet's military sponsors. It was intended to solve technical problems in time sharing on mainframe computers and to transmit official information between the government and contractors on university campuses. In addition it may have also played a role in plans for a redundant communication system able to survive nuclear warfare. But early in its history a graduate student placed a small email program on the system and soon human communication became one of its most important features. His intervention responded to an interpretation of the system different from that of the military. He looked beyond its use for efficiently

---

① Ibid. , 200.

② Ibid. , 220.

distributing computer time to its communicative potential.

The shift in perception implied in such interventions was explained to me by a vice president of the Digital Equipment Corporation in the early days of personal computing. At that time human communication on computer networks was slowly emerging alongside standard usages, much to the surprise of computer professionals. The vice president said, "We were networking computers and we suddenly realized that we were not just connecting the machines but also the users of the machines".

I encountered several similar moments of realization in the early history of computer networks, enough to describe a pattern. For example, the French Minitel system connected millions of users in the early 1980s, long before the Internet was opened to the public. It was originally intended to be an information system delivering official data and news to dumb terminals, called "Minitels". The purpose of the system was clearly articulated by the government: to promote the entry of France into the "information age". But the system was very quickly hacked by users who converted it into a means of human communication. They added instant messaging to the system which exploded its original purpose and introduced an entirely new one: the pursuit of dates and sex. Of course information was still available, but the meaning of the Minitel was irrevocably transformed by the revelation of its social potential.

This basic reinterpretation of the nature of computer networks made possible a long history of user generated innovation that continues to this day on the Internet. The essential idea behind many of these innovations, so obvious but so difficult to realize, is the purely mediating role of the technology in social applications. Operations that depended on holding inventories of goods failed for the most part, but those that simply connected people to each other and to pre-existing information have been wildly successful.

This mediating role is not, however, transparent. For a connection to be made a context must be supplied. That context positions the users to take specific types of initiatives, for example, to seek personal communication or group communication, dates or information, and so on. Constructing these contexts successfully is not a simple matter and no engineering manual contains the knowledge required. This is because contexts are in effect virtual "lifeworlds", frameworks of meaning within which affordances emerge.

These lifeworlds are of course drastically simplified compared to the real thing. But they are not reducible to pure means. They are not tools but environments within which the user moves and works. Consider, for example, email programs such as Eudora and

Outlook. The division of the interface into three "panes", one for titles, another for content, and a third for "mailboxes" constructs a specific temporality. The user is called on to assign connections and priorities. By classifying messages according to various criteria and placing them in mailboxes, she constructs a usable past. By reviewing the title pane and responding to the important communications, she enters her future. The simplicity of the interface belies the complexity of the practices it invites and facilitates.

Still more interesting are the lifeworlds that emerge around group communication on computer networks. Contrary to journalistic hype, these practices do not date from the creation of Web 2.0. The earliest forms of group communication were asynchronous bulletin boards and computer conferencing programs that enabled users to send messages to a shared file instead of to personal addressees. Long before the Internet was open to the public, people were conducting business meetings, social gatherings, and discussing hobbies, illnesses and politics on various computer networks. Of course these discussions lacked many Web 2.0 features, but the all important connection to a group was available.

Within this framework users employed language to construct identities and virtual worlds oriented toward their interests and concerns. They established a communication model at the outset by stating the kind of meeting in which they were engaged. They bounded the group more or less effectively with software or communication practices. They constructed a past and future through techniques of archiving and mutual response. As a result of these activities, computer networks became environments within which communities form and creative activity goes on. [1]

These examples of human communication on computer networks show that the emphasis of critique should be less on features of technology as such or the ills of consumerism and more on the problem of agency and the norms under which agency is exercised. Recognition of the value of human communication is not merely a matter of opinion but activates individuals in new ways that belie the pattern of "technology" as it has been understood by the critics.

We have seen such a recovery of agency in the environmental movement as well as with the Internet. Environmental protests have led to significant changes in technology and shattered the myth of technological determinism and its associated technocratic

---

[1]　Andrew Feenberg and Darin Barney, eds. , *Community in the Digital Age* (Lanham, MD: Rowman and Littlefield, 2004).

ideology. We know now that we are responsible for our own technology and its consequences. An older understanding of technology as a hidden cornucopia managed by experts is giving way to a new sense of technology as a terrain on which human initiative is exercised in the interests of survival and progress.

These new expressions of agency in the technical sphere respond to experience with technologies. The innovators in the case of computer networks and the environmental protestors react to an unsatisfactory situation in which they find themselves. They reinterpret that situation in terms of insights available only outside the mainstream of technological development among users and victims. In Heideggerian terms one could say that they encounter their "world" in the light of potentialities unsuspected by those who first constructed its elements. Innovation is thus not just the discovery of new uses but also of new worlds in which the uses emerge. Marcuse's notion of a normatively informed perception is relevant to the case. Although aesthetics may not be the right term under which to classify the norms involved, clearly innovation emerges from discontent with the given and the projection of new life affirming possibilities of the sort Marcuse advocated.

5

This brief sketch of the role of agency in information technology and the environment calls into question the starting point of the analysis presented here. Apparently the culture of our society is not dystopian after all, nor is our rationality purely formal; substantive considerations intrude with significant consequences. How are we to understand this complication in the picture with which we began? Two possibilities present themselves.

We can simply dismiss the Marxian-Weberian premise of social differentiation and formalization, and with it the related notion of an emancipatory dereification of rationality and the society dependent on it. This approach conforms to the skepticism with regard to traditional social critique expressed by postmodernists and some scholars in science and technology studies. They regard nostalgia for meaning as a dispensable illusion. But their attempts to reconstruct such essential aspects of social life as ethics and political resistance are unconvincing. These anti-critics usually end up with abstract appeals for tolerance of the other under a variety of fancy labels. One can hardly oppose this benign conclusion but it has little relevance to the politics of technology.

Alternatively, we could add to the notions of differentiation and formalization a complementary dedifferentiating and substantive dimension overlooked by most of the earlier social critics. In this case, the stripping down of complex social forms and meanings to functional residues would co-exist with other social processes tending in the opposite direction, toward a reconstruction of complex meaning systems. This is the position of Habermas, for example. He is sensitive to the dystopian threat of modernity but neither resigns himself to the triumph of a purely instrumental social logic nor projects an apocalyptic transcendence of dystopia. Habermas sees a conflict of opposing tendencies in modern societies, a tendency toward total systematization countered more or less effectively by another tendency promising a recovery of meaning. In Lukács a similar conflict is developed as an immanent dialectic between reified institutions and the lives they fail completely to contain. Although I do not think these positions are entirely successful, this general approach opens up the question of modernity in a new and fruitful way. I believe that something along these lines is in fact a correct description of the social process today.

A hermeneutics of technology is needed to articulate the dimension of meaning of technology and to explain how it relates to functionality. Many approaches to developing this hermeneutic are possible. For example, one could turn to the social histories of various technologies for insight into how they engaged with many aspects of the life of their time. A theoretical approach to the relation of function and meaning might be elaborated through analysis of such examples. My very brief sketch of email clients shows the relation of function to meaning from this angle. What appear to be narrowly functional aspects of an interface in reality open up considerations on time that point to an unsuspected complexity. The apparently banal division of panes organizes more than incoming texts; it organizes the users' life, or a more or less significant part of it. In this concluding section of my paper I will sketch an approach that might guide research into such phenomena.

I am struck by the similarity between my own project and Heidegger's two apparently opposite interpretations of technical action in his early and late periods. The early account is explicitly phenomenological. It abstracts from causality to explain functionality in its relation to worldhood. The latter account is also interpretive although Heidegger no longer makes reference to phenomenology. This account, however, also abstracts from the usual causal interpretation of functionality in order to explain it ontologically. I believe that there is a hidden complementarity uniting these two accounts, but Heidegger presents them independently without reference to each other. Making the hidden connection between them explicit provides a basis for a theory of the double aspects of technology which Heidegger himself missed.

The first part of *Being and Time* lays out a remarkable phenomenology of action. Heidegger traces the links from an original attitude of care through the ordered relations of the various instrumentalities, materials and signs implicated in action. This analysis of "readiness-to-hand" shows that functionality is a complex aspect of networks of persons and objects, a "totality of involvements", and not a property of things taken by themselves. The analysis concludes with the notion that the whole matrix of action constitutes "significance", *Bedeutsamkeit*. The argument thus holds that function and meaning are inseparable aspects of what Heidegger calls "world", an ordered system of internal relations between *Dasein* and the objects of action. This notion of world lies behind the reflections of Simpson and Borgmann discussed above.

Heidegger's account is unclear. He does not explain precisely how meaning emerges in action. Is it a condition of action or a result? The "totality of involvements" is a system of relations between entities caught up in a technical network. Each particular functional relation is defined by an operative meaning of some sort that enables coping. Heidegger carefully defines the form of that narrowly functional meaning which is discovered in circumspection (*Umsicht*). The introduction of the concept of significance at the end of this analysis seems intended to tie the variety of these interconnected practices and meanings together in a wider space of action, a "world".

But is it in fact the case that the specific meanings involved in functional activities are connected in a coherent whole? The point of the social theories with which we began is the differentiation of function and meaning in modern societies. This interpretation of Heidegger's early theory resembles pragmatism in attempting to redefine as one what modernity has divided in practice. In this context a mere definition cannot justify the conceptual leap from function to meaning.

We need to ask why Heidegger felt it necessary to present his analysis of function as essentially related to meaning. Awareness of Heidegger's complicated relation to modernity helps to understand this conundrum. The radical social differentiation occurring around him ran increasingly counter to his understanding of life. As railroads, electrical systems and radio broadcasting constructed a functionally organized mass society, Heidegger reverted to the meaning-granting dimension of action to defend a human world. This world was impoverished to be sure, but it was at least a world, and not a mere causal concatenation in which human beings were caught up.

Lucian Goldmann explored one possible explanation for this approach. According to Goldmann, Heidegger's account of readiness-to-hand is an ontological version of the theory of reification in Lukács. Without adopting Goldmann's hypothesis of actual influence, it is still noteworthy that many categories of Heidegger's theory of worldhood have obvious affinities with contemporary cultural critique. Heidegger's argument, like that of Lukács, recognizes and also places a limit on the dystopian tendency of modernity beyond which the very notion of the human would cease to make sense. The inauthenticity of the mass appears to degrade human beings to the level of things, but contained in germ within inauthentic existence is a connection to meaning. This provides the basis for the possibility of authentic action, transcending any given stereotypical response to experience. Thus in his social context, Heidegger's analysis of action has an implicit critical aspect.

What is the evidence that such problems lay behind Heidegger's early theory? The lecture course entitled *Fundamental Concepts of Metaphysics* contrasts human action with other types of engagement with reality. Heidegger explains that the participation of *Dasein* in a world presupposes what he calls the "as" relation, that is, the ability to relate freely to meanings. Without such a relation to meanings, there is no world in his sense but only reflexive responses to particular stimuli. Heidegger needs the concepts of meaning and world to explain care and the projective capacity of the human being. It is this capacity which brings time to being. Animals and machines do not have it and Heidegger therefore distinguishes *Dasein* from these other types of being.

But did Heidegger relate the limitations of animals and machines to the status of *Dasein* in the emerging mass society? He did not descend to the level of social criticism to make such a point, but he came close in his later work. There "technology" is action planned on the basis of representations. Such action does not follow the logic of readiness-to-hand explained in *Being and Time*. Instead it is oriented by a concept of causality and an associated concept of predictability. A truncated functionalization obliterates meaning and its associated world in a mechanically organized and planned order. Human beings become resources and system components in such an order. When *Dasein* as the site of revealing is obscured by the enframing, meaning is impoverished or blocked altogether. If we reconsider the later Heidegger's analysis of technology in the light of his early work it is clear that what he calls the "danger" threatens *Dasein* with reduction to something very much like an animal or a machine. And this reduction threatens world as well as *Dasein*. How could it not given the unity of being-in-the-world which is defining for *Dasein*?

The contrast between these two modes of action—individual action and enframing—evokes the double aspects of technology, and with it, of the society it structures. Enframing describes an order that privileges power, that is, causal relations, functions, over meanings. In fact meanings become instruments of power in this society, little more than advertising slogans. The recovery of meaning in the full sense of the term through authentic action in relation to artifacts is evoked obscurely in the later works, for example in the essays on the thing and on building and dwelling. But Heidegger seems to be in full flight from modern technology in these interesting essays. If we go back to *Being and Time* for a more analytic concept of authenticity, we are steered away from technical practice toward a heroic conception of historical resistance. This conception has so little concrete content that it justified Nazism for Heidegger and communism for his student Marcuse.

Let us try a different tack. In *Being and Time*, the notion of authentic action involves "precisely the disclosive projection of what is factically possible at the time"[1]. Heidegger interprets this proposition in terms of the concepts of death and a vague notion of historical destiny. But we can apply it to *Dasein* in its technical dealings with entities. Then authenticity suggests technical creativity rather than revolution. Indeed, the free improvisations and resistances of individuals within the enframed system obey the logic of action described in *Being and Time*. These actions generate a meaningful world in the face of the "consummate meaninglessness" imposed by enframing. The dialectic of enframing and action describes the modern experience in all its threat and ambiguity.

I have attempted to work out the implications of this approach in terms of a theory of technology as both functional and meaningful. As functional technology responds to a causal logic and is explained in technical disciplines that are relatively differentiated and autonomous. As meaningful, technology belongs to a way of life and embraces not only a minimal significance directly related to its function but also a wide range of connotations that associate it with other aspects of the human world in which it is involved. The evolution of modern technology takes place in part through the interaction between these dimensions. Modern societies tend to separate them institutionally, for example,

---

[1] Martin Heidegger, *Being and Time*, trans. J. Macquarrie and E. Robinson (New York: Harper & Row, 1962), 345.

distinguishing engineering from everyday understanding, management from working life, and control from communication. But in practice there is constant interchange between the differentiated dimensions. In fact they interact and conflict not only institutionally but within the individuals as they respond in routine or innovative ways to the technological environment in which they live.

Consider once again the examples of communication on computer networks. The hacking of the Minitel network responded to users' perception of unexplored potentialities of the technology. These potentialities were suggested by the connection of the Minitel to the telephone network. The hackers must have been puzzled by the obstacles to communication on a familiar network dedicated precisely to that purpose. In introducing a new communicative functionality, they repositioned the computer in the structure of everyday life. Its meaning was transformed through the addition of this function. Notions of efficiency are not helpful in understanding this phenomenon. Nor is it useful to start out from the purely technical possibilities, quasi-infinite in principle. Rather, to imagine this innovation and to understand it, one must work in from meaning in all its complexity to its displacement in narrow functional terms. The key to this hermeneutic approach is thus a notion of substitution or abstraction that responds to specific technical potentials.

Examples such as this suggest a different notion of democracy from the usual one. We are not dealing here with rights and elections but with the negotiation of the technical framework of everyday life. From this standpoint the notion of liberation must be reformulated to signify a reversal in the relations of dominance between the two dimensions and the modes of action that belong to them. So long as the encounter with the technical in everyday experience is subordinated overwhelmingly to enframing, a one-dimensional universe prevails. But a social world in which the meaning-generating activities of the individuals freely interact with technical disciplines and artifacts would have a radically different character. Such a world would realize Marcuse's hope that the imagination might inform technology with values but it would not do so by creating a new form of scientific-technical rationality. Rather, it would invest disciplines and artifacts with the results of the experiential encounter with technology, that is, with new meanings emerging from human action.

功能和意义：技术的双重面相

安德鲁·芬伯格[*]

一

新买的汽车或计算机都配有一本操作手册，描述并说明了该设备的功能以及如何使用。显然，它应尽可能详尽，即只要你读懂了那些术语，就似乎已完全搞懂了这台设备。当然，关于这台设备还可以有更多甚至多得多的东西来讨论。汽车和计算机属于社会，它们是其中复杂的组成部分。它们和那个世界中的方方面面存在着关联，对此全体进行说明或对各种关联在总体上加以把握是不可能的，这其中一些是象征性的关联，另一些则是因果性的。这种复杂性可举例说明。汽车的应用塑造了城市的规划，它们也标示出其主人的身份和地位，同时它们也是污染的制造者，等等。同样，计算机的应用改变的是知识产权的运作体制，改变了个体的自我表达和大众媒体间的关联方式，也克服了各种形式的脱离社会的现象，等等，类似的例子还可以无限地扩充下去。因此需要一个专门的术语来表征这种广泛存在着的关联，我愿意称之为技术的"意义"或"意涵"。

某设备中由社会因素决定的功能，与决定其运转的物理因果性间的关系，引起了哲学家们的兴趣。他们发现这是一种较之表面看来更为复杂的关系。本文的目的在于对此作一些补充说明，但这仍显得有点复杂。我将试图去理解：关于严格意义上的设备的功能——在操作手册中所描述的——与其社会意义间的关联。这就是本文要回答的问题，但采用的方法将和分析性的方法大相径庭，后者更类似于以往对功能进行研究所采用的路径。支持这种区分的充足理由在于：功能和意义的关系并不主要是一个概念问题，而是指向了作为独特的社会构造的现代性的核心所在。它提出的问题是：现代社会中主流的知识范式如何与意义的维度产生关联，当然这是我理解的所谓意义。为此，我将从（前人）关于该问题的一些思考开始论述。

---

* 安德鲁·芬伯格（Andrew Feenberg），加拿大西蒙·弗雷泽大学教授。本文为作者向 2010 年北美海德格尔年会提交的参会论文。——译者注

二

现代社会对自身及世界的理解是基于一种赤裸裸的功能逻辑，它剔除了情感和目的论的附加因素。一方面，这使它能创造出有效控制自然和市场的技术工具，来维持其前所未有的经济增长；另一方面，人的存在自身也成了功能的聚合，通过或多或少能产生疗效的药物、心理和社会干预，这些功能得以维持而不至失效。

把功能化推广为一种文化，并建立以其为基础的文明形态，这样的企图是如此之怪诞，以至于它才被注意到便引起了人们的关注。立场各异的思想家们，无论是韦伯、卢卡奇，还是海德格尔、马尔库塞、哈贝马斯，抑或是福柯，都对这令人的震惊计划有过批判，这同样值得我们关注。这些批判指出了一条道路，展现了一种全新的社会理论思潮，它是关于科学和技术的批判理论，针对的是把科学技术理解为社会生活的核心，而不仅是一种特殊的人类活动。

功能化企图的核心观念可以概括如下：在主流文化中，用功能化的理解替代丰富的意义。但在阐述这一命题前，我必须回应一个反对意见。新近的一些哲学观点认为：功能化同样具有释义学的维度。对某种功能进行定义，这已经是一个释义的行为。只有锤子被认为是一把锤子的时候，它才是有用的。这主张看似能驳倒那些认为现代社会敌视意义的观点，但这其实是混淆了"意义"这个词的两种不同定义。我的结论可预告如下，使用一件工具所需的最低限度的意义通常是一种抽象，是从一个更广泛的意涵和关联领域来的抽象，任何在社会语境中存在的物品必定具有这些意涵和关联。这种抽象当然是一种有用的方式，但不是全部。把对意义的理解设定在其最低限度上的社会，与承认关联的整体来理解意义的社会，二者是不同的。韦伯对于这种差别提出过一个影响深远的分析，即这是基于所谓的"文化模式的差异"（differentiation of cultural spheres）。当这种差异体现在科学、技术和管理上时，便形成了与认识论上的纯粹合理性概念相对应的社会存在层面的概念。它确证了在现实建制层面，物品的功能性方面与其在更广的社会语境中形成的意义间的实质性的分裂。无论是在概念上还是实践上，目的和手段不再是统一的了，它们早已分离。前现代的"本质"概念通过目的论层面的意义概念表征了广泛的关联性，现在它却让位于狭隘的合理性概念，后者又建基于现代的因果性概念。

韦伯观点的方法论基础最终来自马克思，马克思发现：资本主义的经济交换与传统观念、宗教和政治因素的分离，导致市场具有一种独特的理性形式。使用价值把物品带入了其所属的生活方式之中，但这一宏观性的语境却被狭隘的交换价值替代成了经济活动的效率原则。马克思还指出，明显带有"差异化"和自主性的市场合理性，与某个阶级的诞生紧密地联系在一起，进而一个具有阶级偏见的社会也由此产生。中立的合理性与阶级偏见在市场中联起手来。

韦伯关注于自主性概念而忽视了对它的批判，他称得上是马克思信徒中最具影响力的，却非一个正统的追随者。只有到后来，20世纪20年代，卢卡奇才恢复了马克思的合理性批判理论。与韦伯一样，卢卡奇总结了马克思对市场合理性的批判，并将其运用到现代资本主义社会的所有方面，如技术、管理、媒体等。这就是著名的对马克思主义——如果资本主义不算在内的话——具有革命性影响的物化理论，它在法兰克福学派那里得到了发展，同样也直接或间接地影响了海德格尔和其他类型的现代性批判。卢卡奇给出了本文的第一种论证，他认为，在现代性表面上看似自主的、价值中立的理性体制之下，存在的是一套新的权力关系。文化模式导致的合理性上的差异，同时也是一个社会在整体上向资本主义屈服所要克服的差异。

卢卡奇注意到了科学知识和马克思所批判的市场法则间的相似性。市场法则具有自然法则那样的冷漠无情和数学般的精确，因而市场算得上是"第二自然"。与工厂中工人与机器的遭遇相同，行动者在市场中也只能利用市场法则来获利，而不能改变它们。韦伯关于官僚和法律体制的分析，提供了一个相应的形式化体系运转的范例，卢卡奇继承了这点，进而论证：现代科学和技术所给出的各种抽象的属性，是资本主义认识社会的依据所在。

这一观点的得出，又是基于卢卡奇对数学化的物理学所作的批判，它所针对的是后者在现代知识结构和社会实践中所扮演的范导性角色。自17世纪以来，物理学定律就已经成了所有真知识的范本，而追求效果的理性行动，也被等同于那些基于这些法则所进行的技术性操作。卢卡奇写道：

> 需要明确指出的关键问题是：所有人类的关系（可被视为社会活动的目的所在），不断地表现出自然科学概念系统的抽象因素和自然法的抽象基础才具有的客观性的存在形式。同样，行动的主体也不断地表现出只有那些人为的抽象过程的纯粹观测者才具有的态度，即实验者的态度。①

这些从所有具体的时空坐标中抽象而来，从所有其论述对象的发展过程抽象而来的法则，是形式化的普遍性法则。它们与社会中使用的技术产品的功能是分

---

① 卢卡奇这里提到的实验者的被动性只是表面的，事实上实验者积极地建构了被观测的对象，但在卢卡奇看来实验者并没有意识到这点，还以为实验就是自然本身的表达。当然卢卡奇并没有批判自然科学的这一幻象在认识论上产生的结果，但在社会领域中它被定义成为物化。参见卢卡奇 History and Class Consciousness（Cambridge MA：MIT Press, 1971），131.

离的，尽管正是通过这些法则，产品才能被技术地操控。这类法则在认识上的普遍性，同样承诺了对自然和社会的所有方面进行普遍的技术控制的可能。但就其作为纯粹的形式而言，它们并不能理解社会实践及其不断产生的新的历史内容。对于这类控制的反抗，证明了有关现实的人的内容，不能完全适应这些物化的形式法则。在阶级斗争中，卢卡奇找到了体现着物化和生命过程二者辩证关系的典型事例。

卢卡奇分析这一辩证关系所用的术语，是来自黑格尔对康德形式分析的合理性进行的批判，后者同样是基于物理学的知识模式。卢卡奇把黑格尔式的批判移植到了社会领域，并将其与马克思对资本主义市场合理性的批判相互调和起来。因而，马克思的批判被提升到了最高的抽象层面，成了一个替代性的认知范式的基础。这也是马尔库塞的单向度社会批判的背景所在。但卢卡奇并没有在马尔库塞晚年的著述中被援引，我相信原因在于法兰克福学派对这些理念所作的理解。在马尔库塞第二次世界大战后的学术生涯中，他可能已经把卢卡奇关于资本主义和科学技术之间联系的观点当成理所当然的。不过，在这一时期，他思想上还有另一个潜在的倾向，即后期海德格尔对技术的批判。

海德格尔认为现代世界是资源、原材料和系统零件的总和。它不再有任何事物可以保持其内在的运动规律，即它自身存在的本质核心，所有事物都被转变为在技术系统中担任一个服务性的角色。物品从其存在的背景中被剥离，被简化为某种有用的属性。这种去背景和简化本质上就是单向度的和带有暴力的。在这一点上，现代技术不同于手工技艺，在后者中，各种不同的价值和意义被一个预先存在着的本质形式所蕴涵，而工匠则用物质材料来实现这一形式，在工匠眼中这些材料又仿佛是注定要为这件作品来服务的。相反，现代社会则是在被动的材料上实施各种筹划。海德格尔的这一座架理论，连同古代的技艺和现代技术之间的比较，在马尔库塞的方法中得到了体现，同时也是后者的理论基础所在。

卢卡奇和海德格尔的两种批判方式，在许多基本论题上形成了某种有趣的一致，并在马尔库塞那里集中表现了出来，可归纳如下：

（1）科学技术的合理性是作为主流的文化架构而呈现的；

（2）合理性的形式化范式具有中立性，即它与生活世界中的意义和价值的非中立性不同；

（3）较之人与现实的其他联系，人和技术的联系表现出显著的优势；

（4）相应地，在认知上对于世界意义进行权威性的理解已变得不可能；

（5）倾向于技术操作的主流文化中隐含了导致灾难的潜在可能。

从本质上看，卢卡奇和海德格尔论述这些命题的方式大相径庭，但马尔库塞在自己的处理方式中对二者进行了综合。这一综合的核心是要解决价值中立的悖论，即从表面上看来，科学技术与社会不同，它们与价值无涉，但实质上它们只是以一种全新的方式整合了后者。这构成了马尔库塞所谓的"技术合理性"批判的基础，这种合理性主张用纯粹功能化的解说来把握对象，除了推进自身的应用和扩展外不再预设任何其他的目标。

在《单向度的人》一书的第六章，马尔库塞写道："这种理解方式和整个科学的事业（包括其方法和理论）结合在一起，优先于科学的各种应用——包括在社会中的应用，这一结合产生于科学合理性的内在形式层面……正是其中立的属性把客观性和一个具体的历史中的主体——在社会中占主导的个体意识——联系了起来。"[1] 上述段落中所展现的马尔库塞的方法是基于一个观念，即区分出现代科学技术的合理性，为的是与统治产生关联。较之对所有对象进行量化和控制而言，中立性的说法只是硬币的另一面，它与事物自身内在的潜质无关。其他形式的诸如艺术创作、手工技艺、对人类自身的关怀和对自然的崇拜等活动，由于是与活动对象的潜质发生关联，无法提供对事物进行完全控制的可能，因而在前科学或是非理性的名义下被排除了。

但为什么中立性会和资本主义的统治计划联系到一起？因为技术生产要和过去的事物决裂，并摆脱所有妨碍其追求生产力和利润的限制。资本主义在它的发展进程中必须摧毁生活世界，而传统的知识形式正是生活世界中不可分割的组成部分，汇聚其中的认知和价值上的因素阻挡了技术合理性的前进，例如，它主张限制对劳动者和自然环境的剥削，或是阻止对资源和土地的最大限度的利用。

由于科学知识选择性地关注于对象的量的方面，这有利于对象被分割和改造，所以它适合于对权力的追求。有机的和本质主义的知识范式预设了某种目的论，因而对此没有帮助，只能让位给了基于事物可量化属性的机械化方法。因此，现代知识的中立性既是现实的又是非现实的，它打破了传统的束缚，为的却是步入权力的牢笼。

马尔库塞把源自卢卡奇和海德格尔的物化/座架理论付诸应用，这又驱使他要去改变知识的范式和以之为基础的技术，以恢复意义。马尔库塞提倡对技术进行激进的改造：

> 只有在科学技术加上科学和艺术的想象汇成的巨大力量的指导下，才能去构建一个感性的环境；只有当工作世界失去其异化的特征成为一个人类关系的世界，并且只有当生产成为创造之时，统治的根基才会在个体中枯绝。但这并不意味着回到前资本主义、前工业化的手工技艺，而是相反，是以"美的法则"为依据修正和完善在物质世界中被损毁

---

[1]　Herbert Marcuse, *One-Dimensional Man* (Boston: Beacon Press, 1964), 156-159.

和歪曲了的新科技。"美"在这里设定的是一个本体论的条件——它不是与现实存在相隔绝的艺术作品，而是人和世界之间的和谐，和谐才是构建社会形态的标准。①

由技术建制确保的意义，一方面可以使人的生命再次变得有意义，另一方面也能被作为指导实际经验的原则，以改进那些对人类生存和福祉产生危害的技术发展进程。但马尔库塞坚持认为这种恢复不是返回前现代的思想方式，所谓"质化的物理学"从一开始就被排除在外。作为替代，他预言了艺术和科学的综合，这是一种技术的审美化倾向，即在技术的设计过程中，各种价值能像量化的参数那样产生影响。

这种发展完善有赖于"审美的生活世界"的出现，一种以审美为标准的新的经验结构。作为一种价值的审美知觉与生命产生了关联，以此为背景，审美知觉将包容物体的功能性方面。与那种在现代科学技术合理性基础上对物体所作的纯经验型理解不同，与那种目的论意义上、在传统社会生活中为物体确定相应地位的本质概念不同，以丰富的想象力对物体进行的把握，可以使其在人们的自由选择下置身于和谐完满的生活之中。

虽然，马尔库塞雄辩的批判充满了煽动性，但由于在现实性的结论上过于模糊，因而它注定是令人失望的。虽然他对问题的诊断令人信服，但却没有给出令人满意的替代方案。批判卓有成效，而对未来几乎没什么积极的观点，二者比例失调，导致了马尔库塞的理论贡献给人们留下的是悲观主义的印象和敌托邦式的绝望。

三

当一种文明的合理性范式仅仅把因果和功能当成实在时，如何在其中恢复意义的地位，便成为困扰着马尔库塞的难点所在，同样，技术哲学遭遇的也是同一个问题。

技术哲学家伯格曼（Albert Borgmann）和辛普森（Lorenzo Simpson）在后期海德格尔的影响下，在其著述中也讨论了这一问题。② 在海德格尔后期著作中，倡导一种与技术的"自由关系"。这种对技术文明的替代方案，并不要求马尔库

---

① Herbert Marcuse, *Towards a Critical Theory of Society*, ed. Douglas Kellner（New York：Routledge, 2001），138-139.

② Albert Borgmann, *Technology and the Character of Contemporary Life*（Chicago：University of Chicago Press, 1984）；Lorenzo Simpson, *Technology, Time and the Conversations of Modernity*（New York：Routledge, 1995）.

塞式的对技术的改造，就我们所理解的，它是一种态度上的转变。它提出了一个假设：如果我们拒绝对全部现实进行技术化的解释，同时这又不影响我们对技术的利用，那么由功能产生的效率和丰富的意义二者就能兼得，并造福人类。

辛普森和伯格曼采取的似乎是与海德格尔类似的方案。借由一种广义的现象学式的规划，他们想恢复和重新确立起意义。他们指出了与此相关的两个问题。首先，一般所指的文化上的意义缺乏理性的基础，其次，在后契约社会中，各种意见如何取得一致。他们不赞同用一个绝对的合理性标准来排斥各种局部性的意义，因为这一方式不适用于文化层面。但他们也否认现代社会中的文化差异，可以提升到不同类型的善那样的高度；那些被信以为真的"善"，只是一种观念，而不是对某种生活方式的设定，因此它们可以进行无限制和任意的变更。关于技术发达社会中的生活方式，他们与卢卡奇、海德格尔和马尔库塞一样，认为可以存在一种十分明确的共识，这种共识不是观念上的，而是实践层面的，也正是在实践意义上，技术塑造了一个人类经验和行动的共同架构。辛普森和伯格曼都论证出了一种值得追求的生活方式，它能维护人类的各种关系和社会共同体，并摆脱现代社会对技术的迷恋。

辛普森和伯格曼的批判工作，同样是建基于功能和意义的区分，并批评了过高评价前者而忽略后者的做法。现今，人们对于周围物品的一般性认知，也都基于功能化的判断。甚至这种功能化也扩展到了对人类关系和自身的理解上。这是单向度命题在社会学上引起的困惑。功能化的理解不能算错，但肯定是带有偏向的。

现代社会中分门别类的技术及其产生的影响力，助长了这种纯功能化的理解方式。在一种情况下，汽车的运输功能，是汽车工程师要专门考虑的事，显然它是后者工作的核心和目的所在。但在另一种情况下，一个非专业的普通人，假如他对于汽车仅知道其功能化的理解，因而可能在一个不恰当的语境中仍旧采用工程师的观念来理解汽车。这里就存在着由于超越了工程师观念的合法应用限度而引发的风险。工程师的观念展现的是一个技术知识的领域，它超出限度的误用可能误导并妨碍了一个学者、一个用户或一个公民做出各种不同的理解。在许多关于技术的讨论中都存在类似的混淆，在人们日常生活的态度上也存在同样问题。

事到如今，这种认识上的成见应该被清除了。各种社会间的差异，显示出在技术产品和社会生活的关系上，除了功能化的判断外还有其他各种可能。在宏观上对世界只采用功能化的视角，排除其他理解的意义，导致的将是单向度的结论。这种宏观判断产生的影响不仅在理论上，还表现到了经验领域。在辛普森和伯格曼看来，这种影响，用马尔库塞的话说，足以压制另一种认识的产生，那便是意义。意义才是社会生活以及马尔库塞强调的社会进步中最重要的"维度"，它也是使人类理解世界和理解自我得以可能的条件。

辛普森将这一问题描述为"意义"被还原为"价值"。他把价值定义为简单的目的或目标。价值可以从复杂的意义之网中抽象出来，因而看似可以独立于意义而存在，但正是通过意义，价值才能在实际生活中得以展现。这种抽象有其一定的作用，但当它们作为宏观的意义的替代物出现时，结果就变得令人沮丧了。意义生发自经验和生活之间丰富的关联，意义不是可以被我们任意支配的有限和确定的事物，而是人们居于其中的各种架构，它回答了我们是什么和我们是谁这样的发问。意义被设定在我们的感知和实践中。不是我们选择了某种意义，而是意义"对我们提出要求"。目的只是意义现象的一个方面，但它凭借自身的重要性而独立和凌驾于意义整体之上，这就使为实现目的而用到的技术手段显得尤为重要。于是借用手段来追求目的和偏爱技术上行之有效的工具，替代了对意义的理解。对工具的关注导致了意义结构的丰富性被遗忘，进而各种看似与效率最大化无关的原初性经验也最终被一并剪除了。

辛普森还指出了把意义还原为价值的第二个后果。意义结构属于生活方式。一方面，意义只有在其所属的架构中，通过相互援引和诉诸生活方式的普遍性特质才能得到确证。但另一方面，价值似乎总是只能在理性的诉求下来论证。但这类论证总是会被证伪，因而价值观直接导向了相对主义，并普遍地贬损了善生活的理念。

辛普森认为，价值观预设了一个绝对的超乎世外的观察者。但"上帝已死"这句话同样宣告了一个全能的知者不可能存在，对此任何社会、任何传统都不能例外。辛普森写道：

> 当意义转变为价值时，发生了什么？当意义被作为价值论述时，（经验和生活世界之间）多维度的联系转变成了各种预设。——这些联系在"在我们背后（为我们所不知）"产生着作用，它们引发并塑造了我们的经验，使得经验以某种特定的方式向我们呈现。——这些价值预设，脱离了其本身得以可能的意义参照坐标系，因此其有效性只能以理性评价的方式来确立或取消。但是，当价值从先于其存在的意义中被抽象出来后，我们便无法给这些价值预设奠定一个纯粹理性的基础，这致使我们只能以独断的单一性方式来获得经验。如此仅作为价值的价值，如此脱离了有意义的实践的价值，没有资格对我们提出任何要求。①

如果是参与者，而不是一个外在的观察者，被赋予了裁判权，那么通过诉诸生活中无所不在的意义所具有的内在重要性，就可以避免相对主义出现。意义并

---

① Simpson, *Technology*, *Time*, *and the Conversations of Modernity*, 47.

不显现出强制性的迹象，意义可以被讨论，也可以被批评。但运用知识进行批评同样是生活方式的内容之一，不能脱离和超越于生活中所有其他的因素。一般而言，批评并不一定会贬低意义，而是在特定的情形下，使得参与的个体与意义间重新构建一种更恰当和完善的关系。总之，经验并没有在知识中被超越，而是构成了知识的视域。辛普森称之为"对合理性的伦理说明"，这是对价值概念的黑格尔式的定义，即诉诸共同体内在的生活方式，而不是抽象中脱离于现实事物的建构。①

辛普森坚信价值不能从意义的背景中分离出来，技术也是如此，但不幸的是，他没能给出支持其观点的适当理由。脱离其所属的背景来考察技术，就会造成像脱离了意义架构来谈论目标那样的抽象，正是意义架构使得目标得以设立。辛普森不承认技术和意义的区分，但没能给出令人信服的反驳理由，这最终导致其观点不尽完善。在许多场合，辛普森区分了技术的精神和现实的技术，但他的批判针对的是哪一方，却没有清楚地阐明。② 因而，他承认不同的文化背景将产生不同的技术，而且他还相信，在他所批判的各种技术属性的相互作用下，通常会产生一种"剩余"的技术特征。③ 令人遗憾的是，这种剩余特征往往并不为人们所期待，与善良的出发点往往相悖。于是，意义在技术手段的压制下退缩到了现代生活的边缘，而在当前，意义背景及其实现方式也必须依赖于这些技术手段才得以展现。

伯格曼的论证与辛普森类似。他将消费者的生活方式和另一种围绕海德格尔意义上的"焦点物"所展开的生活方式之间作了比较，后者可以作为前者的替代。在海德格尔那里，物指的不仅是一个存在着的实体，还包括了物所蕴涵的某种聚集的能力，它筹划了日常生活的礼俗。在海德格尔看来，各种物"确立为物"，就是在世界的局部范围内，由物展开出一个架构，在其中人的各种关系和身份得以确立。伯格曼强调，在那个局部世界中，个体的行动不是任意的，而是由各种物所提供的可能性条件引导的，行动是由物组织的。虽然如此，置身其中个体仍是一个积极的参与者，而不是被动的消费者。

伯格曼相信，现代人已经变得执迷于商品的获取，而丧失了与（海德格尔眼中的）物的联系。技术教导人们在目的和手段之间做出截然的区分，却忘了二者原先是紧密地缠结在一起的。早先，由于缺乏足够的技术，人的活动并不如现在这样具有影响力，因而人与他人、与自然之间存在着复杂的相互关联，但现今这

---

① Ibid. , 131.

② Ibid. , 8.

③ Ibid. , 174-175, 182.

种情形变成了空洞的技术控制关系。"人与前技术的物的世界所具有的那种密切的相互关联，被各种设备的应用消解了。随着某台设备投入使用，人和世界的直接关联被一种机械化的机制所取代。这种机制隐而不显，造成的后果是：人可以尽情地享用由设备生产的用品，而不用再担心涉足生产过程而产生的负担。"①

消费社会之所以可能，是源于技术高度发展带来的充足产品。这其中技术并不是被动无辜的参与者，不只是为达成外在目的的手段。某种现成可得的技术手段，能满足一种需求，其存在本身就倾向于促使社会认可这种需求。简单方便的背后隐藏的是一种专制，伯格曼称之为"设备的范型"。因而，整个生活方式就被技术潜移默化地设定了。要批评那些由之形成、为之辩护的一般性常识将十分困难，更不用说去挑战和颠覆它了。伯格曼指出，"技术是一种新的统治，是来自现实事物的统治力量"②。

意义则是来自与焦点物打交道的过程。发挥着聚集性力量的各种物品构成了世界。这些物品在成为人们活动的对象的同时，也构成了某种环境和背景，它们要求人们花费精力与之打交道，正是这种实践过程"是我们生活的核心，也证明了生活本身"③。物品提高了与之打交道的人的技艺，增进了彼此间的联系。它们提供了一个切入点，由此出发去体验整个世界，而不是简单和有效地生产某种用品。伯格曼毫不犹豫地承认，焦点物的存在无须科学意义上的论证或说明。当然，一个"直接的"证明，可以指出我们所关注并投入其中的世界的某些特征。通过直接的证明，我们能证实焦点物在生活中的重要性，并使那些也愿与之打交道的人认识到焦点物的聚集的力量。这个概念与康德的"反思判断"类似，后者同样是以一个内在的人类本质概念为依据，而不是外在的绝对条件为前提的。

与辛普森一样，伯格曼也企图从技术退回到边缘性的活动上，同样，他也用了一些含糊的概念来构建其方案。伯格曼拒绝倒退，并主张对焦点物的重新关注将纠正人和技术的关系。这样的话，似乎不是技术，而是设备的范型才是问题的关键，因为，正是后者把生活设定为运用有效的工具去追求那些脱离了意义背景的目标。④ 在其著作的最后，伯格曼提倡了一种"二元经济"，即在其中作为设备的技术和作为手工技艺的技术可以共存。这似乎又暗示了技术最终还是问题所在，答案则是对技术进行限定。到底是谴责技术的精神，还是技术本身，伯格曼的论证同样摇摆于二者之间。⑤

---

① Borgmann, *Technology and the Character of Contemporary Life*, 47.
② Ibid., 189.
③ Ibid., 4.
④ Ibid., 200.
⑤ Ibid., 220.

四

马尔库塞想要证明，技术的根本变革必定是社会根本变革的一个方面。用他的话来形容，向焦点物的回归，包含的不仅是一种关于技术的新的态度，而且还指向了一种新的技术。尽管马尔库塞对这种可能性的反思太过模糊而不能令人信服，但还是有证据显示类似的转变进程已经在某些细微的方面展开了。这在因特网的例子中表现得尤为明显。因特网不仅是作为新的社交形式得以产生的舞台，而且它还赋予了用户一种前所未有的角色，即可以引导或改变技术的发展方向。技术和底层人群之间新型的辩证互动关系，在这个例子中被展现了出来。对此如何来解读呢，我们不会采用卢卡奇式的用阶级行动来对抗物化的观点，也不去诉诸海德格尔的座架理论，但马尔库塞充满希望的思索似乎可以与此相关。

因特网的方案最初是由军方发起提出的，社交功能并不包含在内。设计因特网是为了解决大型计算机运行中技术性的分时问题，以及在政府和加入联网计划的大学之间传输官方信息。此外，它也被计划作为核战争时备用的通信方式。但在其早期发展历程中，有个研究生把一个不大的电子邮件程序添加到了系统中，很快人类通信方式迎来了一次最重要的发展。这种插手改造的行为，其实是对系统进行了一次不同于军事目的的诠释。在有效地分配计算机时间这一功能外，他还看到了系统在通信方面拥有的发展潜质。

这种改造包含的其实是一种理解方式上的转变，一位数字设备公司（DEC）的副总裁向我指出了这一点，那还是在个人电脑发展的早期。那时候，在计算机的标准化应用外，基于计算机网络的通信开始慢慢出现，这大大出乎计算机从业者们的预料。这位副总裁说："我们正在把计算机进行联网，但忽然大家意识到，联起来的不仅是机器，还有它们的用户。"

意外的功能在瞬间被实现，在计算机发展的早期历史上，类似的例子我还发现过很多，它们足以被归纳为一个模型。比如，20 世纪 80 年代初，早在因特网还未向公共开放之前，法国的微型电传网络已经覆盖了数百万的用户。原本它是作为一个发布官方资料和新闻的信息系统，它连接有无数被称为"微型电传"的哑终端①。法国政府对应用该系统的目的有明确的说明，即把法国早日带进"信息时代"。但该系统很快被黑客破解了，它被转变为一种通信工具。黑客们在系统上添加了传送即时信息的功能，这超越了原先的设计目的，并引入了一种新的用途：约会和性交易。当然官方信息仍能被接收到，但微型电传技术在社交

---

① dumb terminal，指无数据处理能力，仅有屏幕和键盘的终端计算机。——译者注

方面展现的发展潜质，使得该技术的意义不可挽回地改变了。

长时间的用户使用过程，使得对计算机网络的本质进行彻底的重新诠释得以可能。在这些创新背后共有一个基本理念，那就是技术在社会交往中扮演了一个纯粹的中介性角色。该理念显而易见，但实现起来却不容易。那种依赖于获取存货清单式的运营方式，大部分都失败了，那些单纯为了用户间的相互联系，并传送已有信息的方式，却获得了广泛的成功。①

不过，技术的中介性角色并非完全透明。一个联结得以可能，有赖于某种语境的存在。语境设定了用户抱有的特定初衷，例如，为了联系某人或某个群体，为了方便约会或获取信息，等等。卓有成效地构建这一语境不是一桩简单的事，技术手册中并没有这种知识。这是因为，这种语境只有在虚拟的"生活世界"中才有效，即只有在意义架构内有用性（affordance）才会产生。

当然，比起现实中的事物，这种生活世界还是被简化了许多。但它们并没有被简化为单纯的工具。它们不是工具，而是用户活动和工作于其中的环境。例如，想象一下诸如 Eudora 和 Outlook 这类电邮软件。它们的界面往往被划分为三个框，邮件标题框、内容框和"信箱"框，它们共同构成了一个具体的世界。用户则被要求给不同的邮件分配优先级并进行答复。依据不同的标准，邮件被分类归到不同的信箱内，由此用户构建起了一个便于查阅使用的历史。通过浏览标题框和回复那些重要信息，用户又跨入了未来。从界面引向实践，界面又简化了实践，因而实践的复杂性被简单的界面掩盖了起来。

然而，更有趣的是网络上由群体信息交流而产生的生活世界。这种信息交流的实践与现今网络新闻中的大肆炒作不同，它并不始于 Web2.0 的产生。最早的群体信息交流的实现形式是异步的公告牌程序（BBS）和计算机会议程序，用户可以把信息发往一个文件共享系统，而不是个人的邮箱。事实上早在因特网公共化之前，在其他类型的计算机网络上，人们已经可以进行商务会谈和社交聚会，或是谈论爱好、疾病、政治等话题了。当然这些讨论还不具备 Web2.0 时代的许多特征，但所有重要的群体信息交流形式都已经可以实现了。

在这样一个架构中，用户通过语言交流构建起各自的身份，同时构建起的还有一个面向用户的兴趣爱好及其所关注事物的虚拟世界。从一开始，通过设定各自喜爱的聚会方式，用户间形成了一种交往模式。在软件和其他通信手段的帮助下，人们或多或少地与某个群体保持着联系。通过存档技术和应答互动，用户在整理过去的同时也构建了未来。这些活动导致的结果是：计算机网络成了一种环

---

① 前一种方式的代表是 Amazon 网上书店，后一种代表是 Ebay 交易网。由于坚持传统的零售模式，前者曾一度陷入困境，但后者仅是一个用户的信息交流平台，运营者不需要任何存货。——译者注

境，从中孕育出了各种社区，同时创新活动也在其中不断开展。[①]

网络上人际交往的实例表明，技术批判的重点不应过多集中在技术的某些特征或是消费主义的弊病上，而应放在规范及规范指导下的行动上。对于人类交往行动价值的承认，不应仅仅停留在观念层面，而是应使个体积极投入一种新的生活方式，这将证明以往那些对于"技术"范式的批评都是错误的。

在环保运动中，在因特网上的交往中，我们可以发现这种行为的复归。要求环保的抗议声，已经促使技术作出了某些重大的转变，这粉碎了技术决定论的神话及相应的技术统治的意识形态。现在我们认识到，必须对技术及其应用的后果负起责任。过去那种把技术理解为由专家们暗中把持着的聚宝盆的观念，正在让位于新的技术内涵：技术是一种环境，在那里人类积极主动地为了生存和发展不断努力。

在技术领域内，用行动进行表达的新方式，针对的是技术应用中产生的问题。由于发现了现实中的令人不满之处，计算机网络上的发明爱好者和环保主义者采取了相应的行动。他们用充满洞见的创意对现状进行了再诠释，这种创意只有在主流的技术研发立场之外的用户和受众那里才能获得。用海德格尔的话说，他们在潜在可能性的指引下找回了"世界"，这种可能性最初体现在为世界制造着物品的创造者身上。因此，创新不仅是发现新的工具，还包括发现使得工具得以可能的新的世界。马尔库塞以规范为依据的知觉概念也与此相关。虽然在对各种相关规范进行分类上，审美因素也许不是合适的标准，但是从对给予事物的不满足，以及对（马尔库塞所倡导的）实现多种可能性的新生活的憧憬中，还是能看到创新的存在。

五

以上对人们的行动在信息技术发展和推进环保运动中所扮演的角色的简要描述，为接下来要进行的分析提供了出发点。显然我们社会的文化毕竟不是敌托邦式的，合理性也不是纯形式化的，切合实际的思考还是可以带来有意义的结果。那么我们如何去充分地理解将要应付的困难之所在呢？这里有两种可能性。

我们可以先简单地忽略马克思-韦伯式的关于社会差异和形式化的预设，以及合理性与合理性社会通过去物化而获得解放的设想。这一进路表现出某种怀疑主义的特征。这种怀疑主义，与后现代主义者和 STS 研究者们提出的传统形式的社会批判有关。他们提出，对意义的怀念是一种多余的幻想，但他们那种企图重

---

① Andrew Feenberg and Darin Barney, eds., *Community in the Digital Age* (Lanham, MD: Rowman and Littlefield, 2004).

塑社会生活，在伦理政治层面对功能化进行抵抗的尝试，并不令人信服。这些反批判主义者的结论，通常是抽象地诉诸各种名目的对他者的容忍。这类善意的结论很难去反驳，但却与技术的政治学毫无关系。

作为另一种选择，我们可以在社会差异和形式化的概念上增加一个实际的去差异的维度作为互补，早期的社会批判主义者们都忽视了这个维度。如此，那种把复杂的社会形式和各种意义精简为功能剩余物的做法，和其他有着不同指向的社会进程将得以共存，这最终将重建出一个综合的意义系统。例如，哈贝马斯所持的正是这一立场。他敏锐地察觉到了现代性所包含的敌托邦迹象，但没有顺从那种纯功能化的社会逻辑，也没有针对敌托邦去规划一个启示录式的超越方案。在各类现代性社会中，哈贝马斯看到了两股背道而驰的潮流间的冲突，一个是趋向于总体上的系统化，另一个则是希望恢复意义，二者或多或少地产生着对立。同样的对立，在卢卡奇那里被阐释为一种辩证运动，一方面是物化的社会机制，另一方面是无法完全融入前者的鲜活的生命，二者间有着内在的矛盾。虽然，我并不认为这些理论是完全成功的，但这一总体上的思路，为如何理解现代性开启了一条崭新和富有成果的道路。我相信，这些思考线索提供的某些结论，已经构成了一种成功的对当代生活进程的描述。

因此，技术的释义学必须对技术的意义维度作出论证，并解释其与功能之间的关系。不少发展这一释义学思路的方法都是可取的。例如，引入各种技术的社会发展史，以考察技术是如何在时代中与社会生活的方方面面产生关联的。通过对这类例子的解读，可以总结出一条探索功能和意义关系的理论进路。前文对电子邮件用户的概括性描述，正是基于这样的视角来展现功能和意义间关系的。现实中，软件界面看似仅仅体现为功能性的那些方面，揭示出的其实是未曾意识到的时间上的丰富内涵。同样看似普通的程序面板设置，处理的不仅是传入的信息，事实上它或多或少地构建起了用户生活中的一个重要的组成部分，有时甚至是整个生活。在本文以下的总结部分中，我将概括出一种用来指导对这种现象进行研究的方法。

如何理解以技术为中介的行动，在早期和晚期海德格尔那里，分别给出了两套似乎是相反的理论方案，但这两者和我的方案间存在着关联。很显然，海德格尔的早期方案是现象学的，它不依赖于因果性，而是从功能与世界的关系中来解释功能。后期的方案不再以现象学作为参照，但也摆脱了通常对功能的因果解释，转而试图从本体论上进行理解。我相信，存在着一个隐藏的补充理论可以沟通这两者，虽然它们在海德格尔那里是独立和互不依赖的。阐明这一隐藏的联系，将为技术的双重面向提供理论基础，尽管这是为海德格尔本人所忽略的。

在《存在与时间》的第一部分中，海德格尔规划了一种出色的行动的现象学。他从一种原初的操心状态出发，通过此在的行动中包含的与工具、物品和符号间的关系，描绘出了此在与世界之间的关联。对"上手状态"的分析表明，功能并不是物体的一种属性，而是人与物体间的关系网络所包含的一个复杂的方面，即一种"关联的整体"。这一分析的结论就是意义（Bedeutsamkeit）这个概念，整个行动的结构本身构成了意义。因而，该理论认为，功能和意义是"世界"的两个无法分割的方面，在海德格尔看来，"世界"就是此在和行动对象之间内在联系的有机系统。这个世界概念正是上述辛普森和伯格曼观点所依赖的基础。

但海德格尔的说明还不够清楚，对于意义究竟如何在行动中产生没有作细致的解释。意义是作为行动的结果还是条件呢？"关联的整体"被认为是缠结在技术之网中实物间的关系系统。每一种特殊的功能性关系又是由某种操作意义决定的，这种意义使得与事物打交道的方式得以可能。海德格尔对于那种在寻视（Umsicht）中发现的严格的功能意义，进行了细致的规定。在分析的最后引入意义概念，似乎为的是在一个更广的行动空间——世界中，把各类相关的实践贯穿起来。

但事实是否真的如此？与功能化的行为相关的各种具体的意义，是否可以连贯成为一个没有矛盾的整体呢？本文开篇所提到的那些社会理论共有的一个论点是：现代社会中功能和意义的区分。现代性的实践把功能和意义进行了分割，海德格尔则把分裂的东西重新定义为一个整体。就这点来看，他的早期解决方案与实用主义有所类似。但在实践的语境下，一个简单的定义，并不能证明在概念上从功能到意义的跳跃是合理的。

其中包含的问题是，为什么这种理解在海德格尔看来是必要的，即把功能在本质上与意义相关联。注意到海德格尔本人和现代性之间的复杂关系，有助于理解这一难题。在海德格尔身边，发生了剧烈的社会分化，所有的一切不断地与他对生活的理解背道而驰。铁路网、供电系统和无线电广播构建起了一个由功能组织起来的大众社会，但他却坚持用行动中产生的意义来捍卫人类世界。虽然这个世界注定要被耗尽，但它至少还是一个世界，而不仅仅是人们被束缚其中的因果性的串联。

古德曼（Lucian Goldmann）对海德格尔的分析给出了一个可能的解释。他认为，海德格尔对上手状态的分析是卢卡奇的物化理论的本体论版本。暂不去考虑古德曼的解释所产生的实际影响，单就海德格尔的世界性（worldhood）理论中许多范畴与当代文化批判之间明显的亲缘关系而言，就值得我们去关注。和卢卡奇类似，海德格尔也认为应对现代性的敌托邦倾向加以限制，否则它将危及人之为人的概念，并使之不再有意义。大众社会的非原真性把人降格为物，但在非原真性生存的萌芽中却还是具有意义的因素。这就为原真性的行动提供了可能的基础，这种行动从切身的经验出发并超越任何给定的陈规旧俗。因此以社会语境为参照，海德格尔对行动的分析中包含了批判性的因素。

有什么证据能证明这一问题就是海德格尔的早期理论要回答的问题呢？海德格尔的名为"形而上学的基本概念"的课程讲稿也许能提供某些线索，文中他将行动和人与现实间其他类型的打交道方式作了对比。他解释道，此在对世界的投入预设了一种被他称为"作为"（as）的关系，即保持与意义的自由联系的能力。如果缺乏这种与意义的联系，那么也就失去了他眼中的那个世界，剩下的只有对某种刺激的条件反射。海德格尔需要意义和世界的概念来解释操心（care）和人类的筹划（projective）能力。正是这种能力把时间带入了存在。动物和机器没有这种能力，正是由此，海德格尔区分出了人的存在（此在）和其他类型的存在间的差异。

当海德格尔面对正在形成中的大众社会时，是否把动物和机器的存在上的这种局限与此在的地位联系起来考虑呢？海德格尔没有下降到这一社会批判的层面，进而得出这方面的观点。但在后期著作中他还是接近了这一立场。"技术"在后期海德格尔看来，是以对世界的图像为基础展开的活动。这种活动并不遵循《存在与时间》提出的上手状态的逻辑，相反，它针对的是因果性概念及与之相关的可预见概念。删节性的功能化处理，把意义和与之相联的世界淹没在机械化的设计和有组织的秩序中。在这种秩序中，人成了可供利用的资源和系统的部件。当此在在其展现的地方就被座架遮蔽的时候，意义也连同一起被封闭和枯竭了。如果用海德格尔早期的著作来解读其后期对技术的分析，就会清楚地发现：威胁着此在，并把此在减损为物的那种"危险"，指的就是把人当动物和机器的做法。这种减损在威胁此在的同时也威胁到了世界。这威胁又怎么可能不去消解此在"在世界中存在"的统一性呢？

对比个人的行动和置入座架这两种行动模式引出了一个议题，即技术的双重面向及其所构筑的社会的双重面向。在座架所设定的秩序中，因果关系和功能胜过了意义，权力被赋予了优先级。事实上，在现代社会中意义成了被权力支配的工具，它和广告语已没什么区别。在海德格尔后期的著作中，通过与物品相关联的原真行动，来恢复意义这个词包含的所有内涵，这只是被隐晦地谈及，例如，在讨论物和建筑、居住的那些篇章中。但在这些有趣的论述中，海德格尔似乎已经转而分析起现代技术了。如果返回到《存在与时间》，对原真性概念作更为细致的分析，便会发现它并不导向技术的实践，而是一个诉诸历史来对抗功能化的宏大设想。但这设想缺少实质性的内容，因此并不能借此证明海德格尔的纳粹主义倾向或他的学生马尔库塞的共产主义倾向。

· **327** ·

让我们尝试另一条路径。《存在与时间》中，原真行动的概念涉及的"正是对当时已经现实地可能存在的事物进行揭示的规划"①。海德格尔用诸如死亡和模糊的历史命运这类概念来解释这个命题。但我们可以把它应用到此在和实物之间的技术性关系上。那样的话，原真性的行动暗示的是技术的创新而非社会革命。毫无疑问，在座架系统内，个体的抵制行为和自由的即兴创作都将符合《存在与时间》中描绘出的行动的规律。当面对由座架导致的无意义的顶峰时，从这些行动中能产生一个有意义的世界。座架和行动之间的辩证关系，其实包括了现代性经验所具有的各种危险，但其中也模糊地存在着希望。

在把技术理解为具有功能和意义双重面向的技术批判理论中，我已尝试着把这条进路将涉及的内容作了规划。功能化的技术由因果逻辑决定，它可以诉诸各种技术学科加以说明，这些学科又是相互区分并具有独立性的。而意义化的技术属于一种生活方式，它不仅具有直接与功能相关的最低限度的意义成分，而且还拥有一个更广义的内涵，包括了技术所属人类世界中所有与之相关的方面。正是这些维度间的互动，在部分意义上促成了现代技术的发展。现代社会倾向于用建制化的手段对这些原因进行区分，例如，工程操作不同于日常理解，经营管理不同于职业生活，控制不同于交往。但在实践中，这些不同的维度间又有着经常性的角色互换。事实上，不仅在建制层面而且在个体层面，它们之间都存在着相互影响，有时甚至是冲突。当个体面对着个体生活于其中的技术环境，他可以选择用常规方法去服从或是用创新的方式去改变。

再回到上述计算机网络中交往行动的例子。微型电传网络被破解，是源于用户对未被开发的技术潜质的洞察。这种洞察的灵感来自微型电传系统与电话网络间的比较。二者都是为了通信的目的建立起来的网络，为什么还要在前者上面设置障碍，黑客们对此肯定觉得无法理解。因此，他们在微型电传网络上引进了新的通信功能，这使得计算机终端在日常生活被进行了重新定位。由于添加了这一功能，微型电传的功能发生了转变。如何理解这一现象，效率概念不能说明问题。从技术在本质上具有无限可能的纯粹应用出发来考虑，也没有切中要害。相反，用意义所包含的丰富内涵来替代狭隘的功能化理解，或许立足于此，才能构想出这一发明，并理解它的内涵。因此，基于某种具体的技术潜质，进行大胆的想象，进而替代现有的功能，这才是这种理解进路的关键。

这类事例，提示出的是一种与通常理解不同的民主概念。它处理的不是权利或选举这类问题，而是如何与日常生活中的技术架构进行商谈。以此为基础，必

---

① Martin Heidegger, *Being and Time*, trans. J. Macquarrie and E. Robinson（New York：Harper & Row, 1962），345.

须对解放（liberation）这一概念进行重构。因为主导性的功能和意义关系，以及由这二者决定的行动模式，都发生了转变。重构是为了突出这种转变。如果日常生活中与技术打交道的经验，还是被座架压倒性地支配着，那么单向度的社会仍然会大行其道。但只要个体与技术规则和技术产品的自由互动是具有意义的活动，那么这样的一个社会将具有截然不同的性质。无须重建科学技术的合理性，想象力就能赋予技术以价值。马尔库塞的这一希望将在这样的社会中实现。而且，它将为技术规则和技术产品注入人与技术打交道的切身经验，这经验就是从人类行动中产生的新的意义。

（计海庆 译，成素梅 校）

Epistemic Agency

Hilary Kornblith

Adult human beings are intellectually more sophisticated than other animals. Even if we allow, as I've argued we should, that many other animals have beliefs and desires, there is reason to believe that we may be the only species with second-order beliefs and desires. And this makes us far more complex intellectually than other animals.

My dog is moved by his beliefs and desires. When he hears food being poured into his dish in the kitchen, he comes running. He believes that there is food in his dish, and that is just what he wants. So his behavior is produced by a rational interaction of his beliefs and desires. And in that respect, he's just like me. My behavior too is, at times, produced by the rational interaction of my beliefs and desires.

But my dog, it is safe to say, never stops to consider whether he should believe the things he does. He doesn't stop to think, "I wonder if I really have good evidence that there's food in my bowl". He doesn't stop to consider whether he should want the things that he in fact wants, and he doesn't ever stop to consider whether he should do the things that he in fact does. He never stops to ask himself whether he is being the sort of dog he wants to be. He isn't conceptually sophisticated enough to have any of these thoughts. And the result of this is that he never stops to critically assess his beliefs, his desires, his motivations, his actions, or his character. And this is a respect in which he and I are different. I, like other adult human beings, do, at times, stop to assess these things. At a minimum, that makes me a more complex creature than my dog, intellectually speaking.

Some writers believe that this sort of intellectual sophistication which adult human beings have amounts to something far more important than just a difference in degree. The ability to critically assess one's beliefs, desires, motivations, actions, and character is connected, on certain views, with issues about freedom and responsibility. Freedom requires, on certain views, the ability to critically assess one's actions. Dogs may have genuine propositional attitudes, but their inability to reflect on their first-order states, and their attendant inability to critically assess those states, assures, on certain views, that they lack a crucial prerequisite for freedom and responsibility.

In this paper, I will examine views which make the ability to reflect on one's first-order psychological states a prerequisite for freedom, or a particularly important sort of freedom. In the course of discussing a number of issues about agency, we will also be led to discuss a particular sort of agency—epistemic agency—which many have argued is deeply connected to reflection on the content of one's mental states.

1. The Infinite Regress

In his seminal article, "Freedom of the Will and the Concept of a Person", Harry Frankfurt draws a distinction between freedom of action and freedom of the will.

> According to one familiar philosophical tradition, being free is fundamentally a matter of doing what one wants to do. Now the notion of an agent who does what he wants to do is by no means an altogether clear one: both the doing and the wanting, and the appropriate relation between them as well, require elucidation. But although its focus needs to be sharpened and its formulation refined, I believe that this notion does capture at least part of what is implicit in the idea of an agent who acts freely. It misses entirely, however, the peculiar content of the quite different idea of an agent whose will is free. [1]

Frankfurt's account of freedom of the will is best understood by way of a comparison between genuine persons and creatures Frankfurt calls *wantons*:

> The wanton addict cannot or does not care which of his conflicting first-order desires wins out. His lack of concern is not due to his inability to find a convincing basis for a preference. It is due either to his lack of the capacity for reflection or to his mindless indifference to the enterprise of evaluating his own desires and motives. There is only one issue in the struggle to which his first-order conflict may lead: whether the one or the other of his conflicting desires is the stronger. Since he is moved by both desires, he will not be altogether satisfied by what he does no matter which of them is effective. But it makes no difference *to him* whether his craving or his aversion gets the upper hand. He has no stake in the conflict between them and so, unlike the unwilling addict, he can neither win nor lose the struggle in which he is engaged. When a *person*

acts, the desire by which he is moved is either the will he wants or a will he wants to be without. When a *wanton* acts, it is neither.

... It is only because a person has volitions of the second order that he is capable both of enjoying and of lacking freedom of the will. The concept of a person is not only, then, the concept of a type of entity that has both first-order desires and volitions of the second order. It can also be construed as the concept of a type of entity for whom the freedom of its will may be a problem. This concept excludes all wantons, both infrahuman and human, since they fail to satisfy an essential condition for the enjoyment of freedom of the will. [2]

Non-human animals, since they lack both second-order desires and second-order volitions, lack freedom of the will, although they may frequently do as they want, and, in that sense, act freely. Adult human beings, however, who at least typically have both kinds of second-order states, are capable of freedom of the will as well as free action.

There is reason to think that this is an especially important distinction. Notice that we do not hold animals and young children responsible for their behavior, and we do not hold them responsible for their behavior, it seems, because they are not the kinds of individuals that could be responsible for their behavior. When my neighbor's dog runs loose in my garden and destroys the flowers, it is not the dog who is responsible, but my neighbor. The dog should have been trained better, or, failing that, penned up. The respect in which the dog is not responsible is not merely a legal issue. Rather, the dog is not morally responsible for its behavior. The same, of course, is true of young children. But now Frankfurt's account seems to give us a perspicuous bit of terminology for describing what it is that adult human beings typically have, and young children and non-human animals inevitably lack, which explains why it is that adults are typically responsible for their behavior and children and animals are not. Adults may have freedom of the will, while children and non-human animals cannot. Adults have the capacity to reflect on and critically assess their beliefs, desires, motives, characters, and actions; children and non-human animals do not. Freedom of the will, and the capacity to act in ways for which one is morally responsible, requires the ability to reflect on, and critically assess, one's first-order mental states.

Let us consider Frankfurt's wanton addict who has conflicting first-order desires. Let us suppose that our wanton has the desire for heroin, but also, since he is hungry, a desire for food. Unfortunately, he currently has neither food nor heroin, and he doesn't have enough money to buy them both. He thus has conflicting desires. Frankfurt says that since he faces such a conflict, "he will not be altogether satisfied by what he does no matter which of them is effective", and this is surely right. But this in no way distinguishes the wanton from a person who has conflicting first-order desires and resolutely acts on one of them. When one wants two things and can't have them both, one will, inevitably, find oneself not fully satisfied. Frankfurt says that in cases of conflict, the wanton "has no stake in the conflict" between his desires, and so "he can neither win nor lose the struggle in which he is engaged". This contrasts, Frankfurt tells us, with the person who has formed a higher-order preference to act on one or another desire, or who forms a volition to act on one of his desires, for the person wins his struggle if and only if his action conforms to his higher-order volition. On Frankfurt's view, in order to have a stake in the struggle between two first-order desires, one must have a second-order preference to act on one of them rather than the other.

But surely, if one sees things this way, then the same will be true in cases where one has no conflict in one's first-order desires at all: that is, Frankfurt should hold that, even when one's first-order desire is unopposed, one has no stake in whether it is satisfied unless one also has a second-order desire that it should be effective in action. And this is, indeed, Frankfurt's view. Frankfurt remarks,

> The essential characteristic of the wanton is that he does not care about his will. His desires move him to do certain things, without its being true of him either that he wants to be moved by those desires or that he prefers to be moved by other desires. The class of wantons includes all nonhuman animals that have desires and all very young children. Perhaps it also includes some adult human beings as well. In any case, adult human beings may be more or less wanton; they may act wantonly, in response to first-order desires concerning which they have no volitions of the second-order, more or less frequently. [3]

So let us consider what an action which exhibits freedom of the will must look like on Frankfurt's view. Suppose Annie has a desire to go to Paris, and she knows that by buying a ticket on Air France, she can get there. If she buys the ticket straightaway, without reflecting on her first-order desire, then she acts wantonly. She is, in acting this way, acting freely, but she does not display freedom of the will. In particular, she acts in a way which is no different from that of a dog or a very young child. If Annie is to act as a person, if she is to exhibit freedom of the will, then she must reflect on her desire and form a second-order desire: the desire to act on her desire to go to Paris. If she acts without forming that second-order desire, she has, on Frankfurt's view, no stake in the outcome of her action; she is merely the venue in which the interaction of her beliefs and desires takes place, "a helpless bystander to the forces that move [her]"[4].

So let us suppose that Annie reflects on her first-order desire to go to Paris, and she thinks to herself that, yes, she would like to act on this desire. She thus forms the second-order desire to act on her desire to go to Paris. She now has a kind of desire which young children and non-human animals cannot have, for they are incapable of reflecting on their first-order desires. This second-order desire, moreover, is unopposed. Annie is not conflicted about the thought of acting on her first-order desire; she has no second thoughts about it whatever. And for precisely this reason, Annie does not stop to reflect on her second-order desire.

But Frankfurt has told us that when a desire is not reflected upon, we have no stake in whether it is effective in producing action. So unless Annie also reflects on this second-order desire, Frankfurt is committed to holding that she still acts wantonly. And, of course, the same will be true should she reflect on her second-order desire, thereby forming a third-order desire to act on it, but fail to reflect, in turn, on that (third-order) desire. An infinite regress results. So long as one's highest-order desire goes unreflected upon, one has no stake, on Frankfurt's view, in whether one acts on it. But however much one reflects, one will inevitably stop somewhere. There will always be some desire which has not itself been reflected upon. And this means that, on Frankfurt's view, wanton behavior is inevitable, and freedom of the will is an impossibility. [5]

The problem here, of course, is exactly parallel to the problems we saw with similar requirements on knowledge and reasoning. There is a temptation to see reflective assessment as a prerequisite for knowledge, reasoning, and freedom of the will. But on the most straightforward understanding of what this would require, this makes knowledge, reasoning, and freedom of the will impossible to achieve. The temptation is one we will need to resist.

2. Higher-order States and Alien Desires

For reflective people, not all of their desires are on a par. We have all sorts of first-order desires, and some of them are ones we care about a great deal more than others. More than this, some of our first-order desires are one we wish we did not have, and ones we do not want to be moved by. Frankfurt discusses a case of such an alien desire, a case involving an unwilling addict.

> It makes a difference to the unwilling addict, who is a person, which of his conflicting first-order desires wins out. Both desires are his, to be sure; and whether he finally takes the drug or finally succeeds in refraining from taking it, he acts to satisfy what is in a literal sense his own desire. In either case he does something he himself wants to do, and he does it not because of some external influence whose aim happens to coincide with his own but because of his desire to do it. The unwilling addict identifies himself, however, through the formation of a second-order volition, with one rather than with the other of his conflicting first-order desires. He makes one of them more truly his own and, in so doing, he withdraws himself from the other. It is in virtue of this identification and withdrawal, accomplished through the formation of a second-order volition, that the unwilling addict may meaningfully make the analytically puzzling statements that the force moving him to take the drug is a force other than his own, and that it is not of his own free will but rather against his will that this force moves him to take it. [6]

One need not be an addict, of course, to recognize this phenomenon. Anyone who has tried to lose weight, stick to a budget, or simply break themselves of an unwanted habit is familiar with this phenomenon from the inside. We may act on one of our desires -as the unwilling addict does -without that action being a manifestation of free will. When we act on alien desires - desires which are, as Frankfurt rightly insists, nevertheless literally our own -free will does not play a role.

What is definitive of this phenomenon, as Frankfurt sees it, is the having of a second-order state which runs counter to the first-order desire. The unwilling addict wants to have the drug, but also wants not to be moved by that desire, and it is because of this second-order preference that the desire for the drug is seen as alien, and thus any action resulting from the first-order desire fails to manifest free will. First-order desires are thus rightly recognized as ones which do not automatically legitimate action, and Frankfurt's account of what it takes to provide the legitimation involves a second-order state giving its blessing, as it were, to the first-order desire.

This is, I think, quite a natural thought. When we think about the reflective agent, reflecting on his or her many and varied first-order desires, it is tempting to see those first-order desires as separate and apart from the agent, merely accidentally produced within him or her, awaiting legitimation by some act of second-order endorsement. Admittedly, these first-order desires, like the addict's desire for the drug, may be extraordinarily difficult, or even impossible, to get rid of. But how easily one may dispense with a desire tells us nothing about the extent to which it is truly one's own rather than some alien force acting on one. As Frankfurt sees it, the distinction between desires which are alien and those which are truly our own comes down to a distinction between desires which are rejected when we reflect and those which are reflectively endorsed. Actions prompted by first-order desires alone are not free-willed actions. It is actions prompted by those first-order desires which have received a second-order endorsement which count as freely willed.

Once we see that there is a problem of alien desires, however, it should be clear that second-order endorsements cannot possibly confer legitimacy merely in virtue of being higher-order. The problem here is exactly parallel to the problem we confronted about justification and knowledge. If we are worried about the fact that first-order processes of belief acquisition, allowed to operate without the benefit of reflection, may not all be reliable, then it is no solution to this problem to suggest that we stop to reflect on these first-order processes, adopting beliefs only if they have undergone reflective scrutiny. Reflective scrutiny itself, just like unreflective belief acquisition, need not be reliable. So if one is worried about reliability, as one might legitimately be, one cannot give second-order beliefs a free pass. Second-order beliefs are not immune—either in theory or in practice—to the possibility of being unreliably produced. And the same is true of the kind of second-order endorsements which Frankfurt uses to discriminate between alien desires and those desires which may legitimately be the source of freely willed action: if one is worried about the legitimacy of first-order desires, and the extent to which they are, in some important sense, truly one's own, it simply won't do to adopt the perspective of the reflective agent, take second-order states at face-value, and drop all of our worries about legitimacy and ownership merely because we are now dealing with states of a higher order.

Thus, Nomy Arpaly asks us to imagine a reflective agent engaged in the following bit of self-scrutiny.

> I see a piece of cake in the fridge and feel a desire to eat it. But I back
> up and bring that impulse into view and then I have a certain distance. Now
> the impulse doesn't dominate me and now I have a problem. Is this desire

really a reason to act? I consider the action on its merits and decide that eating the cake is not worth the fat and the calories. I walk away from the fridge. . .[7]

If we have been reading Frankfurt, we might see this as a model of reflective agency and free-willed behavior. The agent has conflicting desires—the desire to eat the cake and the desire to cut down on calories—and, after reflecting, forms a second-order preference to act on the desire to diet. This second-order desire reveals the desire to eat the cake as merely alien, and the action which is informed by the second-order preference to act on the desire to diet as the properly free-willed action.

As Arpaly points out, however, this reflective monologue could be that of a self-possessed agent who freely wills to stick to a diet, or, instead, someone deeply in the grips of anorexia nervosa, rationalizing every act of deprivation as she slowly but systematically starves herself to death.[8] The anorexic does not understand her own motivations, and, indeed, sees herself as a self-possessed agent with a high level of self-understanding and self-control. Such an agent is, of course, deeply mistaken about her own psychology, but the truth about herself is not available under conditions of reflection. From the first-person point of view—from the perspective of the reflective agent—there need be no difference at all between the self-controlled dieter who carefully follows a rational plan to lose fifteen pounds and the anorexic who pathologically starves herself to death. But this is just to say, of course, that from the first-person point of view, we are subject to the possibility of profound errors about fundamental features of our own psychology. The fact that action on the basis of a certain desire is endorsed by the reflective agent, that such action is agreeable from the first-person point of view, tells us nothing about which desires are alien and which are truly the agent's own, and it tells us nothing about which actions are freely chosen and which are the product of some sort of compulsion. The first-person point of view, the point of view of the reflective agent, cannot automatically make these distinctions accurately. The mere fact that a psychological state is a higher-order state tells us nothing about its legitimacy or its role in the agent's psychology.

Mental disorders of a great many sorts bring with them a characteristic pattern of self-misunderstanding. Paranoid-schizophrenics, for example, do not, on reflection, recognize that they are victims of paranoia, and when they reflect on their various first-order desires and the actions open to them, the (second-order) preferences they form about which of their first-order desires will be effective in producing action are themselves influenced by their paranoia. [9]When they are unreflective, their behavior can be the product of various compulsions they are subject to, but a reflective turn does not, of course, allow them to bypass these compulsions. Their second-order desires to act on certain first-order desires seem as rational to them as yours and mine seem to us. But this is not to say, of course, that paranoid-schizophrenics have a great deal of self-understanding, or that they act rationally, or that their second-order preferences serve to distinguish alien desires from desires which are truly their own, or that they act in a manner which exhibits freedom of the will when their second-order volitions properly line up with the first-order beliefs which actually move them. None of these things, of course are true. Like the anorexic, the paranoid schizophrenic has an interior monologue which appears quite rational, and which mirrors the interior monologue of the self-possessed reflective agent. But the appearance of being self-possessed is part of the problem that such individuals face. One cannot appeal to features of the first-person perspective, or the reflective agent, in order to distinguish those who are self-possessed, or who are acting freely, or who are exhibiting freedom of the will, from those who are not.

Nor is it necessary to chose examples here which involve mental disorders. We all, at times, fail to fully understand our own motives. We are all, at times, moved in ways we do not fully understand. And we all, at times, engage in sincere acts of rationalization. We may be moved to act by first-order desires in ways which are irrational and not fully free, and yet, at the same time, when we reflect on these desires, like the anorexic or the paranoid schizophrenic, we may form a second-order preference to be moved by them. The very irrationality which prompted the first-order desire may serve as a motive for rationalization when we turn reflective. The psychology of self-deceived reflective agents should not be conflated with the psychology of those who are utterly self-possessed.

Once we see that the concerns about the origins of first-order desires may equally apply to second-order states, it becomes clear not only that second-order endorsements are not sufficient for free-willed action (since the second-order endorsement itself may have a heritage which undermines any legitimating role it might otherwise play). It also becomes clear that a failure of fit between second-and first-order states need not, automatically, be taken to reflect badly on the states of lower order. There are, of course, unwilling addicts, just as Frankfurt describes, who are compelled to act by first-order desires which they do not want to be moved by; such agents, as Frankfurt rightly urges, do not act in a way that exhibits free will. But we should not think that whenever an agent acts on a first-order desire in the face of a second-order desire not to be moved by it that this too must be a case of action which is not freely willed.

Arpaly, again, is instructive here. Consider her reconstruction of the incident in Mark Twain's *The Adventures of Huckleberry Finn* in which Finn fails to turn in Jim, an escaped slave. [10] Although Finn can't bring himself to turn Jim in, he believes that what he is doing is wrong, and he berates himself for behaving immorally. Although Twain (unsurprisingly) does not describe it in these terms, we might even imagine that Finn forms the second-order desire not to be moved by his first-order desire that Jim remain free. As Arpaly describes the case,

> Talking to Jim about his hopes and fears and interacting with him extensively, Huckleberry constantly perceives data (never deliberated upon) that amount to the message that Jim is a person, just like him. Twain makes it very easy for Huckleberry to perceive the similarity between himself and Jim: the two are equally ignorant, share the same language and superstitions, and all in all it does not take the genius of John Stuart Mill to see that there is no particular reason to think of one of them as inferior to the other. While Huckleberry never reflects on these facts, they do prompt him to act toward Jim, more and more, in the same way he would have acted toward any other friend. That Huckleberry begins to perceive Jim as a fellow human being becomes clear when Huckleberry finds himself, to his surprise, apologizing to Jim-an action unthinkable in a society that treats black men as something less than human. [11]

As Arpaly argues, Finn is moved by a moral motivation, and his first-order desire that Jim remain free is not merely due to some sort of personal sympathy in the face of (however misguided) moral reasons to the contrary. Jim's first-order preferences are anything but alien; they are not mere compulsions which move him against his will. It is, instead, his second-order beliefs and preferences which are alien and defective. As Arpaly points out, Finn "is not a very clear abstract thinker" [12]. His second-order belief, about the moral acceptability of his motives, for example, and his second-order preference, to be moved by certain desires, are themselves rather shallow, not deeply rooted in his character in the way that his appreciation of Jim's personhood is. For that very reason, these second-order states are badly suited for playing the role of discriminating between alien and legitimating first-order desires; they cannot be the determinants of which actions are freely willed.

Finn's failure to be moved reliably by abstract reasons when he engages in reflection, despite his ability, at the first-order level, to respond reliably to reasons, is neither mysterious nor uncommon. The ability to articulate one's reasons in any detail, even to oneself on reflection, is a highly specialized skill, one which requires a good deal of education and training. Thinking about reasons *qua* reasons is especially abstract, and it is something which does not come naturally even to many who have a good deal of education. After all, a good deal of education does not focus on the nature of reasons *qua* reasons. But this is not to say that, without such an abstract focus, one's first-order beliefs and preferences cannot be moved by reason. We are, after all, very frequently responsive to factors of which we are only dimly aware, and which we can only articulate in the vaguest terms. Reliably responding to reasons does not require a second-order understanding of those reasons, or even the ability to form the concept of a reason. We may thus succeed in responding to reasons at the first-order level while failing to identify, or even respond to them, in our second-order reflections. We should not think that it is only at the second-order level that we engage with reasons, or even that, on those occasions when we are prompted to engage in second-order reflections, their second-order character gives them some automatic connection with reasons.

The same is true of the connection between second-order reflections and freely willed actions. It cannot be, as the infinite regress shows, that higher-order reflection is a necessary condition of free-willed action, at least if free-willed action is to be a genuine possibility. Nor can it be, given the nature of our psychology, that reflection somehow serves automatically to discriminate alien desires from those which are truly our own. The problems which beset first-order thinking and action are ones which may arise at the second-order level as well. By the same token, the kind of engagement we seek, in both our beliefs and our actions, require no special second-order endorsement.

3. Epistemic Agency

Issues about the role of reflection in agency come to the fore in discussions of epistemic agency. The idea that there is such a thing as epistemic agency deserves, I believe, a good deal of examination.

Most of the time, we form beliefs unreflectively. If our eyes are open and there is a table directly in front of us in good light and perfectly normal circumstances, we will come to believe that a table is there. We don't stop to reflect about whether things are really as they seem, nor do we stop to consider whether the belief we thereby form is genuinely justified. These questions could, of course, occur to us, but in the ordinary course of events, they simply do not occur. Our eyes are open; we see the table; we come to believe that the table is there. In cases like this, beliefs seem to be arrived at passively. They are no more chosen than are the visual sensations which the table causes.

But not all belief is like this. We do, at times, stop to reflect. "Is this what I ought to believe?" we ask ourselves. We deliberate. We consciously entertain alternative views, and we think about which, if any, belief about the situation before us we are justified in holding. In situations like this, we seem to play a more active role. We don't just find ourselves believing things. Rather, we decide what to believe; we make up our minds; we choose to believe one thing rather than another. It is in situations such as this that we may be tempted to talk of *epistemic agency*.

Thus, for example, in *Freedom of the Individual*, Stuart Hampshire remarks,

> "What do I believe?" turns into the question "What ought to be believed?" for the man who asks the question; but not necessarily for his audience, if he has one. The audience may be interested in the fact that I believe so-and-so: but for me this is not a fact that I learn, except in very abnormal cases... it is normally a decision, a making up of one's mind, rather than a discovery, a discovery about one's mind. [13]

This idea that, at least when we reflect, our beliefs are typically formed by way of a decision, and our knowledge of our beliefs in these situations is to be explained by the fact that we decided what to believe rather than by any discovery we might make about our minds, is absolutely central to Richard Moran's account of self-knowledge in *Authority and Estrangement*. [14] In commenting on the passage quoted above, Moran remarks,

> Hampshire is not endorsing a voluntarism about belief here, as if one's beliefs were normally picked out and adopted at will. The agency a person exercises with respect to his beliefs and other attitudes is obviously not like that of overt basic actions like reaching for a glass. [15]

But this, of course, raises a puzzle. If Hampshire, and Moran following him, wish to insist that we are agents with respect to our beliefs, that there is, in short, genuine epistemic agency, then how are we to make sense of this idea if it is not by way of some sort of voluntarism about belief?

Hampshire and Moran are not alone here. Voluntarism about belief is not a very widely held position[16], but the appeal to epistemic agency is, indeed, quite widespread. Thus, Christine Korsgaard claims,

> ... the human mind *is* self-conscious in the sense that it is essentially reflective... A lower animal's attention is fixed on the world. Its perceptions are its beliefs and its desires are its will... But we human animals turn our attention on to our perceptions and desires themselves, on to our own mental activities, and we are conscious *of* them.

> And this sets us a problem no other animal has. It is the problem of the normative. For our capacity to turn our attention on to our own mental activities is also a capacity to distance ourselves from them, and to call them into question. I perceive, and I find myself with a powerful impulse to believe. But I back up and bring that impulse into view and then I have a certain distance. Now the impulse doesn't dominate me and now I have a problem. Shall I believe? Is this perception really a *reason* to believe? ... The reflective mind cannot settle for perception and desire, not just as such. It needs a *reason*. Otherwise, at least as long as it reflects, it cannot commit itself or go forward. [17]

On this picture, we may certainly form beliefs unreflectively, as lower animals do, and then our belief formation is entirely passive. But when we reflect on our situation, we cannot come to form beliefs without making a commitment; some genuine activity on our part is required. It is thus when we reflect that we exercise our epistemic agency. Korsgaard goes on,

> The problem can also be described in terms of freedom. It is because of the reflective character of the mind that we must act, as Kant put it, under the idea of freedom. [18]

Unreflective belief does not involve free choice; it is not active; it is not something we do. But when we reflect, we make a choice as to what to believe. We are not passive. We are epistemic agents.

We see the same themes come to the fore in Ernest Sosa's work. In *A Virtue Epistemology*[19], there is an analogy which runs throughout. Sosa begins the book with the example of an archer shooting at a target. Sosa remarks that the archer's performance, like all performances, may be assessed along three different dimensions. We may ask whether it was accurate (i.e., whether it succeeded in its aim); whether it was adroit

(i. e. , whether it manifested a skill) ; and whether it was apt (i. e. , assuming it was both accurate and adroit, whether it was accurate as a result of being adroit) . Sosa refers to this as the "AAA structure" . He then goes on to say, "Beliefs fall under the AAA structure, as do performances generally "[20] . But there is certainly something puzzling about this. Beliefs don't seem to be performances. They are not actions. They do not seem to be something that we do.

Sosa gives a very brief response to this worry. He comments,

> Some acts are performances, of course, but so are some sustained states. Think of those live motionless statues that one sees at tourist sites. Such performances can linger, and need not be constantly sustained through renewed conscious intentions. The performer's mind could wander, with little effect on the continuation or quality of the performance. [21]

But this does not really respond to the concern. The worry is not about whether there is a *conscious* intention at work here. The worry is about whether there is any intention at all. The performer who remains motionless is clearly doing so as a result of an intention. No one can remain as motionless as these performers do without intending to do so. But believers, at least typically, do not form beliefs as a result of an intention. At least typically, when I look at a table in front of me and come to form the belief that there is a table, I am not moved by any intention, conscious or otherwise, any more than my dog is moved by an intention to form a belief when he comes to believe that there is food in his dish. So the suggestion that we may see belief formation, like the shooting of an arrow, as a kind of performance seems to be just a mistake.

We may, perhaps, however, better see what it is that Sosa has in mind by returning to the issue about the value of reflection. What is it about reflection, and beliefs formed under the guidance of reflection, that makes it so important, according to Sosa? One part of Sosa's answer to this question involves epistemic agency:

> . . . reflection aids agency, control of conduct by the whole person, not just by peripheral modules. When reasons are in conflict, as they so often are, not only in deliberation but in theorizing, and not only in the higher reaches of theoretical science but in the most ordinary reasoning about matters of fact, we need a way holistically to strike a balance, which would seem to import an assessment of the respective weights of pros and cons, all of which evidently is played out through perspective on one's attitudes and the bearing of those various reasons. [22]

So, once again, we see a connection being made between reflective belief formation and agency. My unreflective belief which simply registers the presence of the table, like my dog's unreflective belief which registers the presence of his food, is merely passive. But when I stop to reflect—something my dog cannot do—I become an agent with respect to my beliefs. Like Hampshire, Moran, andKorsgaard, Sosa believes that human beings are epistemic agents, and our agency comes into play when we form beliefs reflectively. [23]

A commitment to the existence of genuine epistemic agency is thus quite central to the work of a number of philosophers, philosophers who differ dramatically in other respects. The notion of epistemic agency, however, deserves more scrutiny than it has thus far received. [24]

4. Mechanism and Epistemic Agency

What, after all, is the view of cognition which is implicit in these suggestions? Consider, first, the case of unreflective belief acquisition. The mechanisms at work in a person which produce beliefs of any sort, including unreflective belief, are extraordinarily complex. They are not, for the most part, available to introspection. When I form perceptual beliefs, for example, my perceptual apparatus engages in a process of edge-detection which is made possible by way of mechanisms which are responsive to sudden changes in illumination across my visual field. [25] These mechanisms operate sub-personally. They are, as Sosa puts it, " peripheral modules " . Edge-detection is not something which I engage in, at least in standard cases of perception; rather, it is something done by sub-personal mechanisms within me. Here, at least, I do not act. Mechanisms within me are at work which simply produce perceptual beliefs.

How then are things supposed to be different when I engage in reflection? It will be best to have a simple example before us. Suppose that I am serving on a jury in which someone is charged with murder. Imagine as well that I don't simply react to the evidence presented. Instead, I stop to reflect. I self-consciously consider whether the evidence presented supports a guilty verdict. Here, when I stop to reflect, is where epistemic agency is supposed to be found. But where, precisely, does my agency come into play?

There certainly are things that I do in the course of reflecting on the evidence presented at trial. I may focus my attention on various pieces of evidence and question their relevance as well as their probity. The focusing of my attention is arguably something that I do, as is the activity of questioning both the relevance and the probity of the evidence. So there is genuine agency at work here, at least if we accept these commonsense accounts of what is going on. [26] But activities of this sort, while they are certainly present when a person reflects on his or her beliefs, are no different in kind from various activities we all engage in when forming unreflective beliefs. Thus, for example, just as I focus my attention on various bits of evidence when I carry out my jury duties, I turn my head in the direction I wish to look when I form various perceptual beliefs. Turning my head is certainly a voluntary activity; it is a manifestation of my agency. But the fact that I turn my head voluntarily does not show that my perceptual belief itself is a manifestation of epistemic agency—as all of the authors under discussion here fully agree. Whether I turn my head is determined by my choice, but once my head is turned in a certain direction, with my eyes open, and the lighting just so, my perceptual mechanisms will simply operate in me in ways which have nothing at all to do with the fact that I am an agent. The fact that I focus my attention, and question the relevance and probity of the evidence, thus show no more agency when I reflect than goes on in unreflective cases. Indeed, these activities not only show no more epistemic agency than goes on in unreflective cases in human beings; they show no more epistemic agency than goes on in lower animals when they form perceptual beliefs. But this is just to say that these features of reflectively formed belief do not exhibit epistemic agency at all.

So, once again, we need to ask, just where are we supposed to find the workings of our epistemic agency? As I've mentioned, there are a great many sub-personal processes at work whenever we form unreflective beliefs. But in this respect too, reflection is no different. We certainly shouldn't think that what goes on in reflection is fully and accurately represented in its phenomenology, any more than it is in unreflective belief acquisition. So when we get done focusing on various bits of evidence, and considering their relevance and probity, a host of sub-personal processes go to work eventuating in the production of a belief. How, indeed, could things possibly be any different? There is, after all, a causal explanation to be had of how it is that beliefs are formed, whether belief acquisition is reflective or unreflective. We should certainly not think that while unreflective belief acquisition takes place within a causally structured series of events, leaving no room for epistemic agency (just the workings of "peripheral modules"), reflective belief acquisition somehow takes place somewhere outside the causally structured network of events. There is, of course, no such location. But now the appeal to epistemic agency seems to be nothing more than a bit of mythology. A demystified view of belief acquisition leaves no room for its operation.

5. Epistemic Agency and the First-person Perspective

When we look at belief acquisition, even reflective belief acquisition, from the third-person perspective, there seems to be no room for epistemic agency. When we look at belief acquisition from the first-person perspective, however, there is, at least, the appearance of agency. What should we make of the difference which these two perspectives offer us?

Richard Moran suggests that the first-person perspective cannot simply be explained away as some sort of illusion. "... a non-empirical or transcendental relation to the self is ineliminable. "[27] There is a special authority which we have over our beliefs that is revealed to us, on Moran's view, from the first-person perspective, and it is here that our epistemic agency may be found. As Moran sees it,

> This is a form of authority tied to the presuppositions of rational agency and is different in kind from the more purely epistemic authority that may attach to the special immediacy of the person's access to his mental life. [28]

We may begin to understand what it is that Moran has in mind here by looking at his discussion of an example from Sartre. Consider the case of a habitual and chronic gambler who vows to give up his gambling. The gambler may examine his own past behavior from the third-person point of view and, given his many unsuccessful attempts to reform, question whether he is really able to follow through on his present resolve. But the gambler must also simply decide what it is that he is going to do. As Moran sees it,

> There is one kind of evasion in the empty denial of one's facticity (e. g. , one's history of weakness and fallibility), as if to say "Don't worry about my actual history of letting you down, for I hereby renounce and transcend all that". But there is also evasion in submerging oneself in facticity, as if to say, "Of course, whether I will in fact disappoint you again is a fully empirical question. You know as much as I do as to what the probabilities are, and so you can plan accordingly"[29].

There is no question that we would feel little confidence in someone who, after telling of his resolve to stop gambling, reminded us that he has resolved to do this many times before, with little success. Someone who says this is not only reminding us of relevant information. Such a person seems to be preparing us for his own failure so as to ensure that, should we depend on him to stop gambling and find that we have been disappointed, we will have no one to blame but ourselves. When he returns to the gaming tables, he is now in a position to say, "I told you so", despite his avowed commitment to quit.

So Moran is surely right that there are two different kinds of evasion to be found here. But what are we to make of this? We might well think that the gambler who both insists that he is going to quit, and, at the same time, reminds us of his history of backsliding on such decisions, is doing more than just undermining our confidence that he will follow through with his decision. Some people in situations like this do in fact follow through on their resolve; many others do not. But surely someone who keeps focusing on his history of failure, rather than focusing on his resolve to quit, makes it more likely that he will fail. We may reasonably think that the gambler who insists, "This time I'm really going to quit", is overly optimistic. But such a person has a degree of resolve which is probably necessary (even if not sufficient) for dealing with the inevitable temptations to follow, and the person who is focused on his past failures is missing this.

Consider the approach of Alcoholics Anonymous (AA) in dealing with similar concerns. [30] Those who wish to give up their problem drinking are encouraged to believe that their resolution to quit may carry them through the various temptations that they will face, and they are encouraged to believe that, in facing such temptations, it is not their resolution alone which will allow for their success; there is no attempt here to deny their "facticity", for example, by suggesting that their prior history of backsliding is irrelevant. Rather, AA encourages their members to believe that, in moments of temptation, they may surrender to a "higher power" who will help them through. It is this that will make the difference over past failed attempts.

Now I don't mean to be suggesting that the success which Alcoholics Anonymous has shown in dealing with the difficulties of giving up alcohol is best explained by the existence of such a higher power. This would not, I acknowledge, be one of the more persuasive arguments for the existence of a deity. But while AA should not be deferred to on matters of theology, it does seem to me that they display a fine understanding of human psychology. The twin temptations for evasion which Moran nicely identifies are both sidestepped in their approach. They do not deny that a knowledge of one's history of backsliding is important for the alcoholic, but they work on focusing the would-be quitter's attention elsewhere, and on developing that person's resolve to quit, while simultaneously instilling beliefs which prepare that person for the challenges and temptations which he or she will inevitably face.

Just as we should not take AA's invocation of a "higher power" at face value, Moran's suggestion that there is some sort of "transcendental or non-empirical relation to the self" at work here need not be taken at face value. The gambler in Sartre's example who can do no more than focus on his past history of failure, even while intermittently avowing that he is going to quit, need not be seen as lacking some non-empirical relationship to his self, a relationship which can only be revealed to the first-person perspective. A third-person perspective is all that is needed to appreciate the difference between such a person and the sincere and successful member of Alcoholics Anonymous, as well as those highly motivated individuals who give up bad habits and addictions by other means. Focusing on a history of failure doesn't alienate one from a transcendental self. It merely undermines some of the more effective empirical mechanisms for changing behavior.

The cases of the gambler and the alcoholic involve actions, rather than beliefs. But, as Moran points out, the same issues arise in the case of belief.

> With respect to beliefs, the parallel asymmetry would be the instability in the idea of trust or mistrust being applied to one's own belief, in the sense of treating the empirical fact of one's *having* the belief as evidence for its truth. If a generally reliable person believes that it's raining out, that fact *can* certainly be treated as evidence for rain. But in my own case, as with the resolution not to gamble, I must recognize that the belief is mine to retain or to abandon... That is, my belief only exists as an empirical psychological fact insofar as I *am* persuaded by the evidence for rain, evidence which (prior to my belief) does not include the fact of my being persuaded. If I am unpersuaded enough to need additional evidence, then by virtue of that psychological fact itself I lose the empirical basis for any inference from a person's belief to the truth about the rain. For someone's unconfident belief about the rain provides much less reason for anyone to take it to be good evidence for rain itself. [31]

Moran is certainly right that this kind of case is interesting, but it is not at all clear what it shows. Let us take an example which is slightly more fleshed out. Suppose that Jane is a physician and, after extensive examination of a patient together with a careful scrutiny of a wide range of test results, she comes to believe, somewhat tentatively, that the patient has a certain disease. If Jane is a reliable diagnostician, then we should take this fact about her diagnosis, i. e. , this fact about what she believes, to be good evidence that the patient has that very disease. As Moran rightly points out, however, there would certainly be something odd about Jane herself using this fact about her belief as evidence for its truth.

But why, precisely, is that so? Suppose that Jane discusses this case with her colleague Mary, someone who knows that Jane has an excellent track record as a diagnostician. Jane has had these talks with Mary before, and Mary has come to two conclusions about these conversations: First, Jane is often a bit tentative in the diagnoses she makes; she does not have the same degree of confidence in her diagnoses that many of their other colleagues do. But, second, Jane is much more frequently right than her other colleagues are; she has a superb record as a diagnostician, even when her diagnoses are tentative. Indeed, although she typically expresses (quite sincerely) a good deal of caution about her diagnoses, she very rarely has reason, subsequently, to modify them.

Now in a situation like this, I believe that Mary should come to the belief that Jane's diagnosis is correct. The track record evidence, including a track record in conditions in which Jane's diagnosis is tentative, provides strong evidence for the correctness of the diagnosis. And Moran would certainly agree. But now imagine Mary giving Jane the following pep talk:

Jane, you're a wonderful diagnostician. Your track record of making accurate diagnoses is remarkable. You've done all the relevant tests and scrutinized them carefully. And when you've done this in the past, you've always been very tentative in your conclusions, but they're almost always right. There's nothing different about this case. You should be a great deal more confident than you are. If you think this patient has that disease, then there's excellent reason to believe that he does.

It seems to me that Jane should be convinced by this bit of reasoning, and, if she is, then what she would be doing is using the fact that she believes as evidence for its own truth, in just the way that Mary, and everyone else, uses the fact that Jane has reached a certain conclusion as reason to believe that very conclusion. This bit of reasoning is perfectly good when others use it. It is no less good reason if Jane is convinced by it as well. Indeed, this third-person approach to her own reasoning may be used, as Mary encourages her to do, as a way of apportioning her beliefs to the evidence. Before Jane considers her own track record of successful diagnosis and her own tendency to be more tentative than she should, she is, in this case, as in others before, underestimating the strength of her evidence. By focusing on the fact that she has some small degree of confidence in her diagnosis, and her track record in such situations, she may raise her own degree of confidence in just the way she ought. This may involve a third-person perspective on her own beliefs, but it is none the worse for that. If this involves being alienated from oneself as a believer, then there's nothing wrong with such alienation. And if taking the first-person perspective on one's beliefs prevents one from viewing them in the way Jane does after the pep talk from Mary, then the first-person perspective can thereby get in the way of good cognitive self-management. Indeed, perhaps it would do us all a bit of good if we were alienated from our own beliefs in just this way a good deal more often. [32]

Once again we are led to ask: Just where is the agency involved in belief acquisition supposed to be found? What reason is there for believing that there is such a thing as epistemic agency? At one point, Moran suggests that we may see the workings of epistemic agency whenever we criticize someone's reasons for belief.

> Without the understanding that the person you're speaking to is in a position to exercise some effective agency here, there would be no point in criticizing his reasoning on some point since otherwise what would *he*, the person you're talking to, have to do with either the process or the outcome? He might be in a superior position to view the results of your intervention ("from the inside", as it were), but both of you would have to simply await the outcome. Instead, it seems clear that the very possibility of ordinary argument (and other discourse) presumes that the reasons he accepts and the conclusion he draws are "up to him" in the relevant sense. [33]

But it is not at all clear why the discussion of reasons, and the offering of arguments, presumes any sort of agency at all. When we offer people reasons for believing some proposition, or reasons for changing one of their beliefs, we certainly do take for granted that they are capable of being moved by reasons. We don't offer reasons to a tree stump. But it is one thing to say that people are capable of being moved by reason or that they are responsive to reason; it is quite another to insist that they are genuine epistemic agents. Moran suggests that if we view belief acquisition in a way which divorces it from agency, then when we offer someone reasons for belief, "we would have to simply await the outcome". But we don't, at least in many cases, need to wait at all simply because the wheels of the reasoning mechanism turn rather quickly. If you and I disagree about the sum of two numbers, and then, when you look over my calculations, you point out a simple error I've made, you needn't wait for me to acknowledge my mistake. But this is not because I acted as an agent upon my beliefs; it is because my reasoning mechanism is sensitive to the point you raised, and, given the obviousness of the error, my belief is quickly adjusted. I don't have to do anything once you point the error out to me. My reasoning mechanism does the work for me.

Of course, not all disagreements are so easily resolved. You may disagree with me on some complex issue, and, although you do in fact point out an error which I've made, the precise upshot of the error is not immediately obvious to me. Sometimes the reasoning mechanism operates more slowly. And in these cases, contrary to what Moran suggests, we do both need to wait to see what I end up believing.

Moran suggests that there is something wrong with any picture of belief acquisition which robs it of agency because, in offering someone reasons to change his belief, we must, somehow, assume that belief is up to him. If we don't assume this, Moran argues, "what would *he*, the person you're talking to, have to do with either the process or the outcome?" But I take it that we offer reasons to individuals precisely because we believe that the individuals themselves don't have anything to do with the outcome. When all is working as it should, our belief acquisition mechanisms are simply responsive to reason. We don't have to engage these mechanisms, or decide what to do with their output. Their output just is a belief, and thus no action on our part is required.

What we believe, when all is going well, has nothing to do with what we want, and this is precisely why the reasoning mechanisms may operate so well. Our wants have nothing to do with the reasons we have for belief, at least in the typical case; belief formation which involved agency, and thus allowed our desires to play a role in the beliefs we form, would thus pervert the process. Thus, when we offer others reasons for belief, we assume just the opposite of what Moran suggests: we assume that, to a first approximation, the desires of the people we are talking to will have nothing to do with what they come to believe. They will only be moved by reasons, and thus, their agency will play no role in the beliefs they acquire.

Of course, individuals are not perfectly responsive to reason, and we know this. When we offer reasons for belief, however confident we may be that we are in the right, we do not just assume that our interlocutors will come to share our views. In simple cases, such as the arithmetic mistake, we can safely assume, at least for most interlocutors, that they will simply respond, and respond appropriately, to the reasons offered. But the more complex the reasoning offered, the more room there is for an otherwise rational individual to fail to respond to reasons. Reasoning mechanisms, like all manner of mechanisms, are subject to interfering factors, and even when they operate without interference, they need not always operate properly. Even in otherwise rational individuals, there may be both performance failures and failures of competence. But whether the reasoning mechanisms are operating well or badly, we need not, and do not, assume that the individual to whom reasons are offered will exert any agency with respect to his or her beliefs.

These considerations, it seems, do not provide us with convincing reasons to believe that there really is such a thing as epistemic agency. But we will need to revisit these issues in a somewhat different guise.

6. Epistemic Agency and Deliberation

There is another idea about where and how our epistemic agency best reveals itself: it arises, again, in the course of deliberation about what to believe. When we stop to think about what to believe, we must, some argue, regard ourselves as agents. Thus, Korsgaard remarks that, "It is because of the reflective character of the mind that we must act, as Kant put it, under the idea of freedom"[34]. And Moran has a similar view:

> The basic point can be expressed in a loosely Kantian style, although the idea is hardly unique to Kant. The stance from which a person speaks with any special authority about his belief or his action is not a stance of causal explanation but the stance of rational agency. . . . It is an expression of the authority of reason here that he can and must answer the question of his belief or action by reflection on the reasons in favor of this belief or action. To do otherwise would be for him to take the course of his belief or his intentional action to be up to something other than his sense of the best reasons, and if he thinks *that*, then there's no point in his deliberating about what to do. Indeed, there is no point in calling it "deliberation" any more, if he takes it to be an open question whether this activity will determine what he actually does or believes. To engage in deliberation in the first place is to hand over the question of one's belief or intentional action to the authority of reason. [35]

So on this view, deliberation itself, if it is to be properly so-called, requires that we regard ourselves as free agents in the course of our deliberation. Now even if this were so, there would be a further question we would need to ask. Perhaps we need to regard ourselves as agents in order to engage in deliberation, but the question we were concerned with was not whether we need to think of ourselves as agents when we deliberate, but whether we genuinely are epistemic agents. So even if there is some necessity in regarding oneself as an epistemic agent, we might still reasonably ask whether that view we have of ourselves, when we deliberate about what to believe, is an accurate one. But we need not pursue this issue here, for, as I will argue, it is not even true that we need to regard ourselves in the way that Kant, Korsgaard and Moran suggest that we do. [36]

There are two suggestions which Moran makes in the passage quoted above, and, although he regards them as linked, I believe it is important to keep them separate. First, he suggests that deliberation involves agency, or, as he puts it in this passage, "rational agency". And second, he suggests that in the course of deliberation, we must regard our beliefs as being handed over to "the authority of reason". Now while Moran regards these claims as complementary, it seems to be that they are, in fact, in tension with one another. In my view, we should accept (a qualified version of) the second claim, but we should reject the first.

We may certainly regard our beliefs, under deliberation, as ones which are given over to the authority of reason. We reflect, after all, in the belief that this will allow us to better determine the truth, and we believe that our faculty of reason, if this is the way to put it, will be instrumental in producing true beliefs, or at least in increasing the likelihood that we arrive at true beliefs. [37] But this is perfectly compatible with the view that there is no such thing as epistemic agency. Indeed, it is compatible with the view that when we deliberate, we do not even believe that we are epistemic agents.

Why should we think, then, that we need even to regard ourselves as epistemic agents when we deliberate about what to believe? Consider, for a moment, an analogy. Think of the behavior of airport security personnel screening prospective passengers before they are allowed to proceed to the departure gates. Passengers are sometimes "wanded", that is, the screener will use a hand-held metal detector, or wand, to determine whether the passenger might be carrying a gun or a knife. The wand is systematically passed over the passenger's body, and if some sufficiently large or dense metal object is detected, the wand emits a loud tone. These wands are quite reliable, and the security personnel come to trust them a great deal.

Now when passengers are screened in this manner, the screeners are engaged in a voluntary, intentional activity. They freely decide where the wand is to be held, and they go out of their way to try to focus on areas where guns or knives might be hidden. The wand will sometimes emit a tentative-sounding tone, either when it is in the vicinity of a small metal object (such as a belt-buckle) or when it is a bit further away from a larger or denser metal object (such as a knife or gun), and when the wand emits this tentative sound, the screener will slow down and focus the wand on the area which provoked the response.

In screening passengers in this way, there is an intentional activity—the manipulation of the wand—and there is as well, as a crucial part of the screening, the thoroughly mechanical action of the wand itself. While the screener has control over where the wand is held, the behavior of the wand is not subject to the screener's will. The wand will beep when it is close to something metal, whatever the screener may have in mind. Indeed, the wand can only work effectively—especially given the inattentive way in which some screeners approach their job—if it does behave in a way which is insulated from the screener's intentions.

Deliberation about what to believe, it seems, works in a similar way. There are, beyond doubt, certain activities which we engage in in the course of deliberation. We focus our attention on certain questions or pieces of evidence; we may decide to look for additional evidence before we proceed further. The possibility of proceeding in these ways may well require not only our agency, but a recognition of our own agency. But we are agents here in the same way that the screeners are agents in deciding where the wand is to be held. And just as the wand has a life of its own, as it were, operating in ways which are insensitive to our intentions, the operation of our inferential mechanisms is similarly insensitive to our agency. We direct our attention in various ways, and our inferential mechanisms then go to work. We are able to have confidence in our deliberation, just as the screeners have confidence in their searches, to the extent that we regard our reasoning mechanisms, like the wands, as reliable. And, as I've emphasized, that reliability can only be had if it is purchased at the price of being insulated from our agency.

Beliefs under deliberation thus seem to be no more a product of epistemic agency than are beliefs formed unreflectively. Our agency plays a part in belief acquisition, but not in a way which legitimizes talk of epistemic agency.

7. Epistemic Agency and Epistemic Responsibility

All of the authors under discussion here connect the idea of epistemic agency with reflection, and we have therefore been considering whether there is reason to believe that reflective belief acquisition is different in kind from unreflective belief acquisition, with the former exhibiting agency while the latter does not. There seems, however, to be no such difference. While unreflective belief acquisition does seem to be, in relevant respects, entirely passive, reflective belief acquisition seems to be no different.

But this may be the wrong way to view how the appeal to reflection ought to be understood. Thus, for example, when Sosa introduces the distinction between animal knowledge and reflective knowledge, it might seem that while non-human animals are capable of nothing but animal knowledge[38], human beings sometimes have mere animal knowledge (for we very often form beliefs unreflectively) and sometimes have reflective knowledge. But this is not what Sosa says. Right after drawing the distinction, he remarks, "Note that no human blessed with reason has merely animal knowledge of the sort attainable by beasts"[39]. So human knowledge is viewed as different in kind from animal knowledge even on those occasions when we form beliefs unreflectively. The important dividing line thus seems to be the one between the beliefs of humans (who are capable of reflection) and the beliefs of non-human animals (who are not), rather than the one between reflective belief acquisition and unreflective belief acquisition.

How might we flesh this out so as to underwrite an account of epistemic agency? The account I offer here is not one which I would attribute to any individual. While it is inspired by a number of comments in Sosa's text, as well as by the remarks of a number of theSellarsian philosophers, it is certainly not something to which anyone explicitly commits himself. At the same time, I think that this view is well worth taking seriously, whether it may be attributed to Sosa, or the Sellarsians, or not. It is, I believe, an account which many will find attractive, and it may well underlie the way at least many of us are tempted to think about some of the differences between human and non-human cognition. Indeed, I believe that the suggestion I offer here may have a good deal more initial plausibility than the variations on Kantian themes we have so far considered, even if, in the end, I will argue that it should be rejected.

Sosa presents his view as a virtue epistemology, and we may usefully begin by asking what it is about human belief acquisition which lends itself to some sort of virtue account in a way that non-human animal belief acquisition does not. Some might suggest, as we have seen that Michael Williams does[40], that one important difference here is that, even if animals are properly regarded as having beliefs, the manner in which they form beliefs simply does not change over time. Their processes of belief acquisition are hard-wired, and, even if these processes are, for many purposes, quite reliable, and thus capable of producing (at least a kind of) knowledge, this marks an important difference from human belief acquisition. In our case, the very manner in which we form our beliefs, unlike non-human animals, does change over time. More than this, the manner in which our processes of belief acquisition and revision are modified over time is intimately connected both to reflection and to agency. We human beings periodically stop to reflect on our own beliefs and the manner in which they are acquired and revised, as well as on our own past track record of success and failure. At times, we take steps self-consciously to modify the ways in which we arrive at and revise our beliefs. Thus, for example, when students in an introductory logic class discover that they have been regularly affirming the consequent, and that this inferential strategy is wildly unreliable, at least some of them try to train themselves to stop affirming the consequent; that is, they try to break themselves of a bad inferential habit. This kind of cognitive self-management, it seems, is unique to human beings, and it is a power we have as a result of our ability to reflect on our own mental states.

The suggestion, then, is that reflection allows us to engage in cognitive self-management, unlike other animals, and our ability to monitor our own cognition, and to actively retrain ourselves so as to more accurately form beliefs lends a dimension to our cognition which other animals lack. By the time human beings are adults—and, indeed, arguably well before that—the manner in which we arrive at our beliefs can no longer be accounted for by the direct operation of our native inferential processes. Rather, our (admittedly periodic) scrutiny of our own cognitive performance prompts self-conscious action aimed at cognitive retraining. Many of the processes of belief acquisition and revision in any given adult will thus be the direct result of such active self-modification, just as many of the habitual actions which any adult performs are due to self-conscious activities designed to bring about the habits which produced them. More than this, even in the case of the very many processes of belief acquisition which are not the product of such self-conscious modification, their continued presence in us is explained, in part, by the fact that they have survived our periodic self-scrutiny. It is for this reason that we may reasonably think, it seems, that even unreflective human belief acquisition is importantly different from animal cognition. Even when we fail to reflect, the ways in which we arrive at our beliefs are to be explained, at least in part, by the activities which were prompted when we reflected in the past.

It is for this reason, as well, that we may reasonably be held responsible for the beliefs we hold, unlike non-human animals, and also why, unlike non-human animals, we may be said to deserve credit for our beliefs when they are arrived at reliably. It is not that our beliefs themselves are freely chosen. We do not freely choose our beliefs even when we reflect on what it is that we should believe. Rather, we, unlike other animals, may form our beliefs in ways which are influenced by our self-consciously chosen actions, and thus, we may be credited with a kind of epistemic agency which they lack. When mature human beings arrive at their beliefs in reliable ways, the cognitive mechanisms which produce and retain their beliefs are thus reasonably seen as virtues: not just mechanisms built in by the action of natural selection, but mechanisms whose very presence is due to active intervention by way of self-conscious activity prompted by reflection. [41]

Attractive and commonsensical as this picture is, I believe that it is deeply mistaken about both human and animal cognition. It underestimates the sophistication of animal cognition, and presents an account of the human case which is overly intellectualized. It presents reflection as more deeply involved in our cognition than it really is, and it gives an account of the role of agency in cognition which ties it more closely to reflection than our current understanding of the facts can support. Each of these points requires discussion and elaboration.

First, on the picture just presented, the mechanisms by which humans arrive at and revise their beliefs change over time, while, in the animal case, they do not. But the problem with this suggestion, as we have already seen, is that it is simply untrue. Many non-human animals are quite sophisticated cognitively, and the manner in which they acquire beliefs changes over time in response to new information, just as it does in human beings. Changes of this sort do not require self-conscious reflection. Beliefs, both human and non-human, may change as a result of exposure to new information, even without the intervention of self-conscious reflection. Thus, for example, as soon as I see my car keys on the kitchen table, I come to believe that they are there even if I had believed that I left them in the dining room prior to seeing them in the kitchen. This does not require reflecting on my earlier belief, or my newly acquired sense experience, or the way in which my new sense experience should be integrated into my total body of beliefs. I see the keys on the table and my cognitive machinery does the work of updating my beliefs without any reflection or activity on my part. [42] The same is true in the non-human case. My dog eats his food and fully recognizes when his food dish is emptied. If after arriving at the belief that his dish is empty, he finds that I have refilled the dish, his belief about the contents of the dish are updated by his cognitive mechanisms, without any need for reflection or activity of any sort.

But just as beliefs, both human and non-human, may be updated without the need for reflection or activity, the manner in which beliefs are arrived at may be updated and revised without the need for reflection or activity. If whenever I see a fox approach, I come to believe that it is dangerous, the discovery that a particular fox is harmless will not be something that I simply register atomistically; it will bring about a change in the inferences I draw when I am confronted with this particular fox-assuming, of course, that I can recognize it when I see it again. But the same sort of inferential integration, and change in inferential tendencies, can be found in many non-human animals, as we have already noted. One needn't have anything like the cognitive sophistication of a primate, let alone a human being, in order to integrate information in this sort of way. [43] The suggestion that this ability is a by-product of the ability to reflect, and thus, unique to human beings, is mistaken.

It is important to note that this error misrepresents cognition in both humans and non-human animals. It presents non-human animals as incapable of integrating new information in ways which will inform their subsequent information processing. And it presents human beings as having the ability to integrate new information in this way only by virtue of their ability to reflect on their own mental states. Both of these claims are incorrect. But these are not the only problems with the proposed view.

Reflection is presented, on the proposed picture, as the driving agent of cognitive improvement. According to this view, when we reflect on our beliefs and how they came about, on their logical relations to one another, and on our own past cognitive successes and failures, we come to initiate actions which will allow us to arrive at beliefs more accurately in the future. I have no doubt that we do, indeed, sometimes behave in just this way, and that this sort of behavior is something which no other animal can engage in. This is the important grain of truth in the suggested view. But the importance of this point is greatly exaggerated on this picture. We have already seen that animals incapable of reflection may integrate new information they obtain in ways that change the manner in which they process information. So we should not think that the highly sophisticated reflective strategy available to humans is the only manner in which cognitive improvement may come about. Non-human animals are not doomed to repeat the same cognitive errors throughout their lives simply in virtue of their inability to reflect on their mental states.

Even apart from this point, however, we have also seen that reflecting on our beliefs is not nearly so efficacious in producing cognitive improvement as the commonsense picture would have us believe. The act of reflection is often epiphenomenal with respect to the fixation of belief. [44] Here, as in many other things, phenomenology is a terribly inaccurate guide to the workings of the mind.

Because of these distortions in self-understanding, the commonsense picture of the extent of our influence over the ways in which we process information grossly exaggerates our own efficacy. [45] Thus, it is simply untrue that, as the story we are considering suggests, by the time we are adults, the processes by which we arrive at our beliefs are all either a product of actions self-consciously undertaken with the goal of improving our cognition, or, alternatively, processes which have been left as they were found under reflection only because they passed muster when they were self-consciously scrutinized. A great many cognitive processes are informationally encapsulated in cognitive modules. [46] The workings of these cognitive modules will inevitably contravene the commonsense picture in two different ways: the operations of these modules are invisible to introspection, so they can never be scrutinized by reflection in the manner proposed; and, in addition, the mechanisms by which they work are simply hard-wired, so any defect that the reflective mind might detect in them would be immune to change in any case. It is, of course, a good thing that we are endowed with many cognitive processes which have these features. Our ability to respond both quickly and reliably to much of the complex character of the environment is deeply dependent on the workings of such processes. But this is just to say that the commonsense picture, which presents the workings of our own minds as very much a product of our own activity, grossly mischaracterizes the extent to which our cognitive operations are genuinely malleable. Natural selection has organized the mind in such a way as to make many of its most important features tamper-proof: they cannot be restructured by the action of a well-meaning but frequently uninformed or misinformed agent. The mind doesn't work the way the commonsense picture portrays it, and it's a good thing that it doesn't.

The attempt to portray our cognitive successes as, one and all, a product of the actions and endorsements which flow from our own reflective self-scrutiny gives us far more credit for the workings of our minds than we deserve. This puts an end, I believe, to any project which would view us as deserving credit whenever we have beliefs which are aptly formed, or would see our intellectual capacities as virtues, whose very presence is to be explained by our acts of intellectual self-cultivation. [47]

8. Conclusion

The view that freedom of the will requires some sort of higher-order critical assessment of one's first-order states leads to an infinite regress. As we have seen, however, the regress problem is not the only difficulty with such a view. The attempt to vet one's first-order states by way of reflective higher-order review presupposes a kind of default legitimacy for higher-order states which they simply do not have. There is a certain irony here. The move to critical assessment of first-order states is motivated by the recognition that one's first-order desires, for example, may fail to reflect one's true self; they may, in an important sense, be alien influences. But the same point applies equally to higher-order states, and so we cannot distinguish between desires which are truly our own and those which are alien influences by looking to see which first-order states are backed by second-order endorsements. Second-order states, like their first-order cousins, are not self-legitimating.

We next turned to an examination of a special case of agency. A very wide range of philosophers have presented views of belief acquisition which appeal to some notion of epistemic agency. These authors do not typically endorse the view that belief acquisition is a voluntary activity, and so the appeal to agency of any sort here is prima facie puzzling. Several different motivations for taking talk of epistemic agency seriously were examined, and in each of these cases, appeals to reflection and the higher-order states it produces play a crucial role. Here too, however, we saw that these views presuppose a variety of empirical claims about reflection and higher-order states which run counter to our best available evidence. If there is any legitimate notion of epistemic agency, it cannot do the work which these authors require of it.

Notes

1. "Freedom of the Will and the Concept of a Person", repr. in *The Importance of What We Care About: Philosophical Essays* (Cambridge University Press, 1988),

19-20.

2. "Freedom of the Will and the Concept of a Person", 18-19.

3. "Freedom of the Will and the Concept of a Person", 16-17.

4. "Freedom of the Will and the Concept of a Person", 21.

5. A similar point is made by Gary Watson, "Free Agency", *Journal of Philosophy*, 72 (1975), 218.

6. "Freedom of the Will and the Concept of a Person", 18.

7. *Unprincipled Virtue: An Inquiry into Moral Agency* (Oxford University Press, 2003), 17-18. This passage is modeled, of course, on a passage in Christine Korsgaard, *The Sources of Normativity* (Cambridge University Press, 1996), 93.

8. *Unprincipled Virtue: An Inquiry into Moral Agency*, 18.

9. See *Diagnostic and Statistical Manual of Mental Disorders* (*DSM-IV*), 4th edition (American Psychiatric Association, 1994), 274-288.

10. For a different, and, to my mind, far less plausible, reconstruction of this story, see Jonathan Bennett, "The Conscience of Huck Finn", *Philosophy*, 49 (1974), 123-134.

11. *Unprincipled Virtue: An Inquiry into Moral Agency*, 77.

12. *Unprincipled Virtue: An Inquiry into Moral Agency*, 77.

13. (Chatto and Windus, 1965), 75-76.

14. (Princeton University Press, 2001).

15. *Authority and Estrangement*, 114.

16. But see Carl Ginet, "Deciding to Believe", in Matthias Steup (ed), *Knowledge, Truth, and Duty*, (Oxford University Press, 2001), 63-76; Sharon Ryan, "Doxastic Compatibilism and the Ethics of Belief", *Philosophical Studies*, 114 (2003), 47-79; Matthias Steup, "Doxastic Freedom", *Synthese*, 161 (2008), 375-392.

17. *The Sources of Normativity* (Cambridge University Press, 1996), 92-93.

18. *The Sources of Normativity*, 94.

19. (Oxford University Press, 2007).

20. *A Virtue Epistemology*, 23.

21. *A Virtue Epistemology*, 23.

22. "Replies", in John Greco (ed), *Ernest Sosa and his Critics* (Blackwell, 2004), 292.

23. It is for this reason, no doubt, that Sosa thinks that belief formation may, at least at

times, when it is apt, be something a person deserves credit for. (See *A Virtue Epistemology*, chapter five.) The notion of credit at work here is not merely the notion of causal responsibility, as when the proper spelling of a certain word in some text is "credited" to the automatic operation of a spell checking program. Talk of credit in this context is far more substantive, as befits a virtue-theoretic approach.

24. An exception here lies in two important papers by John Heil: "Doxastic Agency", *Philosophical Studies*, 43 (1983), 355-364, and "Doxastic Incontinence", *Mind*, 93 (1984), 56-70.

25. See, for example, David Marr, *Vision* (W. H. Freeman, 1982) .

26. I don't believe that we should accept these commonsense accounts of our mental lives. Indeed, the history of the cognitive sciences over the last fifty years seems to me to show very clearly that the phenomenology of mental processes is not even roughly reliable in producing an understanding of the mechanisms which actually operate. It is not just that the phenomenology leaves out important features of those mechanisms. Rather, even when it comes to those features of the mechanisms which the phenomenology represents, it very often misrepresents their role. For a recent defense of this view, see Timothy Wilson, *Strangers to Ourselves: Discovering the Adaptive Unconscious*, (Harvard University Press, 2002). I take the commonsensical view at face value in the text here, however, not because I believe it to be correct, but rather because I believe that it presents the best possible case in favor of epistemic agency. Certainly none of the authors under discussion here defend their views on the basis of empirical work in psychology.

27. *Authority and Estrangement*, 90.

28. *Authority and Estrangement*, 92.

29. *Authority and Estrangement*, 81.

30. See http: //www. alcoholics anonymous. org/en_ information_ aa. cfm.

31. *Authority and Estrangement*, 83.

32. Needless to say, the need for adjustment due to overconfidence is almost certainly a larger problem in practice. For discussion of this issue within a Bayesian framework, see Sherrilyn Roush, " Second Guessing-A Self-Help Manual ", *Episteme*, *forthcoming*.

33. *Authority and Estrangement*, 119-120.

34. *The Sources of Normativity*, 94.

35. *Authority and Estrangement*, 127.

36. This argument for epistemic agency is, of course, quite similar to the widely made argument for the claim that, insofar as we deliberate about what to do, we must regard ourselves as free agents, and, on some views, that the very fact of deliberation therefore shows that we are free agents. For a useful discussion of this argument, see Derk Pereboom, *Living without Free Will* (Cambridge University Press, 2001), 135-139.

37. This is an overly optimistic view of what actually happens during deliberation, but I don't deny that it is a view which is widely held.

38. At least if we assume, plausibly, that they are incapable of reflection.

39. "Knowledge and Intellectual Virtue", repr. in *Knowledge in Perspective: Selected Essays in Epistemology* (Cambridge University Press, 1991), 240.

40. "Is Knowledge a Natural Phenomenon?" in R. Schantz, ed., *The Externalist Challenge* (de Gruyter, 2004), 209.

41. There are certainly many points of contact between this view and Sosa's. As mentioned above, Sosa does, at least at times, stress the importance of the distinction between human belief-whether reflectively arrived at or unreflectively arrived at-and the beliefs of non-human animals, rather than the distinction between reflectively arrived at belief and unreflectively arrived at belief. He does, in addition, focus a good deal of attention on the suggestion that we deserve credit for beliefs aptly formed (although without connecting this to self-scrutiny in the way suggested here).

42. Sosa, at times, suggests that the integration of new information, at least when there are considerations at work which might individually pull in more than one direction, requires reflection on the content of one's beliefs and their logical relations, thereby making it a uniquely human possibility.

43. Alcock, *Animal Behavior: An Evolutionary Approach*, is particularly instructive here.

44. See Alison Gopnik, "How We Know our Minds: The Illusion of First Person Knowledge of Intentionality", *Behavioral and Brain Sciences*, 16 (1993), 1-15, and 90 101, and the discussion in section 1.3 above.

45. This is part and parcel of the way in which the commonsense picture exaggerates our own efficacy across the board. See, for example, Shelley Taylor and Jonathan Brown, "Illusion and Well-Being: A Social Psychological Perspective on Mental Health", *Psychological Bulletin*, 103 (1988), 193-210, and Shelly Taylor,

Positive Illusions: *Creative Self-Deception and the Healthy Mind* (Basic Books, 1989) .

46. The classic presentation of this account may be found in Jerry Fodor, *The Modularity of Mind*, (MIT Press, 1983) .

47. By the same token, those who present pictures of the self as " self-constituting " (see especially Christine Korsgaard's *The Sources of Normativity*; *The Constitution of Agency*: *Essays on Practical Reason and Moral Psychology* (Oxford University Press, 2008) ; and *Self-Constitution*: *Agency*, *Identity*, *and Integrity* (Oxford University Press, 2009) ; and, for a similar idea, Robert Kane, *The Significance of Free Will* (Oxford University Press, 1996)) exaggerate the role of reflective acts of character building and creation in determining the nature of the self.

认识的能动性

希拉里·科恩布利斯[*]

　　成年人在智力上比其他动物更复杂。即使我认为，我们应该承认许多其他动物也具有信念和欲望，但仍有理由相信，我们可能是唯一具有二阶信念和欲望的物种。正是这点使得我们在智力上比其他动物复杂得多。

　　我的狗受它的信念和欲望的驱使。当它听到主人在厨房里把食物放入它的盘子里时，它就会跑过去。它相信，它的盘子里有食物，这恰好是它想要的。因而，它的行为是由它的信念与欲望的理性互动产生的。在这方面，它恰好像我一样。有时，我的行为也是由我的信念和欲望的理性互动产生的。

　　但完全可以说，我的狗从未停下来考虑，它是否应该相信它做的事情。它不会停下来思考："不知我是否真有好的证据证明我的碗里有食物。"它不会停下来考虑，它实际上想要的东西，是否应该想要；也从来不会停下来考虑，它实际上做的事情，是否应该去做。它从未停下来反问自己，它是否是自己希望成为的那种狗。从理论上讲，它还没有复杂到足以有这些想法。这样的后果是，它从未停下来批判性地评价它的信念、欲望、动机或性格。这正是我和它的差别所在。我像其他成年人一样，有时会停下来评价这些问题。理智地讲，在最低限度上，这使得我成为比我的狗更复杂的动物。

　　有些人认为，人类所具有的这种智力上的复杂性极为重要，绝非仅仅是程度上的差别。根据某些观点，对信念、欲望、动机、行为、性格进行批判性评价的能力，与自由和责任的问题相关。根据某些观点，自由要求人有能力对自己的行动作出批判性的评价。狗可能具有真正的命题态度，但按照某些观点，它们没有能力反思其一阶状态，随之它们也没有能力批判性地评价那些状态，这必然使它们缺少自由和责任所需要的关键前提。

　　我在本文中将要考察的观点使得对人的一阶心理状态的反思能力成为自由的前提，或某种特别重要的自由的前提。在讨论许多能动性（agency）问题的过程

---

　　* 希拉里·科恩布利斯（Hilary Kornblith），美国马萨诸塞大学阿姆赫斯特分校哲学系教授，浙江大学客座教授。本文为作者 2010 年 6 月 27 日在浙江大学客座教授受聘仪式上的演讲稿之一，是首发稿。——译者注

中，也引导我们讨论一种特殊的能动性——认识的能动性（epistemic agency），许多人认为，这与对人的心理状态内容的反思有深刻的联系。

一、无 穷 回 归

哈瑞·法兰克福（Harry Frankfurt）在他具有开创性的《意志自由和人的概念》（*Freedom of the Will and the Concept of a Person*）一文中，对行动自由（freedom of action）和意志自由作出了区分：

> 根据一种熟悉的哲学传统，自由（being free）从根本上说就是一个人能够随心所欲。现在一位随心所欲的能动者（agent）① 的概念绝不是一个完全明确的概念：做和想做，还有它们之间恰当的关系，都需要阐明。可是，尽管需要突出其焦点，精炼其表述，但我相信，这个概念的确抓住了一位自由行动的能动者观念中所蕴涵的最起码的内容。然而，它完全没有抓住一位意志自由的能动者的这一不同理念中的特别内涵。②

法兰克福对意志自由的解释，最好通过他对真人和被他称为随性者（wanton）的生物之间的比较方式来理解。

> 随性的上瘾者（wanton addict）不能或不关心他的相互冲突的一阶欲望中哪一个欲望能成功。他缺少关心，不应归咎为他没有能力为一种爱好找到一个令人信服的基础，这是因为，要么，他缺乏反思能力；要么，他对评价自己欲望和动机的事业漠不关心。在他的一阶欲望冲突可能导致的斗争中，只有一个问题：哪个欲望更强烈呢？因为他受两种欲望的驱使，不管哪种欲望起作用，他都不会对他做的事情感到完全满足。但不管是他渴求的欲望占上风，还是他反感的欲望占上风，对他来说，根本没有任何差别。他无法把握两种欲望的冲突，因此，与无意志的上瘾者（unwilling addict）不同，在他卷入的斗争中，他分不出胜负。当一个人行动时，驱使他的欲望，要么是他希望的意志，要么是他希望缺少的意志。当随性者行动时，这二者都不是。
>
> ……这只是因为，人具有下列二阶意愿（volitions of the second order）：他既有享受意志自由的能力，也有缺少意志自由的能力。于是，人的概念不仅是既有一阶欲望也有二阶意愿的实体型（type of entity）

---

① "agent" 一词的国内通常译法是"行动者"，但在此篇文章的语境中，它试图表达的是"拥有能动性（agency）的认识主体"，而非"采取行动的认识主体"，故译为能动者。——译者注

② Harry Frankfurt，"Freedom of the Will and the Concept of a Person"，repr. in *The Importance of What We Care About：Philosophical Essays*，Cambridge：Cambridge University Press，1988，pp. 19-20.

概念。也能把人的概念建构成这样的实体型概念：对人而言，人的意志自由可能是有问题的。这个概念不包括所有的随性者，即类人猿和人类的随性者，因为他们没能满足享受意志自由的必要条件。[①]

尽管非人类的动物经常做它们想做的事，并且在这个意义上，它们能够自由行动，但是，由于它们缺少二阶欲望和二阶意志，因而缺少意志自由；然而，至少在典型的意义上，具有两种二阶状态的成年人，既有意志自由的能力，也有行动自由的能力。

有理由认为，这是一种特别重要的区分。请注意，我们不认为动物和幼儿要对他们的行为负责，这似乎是因为，他们是不能对其行为负责的那些个体类型。当我邻居的狗在我的花园里乱跑，破坏了花朵时，为此负责的不是狗，而是我的邻居。这条狗本应该受到更好的训练，否则，就应该被关起来。关于狗不负有责任的问题，不只是一个法律问题。更确切地说，狗对其行为不负有道德责任。当然，幼儿也同样如此。但现在，法兰克福的解释似乎为我们提供了一些明白易懂的术语，来描述成年人一般拥有什么，孩子和动物必然缺少什么，这说明了，为何正是成人一般要为他们的行为负责，而孩子和非人类的动物则不需要。成年人可能具有意志自由，而孩子和非人类的动物则不可能有。成年人有能力反思和批判性地评价他们的信念、欲望、动机、性格和行动；孩子和非人类的动物则没有。意志自由和人们以负有道义责任的方式行动的能力，有赖于反思和批判性地评价人的一阶心理状态的能力。

让我们考虑法兰克福描述的一阶欲望相冲突的随性上瘾者。我们假设，我们的随性者想要海洛因，但同时，他又饿了，因此又想要食物。不幸的是，他当下既没有食物，也没有海洛因，也没有足够的钱去买这两样东西。他因此就有了相互冲突的欲望。法兰克福说，既然他面临着这种冲突，所以，"不管哪一种欲望起作用，他将都不会对他做的事情感到完全满足"，这无疑是正确的。但这不能把随性者与这样一个人区分开来：这个人有相冲突的一阶欲望，并决心按照其中的一个欲望去行动。当一个人想要两样东西，却不能同时拥有两者时，他将必然发现自己没有得到完全满足。法兰克福说，在欲望冲突的情况下，随性者"无法把握他的两种欲望之间的冲突"，因此，"在他卷入的斗争中，他分不出胜负"。法兰克福告诉我们，这与这样的人形成了对比：他已经形成了按照某种欲望行事的高阶偏好，或者，他形成了按照他的欲望之一行动的意愿，因为这个人赢得他的斗争，当且仅当他的行动与他的高阶意愿相一致时。按照法兰克福的观点，人们为了在两个一阶欲望之间的斗争中有立场，他必须有按照某一种欲望行动的二

---

① "Freedom of the Will and the Concept of a Person", pp. 18-19.

阶偏好。

但毫无疑问，如果人们以这种方式看问题，那么，当一阶欲望根本没有冲突时，情况同样如此，即法兰克福应该认为，即使某人的一阶欲望没有遭到反对，他对这个欲望是否能得到满足，也没有把握，除非他拥有在行动中应该起作用的二阶欲望。这确实是法兰克福的观点。他评论道：

> 随性者的基本特点是，他不关心他的意志。他的种种欲望驱使他去做某些事情，但他真的不希望或不喜欢被其中的这些或那些欲望所驱使。随性者的范围包括所有具有欲望的非人类的动物和所有的幼儿，或许还包括某些成年人。无论如何，成年人可能或多或少都是随性者；就他们大概经常没有二阶意愿而言，他们的一阶欲望会使他们随意行动。[①]

因此，让我们考虑一下，根据法兰克福的观点，呈现意志自由的行动必须看起来像什么样的行动呢？假设安妮有去巴黎的欲望，而且，她知道，只要买一张法国航空的机票，她就能到达那里。如果她直接买票，没有反思她的一阶欲望，那么，她就是随意地行动。她以这种方式行动时，是在自由行动，但她没有体现出意志自由。尤其是，她行动的方式与一条狗或一个幼儿一样。如果安妮是作为一个人来行动，如果她呈现出意志自由，那么，她就必须反思自己的欲望，并形成二阶欲望：她按照去巴黎的欲望行动的欲望。如果她在没有形成那种二阶欲望的前提下行动，按照法兰克福的观点，她就无法把握其行动结果；她只是自己的信念和欲望互动的场所，"一位任凭外力驱使（她）的无助的旁观者"[②]。

因此，让我们假设，安妮反思了她要去巴黎的一阶欲望并且暗想，是的，她喜欢按照这个欲望行动。她因而形成了按照她去巴黎的欲望行动的二阶欲望。她现在拥有了一种欲望，这种欲望是幼儿和非人类的动物不可能有的，因为他们没有反思其一阶欲望的能力。此外，这个二阶欲望没有遭到反对。安妮没有抵触按照她的一阶欲望行动的想法；她根本没有重新考虑此事。正因如此，安妮没有停下来反思她的二阶欲望。

但是，法兰克福已经告诉我们，当一种欲望没有被反思时，我们对它在产生行动的过程中是否起作用没有把握。因此，除非安妮也反思了她的二阶欲望，否则，法兰克福就一定认为，她还是在随意地行动。当然，她反思她的二阶欲望，因此而形成按照二阶欲望行动的三阶欲望，但没有依次反思（三阶）欲望，应该同样是这种情况。结果产生了一种无穷回归（infinite regress）。根据法兰克福的观点，只要人的最高阶的欲望没有被反思，人们对他是否按照这个欲望行动，

---

① "Freedom of the Will and the Concept of a Person", pp. 16-17.

② "Freedom of the Will and the Concept of a Person", p. 21.

就没有把握。但无论一个人如何反思，总要在某个地方停下来。总是有某种欲望本身没有被反思。根据法兰克福的观点，这意味着，随意的行为是不可避免的，意志自由是不可能的事。[①]

当然，这里的问题与我们根据对知识与推理的类似要求所看到的问题完全一样。把反思评价视为知识、推理和意志自由的前提很诱人。但对这会提出什么要求的最直接的理解，使得知识、推理和意志自由成为不可能的事。这种诱惑是我们需要抵制的。

二、高阶状态和异己欲望

对反思的人来说，他们的欲望并非都能被相提并论。我们有各种各样的一阶欲望，我们更在乎其中的某些欲望。不仅如此，我们的有些一阶欲望是我们不愿意拥有的，我们也不想受其驱使。法兰克福探讨了这种异己欲望的一个例子，这个例子涉及无意志的上瘾者。

> 无意志的上瘾者是人，他在相冲突的一阶欲望中，最终哪个欲望得到满足，是不同的。无疑，两种欲望都是他的；比如，他最终是吸毒，还是成功地戒毒，取决于他如何满足自己的真实欲望。在任何一种情况下，他都做了自己想做的事，而且，他这么做，不是因为某种目的碰巧与他的自己的目的相一致的外在影响，而是因为他有这么做的欲望。然而，无意志的上瘾者通过形成二阶意愿，在他的相冲突的一阶欲望中认同其中的一种欲望。他使得其中的一种欲望更真实地成为自己的欲望，并且，在这么做时，他放弃了自己的另一种欲望。正是由于通过形成二阶意愿的这种认同和放弃，无意志的上瘾者才可能有意义地作出令人困惑的分析性陈述：驱使他吸毒的力量，是他自己之外的一种力量，并且，这不是他自己的自由意志，而是与他的意志相反的力量，这种力量驱使他吸毒。[②]

我们当然没有必要成为一位上瘾者，才会承认这种现象。试图减肥、坚持预算方案或只是放弃一种不良习惯的任何一个人，都熟悉这种现象的内幕。像无意志的上瘾者所做的那样，我们可以按照我们的各种欲望来行动，但这不是体现自由意志的行动。当我们按照异己欲望行动时，自由意志将不起作用，但正如法兰

---

① 加里·沃森（Gary Watson）得出了类似的结论，参见"Free Agency"，*Journal of Philosophy*，Vol. 72，No. 8，1975，p. 218.

② "Freedom of the Will and the Concept of a Person"，p. 18.

克福正确地坚持的那样，异己欲望不管怎么说确实是我们自己的欲望。

正如法兰克福所认为的，这一现象的特征是其中包含二阶状态与一阶欲望相违背。无意志的上瘾者想吸毒，并且不想被那种欲望所驱使；正是因为吸毒的欲望被视为是异己的这种二阶偏好，因而由一阶欲望引起的任何行动都没有体现自由意志。这样，一阶欲望被正确地认为是不能自动地合法的行动（automatically legitimate action），可以说，法兰克福对是什么提供了合法性的解释里，包括了二阶状态支持一阶欲望。

我认为，这是相当自然的想法。当我们考虑这位反思能动者，即反思他或她的许多不同的一阶欲望时，令人注目的是，把这些一阶欲望看成是与这位能动者相分离的，即只是在他或她的内心里偶然产生的，等待通过某种二阶认可的行动使其合法化。诚然，这些一阶欲望，像上瘾者渴望吸毒那样，可能特别难以摆脱或不可能除掉。但是，不管人们多么容易摒弃一种欲望，都没有告诉我们在多大程度上人们的欲望是真实的，而不是异己力量作用于他。正如法兰克福所看到的，区分异己欲望和我们自己的真实欲望，结果是区分我们反思时被拒绝的欲望和被反思认可的欲望。只受一阶欲望推动的行动，不是意志自由的行动。受已经接受了一种二阶认可的那些一阶欲望推动的行动，才算是自由意志的行动。

然而，一旦我们明白了异己欲望的问题，就应该清楚，二阶认可不可能只凭借更高阶的认可被赋予合法性。这里的问题恰好类似于我们在第一部分关于辩护与知识所面临的问题。如果我们担心的事实是，在没有事先反思就允许信念起作用的前提下，信念获得（belief acquisition）的一阶过程不可能全是可靠的，那么，建议我们停下来反思这些一阶过程，只接受经得起反思审查的信念，并没有解决这个问题。反思审查本身，如同获得非反思的信念一样，不一定是可靠的。因此，正如人可以合法地存在那样，如果他担心可靠性，他就无法接受二阶信念。不论是在理论中，还是在实践中，二阶信念都会受到不可靠地产生的可能性的影响。而且，法兰克福用来辨别异己欲望和可能合法地成为意志自由行动来源的那些欲望的二阶认可，也同样如此：如果人们担心一阶欲望的合法性，担心在某种重要意义上，它们在多大程度上真的是人们自己的欲望，那么，就完全不会采纳反思能动者的视角，对二阶状态信以为真，并放弃对合法性和归属权的担心，只是因为我们现在处理的是高阶状态。

因此，诺米·阿帕利（Nomy Arpaly）要求我们想象从事下列自我审查（self-scrutiny）的一位反思能动者。

> 我看到冰箱里有一块蛋糕，产生了吃掉它的欲望。但我往后退了一步，开始考虑这种冲动，然后，我有了一定的距离。现在，这个冲动没有支配我，而且，我有一个问题。这种欲望真的是行动的理由吗？我考

虑了这种行动的优点，并确定吃这块蛋糕是摄入毫无价值的脂肪和卡路里。我离开了冰箱……①

如果我们正在阅读法兰克福的作品，我们也许会把这一点看成是一种反思能动性（reflective agency）的模型和意志自由的行为。能动者具有相冲突的欲望——吃蛋糕的欲望和减少卡路里的欲望——经过反思之后，形成了按照节食欲望行动的二阶偏好。这个二阶欲望把吃蛋糕的欲望揭示成只是异己的，把受依照节食欲望行动的二阶偏好影响的行动揭示成是适当的自由意志的行动。

然而，正如阿帕利所指出的那样，这种反思独白（monologue）可能是一位从自由意志出发坚持节食的有自制力的能动者的独白，或者相反，是患有严重神经性厌食症拒食（即把每一次拒食行为合理化为她缓慢而有条不紊地使自己饿死）之人的独白。② 厌食症患者不理解她自己的动机，而且确实把她自己看做是具有高度自知之明和能自我控制的有自制力的能动者。当然，这样一位能动者极大地误会了她自己的心理，但她自己的实情在反思条件下是无用的。从第一人称观点来看，即从反思能动者的视角来看，在认真地遵照理性的计划减掉15磅的自我控制的节食者和病态地使自己饿死的厌食者之间，不需有任何差别。但这当然恰好是说，从第一人称的视角来看，我们有可能极大地误会我们自己心理的基本特征。反思能动者认可基于特定欲望的行动，这样的行动从第一人称观点来看是合意的，这个事实没有告诉我们，哪些欲望是异己的，哪些欲望真的是能动者自己的；也没有告诉我们，哪些行动是自由选择的，哪些是强迫产生的。第一人称的观点，即反思能动者的观点，无法自动地作出这些准确的区分。一种心理状态是更高阶的状态，只有这个事实，没有告诉我们它的合法性或它在能动者的心理上所起到的作用。

许多类型的心理异常都伴随有一种自我误解的特征模式。例如，妄想型精神分裂症患者，经过再三考虑都不会承认，他们是妄想症的受害者；而且，当他们反思他们的各种不同的一阶欲望和他们的行动时，他们形成的关于他们的哪些一阶欲望将会导致行动的（二阶）偏好本身受到了妄想症的影响。③ 当他们不反思时，他们的行动可能是他们受到的各种强迫的结果，但是，反思转向当然不允许他们不顾这些强迫。他们凭一阶欲望行动的二阶欲望似乎像你的欲望一样理性，

---

① Nomy Arpaly, *Unprincipled Virtue: An Inquiry into Moral Agency*, New York: Oxford University Press, 2003, pp. 17-18. 当然，这一段是模仿克里斯丁·科斯高（Christine Korsgaard）的一个段落，参见 Christine Korsgaard, *The Sources of Normativity*, Cambridge: Cambridge University Press, 1996, p. 93.

② *Unprincipled Virtue: An Inquiry into Moral Agency*, p. 18.

③ 参见 *Diagnostic and Statistical Manual of Mental Disorders* (*DSM-IV*), 4th edition, American Psychiatric Association, 1994, pp. 274-288.

而且，我的欲望与我们的欲望一样理性。但这当然不是说，妄想型精神分裂症患者很有自知之明；或者，他们理性地行动；或者，他们的二阶偏好能用来把异己的欲望与他们自己的真实欲望区分开来；或者，当他们的二阶欲望真的支持了实际上支配他们的一阶欲望时，他们就以体现意志自由的方式行动。所有这些事情当然没有一件是真的。像厌食症患者一样，妄想型精神分裂症患者有一种内在独白，这种独白显得相当理性，而且，反映了有自制力的反思能动者的内在独白。但有自制力的表象是这些人面临的部分问题。人们为了把有自制力的那些人或自由行动的那些人或体现意志自由的那些人与并非如此的那些人区分开来，不可能求助于第一人称视角或反思能动者的特征。

这里也不必非选择精神异常的例子不可。我们大家有时不能完全理解自己的动机。我们有时都会被我们没有完全理解的方面所驱使。而且，我们大家有时忙于理性化的真实行动。我们可能受非理性的和不完全自由的一阶欲望的驱使而行动，然而同时，当我们反思这些欲望时，像厌食症或妄想型精神分裂症患者一样，我们可能形成受这些欲望驱使的二阶偏好。当我们转向反思时，正是引发一阶欲望的非理性可能充当理性化的动机。自我欺骗的反思能动者的心理不应该与完全有自制力的那些人的心理相混同。

一旦我们明白，对一阶欲望起源的担心，可能同样适用于二阶状态，那么，很显然，不仅二阶认可对意志自由的行动来说是不充分的（因为二阶认可本身可能具有的遗留问题削弱了它本该在其他方面所起的合法作用），而且二阶状态与一阶状态之间的不一致不必自动地被当作对低阶状态的反思不当。当然，正如法兰克福描述的那样，存在着无意志的上瘾者，他们的行动是由一阶欲望强迫的，而他们主观上并不希望受这样的一阶欲望所驱使；正如法兰克福正确地强调的那样，这样的能动者没有以体现自由意志的方式来行动。但我们不应该认为，每当一位能动者不顾不受一阶欲望驱使的二阶欲望，只按照一阶欲望行动时，这也一定不是受自由意志支配的行动情况。

这里，阿帕利对我们再一次有所助益。考虑她在马克·吐温（Mark Twain）的《哈克贝利·费恩历险记》（*The Adventures of Huckleberry Finn*）中重构费恩没有告发吉姆（一个逃亡中的奴隶）这个小插曲。①尽管费恩不能说服自己告发吉姆，但他相信，他所做的事情是错误的，并且痛斥自己不道德的行为。尽管马克·吐温（不出意料地）没有用这些术语来描述情况，但我们甚至可以想象，费恩的一阶欲望是吉姆仍然是自由的；可他形成的二阶欲望并不受这种一阶欲望

---

① 在我看来，对这个故事的不同的、不太可能的重构，参见 Jonathan Bennett，"The Conscience of Huck Finn"，*Philosophy*，Vol. 49，No. 188，1974，pp. 123-134.

的驱使。正如阿帕利对这种情况所描述的那样：

> 哈克贝利通过和吉姆谈论他的期望和恐惧，以及通过与他广泛的互动，持续感知到（从未仔细考虑过）的事实相当于这样的信息：像他一样，吉姆也是人。吐温使哈克贝利轻易地感觉到他自己和吉姆之间的相似性：两个人都同样无知，共享同样的语言和迷信，总的来说，他不需要有约翰·斯图尔特·穆勒（John Stuart Mill）的天赋，就能明白，根本没有特殊的理由认为，他们中的一个人比另一个人下等。尽管哈克贝利从未对这些事实进行过反思，但这些事实的确促使他对待吉姆的行为方式，越来越像他对待其他朋友的行为方式。当哈克贝利惊讶地发现，他自己在向吉姆道歉时——在一个把黑人看成是低人一等的社会里，这种行为是不可想象的——他开始把吉姆看作一个同伴，这一点是明确的。[1]

正如阿帕利所认为的那样，费恩受到了一种道德动机的驱使，而且，吉姆仍然是自由的，他的这个一阶欲望，面对相反的（无论如何是误导的）道德理由，不只是出于某种个人同情。吉姆的一阶偏好绝不是异己的；这些偏好不只是促使他违背自己意志的冲动，相反，他的二阶信念和偏好才是异己的、有缺陷的。正如阿帕利所指出的，费恩"不是一个头脑很清晰的抽象思考者"[2]。他的二阶信念（比如，关于他的动机在道德上的可接受性）和他的受某些欲望驱使的二阶偏好本身是相当肤浅的，在他的个性中不是根深蒂固的，这不同于他对吉姆人格的欣赏。由于这个原因，这些二阶状态在区分异己的一阶欲望和合法的一阶欲望时起作用，是不适当的；它们不能成为哪些行动是自由意志的决定因素。

尽管在一阶层面上，费恩有能力对理由作出可靠的回应，但他在反思时，并不能可靠地受抽象理由的驱使，这一点既不神秘，也很正常。人们极其详尽地清楚表达理由甚至是自我反思的能力，是一项高度专业化的技能，这种技能需要大量的教育和训练。把理由作为理由来思考尤其抽象，甚至对于受过很多教育的人来说，也非轻而易举。毕竟，很多教育不关注理由作为理由的本质。但这并不是说，没有这样一个抽象的焦点，一个人的一阶信念和偏好就不会受理由的驱使。我们毕竟很频繁地对我们仅是隐约意识到的并且只能用最模糊的方式表达的因素作出回应。对理由的成功回应不需要对这些理由的二阶理解，乃至不需要形成理由概念的能力。我们可以这样成功地在一阶层面上对理由作出回应，尽管在我们

---

[1]　*Unprincipled Virtue：An Inquiry into Moral Agency*，p. 77.

[2]　*Unprincipled Virtue：An Inquiry into Moral Agency*，p. 77.

的二阶反思中，不能识别这些理由，乃至不能对它们作出回应。我们不应该认为，只有在二阶层面上，我们才会与各种理由打交道，乃至在那些情况下，当我们马上进行二阶反思时，它们的二阶特征使它们自动地与理由有关。

二阶反思和自由意志的行动之间的关联也是如此。正如无穷回归所表明的那样，至少如果意志自由的行动真的有可能，那么，高阶反思成为意志自由行动的必要条件是不可能的。考虑到我们心理的本性，反思以某种方式自动地把异己欲望和我们自己的那些真实欲望区分开来也是不可能的。困扰一阶思维和行动的问题是也可能在二阶层面产生的问题。出于同样原因，我们在信念和行动中所寻找的这种认可不需要专门的二阶认可。

三、认识的能动性

在讨论认识的能动性时，反思在能动性中的作用问题凸显出来。我相信，存在着认识的能动性这样一种活动的观点应受到认真的考察。

大多数时候，我们形成的信念是没有经过反思的。如果我们在光线充足和完全正常的情况下，睁开眼，看到正前方有一张桌子，那么，我们就会认为那里有一张桌子。我们不会停下来反思，事情是否真的如此，也不会停下来思考，我们由此而形成的信念是否获得了真正的辩护。当然，我们能想到这些问题，但在通常的事件中，不会提出这些问题。我们睁开眼，看见一张桌子，于是，就认为那里有一张桌子。在类似的例子中，信念似乎是被动地获得的。信念也不是精选出来的，只是桌子造成的视觉效果。

但并非所有的信念都这样。有时，我们的确会停下来反思："这是我应该相信的吗？"我们反问自己，谨慎思考。我们有意识地想到可替代的观点，并且考虑到，我们能合理地拥有关于眼前情况的哪个信念（如果真有的话）。在这类情况中，我们似乎起到了更积极的作用。我们不只是发现自己相信各种信念。恰恰相反，我们决定要相信什么；我们作出决定；我们选择相信这个信念，而不是那个信念。正是在像这样的情况中，我们想要探讨认识的能动性。

例如，斯图尔特·汉普希尔（Stuart Hampshire）在《个体的自由》（*Freedom of the Individual*）一书中评论道：

> 对于提问者而言，"我相信什么"变成了"我应该相信什么"的问题；但如果他有一位听众，那么，对于他的听众来说，未必如此。听众可能对我相信如此这般（so-and-so）的事实感兴趣；但对我来说，这不是我得知的事实，很反常的情况除外……这通常是一个决定，即下决

心，而不是一个发现，即关于人的思维（one's mind）的发现。①

至少当我们反思时，我们通过一种决定形成特有的信念，而且，我们在这种情况下关于信念的知识，是由我们决定相信什么的事实来说明的，而不是由我们形成的思维的发现来说明的，理查德·莫兰（Richard Moran）在《权威与隔离》（*Authority and Estrangement*）②一书中解释自我知识（self-knowledge）时，这种观念绝对是至关重要的。莫兰在发表对上面引用的一段的看法时评论说：

> 汉普希尔这里并不是支持信念的唯意志论（voluntarism），好像人的信念通常是被随意挑选的和采纳的。一个人关于他的信念和其他态度所发挥的能动性，显然不同于像伸手拿一个玻璃杯这种明白基本的动作。③

但这当然提出了一个难题。如果汉普希尔和追随他的莫兰希望坚持，我们是关于我们信念的能动者，简而言之，真的存在认识的能动性，那么，若这种观念不是作为信念的某种唯意志论，我们又如何使这种观念有意义呢？

这里持这种观点的人不是只有汉普希尔和莫兰。信念的唯意志论并不是一个被广泛接受的立场④，但诉诸认识的能动性确实是相当普遍的。因此，克里斯丁·科斯高（Christine Korsgaard）主张：

> ……人类的思维在进行本质反思的意义上是自我意识（self-conscious）……低等动物的注意力是集中于世界。它的知觉就是它的信念，它的欲望就是它的意志……但是，我们人类把我们的注意力转向了我们的知觉和欲望本身，转向了我们自己的心理活动，而且，我们能意识到心理活动。
>
> 这就为我们带来了一个其他动物所没有的问题。那就是规范性问题。因为我们把注意力转向我们自己的心理活动的能力，也是使我们自己疏离心理活动并对其发问的能力。我感觉到并发现自己有极强的冲动去相信。但我退回来，开始考虑这种冲动，然后，我有了一定的疏离。现在冲动没有主宰我，而且现在我有了一个问题。我相信吗？这个知觉真的是我相信的理由吗？……反思的思维不能满足于知觉和欲望，不仅

---

① Stuart Hampshire, *Freedom of the Individual*, London: Chatto and Windus, 1965, pp. 75-76.

② Richard Moran, *Authority and Estrangement*, Princeton : Princeton University Press, 2001.

③ *Authority and Estrangement*, p. 114.

④ 但请参见 Carl Ginet, "Deciding to Believe", In *Knowledge, Truth, and Duty*, edited by Matthias Steup, New York: Oxford University Press, 2001, pp. 6376; Sharon Ryan, "Doxastic Compatibilism and the Ethics of Belief", *Philosophical Studies*, Vol. 114, No. 1/2, 2003, pp. 4779; Matthias Steup, "Doxastic Freedom", *Synthese*, Vol. 161, No. 3, 2008, pp. 375-392.

如此，它还需要一个理由。否则，至少只要它反思，它就不能作出承诺或取得进展。①

根据这种图像，我们可能无疑像低等动物那样，形成了未反思的信念，于是，我们的信念形成（belief formation）完全是被动的。但当我们反思我们的处境时，我们只有在作出承诺时，才能形成信念；我们需要有某项真实活动。因此，当我们反思时，就发挥了我们的认识的能动性。科斯高继续指出：

> 这个问题也能用自由来描述。正如康德指出的那样，正是因为思维的反思特点，我们必须在自由观念的影响下行动。②

未反思的信念不涉及自由选择；它不是主动的；它不是我们做的事。但当我们反思时，我们要对相信什么作出选择。我们是主动的，是认识的能动者。

我们在欧内斯特·索萨（Ernest Sosa）的著作中看到了引人注目的同样的主题。在《德性认识论》（*A Virtue Epistemology*）一书中③，有一个贯穿始终的类比。索萨在这本书的开头举了一个射手把箭射向靶子的例子。索萨评论说，射手的行为表现，像所有的行为表现一样，可以从三个不同的维度予以评价。我们可以提问，它是否精准（accurate）（即是否能成功射中目标）；它是否灵巧（adroit）（即是否显示出一项技能）；它是否恰当（apt）（即假设它既精准又灵巧，它的精准是否是由于灵巧的结果）。索萨称之为"三 A 结构"。然后他接着说："正如一般的行为表现一样，信念也属于'三 A 结构'。"④但对此的一些疑惑无疑是存在的。信念似乎不是行为表现，不是动作，它似乎不是我们做的事。

索萨对这种担忧给予了非常简要的回应。他评论道：

> 当然，某些动作是行为表现，但一些持续的状态也是如此。想一想人们在旅游景区看到的那些活生生的一动不动的真人雕像。这样的行为能持续，并且不需要一直通过不断重复的有意识的意向来维持。行为者（performer）可能心不在焉，但很少影响行为表现的持续性或质量。⑤

但这不是对这种担忧的真正回应。我们不担忧这里是否是一种有意识的意向在起作用，而是担忧这里是否有意向。行为者一直不动，显然是故意这么做。没有人能在不是故意这么做时，像这些行为者那样一直不动。但至少典型的相信者并不是由于意向而形成信念。至少在典型意义上，当我看见前面有一张桌子，然

---

① *The Sources of Normativity*, pp. 92-93.

② *The Sources of Normativity*, p. 94.

③ Ernest Sosa, *A Virtue Epistemology*: *Apt Belief and Reflective Knowledge*, *Volume I*, New York: Oxford University Press, 2007.

④ *A Virtue Epistemology*, p. 23.

⑤ *A Virtue Epistemology*, p. 23.

后，形成有一张桌子的信念时，我没有受任何意图、意识的驱使；或相反，当我的狗相信他的盘子里有食物时，他受到了形成一种信念的意向的驱使，我受到的驱使不多于我的狗受到的驱使。因此，像射箭一样，建议我们把信念的形成看成是一种行为表现，似乎只是一种误解。

然而，我们或许只有回到反思的价值问题，才能更好地理解索萨的想法。依据索萨的观点，难道是反思的理由和在反思的引导下形成的信念使反思成为如此重要的吗？索萨在回答这个问题时，有一部分包含了认识的能动性：

> ……反思有助于能动性，有助于控制整个人的行为，不只是次要模块（peripheral modules）的行为。不仅在谨慎思考时，而且在理论化的过程中，不仅在理论科学的前沿领域内，而且在关于事实的最普通的推理过程中，当各种理由，像它们经常所是的那样，是相互矛盾时，我们需要公平处理问题的一种整体性的方式，这似乎意味着评价赞成与反对理由的反思权重（the respective weights），所有这些显然都是通过关于人的态度和具有的那些各种理由的视角进行的。①

因此，我们再一次看到了反思的信念形成和能动性之间具有的关联。我只注意到有一张桌子的未反思的信念（unreflective belief），像我的狗注意到有食物的未反思的信念一样，都只不过是被动的。但当我停下反思时——这是我的狗不能做的事——我变成了关于我的信念的能动者。像汉普希尔、莫兰和科斯高一样，索萨相信，人类是认识的能动者，并且，当我们在反思意义上形成信念时，我们的能动性就发挥了作用。②

这样，承诺真的存在认识的能动性，成为许多哲学家的著作中相当核心的部分。而在其他方面，这些哲学家意见分歧很大。无论如何，认识的能动性概念应受到比迄今更加详尽的审查。③

四、机制和认识的能动性

在这些建议中究竟蕴涵着什么样的认知观呢？首先考虑获得未反思信念的例

---

① "Replies", In *Ernest Sosa and his Critics*, edited by John Greco, Mass. : Wiley-Blackwell, 2004, p. 292.

② 无疑，正是由于这个原因，索萨认为，信念形成，当它恰当时，至少有时可能是人们值得信任的东西（参见 *A Virtue Epistemology*，第五章）。例如，当拼写检查程序的自动运行信任某个文本中对某个词的正确拼写时，这里起作用的信任概念（the notion of credit）不只是因果责任（causal responsibility）概念。在这种语境中谈论信任更加真实，适合于一种德性理论的进路。

③ 这里的一个例外参见两篇重要的论文：John Heil, "Doxastic Agency", *Philosophical Studies*, Vol. 43, No. 3, 1983, pp. 355-364, 以及 "Doxastic Incontinence", *Mind*, Vol. 93, No. 369, 1984, pp. 56-70.

子。人产生任何信念（包括未反思的信念在内）的直接运行机制，是异常复杂的。这些机制多半不是可内省的。例如，当我形成知觉信念时，我的知觉器官进行了边界检测（edge-detection），这个过程得以实现有赖于对在我视域中发生的突然变化作出回应的机制。①这些机制以亚个体（sub-personally）的方式起作用。正如索萨所指出的那样，它们是"次要模块"。至少在标准的知觉情况下，边界检测不是我的事；而是我体内的亚个体机制要做的事。在这里，至少我没有行动。我体内的机制只在产生知觉信念时起作用。

那么，当我进行反思时，事情应该如何不同呢？我最好举一个摆在我们面前的简单例子。假设我是陪审团成员，有人被指控谋杀。最好设想，我不只是对呈堂证据作出反应，相反，我停下来反思。我有意识地考虑，所呈现的证据是否支持有罪裁定。当我停下来反思时，这里应该是获得认识的能动性的地方。但准确地讲，我的能动性到底在哪里发挥作用呢？

在审讯时，我肯定要做的事情是反思向法院提交的证据。我会把我的注意力集中于各种各样的证据，并质疑这些证据的相关性和真实性。我可能调整我的关注点，就像质疑证据的相关性和真实性一样。因此，至少如果我们接受对案情进展的常识性解释，那么，此时就发挥了真正的能动性。但尽管当一个人在反思他或她的信念时，这类活动一定会出现，可是，它们与我们大家在形成未反思的信念时进行的各种活动类型，没有什么不同。②例如，正像当我履行我的陪审员的本职工作时，我会把注意力集中于各种证据那样，当我形成各种知觉信念时，我会把头转向我希望注视的方向。我转头无疑是一项自愿的活动；这体现了我的能动性。但我自愿转头的事实没有表明，我的知觉信念本身体现了认识的能动性——正如这里讨论的所有作者都完全同意的那样。我的选择决定了我是否转头，但是，一旦我胸有成竹地把头转向某一方向，并且光线正常，我的知觉机制就会起作用，而其工作方式与我是能动者的事实完全无关。因此，我集中注意力，质疑证据的相关性和真实性的事实表明，我在反思时的能动性，只是在未反思情况下进行的。的确，这些活动不仅表明，人的认识的能动性，只是在非反思

---

① 例如，参见 David Marr, *Vision: A Computational Investigation into the Human Representation and Processing of Visual Information*, San Francisco: W. H. Freeman, 1982.

② 我不相信，我们应该接受这些关于我们心理活动的常识解释。的确，在我看来，过去近 50 年的认知科学史似乎很明显地表明，心理过程的现象学，对实际运行机制的理解，即使粗略地看，也是不可靠的。现象学不仅忽略了那些机制的重要特征；而且，即使对于那些现象学所涉及的特征，其作用也被错误地表征了。对这种观点的新近辩护，参见 Timothy Wilson, *Strangers to Ourselves: Discovering the Adaptive Unconscious*, Mass: Belknap Press of Harvard University Press, 2002。然而，我在这里的文本中信以为真地接受这种基本常识的观点，并不是因为我相信它是正确的，而是因为我相信它提供了赞成认识的能动性的最有可能的例子。无疑，这里讨论的这些作者都不会在心理学的经验工作的基础上辩护他们的观点。

情况下进行的；而且表明，认识的能动性只是在低级动物形成知觉信念时进行的。但这只是说，在反思意义上形成信念的这些特征，丝毫没有体现出认识的能动性。

所以，我们有必要再一次质问，究竟应该在哪里发现认识的能动性的运行机制呢？正如我所提到的那样，无论我们何时形成未反思的信念，都有大量亚个体过程在发挥作用。但是反思在这方面同样并无二致。我们无疑不应该认为，相比于获得未反思的信念，反思的内在过程能更完整、准确地在反思现象中有所表征。因此，当我们在关注各种证据，并考虑其相关性和真实性时，许多亚个体的过程开始起作用，结果产生了信念。确实，事情可能有所不同吗？毕竟，不论是反思获得的信念，还是未反思获得的信念，对信念是如何形成的这个问题，都存在着一种因果性说明。我们无疑不应该认为，在因果相联的一系列事件中所产生的未反思的信念，没有为认识的能动性留下任何余地（只是"次要模块"的运行），而反思却有办法发生在因果相联的事件之网以外的某个地方。当然，根本没有这样的地方。但现在，诉诸认识的能动性似乎只不过是一个神话。去神秘化的信念获得的观点，没有为其运作留有余地。

五、认识的能动性和第一人称视角

当我们从第三人称的视角审视信念获得，甚至反思信念获得时，似乎没有为认识的能动性留下余地。然而，当我们从第一人称视角审视信念获得时，至少有能动性的迹象。我们应该弄清，这两种视角向我们呈现了什么样的区别呢？

理查德·莫兰建议，第一人称视角不能简单地作为某种幻想来说明。"……与自我的非经验的或超验的（transcendental）关系是不可消除的。"[1]根据莫兰的观点，从第一人称视角看，我们对于向我们自身所揭示的信念有一种特权，我们正是在这里找到了认识的能动性。在莫兰看来：

这是一种预设了理性能动性的权威形式，在种类上与更加纯粹的认知权威不同，后者可能与对自身心理活动获取的特殊直接性有所联系。[2]

他向萨特举了一个例子，我们通过审视他的讨论，可以开始理解莫兰在这里的想法是什么。让我们考虑发誓戒赌的一位嗜赌成性的赌徒的情况。这个赌徒可以从第三人称视角考察他自己过去的行为，基于他多次设法戒赌却都没有成功，因而质疑自己是否真能将现在的决定坚持到底。但这个赌徒也必须要决定，他要

---

① *Authority and Estrangement*, p. 90.

② *Authority and Estrangement*, p. 92.

做些什么。在莫兰看来：

> 在凭空否定某人的事实性（facticity）（比如，某人的弱点和易犯错的历史）中有一种逃避，就好比说"不要为我让你失望的实际历史担忧，因为我兹宣布放弃和超越这一切"。但使自己沉浸在事实性中也是逃避，就好比说"当然了，我实际上是否会再次让你失望，完全是一个经验问题。让你再次失望的概率有多大，我并不比你知道得更多，因此，你可以据此作出预期"①。

毫无疑问，如果一个人说他决心要戒赌，这使我们想起，他之前曾多次下过决心，但都没有成功，我们就会觉得不能轻信这种人。说这种话的人不只是使我们想起了相关信息。这种人似乎在让我们为他的失败有所准备，从而确保当我们对他戒赌的期待破灭时，我们只能责怪自己。当他再次回到赌桌上时，他马上会说"我早就告诉过你们会如此"，尽管他公开承诺过要戒赌。

因此，这里找到了两种不同的逃避，莫兰无疑是正确的。但我们对此有何看法呢？我们可能会认为，如果赌徒既坚持他会戒赌，同时又提醒我们，他对这种决定有反悔的记录，他就不只是削弱了我们对他将会坚持戒赌的信心。有些人在类似的情况下的确坚持了他们的决心；许多其他人则没有。但可以肯定，一个关注自己的失败史，而不是关注自己戒赌决心的人，更有可能会失败。我们可以合理地认为，坚持认为"这次我真的会戒掉"的赌徒是过分乐观的。但是这类人拥有某种程度的决心，这对于抵制随后必将出现的诱惑而言多少是必要的（即便不是充分的），而关注于自己过去的失败史的人则缺乏这一点。

考虑一下嗜酒者互戒会（Alcoholics Anonymous，AA）在处理类似的担忧时的进路。②他们鼓励希望摆脱酗酒问题的那些人相信，他们戒酒的决心能有助于他们抵制将要面临的种种诱惑，而且，鼓励他们相信，在面对这些诱惑时他们仅靠决心是无法成功的；这里没有否定他们的"事实性"，例如，建议说，他们先前的不履行承诺的记录，是无关紧要的。相反，嗜酒者互戒会鼓励他们的成员相信，在受到诱惑的时刻，他们可以屈从于（surrender）将有助于他们抵制诱惑的一种"更强大的力量"。正是这一点，造成了能够超越以往失败尝试的不同结果。

现在，我并不意味着建议，最好通过这种更强大力量的存在来说明，嗜酒者互戒会在处理戒酒难题时显示出的成功。我承认，对于神的存在来说，这并不是更有说服力的论证之一。但尽管不应该使嗜酒者互戒会遵从神学的内容，可是，在我看来，他们揭示了对人的心理的精致理解。莫兰恰当指出的两类要逃避的诱

---

① *Authority and Estrangement*, p. 81.

② 参见 http：//www.alcoholicsanonymous.org/en_ information_ aa.cfm.

惑在这一方法中都被抵制掉了。他们不否认，在酗酒问题上，了解某人曾经出尔反尔的历史是重要的，但他们致力于关注使戒酒懦夫的注意力关注在其他的地方和强化他们戒酒的决心，同时，向戒酒者逐步灌输信念，使他准备迎接将要必然面对的挑战和诱惑。

正如我们不应该轻信嗜酒者互戒会乞求"更强大的力量"的决心一样，我们也不必轻信莫兰提出的这种建议：在这里，有某种"与自我的超验关系或非经验关系"起作用。在萨特举的例子中，即使赌徒间或承认宣称他将戒赌，但他能做到也只是关注他过去的失败史，我们不必把这样的赌徒看成是缺乏了他自我的非经验关系，即只能从第一人称视角才能揭示出来的一种关系。第三人称视角足以揭示这类人与嗜酒者互戒会中真心实意的成功戒酒者之间的不同，还有与很积极地通过其他手段戒掉不良嗜好的那些人之间的不同。关注失败史，不会使人脱离超验的自我，只是削弱了某些更有效改变行为的经验机制发挥作用。

赌徒和酗酒者的例子涉及的是行动，而不是信念。但是，正如莫兰所指出的，同样的问题就信念而言也会产生。

> 就信念而言，在把人们拥有信念的经验事实视为此信念为真的证据的意义上，当用是否信任来表示人们自己信念的观念时，类似的不对称性是不稳定的。如果一个大体上可靠的人相信外面在下雨，这一事实当然可以被视为下雨的证据。但是在我自己的例子中，我必须意识到，我是坚持还是放弃不赌的决心，是我的信念……也就是说，就我相信下雨的证据而言，我的信念只是作为一个经验的心理事实存在的，这个证据（先于我的信念）不涉及使我信服的事实。如果不能充分地说服我需要额外的证据，那么，凭借这个心理事实本身，我就失去了从一个人的信念推出下雨真相的经验基础。因为某人对下雨的信念无把握，就没有足够的理由使任何人把这个信念当做下雨的有效证据。[①]

莫兰无疑是对的，这类例子是有趣的，但它表明了什么却毫不清晰。让我们举一个稍微具体一些的例子。假设简（Jane）是一个外科医生，在对病人做了大量检查并详细分析了很多检查结果后，她有些犹豫地相信，这位病人患了某种疾病。如果简是一位可靠的诊断专家，那么，我们应该接受她的诊断结果这个事实，即她所相信的这个事实是病人确实患了这种病的有效证据。然而，正如莫兰正确地指出的那样，简自己用她的信念作为信念为真的证据，这无疑有些奇怪。

但为什么这恰恰如此呢？假如简与同事玛丽（Mary）讨论这个病例，而玛丽知道，简作为诊断专家有杰出的诊断记录。简之前与玛丽一直有交流，玛丽对

---

① *Authority and Estrangement*, p. 83.

这些交谈有两个结论：第一，简经常对她做出的诊断有些犹豫；她在诊断时的自信度不如其他同事的自信度高。但是，第二，她也比其他同事更经常地给出正确的诊断结果，甚至在她的诊断没有把握时，她作为诊断专家也有极好的记录。的确，尽管她独特地（非常真诚地）表现出对她的诊断很慎重，但后来她很少有理由去更改这些诊断结果。

现在，在像这样的情况中，我相信，玛丽应该产生的信念是，简的诊断结果是正确的。记录的证据，包括简对她的诊断没有把握的条件下的记录在内，为诊断的正确性提供了强有力的证据。而且，莫兰也会肯定赞同。但现在假设玛丽对简说了以下激励的话：

> 简，你是一位优秀的诊断专家。你做出精确诊断的记录是引人注目的。你已经做了所有的相关化验，并对化验结果做了详细的审查。当你以前这样做时，你总是对你的结论很没有把握，但它们几乎总是对的。这个病例也没有什么不同。你应该比现在更加自信得多。如果你认为这个病人患了那种病，那么就有极好的理由相信他的确如此。

在我看来，通过这段推理应该使人相信简，而且，如果她的确如此，那么，她所要做的是，用她相信这一事实作为推理信念为真的证据，正是在这方面，玛丽和其他每个人用简得出某种结论的事实作为相信这个结论的理由。这段推理当被其他人所用时是充分的。即使简也被这段推理所说服，它不失是个好理由。的确，正如玛丽鼓励她去做的那样，对自己推理的第三人称把握可被用作把自身信念归为证据的一个途径。在简考虑她自己的成功诊断记录以前以及她对自身怀疑的倾向超出了应有的程度的时候，与其他情况一样，她在这个病例中低估了自己所持证据的强度。她在诊断时自信度不高，通过关注这个事实和她在这些情况下的成功记录，她恰好应该在这方面提高自己的自信度。这会牵涉到对她自身信念的第三人称视角，但这没有什么不好。如果这包含了与作为相信者的自我的疏离，那么，这种疏离并没有错。并且，如果第一人称视角妨碍了一个人像简在受到玛丽激励之后那样对自我的认识的话，那么，第一人称视角由此会妨碍良好的认知自治（self-management）。的确，如果我们恰好在更加频繁地像这样疏离于我们自己的信念，那么，我们将会受益匪浅。①

由此我们再次质问：究竟应该在哪里才能找到参与信念获得的能动性呢？在那里有什么理由相信，存在诸如认识的能动性之类的东西呢？莫兰曾一度建议说，

---

① 不必说，由于过分自信，需要调整，在实践中，肯定是一个更大的问题。在关于贝叶斯框架内对这个问题的讨论，参见 Sherrilyn Roush, "Second Guessing: A Self-Help Manual", *Episteme*, Vol. 6, forthcoming.

| 认识的能动性 |

每当我们批评某人持有信念的理由时，我们就可以看到认识的能动性的活动方式。

与你说话的那个人在这里能够发挥某种有效的能动性，如果不理解这一点，那么，就没有道理去批评他关于某个观点的推理，因为要不然的话，他，这个与你正在对话的人，与过程或结果有什么关系呢？他能够很好地审视你介入的结果（可以说，是"从内部"），但是，你们两人都只好等待结果。相反，看来很显然的是，普通的论证（和其他论述）很可能假定，在相关意义上，他接受的理由和他所得出的结论都"取决于他"①。

但是，为何对理由的讨论及所给出的论证需要假定某种能动性的存在却绝非显而易见。当我们为人们提供相信某个命题的理由，或改变他们的信念之一的理由时，我们肯定理所当然地认为，他们能够受理由的驱使。我们不会对牛弹琴。但是，说人们能够受理由的驱使或者他们对理由做出回应是一回事；坚持认为他们是真正的认识的能动者则完全是另一回事。莫兰建议说，如果我们在某种程度上区别对待信念获得和能动性，那么，当我们向某人提供信念的理由时，"我们只好等待结果"。但至少在许多情况下，我们完全不需要等待，仅仅因为推理机制系统行动迅速。如果你和我对两数之和的看法不一致，那么，当你仔细检查我的计算，指出我犯的低级错误时，你不必等我认错。但这并不是因为我作为一个能动者奉行我的信念；而是因为我的推理机制对你提出的观点很敏感，并且，考虑到错误是显而易见的，就会很快调整我的信念。一旦你向我指出了错误，我不需要做任何事情。我的推理机制为我完成这项工作。

当然，不是所有的分歧都能被这样轻易地解决。你和我也许对某个复杂问题持有不同意见，并且，虽然事实上你的确指出了我所犯的错误，我却不能很快看出那个错误的准确结果。有时推理机制运行得更缓慢些。那么在这些情况下，与莫兰所建议的相反，我们两人确实需要等着看我最终相信什么。

莫兰建议，剥夺了能动性的信念获得的任何图像都是有问题的，因为在为某人改变他的信念提供理由时，我们必须以某种方式假设，信念取决于他。莫兰认为，如果我们不这么假设，"他，正在与你对话的人，与过程或结果有什么关系呢？"但我以为，我们为人们提供理由，恰恰是因为我们相信这些人自己与结果没有任何关系。当一切都内在地进行时，我们的信念获得机制只对理由作出回应。我们不需要雇佣这些机制，或决定与它们的结果多么相关。它们的结果恰是一个信念，因此，我们不需要采取行动。

当一切进行顺利时，我们所相信的与我们所期待的没有任何关系，而这恰是

---

① *Authority and Estrangement*, pp. 119-120.

推理机制可以运作得这么好的原因。至少在典型情况下，我们的欲求与我们所持信念的理由没有任何关系；包含了能动性，从而允许我们的欲望起作用的信息形成过程，其自身反而会阻止欲望有所作用。因此，当我们为信念提供其他理由时，我们所认为的恰好与莫兰所主张的相反；粗略而言，我们认为，跟我们对话的那些人自身的欲望与他们所相信的并不相干。他们将只受理由的驱使，因而，他们的能动性在他们获得信念时没有发挥作用。

当然，人们不会对理由做出完美的回应，我们也清楚这一点。当我们为信念提供理由时，不管我们多么自信，我们都是正确的，我们没有只假定我们的对话者会共享我们的观点。在简单情况下，比如，算术错误，至少对于大多数谈话者而言，我们可以安全地假设，对所提供的理由他们只会回应，而且是适当的回应。但所提供的理由越复杂，就越有余地使得另外一个理性的人无法对理由做出回应。像形形色色的机制一样，推理机制易受干扰因素的影响，即使当它们在不受干扰的前提下运行，它们也不必总是正确地运行。甚至在另外一些理性的人中，也可能有行为失败和不能胜任。但不管推理机制运作得好与坏，我们都不必假定，也没有假定，被提供了理由的人，就他或她的信念而言，将发挥能动性。

似乎这些考虑没有向我们提供令人信服的理由来相信，真的有认识的能动性这种东西。但我们还需要对以不同形式出现的这些问题有所重新考量。

六、认识的能动性和谨慎思考

关于我们认识的能动性在哪里及如何最好地揭示了自身，还有另外一种观点：在我们谨慎思考（deliberation）相信什么的过程中，能动性再次浮现出来。当我们停下来思考要相信什么的时候，有些人认为，我们必须把自己当成是能动者。因此，科斯高评论道："正如康德指出的那样，正是因为心智的反思特点，我们的行动才会被视作自由的。"①莫兰也持有相似的观点：

> 我们能用不严格的康德式风格表达这个基本要点，尽管这种想法很难说是康德特有的。某人在谈及关于他的信念或行动所拥有的特殊权威时的立场，不是因果解释的立场，而是理性能动性的立场……这里表达了理由的权威性，他能够而且必须通过反思支持他的信念或行动的理由，来回答这种信念或行动的问题。否则，对他来说，他的信念或意向行动的形成就取决于他对于最好的理由的识别力之外的东西，而且，如果他想到这一点，那么，他谨慎思考他要做什么就失去了意义。的确，

---

① *The Sources of Normativity*, p. 94.

如果他把这种行动是否会决定他实际上做什么或相信什么看成一个有争议的问题，那么，更没有理由称之为"谨慎思考"。要进行谨慎思考，首先要把有关信念或意向行动的问题交付给理由的权威性来解决。①

因此，根据这种观点，谨慎思考本身（如果这样说恰当的话）要求，我们在谨慎思考过程中把我们自己看做是自由的能动者。现在，即便是这样，我们还有进一步的问题需要追问。为了进行谨慎思考，也许我们需要把自己看做是能动者，但我们关心的问题并不是，当我们谨慎思考时，我们是否需要把自己看成是能动者，而是我们是否真的是认识的能动者。因此，即使有某种必要把我们自己看做是认识的能动者，我们也可能仍然有理由质问，当我们谨慎思考该相信什么时，我们自己的观点是否是一个准确的观点。但我们在这里不必追究这个问题，因为，正如我将论证的那样，我们需要以康德、科斯高和莫兰建议我们的那样看待自身，对自身的看法是否准确甚至不是真的。②

莫兰在上面的引文中提出了两个建议，并且，虽然他认为这两个建议是相联系的，但我相信，重要的是使它们保持独立。第一，他建议，谨慎思考包含能动性，或者，正如他在这段话中指出的"理性能动性"。第二，他建议，在谨慎思考过程中，我们必须根据"理由的权威性"看待我们的信念。现在，尽管莫兰把这些主张看成是互补的，事实上，它们之间似乎存在着张力。在我看来，我们应该接受第二个主张（的有所限制的看法），拒绝第一个主张。

无疑，我们可以把我们谨慎思考的信念看做是托付给理由的权威性的信念。毕竟，我们在反思信念时允许我们更好地确定真理，而且，我们相信（如果能这么说）在生产真信念时，或者，至少在增加我们获得真信念的可能性时，我们的推理能力是起作用的。③但这与没有像认识的能动性之类的东西这种观点很一致。的确，它与以下观点相一致：当我们谨慎思考时，我们甚至不相信自己是认识的能动者。

那么，我们为什么应该认为，当我们谨慎思考要相信什么时我们甚至需要把自己看做是认识的能动者呢？此时让我们思考一个类比。想一想机场安检人员在准许乘客进入登机口之前，对乘客进行安检的行为。乘客有时接受"棒"的检

---

① *Authority and Estrangement*, p. 127.

② 当然，关于认识的能动性的这个论证，与对下列主张的广泛论证相当类似：这种主张是，就我们谨慎思考要做什么而言，我们一定把自己视为自由的能动者，并且，基于某些观点，谨慎思考的事实本身因此而表明，我们是自由的能动者。关于这种论证的一个有用的讨论，参见 Derk Pereboom, *Living without Free Will*, Cambridge：Cambridge University Press, 2001, pp. 135-139.

③ 这是一种在谨慎思考时实际发生什么的过于乐观的观点，但我并不否认，它是一种被广泛接受的观点。

查，也就是说，安检查人员用手持的金属探测器或探测棒进行检查，来确定他是否携带了枪支或刀具。这个探测棒有条不紊地扫过乘客的身体，如果检测到某个足够大或密实的金属物体，探测棒会发出鸣叫。这些探测棒是相当可靠的，并且，安检人员十分信赖它们。

现在，当乘客以这种方式接受检查时，安检人员进行了一项自愿的意向性活动。他们自由地决定在哪里停下探测棒，然后，他们耐心地设法集中在有可能隐藏枪支或刀具的地方进行检查。当探测棒接近小金属物体（比如皮带扣）或稍微更远离较大的或较密实的金属物体（比如刀具或枪支）时，探测棒有时会发出迟疑的声音，而且，当探测棒发出迟疑的声音时，安检人员会放慢动作，把探测棒集中在引发反应的地方。

在以这种方式检查乘客时，有一种意向性活动——操纵探测棒——作为检查的关键部分，也有探测棒自身的完全机械化的运作。当安检人员控制在哪里停下探测棒时，探测棒的行为不受安检人员的意志的支配。不论检查者在想些什么，当探棒接近金属物体时就会嘟嘟响。事实上，只有探测棒的确不受安检人员意向的影响时才能有效地工作——尤其考虑到一些安检人员对待工作的漫不经心。

关于相信什么的谨慎思考似乎以同样的方式起作用。毫无疑问，有些活动是我们在谨慎思考过程中进行的。我们把注意力集中于某些问题或某个证据；我们在进一步进行下去之外，可能决定寻找另外的证据。以这些方式进行下去的可能性不光需要我们的能动性，还需要承认我们自己的能动性。但我们在这里是能动者，同样，安检人员在决定探测棒在哪里停下时也是能动者。于是，可以说，正像探测棒有自己的使用寿命一样，其操作方式不受我们意向的影响，我们的推理机制活动同样也不受我们的能动性的影响。我们以多种方式支配我们的注意力，然后，推理机制开始发挥作用。我们把我们的推理机制看成是可靠的，就像探测棒一样，在这种程度上，正像安检查人员信任他们的搜查活动一样，我们也能信任我们的谨慎思考的活动。并且，正如我所强调的，我们只有以不受能动性的影响为代价，才能够获得可靠性。

因此，谨慎思考的信念似乎与未经反思形成的信念一样，也不是认识的能动性的产物。在信念获得的过程中，我们的能动性发挥了部分作用，但在某种程度上没有使认识的能动性的讨论合法化。

七、认识的能动性和认识的责任感

这里所讨论到的所有作者都把认识的能动性观念与反思相关联，因此，我们一直在考虑是否有理由相信，反思的信念获得与未反思的信念获得在性质上有所不同，前者体现了能动性，而后者却没有。然而，似乎并没有这样的差别。未反思的信念获得，在相关方面似乎确实是完全被动的，而反思的信念获得似乎并无不同。

但是，在考虑应该如何理解诉诸反思的问题上，这种方式可能是错误的。例如，当索萨引入了动物知识和反思知识的区分时，似乎非人类的动物只能有动物知识①，而人类则有时只有动物知识（由于我们经常非反思地形成信念），有时有反思知识。但这不是索萨所说的全部。索萨在作出区分之后，立刻指出："注意，有幸拥有理由的人类不只有野兽可获得的那类动物知识。"②因此，即使在我们未反思地形成信念的那些场合，也能把人类的知识看成在性质上不同于动物的知识。这样，重要的分界线似乎是在（能够反思的）人类的信念与（不能够反思的）非人类的动物的信念之间，而不是在反思的信念获得和非反思的信念获得之间。

我们如何能充实这一点来支持对认识的能动性的解释呢？我这里提供的解释不能归属到某一个人。尽管这种解释受了索萨文本中的许多评论的启迪，也受到了塞拉斯派的哲学家的许多评论的激发，但肯定不是某人明显地承诺的某种解释。与此同时，我认为，这个观点很值得认真对待，不管它是否可能属于索萨或塞拉斯派。我认为，许多人会发现，这是一种很有魅力的解释，它至少能引起我们中的许多人想要思考人的认知和非人的认知之间的某些差别。的确，我相信，我在此所提供的建议，比迄今为止一直考虑的康德论题的变种似乎从一开始就更有道理，即使最后，我将论证应当拒绝这种解释。

索萨把他的观点描述为德性认识论，我们从以下问题开始提问也许有所帮助，即人类信念获得的什么事实使得其适合某种德性的解释，而非人类动物信念的获得则不适合。正如我们看到迈克尔·威廉姆斯（Michael Williams）所做的那样③，有些人建议，这里的一个重要区别是，即使恰当地认为动物拥有信念，它们形成信念的方式也完全不会随着时间而改变。它们的信念获得过程是天生固定的，而且，即使这些过程对于许多目的而言是相当可靠的，因而能够产生（至少一种）知识，这也标志着与人的信念获得有重大不同。对于我们，我们形成信念的方式不同于动物，是随着时间而改变的。不仅如此，我们的信念获得和修正的过程是随时间而改变的，这种方式与反思和能动性都密切相关。我们人类定时地停下来反思我们自己的信念，反思获得与修正这些信念的方式，也反思我们自己

---

① 至少如果我们假定，他们似乎真的没有反思能力。

② Ernest Sosa, "Knowledge and Intellectual Virtue", repr. In *Knowledge in Perspective: Selected Essays in Epistemology*, New York: Cambridge University Press, 1991, p. 240.

③ Michael Williams, "Is Knowledge a Natural Phenomenon?" In *The Externalist Challenge*, edited by R. Schantz, Berlin: de Gruyter, 2004, p. 209.

过去的成败记录。有时，我们自觉地（self-consciously）采取措施改变我们获得和修正信念的方式。因而，例如，当学生在逻辑导论的课堂上发现，他们经常进行肯定后件推理并且这种推理策略是非常不可靠时，至少部分学生会尝试训练自己停止肯定后件推理；也就是说，他们设法摆脱这个坏的推理习惯。这种认知的自治似乎是人类独有的，我们拥有这种力量，是因为我们有能力对自己的心理状态进行反思。

那么，这种建议是，与其他动物不同，反思允许我们进行认知的自治，使我们有能力检测我们自己的认知，为了更准确地形成信念，主动接受再教育，为我们的认知（即其他动物缺乏的认识）提供一个维度。当人成年之后——甚至大概在此之前——我们获得信念的方式，不能再被内在推理过程的直接运作所解释。相反，我们对自己认知行为的（确实是定期的）审查，引发了旨在重塑认知的自我意识的行为的自我意识的行动。因此，对任何一位成人而言，信念获得和修正的许多过程就是这种主动的自我修正的直接结果，正如成人所表现出的许多习惯性行为应归功于有意识的活动一样，这些有意识的活动旨在导致能引发那些行为的习惯的出现。不仅如此，即使信念获得的许多过程不是这种有意识的修正的产物，即使在这种情况下，这些过程的持续存在部分地被它们通过了我们定期自我审查这一事实所解释。正是由于这个原因，我们似乎可以有理由认为，即便是非反思的人的信念获得，也与动物的认知具有重要的不同。甚至当我们没能进行反思时，我们获得信念的方式，也会通过我们过去反思时引起的活动至少部分地得到说明。

也正是这个原因，与非人的动物不同，我们有理由对所持有的信念负责；同样，与非人类的动物不同，当我们的信念是可靠地获得时，已经表明了我们理应信任我们的信念。我们不是自由地选择我们的信念。即使当我们反思应该相信什么时，我们也不是自由地选择我们的信念。相反，与其他动物所不同，我们形成信念的方式受到了有意识地选择行动的影响，因此，可以认为，只有我们才有其他动物缺乏的这种认识的能动性。当成年人以可靠的方式获得信念时，产生和保留他们信念的认识机制就有理由被视作德性：不仅是通过自然选择行为而内在建构的机制，而且包括由于被反思引起的有意识活动的主动介入才得以存在的机制。①

我相信，尽管这种图像是吸引人且符合常识的，但它极大地误解了人和动物

---

① 索萨的观点和这一观点之间肯定有许多联系点。如前所述，索萨至少有时强调下列区分的重要性：在人类的信念（不论是反思获得的，还是非反思获得的）与非人类的动物信念之间的区分，而不是反思获得的信念和非反思获得的信念之间的区分。此外，他确实非常关注这样的建议：我们值得信任恰当地形成的信念（尽管在这里的建议方式中这与自我审查无关）。

的认知，低估了动物认知的复杂性，对人类认知的阐释显得过于智能化。它过分地夸大了反思在我们的认知中的参与程度，但实际情况并非如此；它提供了对认知中的能动性作用的一种解释，这种解释使它与反思的关系比与我们当前理解的这些事实能支持的关系更密切。这些观点都需要探讨和详尽阐述。

根据刚才呈现的图像，人类获得和修正信念的机制是随着时间而改变的，而动物则不是。但正如我们已经看到的那样，这个建议完全是错误的。许多非人类的动物的认知是相当复杂的，它们获得信念的方式像人类一样，在对新信息作出反应时，随时间而改变。这类改变并不需要有自我意识的反思。即使没有自我意识反思的介入，人类和非人类的信念也都可以因揭示新的信息而改变。例如，一旦我看到我的车钥匙在厨房的台子上，我就相信它们在那里，即便在我看到它们在厨房之前我也以为我把它们放在了餐厅。这并不需要对我之前的信念，或我新获得的感觉经验，或是应该把我的新感觉经验整合到我的整体信念中的方式进行反思。我看到钥匙在台子上，在我没有反思或活动的前提下，我的认知系统完成了信念的更新工作。①动物也是如此。我的狗在吃东西时，完全能识别出什么候吃完。如果获得盘子空了的信念后，它发现，我重新把盘子盛满，那么，在不需要任何一种反思或行动的前提下，它的认知机制更新了它关于盘子里的东西的信念。

但正像人和动物的信念更新不需要经过反思或行动那样，更新和修正获得信念的方式也不需要经过反思或行动。如果每当我看见一只狐狸走近时，我就相信这是危险的，那么，我发现了一只无害的特殊狐狸，不是我能简单自动载入的信息；当我遇到这只特殊假定的狐狸时，将使我改变我做出的推理，当然，当我再看到它时，我能认识到这个改变。但正如我们已经注意到的那样，在非人类的动物身上，也能发现相同类型的推论整合和推论倾向的改变。为了以这种方式整合信息，不需要有类似于灵长类动物（更不用说人类）的认知成熟度（cognitive sophistication）。②这种能力是反思能力的副产品，因此，是人类独有的，这样的建议是错误的。

注意到这种错误歪曲了人和动物的认知是重要的。它把动物描述成，在告知它们的后续信息处理方面，是没有能力整合新的信息的。它把人类描述成，在这方面，只凭借对心理状态的反思能力，就有能力整合新的信息。这两种主张都是不正确的。但这些观点并不是所提出的观点的唯一问题。

---

① 索萨有时建议，至少当人们不止考虑一个目标时，新信息的整合需要反思人的信念内容及其信念之间的逻辑关系，由此使得反思只有人类才有可能。

② Alcock, *Animal Behavior: An Evolutionary Approach*，在这里尤其有启发意义。

基于所提出的图景，反思被描述成是认知进步的驱动力。根据这种观点，当我们对我们的信念、获得信念的方法、信念之间的逻辑联系，以及我们过去认知的成败进行反思时，我们就开始了能使我们今后获得更精确信念的行动。我毫不怀疑，我们有时确实恰好是这样做的，这种行为是其他动物不可能表现出来的。从所建议的观点来看，这是真理的重要性质。但基于这种图景，这种观点的重要性被极大地夸大了。我们已经看到，不能反思的动物可以通过改变它们处理信息的方式，来整合它们获得的新信息。因此，我们不应该认为，人类享用的高度成熟的反思策略，是可能导致认知进步的唯一方式。非人类的动物只是由于不能对其心理状态进行反思，所以，注定在它们的存活期间会重复同样的认知错误。

然而，即使除去这一点，我们也已经看到，为了促进认知进步，反思我们的信念完全不如我们相信的常识图像那么有效。反思行为通常伴有信念的固化（the fixation of belief）现象。①这里，像许多其他状况一样，现象学是对心智运作的极不准确的反映。

由于自我理解中的这些扭曲，常识的图像在很大程度上影响了我们处理信息的方式，这极大地夸大了我们自身的效力。②因此，我们正在讨论的故事使我们想到，当我们是成年人时，我们获得信念的过程，要么是以促进认知为目的的自我有意识的行为的产物，或者要么当它们被反思所寻获时，它们能保留下来完全是因为它们通过了自我有意识的审查。大量的认知过程的信息被囊括在认知模块（cognitive modules）中，这完全不是真的。③这些认知模块的运行机制必然以两种不同的方式违反常识图景：这些模块的运行是无从觉察的内省，因此，绝对不能以所提出的方式通过反思来审查它们；此外，它们的运行机制是天生固定的，因此，反思思维检测到的任何缺陷，无论如何都不会改变。当然，我们被赋予了具有这些特征的许多认知过程，这是一件好事。我们有能力对环境的许多复杂特征做出既快又可靠的反应，这种能力非常依赖于这些过程的运行。但这只是说，常识图景把我们自己的思维机制描述成完全是我们自己活动的产物，这种图景非常错误地刻画了认知运作的真正的可塑度。自然选择以这样一种方式使我们的思维条理化，使得思维的许多最重要的特征成为抗干扰的：这些特征不可能通过善意

---

① 参见 Alison Gopnik，"How We Know our Minds：The Illusion of First—Person Knowledge of Intentionality"，*Behavioral and Brain Sciences*，Vol. 16，No. 1，1993，pp. 115，and pp. 90-101.

② 这是常识图景全面夸大了我们自己的效力的部分方式。例如，参见 Shelley Taylor and Jonathan Brown，"Illusion and Well-Being：A Social Psychological Perspective on Mental Health"，*Psychological Bulletin*，Vol. 103，No. 2，1988，pp. 193-210，以及 Shelly Taylor，*Positive Illusions：Creative Self-Deception and the Healthy Mind*，New York：Basic Books，1989.

③ 对这一解释的经典描述可以在下列文献中找到：Jerry Fodor，*The Modularity of Mind*，Cambridge，Mass.：MIT Press，1983.

的但通常是无知或误导的能动者的行动来重构。思维方式不像常识图景描述它的方式那样发挥作用，并且，不能这么做是一件好事。

人人都企图把我们的认知成功描述为是起因于反思的自我审查的认可和活动的产物，这种企图就心智运作归功于我们的程度远超出了我们应得的部分。我相信，这终结了任何把不论何时只要是恰当形成的信念都归功于我们的方案，或者把我们的智能看成是德性，且这种德性的存在是经由理智的自身教化行为得以解释的方案。①

八、结　　论

意志自由要求对人的一阶状态的某种高阶的批判性评价，这种观点导致了无穷回归。然而，正如我们已经看到的那样，回归问题不是这种观点的唯一困难。通过反思高阶审查的方式诊断一阶状态的试图，为高阶状态预设了根本不存在的一种默认的合法性。这里有某种讽刺意味。对一阶状态进行批判性评价的措施是为了承认，例如，人的一阶欲望也许不能反映真正的自我；它们在某种重要意义上也许是异己的影响。但相同的观点同样适用于高阶状态，因此，我们查明二阶认可支持哪些一阶状态，不能区分下列两类欲望：一类是我们自己的真实欲望，另一类是受异己影响的那些欲望。像一阶状态一样，二阶状态不是自动合法的。

接下来，我们转向了对能动性的特殊情况的考察。许多不同的哲学家提出的信念获得的观点，都诉诸某个认识的能动性概念。这些作者并没有一贯地支持信念获得是一种有意识的行为的观点，因此，这里诉诸任何类型的能动性，初看起来都是令人困惑的。本文考察了认真讨论认识的能动性的几种不同动机，在这些情况下，每一种情况都诉诸反思，而且，反思所导致的高阶状态都起到了关键的作用。然而，我们这里也看到，这些观点预设了关于反思和高阶状态的各种经验主张，这些主张与我们最有用的证据相左。即使存在合法的能动性概念，它也不能胜任这些作者所要求的工作。

（高洁　戚陈炯　丛杭青 译，成素梅 校）

---

① 出于同样原因，那些把自我的图像描述为"自我构成"（self-constituting）（特别参见克里斯丁·克斯高的 *The Sources of Normativity*；*The Constitution of Agency*：*Essays on Practical Reason and Moral Psychology*，New York：Oxford University Press，2008；和 *Self-Constitution*：*Agency*，*Identity*，*and Integrity*，New York：Oxford University Press，2009；关于一个类似的观点，参见 Robert Kane，*The Significance of Free Will*，New York：Oxford University Press，1996）夸大了反思行为在确定自我本性时塑造性格的作用。

The Concept of Normativity from Philosophy to Medicine: An Overview

Claude Debru

1. On Some Twentieth Century Philosophical Ideas about Normativity

The concept of normativity is a transdisciplinary concept which is found in every branch of philosophy and human sciences and pertains to every aspect of human life. However, the terms normative, normativity are rather recent. The term normativity seems to be an invention of twentieth century philosohy. In the nineteen thirties and forties, philosophers from very different backgrounds, including Edmund Husserl in Germany and Georges Canguilhem in France, developed the idea of normativity defined as the power of creating and changing norms. In his 1935 Vienna lecture, *The crisis of European mankind and philosophy*, and in the framework of his phenomenological reflections based on the philosophy of logic and mathematics, Husserl conceived normativity as the creation of a world of ideas put under the control of the norm of unconditional and objective truth. This world of ideas has the property of developing itself in infinity by the continuous creation of new ideas which become themselves the-matter of further creations. Thus, the intentional life of individual persons and of communities is directed towards goals which are subjected to norms and go on in infinity. " Mankind, considered in its soul, has never been and will never be accomplished. The spiritual goal of European mankind (···) is situated at the infinite: it is an infinite idea towards which the spiritual process as a whole seeks, if I may say so, to transcend itself. Not only consciousness in this process of becoming grasps this term as a telos in proportion to and within this development ; but consciousness sets this term also in a practical way as a goal for the will, and erects it in a new form of development, put under the control of norms, of normative ideas. " (Husserl 1950, p. 236) Husserl puts the normative idea of mankind as a spiritual development in sharp contrast with a naturalistic and even "zoological" view of mankind. All aspects of cultural life are involved here, but mathematics as an infinite construction is the best example of human normativity defined as the continuous pursuit of objective and possibly unconditional truths (Husserl 1950, p. 240).

In the very different context of medical thinking Georges Canguilhem introduced the idea of a biological normativity in his 1943 MD thesis *The normal and the pathological* (Canguilhem 1972, p. 77). In a general sense, normativity may be defined as the power of establishing norms. From the particular standpoint of medicine, normativity was defined by Canguilhem as the organism's power to create different and more or less stable ways of functioning according to various normal or pathological states. Canguilhems ideas were formed in the context of pathophysiology. They were confirmed by the reading of Goldstein's neurological interpretations as exposed in his book *The structure of the organism.* Canguilhem was deeply in harmony with the medical theoretical thinking which was very much developed in Germany between the two world wars, and which could be characterized as a philosophical theory of the living being. In the true spirit of this holistic philosophy, Canguilhem wrote: "life is polarity and as such unconscious position of value-in short, life is in fact a normative activity (⋯) In the full sense of the word, normative is what sets up norms. In this sense, we propose to speak of a biological normativity. " (Canguilhem 1972, p. 77) One of the consequences of this thesis for medicine defined as a technique rather than as a science is that "physiology, rather than looking for objective definitions of the normal, should recognize life's original normativity" (Canguilhem 1972, p. 116). Being "normal", for humans, is not a matter of fact, but an aspect of life's original normativity. "If one may speak of a normal man, as determined by the physiologist, this is because there exist normative men, men for whom it is normal to break norms and to establish new ones. " (Canguilhem 1972, p. 106) In the same vein, Canguilhem quoted Reiningers *Wertphilosophie und Ethik*: "Unser Weltbild ist immer zugleich ein Wertbild. " (our view of the world is always at the same time a view of value) (Canguilhem 1972, p. 117) The science (and philosophy) of the living being should include the idea of the organism having not only a world view but mainly a world view which is fully value-laden. According to Canguilhem's fundamental thesis, this matter of fact is indeed the case for normal physiology as well as for pathology. Pathological states are characterized by a remaining, although diminished, physiological normativity. The pathological is not the opposite of the normal ; it keeps some normative character. Pathology does not mean necessarily a chaotic, catastrophic or irrational course. In pathological states, the organism tries to establish a new functional norm certainly different from the normal one but more or less viable. There is some kind of rationality in pathology, in the sense of a holistic concept of rationality. This philosophical thesis may apply to many fields of human experience, not only to medicine.

More recently, the idea of normativity arose much interest in the fields of moral philosophy, of epistemology (with the discussions on epistemic values), and of analytical philosophy of language. Indeed, the terms, normative and normativity, do apply to many different things, be it statements of any kind, from logic to law, to grammar and to language, and to any kind of rule including technological rules. But most importantly, normativity does not only designate the power of the rule which is going to be applied, but also and mainly the power of the human subject who formulates these rules, and applies them thanks to his personal authority, social status or whatever. The ability of the individual who creates a new order in his social environment may be thus described as "normativity", and here medicine comes again into the picture.

Indeed, the medical meaning of normativity remains an essential part of the picture for two very special reasons: the increasing creation of social norms for medical practice and the increasing concern for the patient's autonomy or rather personal normativity (a feature which is particularly desirable for patients in danger of loosing it due to severe neurodegenerative diseases for instance: how to allow these patients to express themselves and interact with their environment, so that they keep most of their normative power). In this paper, we will try to give an overview of the philosophical discussions on the nature and sources of human normativity, and to show their relevance for medical practice in the interaction between patients, health practitioners, and their human environment.

2. Some Philosophical Remarks on Logic and Mathematics

The paramount power of human normativity and its anthropological basis are best described in classics of XXth century philosophy and in more recent developments in analytical philosophy. I will make use of some commentaries on Wittgenstein by the philosopher Jacques Bouveresse in his book *La force de la règle. Wittgenstein et l'invention de la nécessité*. In order to understand what it is to follow a rule, the examples of grammatical and logical rules are basic. They allow to ask a particular question: does the ability to speak a given language rest upon an implicit knowledge of its grammatical rules, or does the ability to speak a language involve also a creative participation of the speaker ? (Bouveresse 1987, p. 14) Indeed rules are not just a matter of external necessity, but rather a matter of choice and perhaps of conventions. Mathematics are the best example of a creative process based on rules and able to create new mental objects which may be used as new rules etc. Now we meet a question about the epistemological status of these rules. Are they similar to ordinary propositions in the logical sense, which are endowed with a truth value, or are they something else, constituting the frame for formulating meaningful sentences without being themselves either true or false ? Are they just conventions, or are they endowed with a kind of necessity which is very likely not of a platonic, external, realistic kind but rather the result of an internal process of creating rules according to rules—a thesis which represents in its essence Wittgenstein's grammatical, non conventionalist, conception of mathematics ? (Bouveresse 1987, p. 23) According to Wittgenstein, there is no contradiction for a mathematical proposition to be a rule, not simply stipulated by convention, but generated according to other rules. There is rather a kind of circle, well defined by Quine who noticed that if logic must result from conventions, logic is necessary for the inference of logic from conventions. Indeed, some kind of logical rules of deduction are necessary there in order to build logic. How can we justify these rules of deduction ? According to Wittgenstein, there is nothing like a preexisting meaning of a grammatical proposition which would force us to accept other grammatical propositions which are logically deducible from it. The new connections which we discover were not already present in any sense. They are the result of a construction which has to be performed and accepted at each step. The new connections are an additional détermination of the meaning and an extension of the grammar, not simply an explanation of any concealed content. The meaning of a word is the rules of its uses ; if we change the rules, the meaning is different. This is a constructivist view of logic and mathematics. In this view, we have to give an account of the relationship between meaning and use (Bouveresse 1987, pp. 35-36) .

In this enquiry, we encounter an unexpected thing. Even in mathematics, which has the most precise and explicit rules, it seems that no anticipation can predict in advance a case which has not been previously encountered. Every application of a rule to a new case is in fact a new application. According to Wittgenstein commenting on the individuality of the numbers in his *Philosophical Grammar*, if a universal rule is given to me, I must know each time anew that this rule can be applied in this particular case. No act of prevision can spare me this act of intuition. Then indeed the form to which a rule is applied is at each step a new form (Bouveresse 1987, p. 36). The application is not a matter of an act of intuition, but of an act of decision. Similar remarks were made by the French mathematician Emile Borel, who asked the question, whether the kind of operation by which we go from one integer to its immediate successor is really the same at each time, or not. From a certain point of view, it is the same, and from another, it is not, since adding one to one in order to get two, and adding one to two in order to get three is not the same operation. Even in such a regular system as arithmetics, a new application involves at each time a decision regarding it. So that, what characterizes best mathematics is its normative character. Mathematics is not descriptive, it is rather prescriptive. Mathematics, in the Wittgensteinian sense, is normative in a way which has nothing to do with any truth of a logical norm. It creates a special kind of necessity, the necessity of a newly created fact, the factual necessity of our special way of doing things, which is ours and does not seem to be possibly of another kind. In other words, this necessity is an anthropological necessity, it depends on a fact of nature, which appears necessary because there is no other alternative-and moreover which appears necessary from within (it could appear as contingent rather than necessary from an external point of view). If we try to summarize the most significant results, from the standpoint of the idea of normativity, of this sketchy investigation in the field of the philosophy of mathematics, it appears that the idea that the recursive application of the same rule is at each step a new situation different from the previous ones, and that there is something unforeseeable in this application. This rather unexpected result seems to be of great relevance for any aspect of human thinking and action, including certainly the ones which are most remote from pure mathematics, like medicine.

3. The Nature of Normativity According to the Analytical Philosophy of Language

Let us try to get now into another field of contemporary philosophical enquiry, the analytical philosophy of language, and ask the following questions: is there something irreducible in normative statements (statements of the "ought to" kind) compared with factual statements ? Can we give an account of this difference by using the logical tools of analytical philosophy ? This problem was dealt with by Ralph Wedgwood in a recent book on the nature of normativity. As an analytical philosopher, Wedgwood considers the existence of normative statements or judgments as a fact, which needs explanations, but refuses to propose any definition of the term "normativity" as such, perhaps fearing that defining concepts such as normativity would lead us into a kind of mystical realism. He defines his task as giving a theoretical account of the ordinary understanding of normative terms by analysing their truth conditions in a logical and semantic way. This task is very different from more naturalistic approaches of the normativity problem, which are more and more developed in natural sciences like cognitive neuroscience and psychology, which deal with the different problem of understanding how we as human beings create and possibly change norms, and not only follow them in our language.

From the standpoint of the analytical philosophy of language, normative thinking, normative knowledge, normative truths, normative judgments, normative intuitions, normative beliefs etc. are matters of fact—facts of discourse, language and thinking which bear on non existent realities. What is the semantic content of these very peculiar kinds of statements ? Does the semantic explanation of normative statements need statements or concepts which are themselves non normative, or not ? In other words, is normativity in language and thinking reducible or not reducible to factual properties ? In case it would turn out to be irreducible, we would face a logical circle. Wedgwood's own thesis, is that we can avoid this circle. Other philosophers, like John Mac Dowell, Ronald Dowrkin or Derek Parfit claim on the contrary, in their own "quietist" thesis, that it is impossible to give a non-circular explanation of normative discourse. Wedgwood relies on an argument of Saul Kripke regarding reference, in order to justify his own claim: "I shall give an accont of what it is for a thought or a statement to be about what ought to be the case without making any use of (···) a notion of a thought's or a statement's being about what ought to be the case. " (Wedgwood 2007, p. 20) Only very general normative properties like truth, proof, belief, decision can be used in order to analyse normative statements exactly in the same way as any other kind of statements. Indeed, normativity can never be absent from statements of the sort "Necessarily, if one is rational, if one judges 'I ought to do this', one also intends to do this" . Necessity and rationality are normative properties, so that the project of an explanation of normativity in totally non normative terms makes non sense.

However, Wedgwood claims that particular normative statements of the form " I ought to do this" can beanalysed without referring to any normative property of the same kind. As a philosopher of language, he notices that normative terms are extremely context-sensitive and express different concepts in different contexts (Wedgwood 2007, p. 23). Under such conditions, the philosophical theme of intentionality enters the discussion. Indeed, Wedgwood's own theory of the nature of normativity, which he pictures as "normative judgment internalism" as opposed to externalism, stresses the subjective, first person dimension of normative judgment, whereas externalism stresses the external, objective source of normativity. According to Wedgwood's internalist theory, there is an essential link between normative judgment and practical argumentation and motivation for action. Normative judgments or statements can be pictured as first-person judgments, with the qualification, that action is strongly context-dependent, and should be appropriate and reasonable. Normative judgments can be understood in terms of their regulative roles in practical reasoning. The concept of a right planning for action is introduced by Wedgwood under conditions, which do not allow any uncertainty. Under the specific conditions of certainty it is clear that the role of normative concepts or judgments can be easily described as commitment or obligation, which necessitates the realisation of the planned action. This is a rather ideal model, which should be totally different under conditions of uncertainty.

Then the concepts of intentionality and first-person viewpoint come necessarily into the picture. The slogan "the intentional is normative" has been much discussed by philosophers recently. From the previous analyses it is clear that normativity represents a special form of intentionality, which has to do with practical judgment, disposition, and commitment for action. Does intentionality in general and essentially involve some for of normativity ? Many philosophers, including Husserl and more recently Davidson, defended already this idea. According to Wedgwood, "intentional facts are partially constituted by normative facts" (Wedgwood 2007, p. 159). The reason why this seems to be the case is that all intentional states or properties involve two elements: "(i) a content, which is composed out of concepts, and (ii) a mental relation or attitude (such as belief, desire, hope, fear, and so on) towards that concept." (Wedgwood 2007, p. 161) For instance, believing can be pictured in two ways. The belief is right if and only if its content is true, and second, the belief is rational (under certain circumstances) only if these circumstances determine its probability so that its content is true (Wedgwood 2007, p. 162). The normative element of the intentional is found in the relationship between the intentional conceptual content and the intentional attitude regarding this particular concept. With his theory of "normative judgment internalism", Wedgwood wants to avoid to different dangers: the danger of a total reduction of the normative to natural properties (naturalism), and the danger of a purely metaphysical and mysterious picture of human normativity.

4. Some Empirical Studies on the Sources of Human Normativity

However, internalism leaves some room for an explanation of normativity in naturalistic, psychological and even neurobiological terms, as we may observe more and more frequently in some research programmes and reports which are worth mentioning. It is rather striking to notice that in the naturalistic approach of normativity in developmental cognitive psychology (dealing with young children) normativity is considered as a power as well as as a fact. The description of normativity as a fact which is found in language and whose logical and analytical properties can be looked for (with the not entirely unexpected result that normativity involves intentionality and rationality) —this description is not enough. How can these normative rules be created, how can they be perceived and accepted ? These questions were investigated by reseachers at the Max Planck Institute for evolutionary anthropology in Leipzig. The children who participated in the study were two and three years old. These children can understand the rules of a game and follow them. Moreover, they can notice situations in which the rules are not followed and they can protest. A "normative awareness" is found in very young children (Eakoczy 2008, p. 875-881). Similar empirical studies can be found on the inhibition of violence among children, showing the existence of normative behaviors. In a recent book, the neurobiologist Jean-Pierre Changeux tries to show how factual knowledge about the human brain may be useful for the scientific understanding of the sources of norms and values and for the practical development of normative behavior in society. Examples of empirical studies on normative behaviors and moral emotions include the inhibition of violence in children, and the distinction between social conventions and moral imperatives. The study of children raised in different cultures and in different social conventions shows that after the age of three, children judge as acceptable not to follow religious rules, but they consider as unacceptable the transgression of essential moral rules (Changeux 2009). However, and not surprisingly, the question of the sources of human normativity was dealt with mainly in moral philosophy, as we shall see.

5. Moral Philosophy on the Sources of Human Normativity

In her book *The sources of normativity*, Christine Korsgaard wrote: "It is the most striking fact about human life that we have values. We think of ways that things would be better, more perfect, and so of course different than they are ; and of ways that we ourselves could be better, more perfect, and so of course different than we are. Why should this be so ? Where do we get these ideas that outstrip the world of experience and seem to call it into question, to render judgment on it, to say that it does not measure up, that it is not what it ought to be ? Clearly we do not get them from experience, at least not by any simple route. And it is puzzling too that these ideas of a world different from our own call out to us, telling us that things should be like them rather than the way they are, and that we should make them so. " (Korsgaard 1996, p. 1) According to Korsgaard, "the fact of value is a mystery", a mystery which is in need of a philosophical enquiry. Four kinds of solutions were proposed by philosophers: (i) realism-norms and value are real things (Plato, Aristotle) ; (ii) voluntarism - there is no right or wrong in a state of nature, norms and values are imposed on us by an authority, obligation derives from the command of someone who has legitimate authority over the moral agent (Hobbes) ; (iii) norms and values are discovered by an internal reflection-theory of the "reflective endorsement", meaning personal approval (Hume) ; (iv) the content of norms and values is potentially universal, it is found by an autonomous act of free will—the laws of morality are the laws of the agent's own will and its claims are the ones he is prepared to make on himself (Korsgaard 1996, p. 19). Kant and contemporary kantians like John Rawls are representatives of this trend—Kant's formal universalism is well known.

These four basic philosophical solutions are answers given to a very special question, which Korsgaard defines as the "normative question", which may be posed in the following way: "Ethical standards are normative. They do not merely describe a way in which we in fact regulate our conduct. They make claims on us ; they commnd, oblige, recommend, or guide. (···) And it is the force of these normative claims—the right of these concepts to give laws to us—that we want to understand. " (Korsgaard 1996, p. 8-9) "When we seek a philosophical foundation for morality, we are not merely looking for an explanation of moral practices. We are asking what justifies the claims that morality makes on us. This is what I am calling the normative question. Most moral philosophers have aspired to give an account of morality which will answer the normative question. But the issue of how normativity can be established has seldom been directly or separately addressed, as a topic of its own right. " (Korsgaard 1996, p. 10) Where could we find, then, the answer to the normative question ? According to Korsgaard, the answer "must appeal, in a deep way, to our sense of who we are, to our sense of our identity" (Korsgaard 1996, p. 17) . If we go back into ourselves, in difficult situations, we should be able to find an answer. Korsgaard's own answer to the normative question is eventually Kantian (and some hints to Kantianism are found also in Wedgwood's discussion) —with the qualification that the content of formal universalism is given by reflective endorsement. " The reflective structure of human consciousness requires that you identify yourself with some law or principle which will govern your choices. It requires you to be a law to yourself. And that is the source of normativity. So the argument shows just what Kant said that it did: that our autonomy is the source of obligation. " (Korsgaard 1996, p. 103-104) In the end, and as a possible consequence of this mixture of Kantianism and reflective endorsement, we are lead to conclude that all four philosophical theories of the sources of normativity are true, they are all parts of the overall picture of human normativity (Korsgaard 1996, p. 165) . [1]

6. Normativity versus Autonomy: Medical Implications

The concept of autonomy is a philosophical concept which underwent a change of meaning from the collective and political to the more individual level. Christian Wolff used this term in 1757 in his book on moral and political philosophy in the sense of the independance or self-determination of a state. Although Jean-Jacques Rousseau did not use the term, he is considered as responsible for the introduction of the idea of autonomy in the sense of obeying a law which the individual prescribes for himself. This move towards moral philosophy is even stronger in Kant's work. Autonomy in the Kantian sense means self-rule of the will which discovers in itself universal laws. Much more recently, the idea of autonomy was introduced in medical ethics, in much more pragmatic discussions regarding health care and health economy. The idea of autonomy

---

[1] Ibid. , p. 165.

was meant as the ability of the patient to sustain him or herself, to rely on him or herself rather than on the others. The concept of normativity, as we did see, has quite a different background. As a matter of fact, it involves the collective horizon of mankind, the essential relationship between individuals. It means the ability of the individual even in diminished conditions to establish new norms for himself and for the others. Stephen Hawking is an extreme case, since he keeps his normative, creative power, his ability as a man not only to rule himself (or at least his mind), but to communicate with the others and to enrich their lives.

It is worth introducing here the concept of health as a dynamical concept. It has been stressed many times that recovery does not mean a full return to the previous health state, because the organism's whole functionality has changed. The idea of normativity is in harmony with this dynamical concept of health. In the case of severe chronic, degenerative or genetical diseases, for which there is no cure, normativity has a much more serious and interesting content—a dynamical content, obviously. Let us go to clinics to illustrate this point. Between two bouts of multiple sclerosis, the patient may have the time to adjust to his or her pathology, thus showing his or her remaining normativity. The Freireich disease is a neurodegenerative disease of the motor command. There is no cure, and no hope. Handicapped students suffering from this disease may keep however an extremely strong desire for work, recognition, scientific contribution, and interaction with the others. Patients suffering of amyotrophic lateral sclerosis, a degenerative disease of the spinal chord motoneurones, ending in a paralysis of the respiratory muscles, may see their lives improved by technical devices which allow them to communicate with their environment even in a still creative way, showing an incredibly rich human expérience, keeping their normativity, in the sense of their ability to contribute to the human experience of their immediate environment, in spite of having lost most of their autonomy (Le Forestier, forthcoming in *Medicine Studies*). The same is true for very elderly people, suffering from the syndrom of frailty, of diminished physiological functions making them more sensitive to unexpected external perturbations. Their autonomy is reduced, their normativity may be even enhanced, depending on the quality of their interaction with their human environment.

Normativity is a philosophical term which describes very well almost every aspect of human experience. The new awareness of human normativity may be useful to improve many social and technical aspects of medical practice and care.

References

Bouveresse, Jacques, 1987, *La force de la règle. Wittgenstein et l'invention de la nécessité*, Paris: Editions de Minuit

Canguilhem, Georges, 1972, *Le normal et le pathologique*, Paris: Presses universitaires de France.

Changeux, Jean-Pierre, 2009, *Du vrai, du beau, du bien. Une nouvelle approche neuronale*, Paris: Odile Jacob.

Husserl, Edmund, 1950, " La crise de l'humanité européenne et la philosophie", *Revue de métaphysique et de morale* No. 3: 229-258.

Korsgaard, Christine, 1996, *The Sources of Normativity*, Cambridge: Cambridge University Press.

Rakoczy H. , F. Warnecken and M. Tomasello, 2008, "The sources of normativity: young children's awareness of the normative structure of games ", *Developmental Psychology* 44 (3): 875-881.

Wedgwood, Ralph, 2007, *The Nature of Normativity*, Oxford: Clarendon Press.

从哲学到医学的规范性概念的概述

克劳德·德布鲁[*]

一、20 世纪关于规范性的各种哲学观

规范性概念（concept of normativity）是一个跨学科的概念，它存在于哲学和人类科学的每个分支学科中，并与人类生命的每个方面都密切相关。然而，术语"规范的"（normative）和"规范性"（normativity）是最近的概念。规范性术语似乎是 20 世纪哲学的一个发明。在 19 世纪 30 年代和 40 年代，包括德国的胡塞尔和法国的乔治·康吉兰（Georges Canguilhem）在内的具有完全不同背景的哲学家发展了规范性的观念，把规范性定义为创造与改变各种规范（norms）的力量（power）。胡塞尔在 1935 年的"欧洲人的危机与哲学"的维也纳讲座中和他基于逻辑学与数学进行现象学反思的框架内，认为规范性是在绝对的客观真理的规范控制下创造出来的一个观念世界。这个观念世界具有的属性是，不断创造新观念，新观念本身又成为进一步创造的根据，由此无限地扩展自身。因此，个人和群体的意向生活是努力不断地服从规范的目标。"人类，从其灵魂深处来考虑的话，从来就不是完美的，以后也不会是完美的。欧洲人的精神目标……定位于（追求）无限：一个无限的观念（an infinite idea）是，以作为一个整体的精神过程设法（如果我可以这样说的话）超越自身为目标的。不仅意识在这种形式过程中把无限作为正比于这种发展和内在于这种发展的一个目的来把握，而且意识以实践的方式把无限设定为是意志的目标，并以新的发展形式创立无限，即受到规范（即规范观念）的控制。"[①]胡塞尔把人类的规范观念视为精神发展，与自然主义的乃至人类的"动物"观形成了明显的对比。这里包括文化生活的方方面面，但是，作为一种无限构造的数学是把人类的规范性界定为不断追求客观真理

---

[*] 克劳德·德布鲁（Claude Debru），法国巴黎高等师范大学哲学系教授。本文为作者 2010 年 7 月访问上海社会科学院哲学研究所时的演讲稿。——译者注

[①] Husserl, Edmund, "La crise de l'humanité européenne et la philosophie", *Revue de métaphysique et de morale* No. 3, 1950, p. 236.

和可能的绝对真理的一个最好事例。①

　　在完全不同的医学思维的语境中，乔治·康吉兰于 1943 年在他的《规范和病理学》② 的医学博士论文中提出了生物规范性的观念。从一般意义上看，规范性可以被定义为确立规范的力量。从特殊的医学观点来看，康吉兰把规范性定义为，机体根据各种不同的规范状态或病理状态创造几乎稳定的不同功能方式的力量。康吉兰的观念是在病理生理学的语境中形成的。正如在他的《有机体的结构》一书中揭示的那样，对古尔德斯坦（Goldstein）的神经学解释的解读，确证了这些观念。康吉兰的思维与德国的医学理论思维非常一致，德国的医学理论思维是在两次世界大战之间充分发展起来的，而且，能够被视为有机体的哲学理论。在这种真正的整体论的哲学精神鼓舞下，康吉兰写道：“生命是两极对立的（polarity），简而言之，就无意识的价值定位而论，生命事实上是一项规范的活动……在这个词的全部意义上，规范的（normative）就是建立各种规范（norms）。我们建议在这种意义上说生物学的规范性。”③ 对于被定义为是一种技巧（technique）而不是科学的医学来说，这篇论文的结论之一是，“生理学应该承认生命的原始规范性，而不是寻找正常的客观定义”④。对于人类而言，“正常”不是一个事实问题，而是生命的原始规范性的一个方面。“如果人们可以像生理学家所确定的那样来谈论一个正常人，这是因为，存在着规范的人（normative men），对于这些人来说，打破旧的规范和确立新的规范是正常的。”⑤ 沿着同样的思路，康吉兰引证瑞宁格斯（Reiningers）的《价值哲学与伦理学》（*Wertphilosophie und Ethik*）中的话说，“我们的世界观总是同时也是一种价值观”（Unser Weltbild ist immer zugleich ein Wertbild)⑥。生命体的科学（和哲学）应该包括不仅具有一种世界观而且主要是一种完全负载价值的世界观的机体观。按照康吉兰的基本论点，这个事实问题确实是正常生理和病理的情况。病理状态的特征在于，继续保持（尽管有所减弱）生理的规范性。病理状态不是与正常状态

---

① Husserl, Edmund, "La crise de l'humanité européenne et la philosophie", *Revue de métaphysique et de morale* No. 3, 1950, p. 240.

② Canguilhem, Georges, *Le normal et le pathologique*, Paris: Presses universitaires de France, 1972, p. 77.

③ Canguilhem, Georges, *Le normal et le pathologique*, Paris: Presses universitaires de France, 1972, p. 77.

④ Canguilhem, Georges, *Le normal et le pathologique*, Paris: Presses universitaires de France, 1972, p. 116.

⑤ Canguilhem, Georges, *Le normal et le pathologique*, Paris: Presses universitaires de France, 1972, p. 106.

⑥ Canguilhem, Georges, *Le normal et le pathologique*, Paris: Presses universitaires de France, 1972, p. 117.

相对立，而是维持了某些规范的品质。病理不一定意味着是一个混沌的、灾难性的或非理性的过程。在病理状态下，机体尝试着确立新的功能规范，这种新的功能规范肯定不同于正常的功能规范，但或多或少是可行的。在整体论的合理性（rationality）概念的意义上，病理中存在着某种合理性。这种哲学论点可以应用于人类经验的许多领域，不只是应用于医学领域。

最近以来，规范性的观念引起了对道德哲学、认识论（包括对认知价值的讨论）和语言分析哲学领域的很大兴趣。规范的和规范性这两个术语确实能真正地应用于许多不同的事态，即从逻辑到定律、语法和语言，再到包括技术规则在内的各类规则的任何一种陈述。但最重要的是，规范性不仅指明了将被应用的规则的力量，而且主要是表述这些规则并由于其个人权威、社会地位等应用这些规则的人类主体的力量。一个人在他的社会环境中创造新秩序的能力就可以被描述为是"规范性"，在这里，医学再一次体现了这个图像。

的确，由于下列两种非常特殊的原因，规范性的医学意义依然是这个图像的一部分。这两种特殊原因是，不断地创建医学实践的社会规范和不断地关注病人的自主权或相当个人的规范性（由于重度神经退行性病变，在失去控制的危险中，病人特别期望的一个特征，比如，如何允许这些病人表达自我，如何允许他们与其环境互动）。在本文中我们将试图概述关于人的规范性的本质和来源的哲学讨论，并表明，在病人、理疗师及其环境之间的互动中这些哲学讨论与医学实践的相关性。

二、关于逻辑学与数学的某些哲学评论

人类的规范性及其人类学基础的首要力量，在 20 世纪的哲学典籍和分析哲学的近期发展中得到了最好的描述。我将利用哲学家雅克·布弗雷斯（Jacques Bouveresse）在他的《规则的力量：维特根斯坦与必要性的创造》（*La force de la règle. Wittgenstein et l'invention de la nécessité*）一书中对维特根斯坦的某些评论。为了理解遵循规则是什么意思，我举的两个基本事例是语法规则和逻辑规则。这些例子允许提问的特殊问题是：一种特定语言的能力取决于其语法规则隐含的知识吗？或者，一种语言的能力也包括说话者的富于创造性的参与吗？①的确，规则不仅仅是一个外在必然的问题，反而是一个选择也许是约定的问题。一个最佳事例是数学：数学是一个基于规则的创造过程，并且能创造出新的心智对象

---

① Bouveresse, Jacques, *La force de la règle. Wittgenstein et l'invention de la nécessité*, Paris: Editions de Minuit, 1987, p. 14.

(mental objects)，而这些心智对象又可以被作为新的规则等来使用。现在，我们面临着一个关于这些规则的认识论地位的问题。这些规则是类似于被赋予真值的逻辑意义上的普通命题呢？还是类似于自身没有真假只是为表述有意义的语句构造框架的其他命题呢？它们只是各种约定吗？还是它们被赋予了一种必然性呢？即，它们很可能不是柏拉图式的外在实在论的类型，而是根据规则来创造规则的内在过程的结果——其实，代表了维特根斯坦的非约定主义的数学语法概念的一个论点。① 根据维特根斯坦的观点，数学命题成为一条规则，不只是靠约定来规定，而且还根据其他规则来产生，这样的数学命题是没有矛盾的。奎因注意到，如果逻辑学必须产生于约定，那么，逻辑学必须是源于约定的逻辑推理，因此，他明确指出，倒是存在着一种循环论证。的确，在那里，为了建立逻辑学，某种演绎的逻辑规则是必要的。我们如何能够为这些演绎规则辩护呢？根据维特根斯坦的观点，一个语法命题，如果迫使我们接受在逻辑上由它推断出的其他语法命题，那么，最好是预先给定这个语法命题的意义。从任何一个方面来说，我们发现的这些新的联系都不是现存的。它们是在每一步不得不被完成和接受的构造的结果。这些新的联系补充决定了意义和扩展了语法，不只是对任何隐藏内容的一种说明。一个词的意义就是它的使用规则；如果我们改变规则，意义就会不同。这是一种构造主义的逻辑观和数学观。从这种观点来看，我们不得不对意义与用法之间的关系作出说明。②

在这种探究中，我们遇到了一个意想不到的问题。即使在拥有最精确的清晰规则的数学中，似乎也没有任何一种预期能够事先预言以前从未遇到过的情况。每次把一个规则应用于一种新的情况，事实上，都是新的应用。根据维特根斯坦在他的《哲学语法》（*Philosophical Grammar*）一书中对数的个体性（individuality）的评论，假如在我看来一个普遍规则是已知的，我必须每次都重新了解这个规则能够被应用于这种特殊情况。所有的预见行为都不能为我提供这种直觉行为。因此，应用了一种规则的形式确实在每一步都是一种新的形式。③ 这种应用不是一个直觉的行为问题，而是一种决定的行为。法国数学家博雷尔（Emile Borel）作出了同样的评论，他提问的问题是，我们从一个整数到与其相邻的另一个整数进行的运算是否每次都真的相同呢？从一种特定的观点来看，是相同的，但从另一种观

① Bouveresse, Jacques, *La force de la règle. Wittgenstein et l'invention de la nécessité*, Paris: Editions de Minuit, 1987, p. 23.

② Bouveresse, Jacques, *La force de la règle. Wittgenstein et l'invention de la nécessité*, Paris: Editions de Minuit, 1987, pp. 35-36.

③ Bouveresse, Jacques, *La force de la règle. Wittgenstein et l'invention de la nécessité*, Paris: Editions de Minuit, 1987, p. 36.

点来看，是不同的，因为 1 加 1 得到 2，1 加 2 得到 3，所以，没有相同的运算。即使在像算术运算这样一个有规则的体系中，一次新的应用也每次都包括了对它的一种决定。因此，最佳数学具有的特征是它的规范特性。数学不是描述的（descriptive），而是规定的（prescriptive）。在维特根斯坦的意义上，数学是规范的，在某种程度上与符合逻辑规范的任何真理无关。数学创造了一种特殊类型的必然性，即创造新事实的必然性，我们做事的特殊方式的事实的必然性，这种做事方式是我们的，似乎不可能是其他种类的。换言之，这种必然性是人类学的必然性，它取决于自然界的事实，自然界的事实呈现出必然性，因为没有其他的替代选择——而且它呈现出内在的必然性（从外在的观点来看，它可能扮演了偶然性的角色，而不是必然性的角色）。如果我们站在规范性观念的立场上，试图在数学哲学的领域内总结这种概述性研究的最有意义的结果，那么，好像是这样的观念：相同规则的递归应用（recursive application）在每一步都处于不同于前面步骤的新境情（new situation），并且，在这种应用中有些事情是无法预见的。这种相当无法预见的结果，似乎与人类思维和行动（无疑包括诸如医学之类的最远离数学的情况在内）的任何方面都很相关。

三、从语言的分析哲学来看规范性的本质

现在我们试图进入当代哲学研究的另一个领域——语言的分析哲学（the analytical philosophy of language），并提出下列问题：与事实陈述相比，规范的陈述（即"应当"型的各种陈述）有时是不可还原的吗？我们运用分析哲学的逻辑方法能对这种不同之处作出说明吗？威奇伍德（Ralph Wedgwood）在近期出版的关于规范性本质的书中研究过这个问题。作为一名分析哲学家，威奇伍德考虑了作为一种事实的规范陈述或判断的存在性，事实是需要说明的，但拒绝提出这样的"规范性"术语的任何定义，也许害怕定义像规范性之类的概念会把我们导向一种神秘的实在论。他把他的任务界定为，通过以逻辑的和语义的方式来分析规范术语的真值条件，对规范术语的普通理解给出理论说明。这项任务与规范性问题的更多自然主义的进路完全不同，在像认知神经科学和心理学之类的自然科学中越来越体现出自然主义的进路，这些自然科学研究不同的理解问题，即我们作为人类如何创造规范，如何可能改变规范，而不仅仅是在我们的语言中遵循这些规范。

站在语言的分析哲学的立场上，规范的思维、规范的知识、规范的真理、规范的判断、规范的直觉、规范的信念等，都是事实问题——与不存在的实在有关的话语、语言和思维的事实。这些很特殊的陈述类型的语义内容是什么呢？规范

陈述的语义说明是否需要那些本身是非规范的陈述或概念呢？换言之，语言和思维中的规范性是否可还原为事实属性呢？假使结果是不可还原的，我们就会面临一个逻辑循环。威奇伍德自己的论点是，我们能够避免这种循环。相反，像麦克道维尔（John McDowell）、德沃金（Ronald Dowrkin）或帕菲特（Derek Parfit）那样的其他哲学家在他们自己的"寂静主义的"论题（quietist thesis）中声称，对规范的论述作出非循环论证的说明是不可能的。威奇伍德信赖克里普克（Saul Kripke）对指称的论证，是为了为他自己的主张辩护："我将在不使用……情况应该是什么的思想或陈述概念的前提下，对情况应该是什么的思想或陈述的认识作出说明。"①为了完全像分析其他类型的陈述一样来分析规范的陈述，只能运用像真理、证明、信念、决定之类的很普遍的规范属性。的确，规范性不能缺少这类陈述："必定，如果人是理性的，如果有人判断，'我应该做这件事'，那么，他也就打算做这件事。"必然性和合理性都是规范的属性，因此，用完全非规范的术语说明规范性的方案是毫无意义的。

然而，威奇伍德断言，人们在不参照相同类型的规范属性的前提下，能够分析"我应该做这件事"这种形式的特殊的规范陈述。作为一名语言哲学家，他注意到，规范的术语是对语境极其敏感的，在不同的语境中表示不同的概念。②在这些条件下，意向性的哲学论题开始加入了这种讨论。的确，鉴于外在论强调规范性的外在的客观来源，所以，威奇伍德把"规范判断的内在论"描绘成与外在论相对立，他自己的这种规范性本质的理论，强调了规范判断的主观的第一人称维度。根据威奇伍德的内在论的理论，在规范判断和对行动的实际论证与动机之间存在着一种根本联系。规范判断或陈述能够被描绘成是第一人称判断的限定条件是，行动是非常依赖于语境的，而且，应该是适当的和合理的。规范判断能够在实际推理过程中根据其发挥的规则作用来理解。一个正确的行动计划的概念是由威奇伍德在不允许任何不确定的条件下提出的。在确定的特殊条件下，规范概念或判断的作用显然很容易能被描述成是承诺或责任，这使得实现有计划的行动成为必要。这是一个相当理想的模型，在不确定的条件下，应该是完全不同的。

于是，意向性概念和第一人称的观点必定进入了这个图像。最近，哲学家们已经非常多地讨论了"意向的就是规范的"这一口号。从前面的分析来看，规范性显然代表了一种特殊形式的意向性，它与对行动的实际判断、倾向和承诺相关。意向性通常从本质上包括某种规范性吗？包括胡塞尔和更新近的戴维森在内

---

① Wedgwood, Ralph, *The nature of normativity*, Oxford：Clarendon Press, 2007, p. 20.

② Wedgwood, Ralph, *The nature of normativity*, Oxford：Clarendon Press, 2007, p. 23.

的许多哲学家已经捍卫了这种观念。根据威奇伍德的观点，"意向的事实有一部分是通过规范的事实建构的"①。之所以是这种情况，似乎原因在于，所有的意向状态或属性都包括两个要素："①由概念组成的内容；②关于那个概念的心智关系或态度（比如，信念、愿望、希望、恐惧等）。"② 例如，相信能用两种方式来描绘。这种信念是正确的，当且仅当它的内容是真的，其次，（在特定的情况下）这种信念是合理的，仅当这些情况决定了它的可能性时，它的内容才为真。③这种意向的规范要素能在意向的概念内容和关于这个特殊概念的意向态度之间的关系中找到。按照他的"规范判断的内在论"的理论，威奇伍德希望避免另外一些危险：把规范的属性完全还原为自然属性（自然主义）的危险和人类规范性的纯形而上学的神秘图像的危险。

四、关于人类规范性来源的某些经验研究

然而，正如我们从值得提到的某些研究纲领和报告中越来越频繁地看到的那样，内在论为用自然主义的、心理学的乃至神经生物学的术语说明规范性留下了余地。相当引人注目的是注意到，在发展认知心理学（研究幼儿）的自然主义的规范性进路中，规范性被认为是一种力量，也是一个事实。我们能期待在语言中发现作为一个事实的规范性描述及其逻辑的和分析的属性（由于规范性包括意向性和合理性，不完全是无法预见的结果）——这种描述是不够的。我们如何创造这些规范的规则呢？如何能感觉到和接受它们呢？德国莱比锡的马克斯·普朗克进化人类学研究所的研究者们研究过这些问题。参与研究的儿童的年龄是 2 ~ 3 岁。这些孩子们能理解和遵守游戏规则。此外，他们能注意到不遵守这些规则的情形，并还会提出抗议。在幼儿身上就能发现"规范的意识"④。在禁止对孩子实施暴力的情况下也能发现类似的经验研究，这表明了规范行为的存在性。在最近的一本书中，神经生物学家琼·皮埃尔·尚热（Jean-Pierre Changeux）试图表明，关于人类大脑的事实的知识，对规范和价值之来源的科学理解和对社会中的规范行为的实际发展，如何有用的问题。对规范行为和道德情感的经验研究的例子包括禁止对孩子实施暴力和社会约定与道德需要之间的区分。在不同文化中和不同社会约定中进行的对孩子的研究表明，3 岁以后，孩子把不遵循宗教规则

① Wedgwood, Ralph, *The nature of normativity*, Oxford: Clarendon Press, 2007, p. 159.

② Wedgwood, Ralph, *The nature of normativity*, Oxford: Clarendon Press, 2007, p. 161.

③ Wedgwood, Ralph, *The nature of normativity*, Oxford: Clarendon Press, 2007, p. 162.

④ Rakoczy H., F. Warnecken and M. Tomasello, "The sources of normativity: young children's awareness of the normative structure of games", *Developmental Psychology*, 2008, Vol. 44, No. 3, pp. 875-881.

判断为是可接受的，但他们认为，违反基本的道德规则是无法接受的。① 然而毫不奇怪的是，正如我们将会看到的那样，人类规范性的来源问题主要在道德哲学中被加以研究。

五、关于人类规范性来源的道德哲学

科尔斯戈德（Christine Korsgaard）在她的《规范性的来源》一书中写道："关于人类生活的最显著的事实是，我们拥有各种价值。我们想方设法让事情更好些、更完善些，因而当然与它们本来的样子不同；我们设法让我们自己更好些、更完美些，因而当然与我们本来的样子不同。我们应该这么做吗？我们从哪里获得超越经验世界，对经验世界提出质疑、作出判断，说它不符合标准，不应该是如此的这些观念呢？显然，我们从经验中得不到这些观念，至少不能通过简单的思路来获得。于是，很令人困惑的是，与我们自己的世界观不同的这些世界观向我们呼吁，告诉我们，事情应该像他们的世界观一样，而不是事情本来的样子，我们应该使事情成为这样的。"② 按照科尔斯戈德的观点，"价值的真相是神秘的"，是需要哲学探究的一个秘密。哲学家提供了四种解答：①实在论者认为，规范与价值是真实情况（柏拉图、亚里士多德）；②唯意志论者认为，在自然状态中，根本没有对错，规范和价值是通过一种权威和义务强加给我们的，这种权威和义务源于有合法权威性来指挥道德行动者（moral agent）③ 的某个人的命令（霍布斯）；③规范和价值是通过一种内在反思——"反思认同"（reflective endorsement）理论，意思是个人赞同——发现的（休谟）；④规范和价值的内容可能是普遍的，是通过自由意志的自主行为发现的——道德律就是行为者自己意志的法则，其主张是使他自己做好准备的那些主张。④ 康德和像罗尔斯那样的当代的康德主义者是这种趋势的代表人——康德的形式普遍主义（formal universalism）是著名的。

这四个基本的哲学解答回答了一个给定的特殊问题，科尔斯戈德把这个问题定义为"规范的问题"，这个问题可能是根据下列方式提出的："伦理标准都是规范的。它们不只描述了我们对自己行为的实际调节方式。它们还对我们提出要

---

① Changeux, Jean-Pierre, *Du vrai, du beau, du bien. Une nouvelle approche neuronale*, Paris: Odile Jacob, 2009.

② Korsgaard, Christine, *The Sources of Normativity*, Cambridge: Cambridge University Press, 1996, p. 1.

③ 目前在国内学术界有的人把"moral agent"译为"道德主体"，有的人译为"道德责任人"，有的人译为"道德行为者"，有的人译为"道德代理人"，有的人译为"道德者"，本文采纳了"道德行动者"的译法是为了与当前科学知识社会学等领域内的译法相一致。——译者注

④ Korsgaard, Christine, *The Sources of Normativity*, Cambridge: Cambridge University Press, 1996, p. 19.

求；它们命令、迫使、建议或指导……因此，我们希望理解的正是这些规范主张（这些概念的权利是为我们提供法则）的威力。"① "当我们为道德寻找哲学基础时，我们不只是寻找对道德实践的一种说明。我们同时提出这样的问题：如何对道德向我们提出的要求作出辩护。这就是我所说的规范的问题。大多数道德哲学家渴望对道德给出一种说明，这也就回答了规范的问题。但如何能确立规范性的论题很少被直接或单独称为一个自身具有的话题。"②那么，我们能从哪里找到关于规范问题的答案呢？按照科尔斯戈德的观点，这个答案 "一定以一种深刻的方式诉诸我们对自己的理解，即诉诸我们对自己的认同感"③。如果我们在不同情况下回到自身，我们应该能找到一种答案。科尔斯戈德自己对规范问题的回答最终是康德主义的（而且，在威奇伍德的讨论中也能找到某些康德主义的迹象）——但限定条件是，形式普遍主义的内容是由反思认同赋予的。"人类意识的反思结构要求你认同自己选择的某个法则或原则。这需要你自己成为自己的法则。那就是规范性的来源。因此，这个论证刚好表明，康德所说的确实是：我们的自主权是义务的源泉。"④ 最后，作为这种康德主义和反思认同相混合的一种可能结果，我们断定，规范性来源的四种哲学理论都是正确的，它们是人类规范性的总图像的所有组成部分。⑤

六、规范性与自主性：医学的含义

自主性概念是一个哲学概念，它经历了从集体和政治层面到更个体层面的意义变化。沃尔夫（Christian Wolff）于 1757 年在他的关于道德和政治哲学的书中，在一种状态的独立或自决（self-determination）的意义上，使用这个术语。尽管卢梭没有用这个术语，但他被认为是在遵循自我治疗法则的意义上负责提出自治权的观念。道德哲学的这种动机甚至比康德的著作中的动机更强。康德意义上的自治权意味着发现普遍法则本身的意志的自治（self-rule）。最近，在关于健康保健和健康经济学的更加实用的讨论中，把自治观（idea of autonomy）引入到医学伦理中。这种自治观是病人靠自己而不是靠别人来承受的能力。正如我们所

① Korsgaard, Christine, *The Sources of Normativity*, Cambridge: Cambridge University Press, 1996, pp. 8-9.

② Korsgaard, Christine, *The Sources of Normativity*, Cambridge: Cambridge University Press, 1996, p. 10.

③ Korsgaard, Christine, *The Sources of Normativity*, Cambridge: Cambridge University Press, 1996, p. 17.

④ Korsgaard, Christine, *The Sources of Normativity*, Cambridge: Cambridge University Press, 1996, pp. 103-104.

⑤ Korsgaard, Christine, *The Sources of Normativity*, Cambridge: Cambridge University Press, 1996, p. 165.

看到的那样，规范性概念有相当不同的背景。事实上，它包括人类的集体见识，即人与人之间的基本关系。它意指一个人即使降低条件来为自己和他人确立新的规范时的能力。霍金是一个极端的事例，因为他保持了自己规范的创造力，他作为一个人，不仅有能力控制自己（至少控制他的内心），而且有能力与他人交流，丰富他人的生活。

这里值得把健康概念作为一个动态的概念提出。人们已经多次强调，康复并不意味着完全回到从前的健康状态，因为机体的整个功能已经发生了改变。规范性的观念与动态的健康概念是一致的。在无法治愈的重度退行性慢性病或遗传病的病例中，规范性有更重要的和更有趣的内容——显然是动态的内容。让我们用诊所的病例说明这一点。多发性硬化症在两次发作期间，病人可能有时间适应他或她的病变，因此表明他或她保持了规范性。弗赖雷克病（Freireich disease）是一种控制运动的神经性退行性疾病，根本无法治愈，也没有希望能治愈。然而，身患这种疾病的残疾人可能一直有极强的愿望去工作、得到承认、作出科学贡献以及与他人合作。身患肌肉萎缩症（一种脊髓运动神经元退行性疾病，最终导致呼吸肌麻痹）的人明白，通过技术装置来改善他们的生活，这些技术装置甚至以一种更加创新的方式允许他们能与其环境互动，在他们有能力为他们的直接环境贡献人类经验的意义上，体现了一种不可思议的人类经验，保持了他们的规范性，尽管他们在很大程度上丧失了自治权。[①]对于对意外的外界干扰更敏感的生理功能退化的脆弱综合征的老年人来说，同样如此。他们的自治权降低了，但他们的规范性却提高了，这取决于他们与自己机体的互动能力的质量。

规范性是一个哲学术语，它很好地描述了几乎每个方面的人类经验。对人的规范性的这种新认知，对改进医学实践和保健方面的许多社会现状和技术现状来说，可能是有用的。

（成素梅 译，童世骏 校）

---

① Le Forestier, forthcoming in *Medicine Studies*.

Power, Trust, and Risk: Some Reflections on an Absent Issue

Harald Grimen

When health professionals in public reflect on themselves and their relations to patients, discussions of power are rare, with some exceptions (psychiatry and empowerment). Power is most conspicuously absent in two (overlapping) discussions, where it should have been discussed: those about trust and about the ideal physician-or nurse-patient relationship. For years, I have as an outsider followed these discussions in medical and nursing journals, and the rarity of discussions of power is striking. This is not unique for health professionals. Power is also conspicuously absent when college teachers—like me—in public discuss their relations to students.

A philosopher reflecting on this situation is led to ask a question that, is probably unanswerable in general terms: Why do not professionals in public discuss power differentials inherent in their work? And what are the consequences of this lack of public discussion? Health professionals (e. g. , physicians, nurses, psychologists, physiotherapists, and rehabilitation specialists) are socialized to see themselves as beneficial helpers and not as powerful gatekeepers or controllers. And they are beneficial helpers. But they are also gatekeepers and controllers. Differentials in knowledge and opportunities of control are, moreover, essential to what it means to be a professional. A modern society could not function without uneven distribution of knowledge and control. Professionals are there because they are assumed to have superior knowledge in an area, for example, medicine or law.

A lack of public reflection on power can hamper a serious moral discussion—with health professionals as important participants—of various institutional forms of their relations to others. This ought to be a cause of concern. It can lead to unrealistic views of how, for example, physician-patient interaction can be organized, which can do more harm than good if implemented. Moreover, it leads to distorted views of trust and risk. Power, trust, and risk are connected, and these concepts are not socially, morally, or politically innocent. It can also lead to blindness to the possibility of certain kinds of change. At the core of modern health care there is a nexus of power, trust, and risk that cannot change if deep-seated structural features of health care do not change first. I am unsure if we should opt for a radical change of these structures. Radical change may not lead to better health care. It may be better to opt for more self-consciousness among professionals about power and find more humane institutional forms of this nexus, not its abolition. Public discussions about power—where health professionals participate—are then crucial.

I first discuss some concepts that are crucial to my argument (Section I). Then I outline the nexus of power, trust, and risk in health care (Section II). I discuss an example of the absence of power (Section III) and how power can be approached (Section IV). Finally I briefly explain why we should not opt for a radical change (Section V).

I

Analyses of trust that neglect power are naive. Analyses of power that neglect trust are shallow. But there are no uncontroversial definitions of trust, risk, and power. I shall not define them, but I shall say something about what someone who trusts, takes a risk, or exercises power de facto does, to show how closely connected trust, risk, and power are. [1] If A trusts B, then:

1. A leaves or has something, X, in B's custody for a period of time.

2. A transfers—always de facto, sometimes de jure—discretionary powers over X to B for this period of time or is in a situation where B has such powers.

3. A values X.

4. A expects that

(a) B is not going to do something that harms A's interests.

(b) B is competent to take care of X according to A's interests.

(c) B has the necessary means to take appropriate care of X.

5. A takes few precautions against B's misuse or careless use of his discretionary powers over X.

Trust has a transactional side. A trustier often leaves something, for example, money, children, property, information, or his body, in someone's custody. By doing this, he transfers discretionary powers to others. When X is in B's custody, B can decide what should happen to it, whether or not A knows or wants it. When the parents are absent, a babysitter has de facto power over the child, regardless of what they believe about this. The trustier expects that the trustees are not going to harm him but has no guarantee that they will not do that. He expects them to be competent and to have adequate means but has no guarantee that this is so.

What he leaves to others can be misused, treated in a careless or incompetent way, or used against him. One person's trust can become another person's power base. The trustee can use what he has got in his hands to make the trustier do what he wants. The ability to make others do what one wants—often against their wills—is a core element in power (Weber 1976: 28). There are many ways of doing that, for example, to threaten to harm or misuse something that they value. Trust is risky and can have adverse consequences for the trustier. By trusting he makes himself vulnerable.

This transactional side of trust is important in interaction with professionals. We leave information, money, wills, children, our own bodies, and other things in the custody of professionals for shorter or longer times. Our willingness to leave our bodies in the custody of health professionals enables them to do their job. If I did not trust my physician, I would not permit him to do what physicians do with people's bodies. As Hall and Berenson say, trust makes beneficial power exist in health care (Hall and Berenson 1998). Beneficial power is necessary to get work done. But trust also creates the structural conditions for power, which need not be beneficial. I call these connections the "nexus of trust, risk, and power". This nexus is found in all interaction between laypeople and professionals. But it can have different forms, dependent on the tasks of the professionals and the institutional settings in which the interaction takes place.

In modern health care systems its forms are determined by professional autonomy combined with the structurally inferior situation of patients. [2] Professional autonomy leads to a massive use of discretion and clinical judgment, which are difficult to make accountable. Nonaccountable discretion is a potent source of power, especially in legal contexts and in trust-based interaction. The patients' structural inferiority has many sources (Grimen 2001). There are, first, knowledge differentials between professionals and patients. The knowledge gap is probably greater in health care than in any other consulting profession, and in my opinion it is increasing, not decreasing. [3] This makes it difficult to challenge professional judgments. Second, patients have often only extreme alternatives to trusting the competence, technology, and professionalism of health care staff. They may be forced to trust what they get. With Hirschmann (1970) we could say that they lack exit options, have a weakened voice, and can be forced to loyalty. In addition, health professionals often act as gatekeepers to goods and services that patients need, for example, specialist services or disability pension. Gatekeepers can have mixed interests, and not all of them need be in accordance with the patients' interests. Finally, when patients interact with health professionals, they often do not function in a normal way physically or mentally. They may be confused, afraid, or in shock, or they may have lost parts of their linguistic and cognitive capacities, as stroke patients often do. They can therefore be easier to manipulate than healthy persons. Professional autonomy is not specific for health care. Nor is structural inferiority of users and forced trust reserved for health care. But these issues are more pronounced in health care than in other contexts. It is therefore important to reflect on power, trust, and risk in health care. This could, for example, lead to a greater sobriety in discussions about the relations between health professionals and patients.

The links among power, trust, and risk in health care in one sense belongs to the human condition. It would not be there if we did not need medical help. But its specific forms stem from the social structure of health care systems and the kind of knowledge and technology applied in the treatment of diseases. If health care were not in the hands of modern-style professionals, with autonomy and specific kinds of knowledge and technology, the nexus would have had another form. It has a different form when people interact with traditional healers. We can rationally discuss if this form is more humane than the present one. It is morally important to discuss whether some institutional forms of the nexus are more humane than others. Institutional forms should be more frequently discussed in professional ethics than it is. The conclusion need not be that the nexus should be abolished. It could also be that some forms of it are more humane than others, and that we should opt for them, rather than for a radically different health care.

An example of what I mean by a discussion of institutional forms is the following one: An editorial in the *British Medical Journal* in 2001 proposed that the NHS should implement the principle of subsidiarity: "Decision making should be located as closely as possible to the place where actions are taken. The performance of organizations is most effectively governed when subsidiarity is applied." (Welsh and Pringle 2001: 177) This principle originated in Catholic social philosophy. It is central in EU jurisprudence. The editorial presented one possible interpretation of it. Implementing it in health care could strengthen the autonomy and power of physicians, nurses, and psychologists. They work closer than any other professionals to the place where medical actions are taken. Subsidiarity can also conflict with new forms of accountability based on profession-external audits. A moral discussion of subsidiarity in health care must to a large degree be a discussion about the institutional forms that could result from implementing it: Should we want health care institutions that give physicians and nurses more power?

II

I shall first discuss the nexus of power, trust, and risk more in depth. It is created by the professional autonomy of consulting professions combined with the manysided structural inferiority of patients. This link is set up by how modern health care institutions work, and it makes patients very dependent on professionals. It can only change if some basic structures of modern health care change (see Section V). Thus, as I argue in Section III, some proposals for how to organize the relations between health care staff and patients are unrealistic and may create more harm than good. There are, for example, limits to how far patients can be dialogue partners with professionals. If one goes too far, the patients' dependence on the professionals can be difficult to recognize and handle in a morally sensitive way. Conversations between patients and health professionals cannot be dialogues between equals. But there is also reason to be suspicious about proposals that may deepen the patients' dependence. Subsidiarity can have this effect. It can make it more difficult to make physicians, nurses, or psychologists accountable, by widening their areas of discretion (Section I).

1. I shall first discuss professional autonomy. According to Freidson, the core of professional autonomy is control over the technical aspects of work, that is, over how to organize work operations and their contexts. An occupational group can have professional autonomy, even if does not control the economic, administrative, or educational aspects of its work. Freidson's example is Soviet physicians, who did not control the economic and administrative aspects of their work, but only the technical aspects. Freidson's view is contested. But it pinpoints one important aspect of what must change if the nexus of power, trust, and risk in health care is to erode in its present form: Health professions must then lose control over the technical aspects of their work.

| Power, Trust, and Risk: Some Reflections on an Absent Issue |

An occupational group in control of the technical aspects of its work has considerable space for discretion and judgment. It can handle contingencies (on behalf of others) according to its view of the world (including a view of its clients' needs) . It has considerable liberty to shape its worldview. It can shape the important routines of daily work according to its worldview and needs. Discretion and judgment are difficult to make accountable. To have discretion in consulting professions (like physicians, nurses, lawyers, or social workers) is to have power to decide for or on behalf of others. Unaccountable discretion is a source of unaccountable power. But it is difficult to imagine what professional autonomy could be, if it did not contain wide spaces for discretion and judgment. I guess that the areas for discretion in health care are widest in two kinds of contexts: When professionals assess patients' future functional abilities for eligibility to welfare benefits allotted according to need (e. g. , disability pensions), and when they assess risk, for example, in surgery, ambulatory medicine, or general practice. There is also much discretion in other areas of health care, but I use these two contexts to illustrate the point.

In the first kind of case there is a wide scope for discretion both because welfare laws are open ended, because physicians and nurses are not particularly competent to assess future functional abilities, and because it is difficult to predict the development of such abilities. Physicians and nurses are not necessarily trained to do this. In the Norwegian system physicians are gatekeepers to welfare benefits distributed according to medical need. In some other countries this is more the domain of other health professionals, for example, physiotherapists, occupational therapists, or rehabilitation specialists. They may be better trained, but there is still a wide space for discretion. Welfare benefits are at the margins of medicine, but at the heart of what T. H. Marshall called citizens' social rights, in contrast to civil and political rights (Marshall 1992: ch. 2) . In social rights cases juridical discretion and clinical judgment meet in unpredictable and uncontrollable ways. Such cases are a growth area in modern welfare states, and health professionals play complex roles in such cases.

Assessment of risk, however, is at the core of medicine, concerning everything from prognostics to side effects of medication. Much of it is routinized. Routines are double-edged swords. They facilitate work but restrict the field of vision. Routines can bring both mental comfort and medical (and juridical) disaster. This is the paradox of routinization: What makes routines helpful also makes them dangerous. In this second kind of case there is a wide scope for judgment mainly for three reasons. First, to be a patient is to delegate to others to assess risks. We all do that when we comply with the prescriptions of health professionals. Even the most well-informed patients must rely on the ability of others to assess what is risky. A caricature of modern health care as I see it as an experienced patient (the caricature is not too far off the point) is that in such systems there is always someone with the autonomy and power to detect, assess, take, conceal, create, or eliminate risks for me. I must trust that they know what they are doing and that they can and will correct their own mistakes, if they make mistakes (as they often do). Second, in many of the situations that may arise in health care, for example, in ambulatory medicine or advanced surgery, it is impossible to regulate health professionals' behavior in detail, and it would be foolish to try to do so. Sometimes regulation can do more harm than good. The professionals must—in a literal sense—handle unpredictable contingencies. Then personal judgments could be the best we can hope for. Third, health care is often based on uncertain and contested knowledge and knowledge that may be more outdated than it ought to be. General practitioners or the average hospital physicians are normally not at the research front.

To trust someone in health care is to delegate power to assess and take risks. It is to transfer (de facto) the right to assess what is risky to professionals whose judgments can—and often do—differ significantly from one's own. A patient leaves the assessments of what is risky for him in the custody of professionals for some time. Being anesthetized in the operating theater, one can only hope that the professionals know what they are doing and have a sound judgment of risk. A space for discretion (Dworkin's "hole in a doughnut" [Dworkin 1978: 31]) may be wide or narrow. It is also often a space for the exercise of power. Moreover, the worldviews, opinions, and judgment of professionals tend to differ significantly from the worldviews, opinions, and judgment of their clients. Professionals are often unaware of this gap between their views and those of their clients. Health care staff tends to think that patients think like themselves. But that is not generally true.

2. I shall then say something about the structural inferiority of patients. Health care contains a set of asymmetric relations in which both power and risk thrive. Patients are in a structurally inferior position vis-a-vis health care staff. This structural imbalance cannot be conjured away by good intentions. Some people need help; other, less helpless and—hopefully—more competent people can provide it. Some people need goods, for which other people are gatekeepers. Those in need have little to offer in return, except gratitude. Some people know much; others know less about the relevant issues. Knowledge

differentials mean that the less knowledgeable cannot challenge the definition of the situation—to use W. I. Thomas's term—which informs interaction. Moreover, because they know less, they must trust others from a position of epistemic inferiority. They do not know the road and must trust that others know it, without being able to challenge the choices made at each crossroad. Finally, patients are in the hands of professionals' ability and will to detect and correct their own mistakes, and they are vulnerable to mixed motives on the side of those whom they trust.

These are some elements of the nexus of power, trust, and risk at the core of modern health care. It is an institutional form of a fact of the human condition, namely the fact that people get ill and need medical help. Health care in modern societies is in the hands of specific kinds of professionals, controlling a kind of knowledge that transcends everyday experience (Fredriksen 2002) and that they can apply according to their autonomous judgment. Patients must trust their competence, professionalism, and technology. But patients trust from a structurally inferior position. They leave their lives in the custody of professionals, whose judgments and actions they often lack opportunity to challenge. This nexus is easy to overlook. An important question is if we should wish it away. I do not think we should. But we should discuss different institutional forms of it. The moral issues that it raises concern power, dependence, and vulnerability. To neglect these issues can create a lot of wishful thinking.

III

There is a lot of wishful thinking when health professionals in public discuss themselves and their relations to patients. In much of the literature where they discuss such issues, the nexus of power, trust, and risk is neglected. I shall, as an illustration, briefly discuss an example that is typical and influential.

One place where discussions of power ought to be central is in the texts where health professionals discuss the proper relations between themselves and patients. This is a kind of texts—found in professional journals—where health professionals, mostly physicians and nurses, discuss themselves and their roles. In the last decades there has been a growing concern about liberating patients from paternalism and finding new models for how health professionals and patients should interact. Key words are patient autonomy, empowerment, free choice, and informed consent. A much-quoted work is Ezekiel and Linda Emanuel's article about models for patient-physician relations

（Emanuel and Emanuel 1992）. If anything is a reference article in these discussions, this one is. They outlined four models: the *paternalistic*, the *informative*, the *interpretative*, and the *deliberative*. I shall not say much about the first three models, but concentrate on the deliberative one.

In this model "the physician acts as a teacher and friend, engaging the patient in dialogue on what course of action would be best" （Emanuel and Emanuel 1992: 2222）. Each model entails a concept of patient autonomy. In the deliberative model, autonomy is a kind of moral self-development: "The patient is empowered not simply to follow unexamined preferences or examined [should it be "unexamined"?[4]] values, but to consider, through dialogue, alternative health values, their worthiness, and their implications for treatment" （Emanuel and Emanuel 1992: 2222）. Here the physician is teacher and the patient student. The Emanuels profess a kind of dialogical pedagogic, which aims at moral self-development through the examination of preferences and values. But it is the patient and not the physician who is to develop morally. In spite of dialogue, there is a teacher and a student. The Emanuels' model does not conform to deliberative models as discussed in theories of democracy, where all participants are assumed to develop morally. It does not conform, for example, to Habermas's practical discourses （Habermas 1991）. There are, I think, four reasons for this. They do not point out a weakness of deliberative models. The weak spot is the claim that such models can conceptualize the physician-patient relationship.

1. The Emanuels claim—fully justified—that physicians are teachers and patients are students. Epistemic asymmetry necessitating teaching is unavoidable in physician-patient interaction. But epistemic asymmetry has no essential place in deliberative models of decision making. To square their model with reality, the Emanuels must—unwittingly—introduce another professional dyad to the physician-patient relationship, namely the teacher and the student. This dyad does not fit deliberative models of decision making. It does not remove but presupposes epistemic asymmetry. Deliberative models presuppose epistemic symmetry between participants. They contain no unavoidable teacher-student dyad. But epistemic asymmetry is a basic feature of the interaction between patients and health professionals.

2. Deliberative models fit only parts of what is important in the relationship between physicians and patients, namely some kinds of value questions (mostly those pertaining to a good life) . They do not fit all kinds of values (e. g. , those pertaining to the goal of health care), or the cognitive aspects of the relationship (e. g. , theories and facts

about the causes of diseases). Such models are better suited to dialogues about some kinds of values than to discussions about diagnoses and treatment. When the Emanuels introduce empowerment in the deliberative model, it is mainly handled as a problem of preference and value examination, not as a problem of decision-making. Patients become empowered if they acquire a more clairvoyant relation to their preferences and values. This is a weak kind of empowerment, which is consistent with leaving the physicians' decision-making power intact. Values count, but physicians decide. [5] This is probably the only kind of empowerment that can be squared with using the deliberative model to conceptualize physician-patient interaction. For some values it makes sense to claim, as the Emanuels implicitly do, that overcoming paternalism creates symmetry between health care staff and patients. Overcoming paternalism about moral issues simply means to remove the idea of moral expertise from the relations: Physicians (and other health professionals) lose their role as experts on the good life, and in this sense patients become empowered. But with diagnoses and treatment, this does not make good sense. There is no problem of paternalism connected to the cognitive aspects of diagnoses and treat-ment. But there is a problem of power and knowledge. Deliberative models are not well adapted to some of the really tricky issues of physician-patient interaction. Such models have an idealist ring. The Emanuels actually discuss the problem of compliance in disguise. [6] They do not seriously entertain the possibility that patients could convince physicians that the basic values of health care or biomedical theories of disease could be incorrect. Although the Emanuels do not say so, the point of the patients' moral self-development is compliance, not moral revolution of health care or scientific revolution of medicine. But in a really symmetrical dialogue, this possibility should be open.

3. Some patients are unable to examine their preferences and values. Even reasonably intelligent people may be unable to do this when they are seriously ill. Under the stress and distraction of pain and illness they may be unable to absorb complex information and make autonomous decisions. This is one of the sources of the structural inferiority of patients (see Section I). Moreover, some people do not want to examine their preferences and values when they are ill. Patients from some cultures may not want to deliberate in such situations. They expect and want physicians to make decisions. This fits with their expectations of the role and purpose of a physician. It may be their way of being autonomous and making free choices. [7] But it does not fit with the Emanuels' claim about the importance of self-scrutiny.

4. The Emanuels more or less neglect the structural imbalances in physician-patient

interaction. The nexus of power, trust, and risk then unwittingly reappear in the deliberative model in the shape of teacher and student, another structurally unbalanced dyad. To the extent that power is given attention, it is connected to one of the models from which the Emanuels distance themselves, namely the paternalistic model. They welcome the move away from paternalism, something with which it is easy to agree. But they seem to think that the move to deliberation will eliminate power differentials in physician-patient interaction. Some of the nursing literature builds on the same assumption to an even stronger degree. Such an assumption can, however, only be defended if one ignores power. If one does not, the assumption is kind of wishful thinking.

When power is discussed in this kind of medical and nursing literature, it is either discussed as power imbalances between health professions, or as power struggles between professionals and management. Typical in this respect is Gray, who, in a comment on managed care, says: "Managed care shifts much control of the flow of dollars and patients away from physicians to large organizations that have strong economic goals and the power to influence patient care in pursuit of those goals. This combination of purpose and power is a major source of doubts about the trustworthiness of these organizations, particularly because they reduce the power of health professionals who profess adherence to the fiduciary ethic. " (Gray 1997: 35) Otherwise, power in health care is mostly discussed by sociologists and anthropologists (and by dissidents within the system). [8] But their views do not seem to seep into the mainstream discussions among health professionals.

The idea that patients and health care staff can become equal in a dialogical relationship or alliance partners, as some physicians suggest, is wishful thinking. Even if the reduction of paternalism is laudable, it does not mean that the relations become symmetrical. Paternalism is less important than many discussants think. Nonpaternalistic relations may soothe our feelings, but feelings of well-being do not remove power imbalances. To make paternalism a main source of power in health care is misleading and can conceal more sinister sources of power. The deliberative model underestimates how dependent patients are on professionals—in spite of free choice, patient autonomy, and informed consent. It underestimates both the structural imbalances in the relationship and the differences between a free and symmetrical discussion of values and a free and symmetrical discussion of diagnoses and treatment. The first is laudable, because it presupposes that there is no moral expertise. The second may lead to disaster, because it presupposes that there is no other kind of expertise either. But there is. I may dislike getting a cancer diagnosis. But its correctness is not an issue of deliberation between my physician and me. I think my physician knows more about this than I do, and I hope I am right. And I do not think we should wish for a health care system where it were an issue of deliberation. I would not like a health care system where I, being incompetent in medicine, had a really fair chance to convince my physician that his diagnoses were wrong. I prefer a system where I must live with bad feelings but have the opportunity of being treated by competent people. But I also prefer a system where the professionals are aware of how dependent I am on them as a patient. Blindness to power can perpetuate morally objectionable institutional forms of the nexus of trust, risk, and power.

IV

I have discussed what the nexus of trust, risk, and power is in health care, why health professionals should discuss it, and I have given an example of what I mean by claiming that it is often neglected when health professionals publicly discuss their roles. It should be discussed because it is a system of structural imbalances between professionals and patients. To neglect it can lead to indulgence in wishful thinking about how health care can be organized and to the perpetuation of more sinister forms of power than paternalism. But one must also ask how it should be discussed. It should be a crucial part of the education of health professionals that they are sensitized to trust, risk, and power. But that is not my concern here. I shall make three other points: First, it should be discussed as an issue about institutional forms. Second, it should be discussed with an awareness of the role, which concepts like "trust", "risk", and "power" can play in social life. Third, it should be discussed with a concept of power that can account for its main features.

1. To suggest that the NHS in Great Britain should be organized according to the principle of subsidiarity (see Section I) is to propose an institutional form. It will probably transfer more power to physicians, psychologists, nurses, and other health professionals close to the sites of medical actions. Managed care is another example. It may, as Gray claims (see Section III), reduce the power of health professionals but also the trustworthiness of health care organizations. The Norwegian system of giving diagnoses a price and rewarding hospitals economically on the basis of the prices of diagnoses is a third example. It may tempt hospitals to concentrate on diseases that bring

money from the state and neglect less profitable diseases. Still another example is the role of general practitioners as gatekeepers to specialist services. In Norway initial contact between patients and specialists is channeled through general practitioners. Institutional forms are important because they determine the terms of interaction. Discussions of institutional forms are almost completely absent in professional ethics. But they can be discussed in moral terms, and reasonable answers can be provided. One can, for example, discuss if it would strengthen the position of patients if they could make the initial contact with specialists directly, without having to rely on gatekeepers. Institutional forms to a large degree determine the shape of the nexus of trust, risk, and power in health care, because they determine how and to what extent patients become dependent on professionals.

2. Concepts like power, trust, and risk are not socially, morally, and politically innocent. They come with stakes and interests, which make a practical difference for institutional forms. Power is a phenomenon of which we all have experiences and an intuitive understanding of what it is. But no definition commands unanimous consent. The literature abounds with definitions. "Power may be defined as the production of intended effects," Russell said in 1938 (Russell 1975: 25). He distinguished between three kinds of power: traditional, revolutionary, and naked. But he did not distinguish between power over things and power over persons. Weber defined power as the opportunity of a social actor to have his will in a social relation, even if he encounters resistance, regardless of the foundation of this opportunity (Weber 1976: 28). He also distinguished between three kinds of legitimate authority (subclasses of power): legal, traditional, and charismatic. According to Foucault, power is just a name given to a complex strategic situation (Foucault 1980: 93). I could have continued this list ad infinitum. It is not easy to say how these definitions are connected. The fact that it is difficult to define a concept does not, however, mean that which we try to grasp by the concept doesn't exist. From difficulties in definition one cannot infer to nonexistence, although it may be tempting to do so.

Much of the same is true for risk and trust. Some people cling to the standard concept of risk from decision theory, where risk is contrasted with certainty, uncertainty, and ignorance. Others, like Douglas and Caplan, emphasize the cultural variability of concepts and the social functions of risk talk (Douglas 1992; Caplan 2000). The rational choice theorist Coleman and the system theorist Luhmann claimed that to trust is risky. But Luhmann's concept of risk was closer to Coleman's concept of uncertainty than to his concept of risk. They used the same word, but their claims were different (Coleman 1990: ch. 5; Luhmann 1968). Definitions of these concepts that will command unanimous consent are unlikely to appear in the near future.

My issue here is how definitions can function in social life. The concept of power denotes what Hacking calls an interactive and not an indifferent kind (Hacking 2001: 104-105). The same is true for risk and trust. An interactive concept can affect that which it is about; an indifferent concept can not. Planet is an indifferent concept. In the history of astronomy there have been many definitions of it. But the heavenly bodies are unaffected by our efforts to define the concept. Democracy is an interactive concept. In political history there have been many definitions of it. It makes good sense to claim— and one can document—that the development of political institutions have been affected by these definitions. Definitions may enter persons' self-understanding and their understanding of the world and thereby have practical effects. Humans think and act on the basis of how they understand themselves, one another, and the world. Planets do not care about the self-understanding of humans, and rightly so. But humans should care about the self-understanding and worldviews of other humans, especially if they depend on these other humans. Power, trust, and risk concern relations among humans. Such relations can be affected by how we understand them.

Power is also what W. B. Gallie called an essentially contested concept (Gallie 1962). By such concepts there are rival—often mutually incommensurable and mu-tually hostile—traditions of interpretation of their meaning. They have no uncon-tested core meaning. This can be contrasted both with the Fregean view, where it is assumed that one can give a precise definition of a concept's meaning, and to concepts that exhibit what Waismann called "open texture" (Waismann 1984). All concepts have "open texture", that is, vague boundaries, and can encounter cases that are difficult to classify. But most concepts have a core meaning, which is not affected by their "open texture". Gallie's article challenges this view for central social scientific and normative concepts. In his view there are stakes attached to the rival interpretations of such concepts. How democracy is defined is not a politically innocent matter. Different definitions can serve different interests. So we should not expect consent about a definition of democracy.

If we combine Hacking's and Gallie's views, it is obvious why this is so. If democracy is not only essentially contested, but also interactive, its definition is not only a theoretical matter. It has practical consequences for the design of democratic institutions, because definitions can enter political actors' self-understandings and worldviews and shape their actions. How we define democracy does not only affect our theoretical understanding of democracy. It also affects what we in practice view as democratic institutions. There is a reason why there are so many different models of democracy (Held 1992). Much the same goes for power. How we understand it is not a purely theoretical matter. It also has practical social and political consequences for how we behave in the face of power relations, what we see as power, what kinds of power we try to constrain, or what instruments for the acquisition, exercise, and maintenance of power we think are legitimate. A part of health professionals' neglect of power stems from an implicit definition that equals it with physical force. If so, power can be seen as a marginal phenomenon in health care, at least in the Western world. One consequence of such a view is that, even if power has a negative ring, kinds of relations and behavior, which by other definitions would be described as the exercise of power, can be seen as innocent and unproblematic and be perpetuated under some other headings. Discussions of trust, risk, and power should be sensitive to such issues. Definitions of essentially contested and interactive concepts can play a part in political struggles. Their definitions are never purely theoretical. They are practical in the sense that they concern how phenomena in the social world are shaped for social actors. Ongoing discussions of different concepts of power, trust, and risk are a vital part of moral discussions of institutional forms of the nexus of power, trust, and risk in health care.

3. But how should power be approached in health care? We need an approach that is suited to moral discussions of institutional forms and their effects. I think that the most fertile framework for discussing power in health care is developed by Lukes (1974). It can account for the nexus of power, trust, and risk and sensitize us to its importance and variations. It also points out what must change if the nexus is to change. One could claim that Foucault's approach to power can do the same. But Lukes's framework has an advantage over Foucault's. In Lukes's approach it is possible to talk about being liberated from certain power relations and of the reduction of power within a relation. Within Foucault's framework, such judgments are difficult to warrant. All reforms of health care need normative standards. Such standards are hard to justify on the basis of Foucault's work.

Lukes proposed a threefold description of power. First-dimensional power concerns contexts where A forces B to do something. This includes physical force. Second-dimensional power concerns contexts where A controls the agenda of the interaction with

B. Agenda control can, but need not, be based on force, that is, on first-dimensional power. It concerns what is up for grabs in an interaction and what is not, and who decides. It is important to understand situations where second-dimensional power is not based on force. Third-dimensional power concerns contexts where A controls the world as B sees it. It means control of the definition of the situation: What is illness, what is normal, and so on. This way of conceiving power has some advantages in the study of power in health care.

First, in modern health care there is, with some exceptions (mainly in psychiatry), not much force. A bias in the literature about proper relationships between health professionals and patients is that the discussants tend to equalize power with force. Thus, if we get rid of force, what is left is a community of equals. There is also much illusion surrounding the idea of informed consent and the patients' free choice. Such illusions are often fostered by a concept of power that equalizes power with force, that is, first-dimensional power. Absence of force can then be interpreted as opportunity for free choice. But this assumption is misguided. Power is more complex. Agenda setting and control of worldviews are even so important for free choice and informed consent as is force.

Second, by emphasizing agenda setting and control of worldviews, Lukes's approach can account for most of the asymmetries in the relationship between health care staff and patients. These asymmetries are normally not based on force. One asymmetry is epistemic asymmetry, where one party knows more than the other. The inferior party then lacks the intellectual resources to challenge the interaction's agenda interaction and the presupposed worldview. Canter says that medical power is not a currency that can be easily transferred from physician to patient (Canter 2001). One reason for this is the fact that medical power is not based on force. If it were, the removal of force would transfer power. But much warrants the claim that health care staff and patients can never become equals who solve all problems and make all decisions by deliberation, even if there is not much use of force. I do not suggest that the deliberative model is morally wrong. I only suggest that to discuss it in a rational way, one cannot ignore unavoidable power differentials even in the absence of force in the relationship between health care staff and patients.

A third advantage of Lukes's approach is that it can account for the connections among power, trust, and risk. When I enter the health care system as a patient, there are people who can detect, assess, take, conceal, create, or eliminate risks for me. It is not their ability to use force that creates such opportunities. Rather, to understand such phenomena, we must apply the two other dimensions in Lukes's approach. The professionals' opportunities are mainly created by my helplessness and my trust in their superior epistemic position in comparison with my own.

V

Health care systems in Western countries change rapidly. I shall pick out a couple of trends that may change the nexus of trust, risk, and power. Such a perspective can say something about how deep-seated it is. If it is unlikely that the reduction of paternalism and the introduction of deliberation will turn health care interaction into symmetrical partnerships, what could change it in radical ways? And would it be a change to something better? The answer to that question is ambiguous. Maybe the best we ought to hope for, if paternalism erodes, is that health professionals become more self-conscious about power and participate more in the moral and political discussions about institutional forms. This could transform the nexus, but not eliminate it. To eliminate it could change health care beyond recognition, in a direction for which we should maybe not opt.

One kind of change is the feminization of medicine. In Denmark more than 60 percent of new physicians entering general practice are women (Holge-Hazelton 2004: 143). Denmark is not unique (Riska 2001: 181). Some people, a few physicians included, think that a new care-based definition of medicine should be developed based on more "feminine" values. But it is difficult to see that this could significantly change the nexus of power, trust, and risk. The question of how to define the profession of medicine is mainly a question of what values the profession should have. Values are important, but they do not by themselves change deep-seated structural phenomena like epistemic imbalances. They can make the helpless feel better but do not by themselves make them less helpless.

A second set of changes concern the new patient and the democratization of knowledge. A new patient culture is developing. Patients are better educated. They have (e. g. , through the Internet) better access to expert opinions. They have better health and live longer. And they pose greater demands on health care staff. Technical development changes the relations between health care staff and patients. One could say that the importance of health-related knowledge grows. But health professionals' control over such knowledge decreases. And although the literature about patient trust is difficult to interpret (because so many different definitions of trust are in play), it seems that patient trust in the health care system and in health care staff is declining in Western countries. All this could lead to an erosion of health professionals' monopoly of defining the situation of the patient and distributing health care benefits according to this definition (Holge-Hazelton 2004: 150, n. 3; Hilden 2003). It could threaten not only the idea of moral expertise, but also the idea of medical expertise. This set of changes can pose a serious challenge to the nexus of power, trust, and risk. It challenges its core, namely the scope and legitimacy of professional autonomy. It challenges the health professionals' control of the technical aspects of their work. The trend toward greater sophistication of health-related knowledge is mostly visible in some segments of modern societies. But it has the potential to spread.

A third set of changes concern structural characteristics of the role as a physician. This role changes along several dimensions. Each physician treats more patients than before and spends shorter time with each patient. Physicians change lifestyle, with a clearer separation between work and leisure time. Being a physician (or a nurse) is more and more an ordinary job and less a vocation. General practice is changing. General practitioners treat trivial cases. Most other cases are referred to specialists. The medical profession is losing influence on the division of labor in hospitals. It is also losing ground in management, partly because of changed relations to other professions (e. g. , nurses, an upcoming profession) and partly because hospitals become more and more stuffed with professional management. Some of these changes can also challenge the nexus of power, trust and risk because they affect professional autonomy. Division of labor may be important for control of the technical aspects of work. [9]

Genuine deliberation between health professionals and patients presupposes such changes, because first the relations between them can become symmetrical. Then there will be no teachers and no students, but equals. Relations between health professionals and patients can only become symmetrical if the nexus of power, trust, and risk really erodes. This, and not paternalism, is the serious moral issue. But this development is ambiguous. It may be that we should not wish much further development in this direction.

I shall briefly illustrate my point by reference to informed patients and the de-mocratization of knowledge. Both are good things, if they come about. But they raise some serious issues.

1. The label "democratization of knowledge" is probably a misnomer. It presupposes that we are witnessing a development in the social distribution of knowledge that is analogous to the political democratization in Western countries in the 19th and 20th centuries. But we are probably not witnessing something like this. The end of a process of democratization of an issue is either that all concerned parties get the right to vote on an equal footing about the issue (one person, one vote), or an equal right to participate in discussions leading to decisions about the issue.

Political democratization starts with an uneven and ends with an even distribution of rights or participation or both. But an analogous claim about knowledge is difficult to warrant. What is true about Western countries is that the level of general education has increased. So has the access to different kinds of information. But do we have good reasons to think that the distribution of knowledge—and of health-related knowledge in particular—has become more even? I do not think so, mainly for three reasons. First, people have different abilities, which are relevant for what knowledge they can acquire. This alone counts heavily against the claim that knowledge can be fully democratized. Diffusion of knowledge does not cancel out natural abilities, whether we like it or not. Second, different kinds of knowledge spread in uneven ways in society, and we do not know much about how. But the diffusion of knowledge is not like the diffusion of rights. Different kinds of knowledge spread in different ways. Even if we could make a case for the claim that theoretical medical knowledge is more evenly distributed than before, a similar case cannot be made about practical medical skills. The fact that someone is informed about some of the causes of cancer does not mean that he can diagnose and treat himself or others. Theoretical knowledge and practical skills spread in very different ways. Third, frontier knowledge is a moving target. When some pieces of knowledge have become more evenly distributed, the frontier can be way ahead and the old unbalanced distribution is maintained on another level. Much warrants the claim that at least some segments of patients have become better informed. But it is implausible to claim that they have become or will become so well informed that we should abandon the professional autonomy of health care staff, that is, really challenge their control of the technical aspects of their work. To become better informed is not the same as acquiring the depth and scope of an expert's theoretical knowledge and practical skills in a field. What we do witness is not a democratization of knowledge, but a new division of epistemic labor in society. In health care, I think we see two tendencies, which it is important not to confound: The knowledge gap between medical knowledge and the knowledge of laypeople (patients) is increasing, not decreasing. At the same time, the profession of medicine is losing control over medical knowledge. The control is not transferred to laypeople, but to other professions. Evidence basing is one interesting example of this. Evidence basing transfers control over knowledge to a completely new kind of expert: those who are specialists in meta-analyses. Many of them are not physicians at all. But it is dubious if evidence basing democratizes knowledge, as some of its defenders claim.

2. There is a difference between challenging the idea of moral expertise and the idea of medical expertise. There are sound arguments against the claim that there is genuine moral expertise. Someone who knows a lot about moral theory, or has accumulated a vast knowledge about moral cases, need not be a better person, and he need not be the best choice as an advisor on moral issues. Knowledge of moral theory and cases does not translate into actions and characters in this way. But there are sound arguments for the claim that there is genuine epistemic, for example medical, expertise. More knowledge about the causes of diseases and their treatment does translate into action, given adequate medical skills. One can claim that a physician who has more knowledge is also a better physician, even if this is not always true, because other abilities are also important for being a good physician. So, even if we cannot claim that someone who knows a lot about moral theory should be allowed to control our moral behavior, it makes sense to claim that health professionals should control the technical aspects of their work, which includes diagnosing and treating us—the patients—in various ways. If "the democratization of knowledge" is largely an illusion, the real alternative to the nexus of power, trust, and risk is to give the "informed patient" the upper hand. But that could have bad side effects. The "informed patient" is informed, but not much more than that. In spite of the informed patient, there is still a deep-seated—although changing—division of epistemic labor in health care. I do not think it is much more democratic than before.

To conclude, we should pray for the maintenance of the nexus of power, trust, and risk and thus for the autonomy of health professionals. It is not a bad thing that some people know more about diseases than I do. Actually, I would be very worried if it were not so. But we should seriously discuss various institutional forms of the nexus. They really matter. And health professionals should participate in such discussions. Actually, they should be the main participants, because their self-understanding and worldviews probably matter more than anything else in such discussions.

Notes

Acknowledgments. I am grateful to Lars Inge Terum, Jan Helge Solbakk, Reidar Pedersen, Stale Fredriksen, and the journal's two anonymous reviewers for comments on earlier drafts of this article.

1. My idea of what the trustier does is mainly based on Warren (1999: 311), Coleman (1990: ch. 5), and Baier (1986, 1993); see also Potter (2002). I have also been inspired by Elster's approach (Elster 2000). I differ from Elster in the sense that I do not define *trust* in terms of behavior, as he does (2000: 5). I simply ask what a trustier does.

2. I do not intend a very precise use of the concept of "health care systems". My use corresponds roughly with Kleinmann's (1980) classical view of the professional part of health care systems.

3. I use Freidson's distinction between consulting and scholarly professions (cf. Freidson 1988: 73-75f).

4. The text says "examined", but that does not make good sense.

5. This is a twist on Stein Rokkan's famous dictum: "Votes count, but resources decide." (Rokkan 1971)

6. This was suggested to me by Per Kristian Hilden.

7. I owe this point to one of the journal's reviewers.

8. Typical in this respect are Freidson (1988) and Fainzang (2002).

9. One could argue that evidence-based medicine will also change the nexus of power, trust, and risk. Evidence-based medicine is not only an issue of knowledge, but also of social relations. But I do not think it will change physician-patient interaction in the direction of more epistemic symmetry.

References

Baier, A., 1986, Trust and Anti-Trust, *Ethics* 96 (3): 231-260.

Baier, A., 1993, "Doing Things with Others: The Mental Commons. Paper presented at the Tenth International Nordic Philosophy Symposium", Turku, Finland, August 20-23.

Canter, R. , 2001, "Patients and Medical Power", *British Medical Journal* 323: 414.

Caplan, P. , ed. , 2000, *Risk Revisited*, London: Pluto.

Coleman, J. S. , 1990, *Foundations of Social Theory*, Cambridge, MA: Belknap Press of the Harvard University Press.

Douglas, M. , 1992, *Risk and Blame: Essays in Cultural Theory*, London: Routledge.

Dworkin, R. , 1978, *Taking Rights Seriously*, London: Duckworth.

Elster, J. , 2000, "Trust and Emotions", *Sosiologi i dag* 30 (3): 5-11.

Emanuel, E. , and L. Emanuel, 1992, "Four Models of the Physician-Patient Relationship", *Journal of the American Medical Association* 267 (16): 2221-2226.

Fainzang, S. , 2002, "Lying, Secrecy and Power within the Doctor-Patient Relationship," *Anthropology and Medicine* 9 (2): 117-133.

Foucault, M. , 1980, *The History of Sexuality*, Vol. 1. New York: Vintage.

Fredriksen, S. , 2002, "Diseases Are Invisible", *Medical Humanities Review* 28 (2): 71-73.

Freidson, E. , 1988, *Profession of Medicine: A Study in the Sociology of Applied Knowledge*, Chicago: University of Chicago Press.

Gallie, W. B. , 1962, "Essentially Contested Concepts", In: M. Black, ed. *The Importance of Language*. Englewood Cliffs, NJ: Prentice Hall. 121-146.

Gray, B. H. , 1997, "Trust and Trustworthy Care in the Managed Care Era", *Health Affairs* 16 (1): 34-49.

Grimen, H. , 2001, "Makt og tillit—tre samanhangar", *Tidsskrift for den norske lægeforening* 121 (30): 3617-3619.

Habermas, J. , 1991, *Erlauterungen zur Diskursethik*, Frankfurt am Main: Suhrkamp.

Hacking, I. , 2001, *The Social Construction of What?* Cambridge: Cambridge University Press.

Hall, M. A. , and R. A. Berenson, 1998, "Ethical Practice in Managed Care: A Dose of Realism", *Annals of Internal Medicine* 28 (5): 395-402.

Held, D. , 1992, *Models of Democracy*. Oxford: Polity.

Hilden, P. K. , 2003, *Risk and Late Modern Health: Socialities of a Crossed-Out Pancreas*. Oslo: Unipub.

Hirschmann, A. O. , 1970, *Exit, Voice, and Loyalty*. Cambridge, MA: Harvard

University Press.

Hølge-Hazelton, B. , 2004, "En klassisk profession skifter køn. " In: K. Hjort, ed. *De professionelle—forskning i professioner og professionsdannelser*. Roskilde: Roskilde Universitets-forlag. 143-151.

Kleinmann, A. , 1980, *Patients and Healers in the Context of Culture: An Exploration of the Borderland between Anthropology, Medicine, and Psychiatry*. Berkeley: University of California Press.

Luhmann, N. , 1968, *Vertrauen. Ein Mechanismus der Reduktion sozialer Komplexitaï*. Stuttgart: Ferdinand Enke.

Lukes, S. 1974. *Power: A Radical View*. Oxford: Blackwell.

Marshall, T. H. 1992. *Citizenship and Social Class*. London: Pluto.

Potter, N. N. 2002. *How Can I Be Trusted? A Virtue Theory of Trustworthiness*. Lanham: Rowman and Littlefield.

Riska, E. 2001. "Towards a Gender Balance: But Will Women Physicians Have an Impact on Medicine?" *Social Science and Medicine* 52 (2): 179-187.

Rokkan, S. 1971. *Citizens, Elections, Parties*. Oslo: Scandinavian University Press.

Russell, B. 1975. *Power*. London: Allen and Unwin.

Waismann, F. 1984. "Verifiability". In *The Theory of Meaning*. G. H. R. Parkinson, ed. pp. 35-60, Oxford: Oxford University Press.

Warren, M. E. 1999. "Democratic Theory and Trust". In *Democracy and Trust*. M. E. Warren, ed. pp. 310-345. Cambridge: Cambridge University Press.

Weber, M. 1976. *Wirtschaft und Gesellschaft*. Tubingen: J. B. C. Mohr.

Welsh, T. , and M. Pringle. 2001. "Social Capital: Trusts Need to Recreate Trust". *British Medical Journal* 323 (28): 177-178.

权力、信任和风险：关于权力问题缺失的一些反思

哈罗德·格里曼[*]

当医护人员公开反思他们自身及他们与患者的关系时，除了一些例外情形（精神病治疗和赋权[①]），关于权力的讨论是罕见的。在下列两类（相互交叉的）讨论中，本来是应该讨论权力的情况，反而最明显缺乏讨论的就是权力。这两类讨论是：关于信任的讨论和关于理想的医生——或护士——与患者关系的讨论。几年来，我作为一名局外人，一直密切注意医学杂志和护理杂志中的这些讨论，令人震惊的是，关于权力的讨论十分罕见。权力问题的缺失并不单单针对医护人员。像我这样的大学教师在公开讨论教师与学生的关系时，明显缺乏讨论的也是权力。

这使得反思这种局面的哲学家去追问一个无法笼统回答的问题：为什么医护人员不公开讨论内在于他们工作之中的权力差异呢？那么，缺乏这种公开讨论会带来什么后果呢？医护人员（例如，医生、护士、心理学家、理疗师和康复专家）在社会生活中把他们自己视为能提供有益帮助的人，而不是掌权者或控制者。的确，他们是能提供有益帮助的人。但是，他们也是掌权者和控制者。此外，知识的差异和控制的机会，对于所谓的专业人员来说，是至关重要的。如果没有知识和控制的不均衡分配，一个现代社会就不可能运行。之所以有专业人员，是因为人们假定他们更好地掌握了某个领域（例如，医学或者法学）内的较多知识。

缺乏对权力的公开反思，会妨碍从道德上认真地讨论医护人员同他人关系的各种制度形式，因为医护人员被看成是这一讨论的重要参与者。这应该是关注的一个理由。缺乏关于权力的公开反思，还会导致互动关系中的不切实际的观点，比如，如何建立医生和患者的互动，这些观点如果付诸实践的话，一定会是弊大

---

[*] 哈罗德·格里曼（Harald Grimen），奥斯陆大学学院职业研究中心教授。本文刊于 Medical Anthropology Quarterly，Vol. 23，Issue1，pp. 16-23，2009.

[①] Empowerment，从字面上理解，即将权力赋予某人，在本文中意指医护人员在对患者实施的医疗行为中赋予患者权力。——译者注

于利。而且，这还导致了曲解信任和风险的观点。权力、信任和风险是相互联系在一起的，从社会、道德或政治意义上来看，这些概念并不是无辜的。缺少对权力的公开反思也会导致对特定类型的变化的可能性视而不见。如果不首先改变医疗保健体制的根深蒂固的结构特征，那么，就不可能改变处于现代医疗保健体制核心的权力、信任和风险的关系。我不确定，我们是否应该选择彻底改变这些结构。彻底改变可能不会导向更好的医疗保健体制。选择对权力更有自我意识的医护人员，并且找到这种关系的更加人性化的制度形式，而不是废除这种关系，也许更好。因此，关于权力的公开讨论——在医护人员参与的情况下——是至关重要的。

我首先讨论对我的论证至关重要的一些概念（第一部分）。然后，我勾勒出医疗保健体制中权力、信任和风险的关系（第二部分）。接下来，我讨论权力缺失的一个例子（第三部分）和如何对待权力的问题（第四部分）。最后，我简要说明为什么我们不应该选择彻底改变的原因（第五部分）。

一

忽视权力来分析信任是幼稚的，忽视信任来分析权力是肤浅的。但是，根本没有无可争议的信任、风险和权力的定义。我不打算给它们下定义，而是打算探讨给予信任、承受风险或行使权力的某个人事实上做什么来表明信任、风险和权力是如何紧密地联系在一起的。① 如果 A 信任 B，那么：

（1）在一段时间内，A 使某物 X 处于 B 的监管之下。

（2）在这段时间内，A 把支配 X 的自由决定权——总是在事实上，有时在法律上——转让给 B，或者，A 处在 B 拥有这些权力的情境中。

（3）A 重视 X。

（4）A 期待：①B 将不会做出损害 A 的利益的事情。②B 有能力根据 A 的利益看管好 X。③B 有适当看管好 X 的必要手段。

---

① 我关于信托人（trustier，经济学词汇，在本文中主要指给予医护人员以信任的患者。——译者注）做什么的思想，主要基于 Warren，M. E.，"Democratic Theory and Trust"，In *Democracy and Trust*，M. E. Warren ed，Cambridge：Cambridge University Press，1999，p. 311，pp. 310- 345；Coleman，J. S.，*Foundations of Social Theory*，Cambridge，MA：Belknap Press of the Harvard，1990，ch. 5；以及 Baier，"Trust and Anti-Trust"，*Ethics*，96（3），pp. 231-260；"Doing Things with Others：The Mental Commons"，第十届国际北欧哲学会议论文，Turku，Finland，August 20-23，1993。我也受到了埃尔斯特（Elster）的进路的启发 [参见 Elster，J.，"Trust and Emotions"，*Sosiologiidag*，2000，30（3），pp. 5-11]。但我不是根据行为来定义信任，在这个意义上，我不同于埃尔斯特，而他是这么做的 [Elster，J.，"Trust and Emotions"，*Sosiologiidag*，2000，30（3），pp. 5-11]。我只是提出问题：信托人做什么呢？

5. A 很少防范 B 对他支配 X 的自由决定权的滥用或草率使用。

信任有交易的一面。一个信托人经常把一些东西，例如，金钱、小孩、财产、信息或者他的身体，交给某人监管。这样做，他把自由决定权转让给他人。当 X 由 B 监管时，B 能决定 X 应该发生什么事，不管 A 是否知道或希望这样。当父母不在场时，保姆具有对小孩的实际支配权，不管对此他们怎么认为。这个信托人期望受托人将不会伤害小孩，但是，他并不能保证受托人不会那么做。他期望受托人是称职的，并有适当的手段，但却不能保证真是这样。他托付给别人的东西会被滥用，被以草率的或不称职的方式来对待，或者，会被用来反对他。一个人的信任能够成为另一个人权力的基础。受托人可以利用自己手中得到的东西使信托人做他想做的事情。使其他人做某人想做的事情——通常与他们的意愿相违背——的能力是权力的核心要素。① 这么做有许多方式，例如，威胁要伤害或滥用他们重视的东西。信托是有风险的，而且对信托人会有不利后果。通过信托，他使自己变得脆弱。

在与专业人员的互动中，信任的交易的这一面很重要。我们把信息、金钱、意愿、小孩、我们自己的身体等交给专业人员监管或长或短一段时期。我们情愿将自己的身体交给医护人员监管，使他们做好本职工作。如果我不信任我的医生，我就不会允许他做医生对病人的身体所做的事。正如霍尔（Hall）和贝伦森（Berenson）所说的，信任使医疗保健体制中有了受益权（beneficial power）。② 受益权对于完成工作来说是必要的。但是，信任也创造了权力的结构条件，这些未必是有益的。我将这些联系称作"信任、风险和权力的关系"。在外行与专业人员的所有互动中都可以发现这种关系。但它可能有不同的形式，取决于专业人员承担的任务和外行与专业人员互动时的制度设置。

在现代医疗保健体制中，这种关系的各种形式是由医护人员的自主权和患者的结构性的劣势地位共同决定的。③ 医护人员的自主权导致了大量使用自由决定权和临床判断，这些很难做出解释。不问责的（nonaccountable）自由决定权是权力的有效来源，尤其是在法律语境中和在基于信任的互动中。患者的结构性劣势有许多根源。④ 首先，医护人员和患者之间有知识差距。在医疗保健体制中的

---

① Weber, M., *Wirtschaft und Gesellschaft*, Tubingen：J. B. C. Mohr, 1976, p. 28.

② Hall, M. A., and R. A. Berenson, "Ethical Practice in Managed Care：A Dose of Realism", *Annals of Internal Medicine*, 28（5），pp. 395-402.

③ 我不打算在非常严格的意义上使用"医疗保健体制"的概念。我的用法大致对应于克莱曼对医疗保健体制的专业部分的经典观点（Kleinmann A., *Patients and Healers in the Context of Culture：An Exploration of the Borderland between Anthropology, Medicine, and Psychiatry*, Berkeley：University of California Press, 1980）。

④ Grimen, H., "Makt og tillit—tre samanhangar". *Tidsskrift for den norske lægeforening*, 121（30），pp. 3617-3619.

知识差距很可能比任何其他咨询业中的知识差距更大，在我看来，这种差距是在加大，而不是在缩小。① 这使得难以对医护人员的判断提出挑战。第二，患者通常只是极端地相信医护人员的胜任能力、技术和专业。他们可能被迫信任对他们的医治。按照赫希曼②的观点，我们会说，他们缺少退出的选择，减少了发言权，会被迫忠诚。此外，医护人员通常充当患者需要的商品和服务（例如，专家服务或伤残抚恤金）的把关人。把关人会利益混杂，并不是所有这些利益都需要与患者的利益保持一致。最后，当患者与医护人员互动时，他们的身心通常不能发挥正常作用。他们可能迷茫、害怕或震惊，或者，他们可能失去了部分语言和认知能力，就像中风病人经常表现的那样。因此，与健康人相比，他们可能更容易被操纵。专业人员的自主权不是医疗保健业特有的。使用者的结构性劣势及被迫信任也不是医疗保健业专享的。但是，这些问题在医疗保健体制中比在其他环境中更加明显。因此，反思医疗保健体制中的权力、信任和风险是重要的。例如，这会导致更加认真地讨论医护人员和患者的关系。

医疗保健体制中权力、信任和风险的联系，在某种意义上，就是人类的生活状况。如果我们不需要医疗救助，那么，这种联系就不复存在。但它的特殊形式根源于医疗保健体制的社会结构，以及在治疗疾病时应用的那种知识和技术。如果医疗保健体制不受具有自主权并且掌握了具体类型的知识和技术的现代风格的专业人员的控制，那么，权力、信任和风险的关系就会有其他形式。当患者与传统的行医者互动时，这种关系表现为不同的形式。我们能够理性地探讨，这种形式是否比现有的形式更加人道。道德上重要的是探讨这种关系的某些制度形式是否比其他形式更加人性化。各种制度形式应该在职业伦理学中得到比现在更加频繁的讨论。其结论未必是应该废除这种关系，也可能是，这种关系的某些形式比另外一些形式更人性化，并且我们应该选择这些更加人性化的形式，而不是选择完全不同的医疗保健体制。

下面举一个例子说明我所说的讨论制度形式的意思：2001 年《英国医学杂志》中的一篇社论提议，国家卫生事业局（NHS）应当执行辅助性原则："决策的制定应该被尽可能接近地定位于采取行动的地方。在辅助性原则得以应用时，组织的行为就得到了最有成效的管理。"③ 这一原则起源于天主教的社会哲学，在欧盟的法律体系中居于核心地位。这篇社论提出了对这一原则的一种可能的解

① 我使用了弗雷德森在咨询业和学术界之间所作的区分，参见 Freidson, E. *Profession of Medicine: A Study in the Sociology of Applied Knowledge*, Chicago: University of Chicago Press, 1988, pp. 73–75f.

② Hirschmann, A. O. *Exit, Voice, and Loyalty*, Cambridge, MA: Harvard University Press, 1970.

③ Welsh, T. and M. Pringle, "Social Capital: Trusts Need to Recreate Trust", *British Medical Journal*, 323 (28), pp. 177–178.

释。在医疗保健体制中贯彻这一原则会强化医生、护士和心理学家的自主权和权力。他们在比任何其他专业人员更接近采取医疗行动的地方工作。辅助性原则也可能与基于职业外聘审计的各种新的问责形式（ forms of accountability） 相冲突。关于医疗保健体制中的辅助性原则的道德讨论，在很大程度上，必须是关于因贯彻这一原则而导致的制度形式的讨论：我们应该期待给予医生和护士更多权力的各种医疗保健制度吗？

二

首先，我将更加深入地探讨权力、信任和风险之间的关系。这种关系是由两个方面的因素促成的：一是咨询业的专业自主权，二是患者多方面的结构性劣势处境。这种联系是通过现代各种医疗保健制度的运作方式建立起来的，而且，它使得患者非常依赖于医护人员。只有当现代医疗保健体制的某些基本结构发生改变，这种关系才可能发生变化（参见第五部分）。因此，正如我在第三部分将要论证的，关于如何组建医护人员和患者关系的一些建议是不切实际的，而且可能弊大于利。例如，患者在多大程度上能够成为医护人员的对话伙伴，是有限度的。如果人们走得太远，就很难以一种道德敏感的方式承认和对待患者对医护人员的依赖性。患者与医护人员之间的对话不可能是平等个体之间的对话。但是，对可能加深患者对医护人员的依赖性的提议，也有理由提出质疑。辅助性原则可以起到这样的作用。这一原则由于拓宽了医生、护士和心理学家的自主决断的范围，使得对他们的问责更加困难（第一部分）。

我将首先探讨职业自主权。按照弗雷德森（Freidson）的观点，职业自主权的核心是有对工作的技术方面的控制权，即对如何组织工作的运作及其环境的控制权。一个职业群体，即便在其工作的经济、管理和教育方面没有控制权，也会具有职业自主权。弗雷德森举例说，原苏联的医生在自己工作的经济和管理方面没有控制权，而只有对工作的技术方面的控制权。弗雷德森的观点是有争议的。但是，它明确指出，如果在当前形式中削弱医疗保健体制中权力、信任和风险的关系，那么，必须要改变的一个重要方面是：医护人员必须失去他们对其工作的技术方面的控制权。

对其工作的技术方面有控制权的一个职业群体拥有相当大的自主决断（discretion and judgment）的空间。它可以根据自己的世界观（包括为满足其顾客的需要在内）处理偶发事件（代表他人）。它有相当大的自由来形塑自己的世界观。它可以根据自己的世界观和需要确定日常工作中的重要事务。自主决断很难被问责。在咨询业中（像医生、护士、律师或社会工作者）拥有自主决断就是有权为他人或者代表他人作决定。不可问责的自主决断是不可问责的权力的来源。但是，如果没有自主决断的广阔空间，很难想象职业自主权会是什么样子。我推测，在医疗保健体制中为自主决断留下的空间在两种情形下是最广阔的：当专业人员对患者未来的功能性能力进行鉴定，以确定其是否有资格享受按需要分配给他们的福利待遇时（例如，伤残抚恤），以及当专业人员评估（例如）手术、急救治疗或者一般的医疗实践中的风险时。在医疗保健体制的其他领域中也存在很大的自主决断的空间，不过我用两种语境来举例说明要点。

在第一种情形中，自主决断的空间很大，这不仅是因为福利法是开放式的，原因是，医生和护士不是特别能胜任对患者未来功能的能力做出鉴定，而且因为预测这些能力的发展变化是非常困难的。医生和护士不一定受过这么做的训练。在挪威的医疗保健体制系中，医生是根据医疗需要所分配的福利待遇的监管人。在其他一些国家，这些更多是其他医护人员的"地盘"，比如理疗师和职业治疗师或康复专家。他们可能接受过较好的培训，但仍然有很大的自主决断的空间。福利待遇位于医疗实践的边缘，但与公民权利和政治权利相比，却是马歇尔（T. M. Marshall）所谓的公民的社会权利的核心。[①] 在社会权利的情形中，司法上的自主裁决和临床诊断不期而遇。这些情形在现代福利国家中的范围在扩大，而且医护人员在这些情形中扮演着复杂的角色。

然而，对风险的评估是医学的核心，涵盖了从疾病预后到药物治疗的副作用等所有环节。其中大部分工作已经成为日常工作。日常工作是把双刃剑，方便工作，但也限制视域。日常工作既能带来精神安逸，又能造成医疗（或司法）灾难。这是工作常规化的悖论：有助于日常工作的事情，也会使日常工作变得很危险。在上面的第二种情形中，自主决断有很大的空间，主要有三个理由。第一，成为一名患者，就是委托他人来评估风险。当我们听从医护人员的建议时，我们大家都是这么做的。即便是知情患者，也必须依赖他人的能力来评估风险是什么。正如我把现代医疗保健体制的漫画看成是一名有经验的患者那样，这幅漫画（它不是很不及要领）是，在这样的体系中，总会有某个人，拥有自主权和权力为我查明、评估、承担、隐瞒、造成或消除诸风险。我必须相信，他们知道他们正在做什么，而且，假如他们犯了错误（因为他们通常会犯错），他们将会纠正自己的错误。第二，在医疗保健中可能出现的许多情境中，例如，在急救治疗或在先进的外科手术中，不可能详细地管制医护人员的行为，而且，试图这么做也很愚蠢。管制有时可能弊大于利。医护专业人员必须——在确切意义上——处理不可预测的偶然情况。因此，个人决断会是我们能指望的最好结果。第三，医疗

---

① Marshall, T. H., *Citizenship and Social Class*, London：Pluto, 1992, ch. 2.

保健总是建立在不确定的和有争议的知识及比应有的情况更加过时的知识之基础上。全科医生或正常医院的医生通常都不处在研究的前沿。

信任某个医护人员就是授权某人评估和承担风险。这是把评估承担什么风险的权利事实上转让给医护人员，这些医护人员的判断可能——并且经常确实——与某个人自己的判断明显不同。一名患者把对他所承担风险的评估委托给医护人员监管一段时间。在手术室里接受麻醉时，他只能期望，麻醉师知道他们正在做什么，并且对麻醉的风险有明智的判断。自主决断的空间［德沃金的"一个面包圈中的洞"（Dworkin's "hole in doughnut"）①］或宽或窄。这通常也是行使权力的空间。此外，医护人员的世界观、意见和判断往往与他们的患者的世界观、意见和判断明显地不同。医护人员通常并没有意识到他们的观点及其患者的观点之间的这种鸿沟。医护人员倾向于认为，患者会和他们一样思考问题。但是，通常并非如此。

接下来，我会就患者的结构性的劣势处境做出说明。医疗保健包含了一系列不对称关系，其中滋生了权力和风险。患者相对于医护人员处于结构性的劣势地位。各种善意不可能消除这种结构性的失衡。一些人需要帮助；另外一些人不那么绝望，并且——但愿——有更称职的人能够提供帮助。一些人需要商品，另外一些人则是这些商品的看护者。那些需要帮助的人，除了感谢之外，几乎无可回报。一些人知情；另外一些人对相关问题知之甚少。知识差距意味着，所知甚少者无法对——用托马斯（W. I. Thomas）的术语来说——告知互动（informs interaction）情境的界定提出挑战。此外，因为他们知之甚少，他们必须站在认知劣势的立场上信任别人。他们不知前行之路，必须相信他人知道，没有能力反对他人在每一个十字路口做出的选择。最后，患者是在医护人员的能力与意愿的掌控下，发现和纠正他们自己的错误，并且，他们很容易受到各种复杂动机的影响，支持他们所信任的那些人。

这些是位于现代医疗保健核心的权力、信任和风险关系的一些元素。这是人类生活状况的一种事实性的制度形式，即这个事实是，人们生病了，需要医疗救助。现代社会中的医疗保健受特定类型的专业人员的控制，他们掌握着超越日常经验的某种知识②，并且，他们可以根据自己的自主判断来运用这些知识。患者必须信任他们的胜任能力、职业素质和技术。但是，患者的信任源于他们的结构性的劣势地位。他们将自己的生命交给那些他们没有机会对其判断和行动提出挑战的专业人员来监管。这种关系很容易被视而不见。一个重要的问题是，我们是

---

① Dworkin, R., *Taking Rights Seriously*, London: Duckworth, 1978.

② Fredriksen, S., "Diseases Are Invisible", *Medical Humanities Review*, 2002, 28 (2), pp. 71-73.

否应该希望这样的关系完全消失。我认为，我们不应该这么做，而是应该探讨它的不同的制度形式。它所提出的道德问题关系到权力、信赖及脆弱性。忽略这些问题会引起许多一厢情愿的想法。

三

当医护人员公开讨论他们自己及他们同患者的关系时，有许多一厢情愿的想法。在他们讨论此类问题的许多文献中，忽略了权力、信任和风险的关系。作为一个例证，我将简要地讨论一个典型的有影响的例子。

在医护人员讨论他们同患者之间的适当关系的文本中，应该以讨论权力为核心。这类文本是（可以在专业期刊中找到）医护人员，即大多数是医生和护士，讨论他们自己及他们所扮演的角色的文本。在过去的几十年中，人们越来越关切将患者从家长制式的统治中解救出来，关切寻找医护人员与患者应该如何互动的新模式。关键术语是患者的自主（autonomy）、赋权（empowerment）、自由选择（free choice）以及知情同意（informed consent）。一篇被广泛引用的文章是伊齐基尔（Ezekiel）和琳达·伊曼纽尔（Linda Emanuel）关于患者-医生关系模式的文章。① 如果在这些讨论中需要有任何参考文章的话，这篇文章就是很好的参考。他们概括出了四种模式：家长制式的、告知式的、解释性的和协商的。我将不再多讨论前三种模式，而是集中讨论协商模式。

在协商模式中，"医生扮演老师和朋友的角色，就什么是最佳治疗方案与患者交换意见"②。每种模式都需要一个患者自主的概念。在协商模式中，患者自主是一种道德的自我发展："患者不仅有权遵从未经检视的那些偏好或经过检视[它应该是'未经检视的'吗?③]的价值，而且有权通过对话考虑其他可供选择的健康价值，即是否值得治疗及其治疗的意义。"④ 在这里，医生是老师，患者是学生。伊曼纽尔夫妇的文章主张一种对话式的教学法，其目的是，通过对偏好和价值的考察，实现道德的自我发展。但在道德上得到发展的是患者，而不是医生。尽管有对话，但对话的一方是老师，另一方是学生。伊曼纽尔夫妇的模式与

---

① Emanuel, E., and L. Emanuel, "Four Models of the Physician-Patient Relationship", *Journal of the American Medical Association*, 1992, 267 (16), pp. 2221-2226.

② Emanuel, E., and L. Emanuel, "Four Models of the Physician-Patient Relationship", *Journal of the American Medical Association*, 1992, 267 (16), pp. 2221-2226.

③ 文本中说"经过检视的"，但这样说并不十分合理。

④ Emanuel, E., and L. Emanuel, "Four Models of the Physician-Patient Relationship", *Journal of the American Medical Association*, 1992, 267 (16), pp. 2221-2226.

民主理论中所讨论的协商模式不相符，在民主理论中，人们假定所有的参与者都在道德上得到了发展。例如，它与哈贝马斯的实践话语不相符。① 我认为，这有四种原因。它们没有指出协商模式的一个弱点。这个弱点是，主张这些模式能使医患关系概念化。

（1）伊曼纽尔夫妇（很合理地）主张，医生是老师，患者是学生。在医生与患者的互动中，认知的不对称迫使讲授成为不可避免的。但是，这种认知的不对称在决策的协商模式中没有重要地位。要使他们的模式与现实情况相一致，伊曼纽尔夫妇必须——无意地——在医患关系中引入另外一个二分体，即老师与学生。这对二分体不适合决策的协商模式。它非但没有消除，反倒假定了认知的不对称。协商模式假定，参与者之间的认知是对称的。这一模式根本没有包括不可避免的老师-学生的二分体。但是，认知的不对称是患者和医护人员互动的一个基本特征。

（2）协商模式只适合医患关系中重要的那一部分，即某些类型的价值问题（主要是与好生活相关的那些价值）。它们不符合所有类型的价值（例如，与医疗保健目标相关的那些价值）或对医患关系的认知（例如，关于病因的理论和事实）。与讨论诊断和治疗相比，这些模式更适合于进行某些类型的价值之间的对话。当伊曼纽尔夫妇在协商模式中引入授权时，这主要是作为一个偏好和价值检视的问题，而不是作为决策问题来处理的。如果患者能更富有洞察力地考虑他们的偏好和价值观，那么，他们就适合被赋予权力。这是一种弱意义上的授权，这种授权与完全维护医生的决策权相一致。价值是值得考虑的，但医生作决定。② 这可能是能够与用协商模式使医生和患者的互动概念化相一致的唯一的一种授权。正如伊曼纽尔夫妇隐含地表明的那样，对于有些价值而言，这种授权使得下列主张成为合理的：克服家长作风能在医护人员和患者之间建立对称关系。克服关于道德问题的家长作风只意味着从这些关系中排除道德专长（moral expertise）的观念：医生（和其他医护人员）在关于什么是好的生活上失去了作为道德专家的地位，而且，在这种意义上，患者适合被赋予权力。但是对于疾病的诊断和治疗来说，这么做不是很有意义。不存在与诊断和治疗的认知方面相关的家长作风问题。但存在一个权力和知识的问题。对于医生与患者互动的一些确实棘手的问题来说，协商模式是很不适合的。这些模式带有理想主义的色彩。伊曼纽尔夫妇实际上是歪曲地讨论遵从问题。③ 他们没有认真地考虑这种可能性：

---

① Habermas J, *Erlauterungen zur Diskursethik. Frankfurt am Main*：Suhrkamp，1991.

② 这是对斯特恩·罗坎（Sterin Rokkan）著名的格言的歪曲："选举是重要的，但资源起决定作用。"（Rokkan S.，*Citizens*，*Elections*，*Parties*，Oslo：Scandinavian University Press，1971. ）

③ 这是 Per Kristian Hilden 向我建议的。

患者能使医生相信，医疗保健或生物医学的疾病理论的基本价值可能是不正确的。尽管伊曼纽尔夫妇并没有这么说，但是患者道德的自我发展的要点是遵从，不是对医疗保健的道德革命或对医学的科学革命。然而，在一种真正对称的对话中，这种可能性应当是开放的。

（3）一些患者没有能力去检视他们的偏好和价值。即便是相当有才智的人，当他们身患重病时，也没有能力这么做。在疼痛和疾病的压力与干扰下，他们或许没有能力去接受复杂的信息并做出自主决定。这是患者的结构性劣势的根源之一（参见第一部分）。此外，一些人在生病时，并不希望检视自己的偏好和价值。在这些情况下，处在某些文化环境中的患者可能不想协商。他们期待和希望医生来作决定。这正好契合了他们对医生的角色和目标的预期。这或许正是他们自主地做出自由选择的方式。① 但是，这不符合伊曼纽尔夫妇关于自我审视的重要性的主张。

（4）伊曼纽尔夫妇或多或少忽略了在医生-患者互动中的结构性失衡。权力、信任和风险的关系于是就不知不觉地在协商模式中再次呈现出老师和学生的形式，即另外一种结构性失衡的二分体。就对权力的关注而言，这与伊曼纽尔夫妇敬而远之的一种模式联系在一起，即家长制式的模式。他们欢迎改变家长作风，同意这一点是一件容易的事情。但他们似乎以为，转向协商将会消除医生-患者互动中的各种权力差异。一些护理文献甚至在更大程度上以同样的假定为基础。然而，人们只有忽略权力，才能对这样一个假定作出辩护。如果人们不忽略权力，那么，这个假定就是一个一厢情愿的想法。

当人们在这类医学和护理文献中讨论权力时，它要么是作为医护人员之间的权力失衡来讨论，要么是作为专业人员和管理者之间的权力斗争来讨论。在这方面的典型人物是格雷（Gray），他在评论管理医疗时说："管理医疗把对美元和患者流量的控制权从医生转让给了具有强烈经济目标和在追求这些目标时有权影响患者医疗的大型机构。目的和权力的这种结合是这些机构的可信度遭到质疑的一个主要根源，尤其是因为它们削弱了表示坚守信托伦理的医护人员的权力。"② 除此之外，社会学家和人类学家（以及医疗保健体制内持不同意见者）也主要

---

① 这一点我要归功于杂志的一位评论人。

② Gray B. H. , "Trust and Trustworthy , Care in the Managed Care Era", *Health Affairs* 1997, 16 (1), pp. 34-49, p. 35.

讨论医疗保健中的权力问题。① 但他们的观点似乎没有渗透到医护人员的主流讨论中。

正如一些医生所建议的那样，患者和医护人员能够在一种对话关系中成为平等的或成为同盟伙伴，这种观念只是一厢情愿的想法。即便削弱家长作风是值得称赞的，这也绝不意味着这些关系变成了对称的。家长作风没有讨论者认为的那么重要。非家长制式的关系可能会抚慰（原文为 soot，疑原文有误，应为 soothe——译者注）我们的情感，但快乐感并不会消除各种权力失衡。认为家长制是医疗保健中权力的主要来源，是误导人的，并且能够遮蔽更加邪恶的权力来源。协商模式低估了患者对医护人员的依赖程度——尽管患者有自由选择、患者自主和知情同意。这不但低估了患者和专业人员之间关系中的结构失衡，而且还低估了关于价值的自由与对等的讨论与关于诊断和治疗的自由与对等的讨论之间的差别。前一种情况是值得称赞的，因为它预先假定了没有道德专长。第二种情形可能会造成灾难，因为它预先假定了也没有其他类型的专长。但是，其他类型的专长是有的。我可能不喜欢被诊断患了癌症，但是这个诊断的正确性并不是我的医生和我协商的一个问题。我认为，对此我的医生比我更了解情况，我希望我是对的。而且我认为我们不应该渴望这样一种医疗体制：在那里，对癌症的诊断是一个协商的问题。我将不喜欢这样一种医疗体制：在那里，尽管我没有行医能力，但我有很公平的机会使我的医生信服，他的诊断是错误的。我喜欢这样一种医疗体制：在那里，我必须忍受痛苦，但有机会得到称职医生的救治。然而，我同样喜欢这样一种医疗体制：在那里，医护人员意识到我作为一名患者是多么依赖他们。无视权力可能会从道德上使权力、信任和风险之间的应该反对的制度形式永久存在。

四

我已经讨论了医疗保健中权力、信任和风险之间的关系是什么，为什么医护人员应该对此展开讨论，而且我举例说明了，我声称当医护人员公开讨论他们的角色时经常忽略这一点意味着什么。医疗保健体制之所以应该加以讨论，是因为它是一个医护人员和患者之间的结构性失衡的体制。忽略它就会导致医护人员沉迷于如何能组建医疗体制和如何能使比家长作风更邪恶的各种权力形式永久存在

---

① 在这方面具有代表性的人物是 Freidson（Freidson E., *Profession of Medicine：A Study in the Sociology of Applied Knowledge.*，Chicago：University of Chicago Press，1988）和 Fainzang［Fainzang, S.，"Lying, Secrecy and Power within the Doctor-Patient Relationship"，*Anthropology and Medicine* 2002，9（2），pp. 117-133］。

的一厢情愿的想法中。但是人们也必须质问：对此应该如何加以讨论呢？医护人员教育中至关重要的一部分应该是，他们对权力、信任和风险很敏感。但这并不是我在此所关切的内容。我将会表明其他三个要点：第一，应该把制度形式作为一个问题来探讨。第二，应该根据角色意识来探讨，像"信任"、"权力"和"风险"之类的概念能在社会生活中起什么作用。第三，应该用能够解释其主要特征的权力概念来讨论。

第一，建议按照辅助性原则组建英国 NHS（参见第一部分）就是提出一种制度形式。这一建议很可能将更多的权力移交给接近于医疗行动场所的医生、心理学家、护士等医护人员。管理医疗是另外一个例子。正如格雷断言的那样（参见第三部分），这不但会削弱医护人员的权力，而且会降低医疗保健机构的可信度。挪威的医疗保健体制是，给诊断定价，然后，基于诊断价格对医院进行经济奖励，这是第三种例子。这样做会诱惑医院集中医治那些能为国家带来经济收益的疾病，忽视那些不太有利可图的疾病。另外还有一个例子是，全科医生扮演着专家服务的监管人的角色。在挪威，患者和专家之间最初的接触是通过全科医生的渠道来决定的。制度形式是非常重要的，因为它们决定着互动的条件。对于制度形式的讨论，在职业伦理学中，几乎是完全缺失的。然而，这些制度形式能用道德术语来讨论，并且能够提供合理答案。例如，人们会讨论，如果它们能使患者从一开始就接触专家，不必依赖于监管人，那么，这是否会强化患者的地位。制度形式在很大的程度上决定着医疗保健体制中权力、信任和风险之间的关系的形成，因为它们决定了患者会如何和在多大程度上依赖于医护人员。

第二，像权力、信任和风险之类的概念，在社会、道德和政治意义上，不是无辜的。它们伴随有风险与利益，风险和利益带来了制度形式的实践差异。权力是这样一种现象：关于它是什么，我们大家都有各种体验和一种直觉的理解。但是，没有一个定义博得一致同意。相关文献中充斥着对权力的各种定义。罗素在1938 年说："权力可以被定义为产生预期的效果。"[1] 他区分了三种不同类型的权力：传统的、革命性的和赤裸裸的。但他没有区分人的权力和物的权力。韦伯把权力定义为是社会行动者在社会关系中各行其是的机会，即便他遇到了阻力，不管这种机会的基础是什么。[2] 韦伯也区分了三种合法的权威（权力的子类）：法律的、传统的和有魅力的。按照福柯的观点，权力仅仅是人们赋予一个复杂的战略态势的名称。[3] 我本可以无穷无尽延续这个清单。但是，说出这些定义是如何

---

① Russell, B. , *Power*, London: Allen and Unwin, 1975, p. 25.

② Weber, M. , *Wirtschaft und Gesellschaft*, Tubingen: J. B. C. Mohr, 1976, p. 28.

③ Foucault, M. , *The History of Sexuality*, vol. 1, New York: Vintage, 1980, p. 93.

联系在一起的是不容易的。然而，难以定义一个概念这一事实并不意味着不存在我们通过这个概念试图要领会的内容。从下定义遇到的困难，人们并不能推断出所要定义的概念所指称的对象不存在，尽管这么做可能很有诱惑力。对于信任和风险而言，大多数情况也是如此。有些人固守决策理论的标准的风险概念，在决策理论中，风险与确定性、不确定性和无知形成了对比。另外一些人，像道格拉斯（Douglas）和卡普兰（Caplan），强调概念的文化可变性和风险洽谈的社会功能。① 理性选择理论家科尔曼（Coleman）和系统论理论家卢曼（Luhmann）主张，信任是有风险的。但是卢曼的风险概念更加接近科尔曼的不确定性概念，而不是他的风险概念。他们用了同一个词，但他们提出的主张却各不相同。②这些概念的定义博得一致同意，在不久的将来是不可能出现的。

我这里的问题是，这些定义如何在社会生活中发挥作用。权力的概念指哈金（Hacking）所谓的一种互动，而不是一类中性（indifferent）的概念。③ 风险和信任也是如此。一个互动的概念能够影响事实真相；一个中性的概念无法做到。行星就是一个中性概念。在天文学史上，行星的定义有许多。但是，天体不受我们定义这个概念的努力的影响。民主是一个互动的概念。在政治史上，民主的定义有许多。有很好的理由断言——人们可以查文献——政治制度的发展受到了这些定义的影响。定义可以进入人们的自我理解和他们对世界的理解中，因而有实际效果。人类的思与行是建立在他们如何理解自己、如何相互理解和如何理解世界之基础上的。行星绝不关心人类的自我理解，而且，的确如此。但是，人类应当关心自我理解和他人的世界观，尤其是当他们有赖于其他人时。权力、信任和风险关乎人与人之间的关系。这些关系受到了我们如何理解它们的影响。

权力也是加利（W. B. Gallie）所谓的本质上相抗争的一个概念。④ 就这些概念而言，对其意义的传统解释是抗争性的——通常是互不通约的和相互敌对的。它们没有无异议的核心含义。这可能既与弗雷格的观点相反，弗雷格假定人们可

① Douglas, M., *Risk and Blame: Essays in Cultural Theory*, London: Routledge, 1992; Caplan, P., ed., *Risk Revisited*, London: Pluto., 2000.

② Coleman, J. S., *Foundations of Social Theory*, Cambridge, MA: Belknap Press of the Harvard University Press, 1990, ch. 5; Luhmann, N., *Vertrauen. Ein Mechanismus der Reduktion sozialer Komplexität*, Stuttgart: FerdinandEnke, 1968.

③ Hacking, I., *The Social Construction of What*? Cambridge: Cambridge University Press, 2001, pp. 104-105.

④ Gallie, W. B., "Essentially Contested Concepts". In *The Importance of Language.*, M. Black, ed., pp. 121-146, Englewood Cliffs, NJ: Prentice Hall, 1962.

以为概念的含义提供精确的定义，又与魏斯曼①所谓的"开放结构"（open texture）所显示的那些概念形成对比。所有的概念都有"开放结构"，也就是说，有模糊的边界，并且会遇到难以分类的情况。但是，大部分概念都有一个不受其"开放结构"影响的核心含义。加利的文章向社会科学的规范概念的核心观念提出了挑战。在他的观点中，对这些概念相抗争的解释都附带有风险。如何定义民主，不是一个单纯的政治问题。不同的定义会服务于不同的利益。因此，我们不应该指望一致赞成民主的定义。

如果将哈金的观点和加利的观点结合起来理解，我们显然会明白为什么是这样。假如民主不仅在本质上有争议，而且是互动的，那么，民主的定义就不只是一个理论问题。就民主制度的设计而言，它具有实践结果，因为这些定义会进入政治行动者的自我理解和世界观中，并进而形塑他们的行动。我们如何定义民主，不仅会影响我们对民主的理论理解，也会影响我们在实践中把什么样的制度看成是民主制度。关于民主为什么会有如此多的不同模式，是有原因的。② 对于权力来说，几乎也一样。我们如何理解权力并非纯粹的理论问题。我们在面对各种权力关系时，如何行事，也有实际的社会和政治后果，即我们把什么看做是权力，我们试图约束哪些类型的权力，或者，我们认为的获得、行使和维持权力的哪些手段是合法的。医护人员对权力的忽视在某种程度上是源于他们把权力等同于物质力量这种隐含的定义。如果是这样，那么，权力可以被看做是医疗保健中的一种边缘现象，至少在西方世界是这样的。这样一种观点的后果之一是，即使权力有负面的特性，按照其他定义被描述为是行使权力的各种不同关系和行为，也会被看做是无辜的和不成问题的，而且会以其他名头被永久化。关于权力、信任和风险的讨论对诸如此类的问题应该是敏感的。为本质上有异议相互竞争的互动的概念下定义，会在政治斗争中起到一定的作用。它们的定义绝不完全是理论上的。它们涉及如何为社会行动者形塑社会世界的各种现象，在这种意义上，它们是实践的。对权力、信任和风险的不同概念的持续讨论，是对医疗保健中权力、信任和风险关系的制度形式进行道德讨论的至关重要的一部分。

第三，在医疗保健中对待权力，我们需要适合于对制度形式及其效果进行道德讨论的一条进路。我认为，讨论医疗保健中的权力的最富有想象力的框架是由卢克斯（Lukes）提出的。③ 这个框架能解释权力、信任和风险之间的关系，并使我们对这种关系的重要性和各种变化很敏感。它也指出，如果要改变这种关

---

① Waismann, F. , "Verifiability. ", In *The Theory of Meaning*, G. H. R. Parkinson, ed. , pp. 35-60, Oxford：Oxford University Press, 1984.

② Held, D. , *Models of Democracy*, Oxford：Polity, 1992.

③ Lukes, S. , *Power：A Radical View*, Oxford：Blackwell, 1974.

系，必须要改变什么。人们可能断言，福柯的权力进路同样能做到。但是卢克斯的框架比福柯框架有一个优势。在卢克斯的进路中，谈论摆脱某些权力关系，或者在某种关系中削弱权力，是可能的。而在福柯的框架中，很难为这些判断提供担保。所有的医疗改革都需要规范标准。基于福柯的工作，难以为这些标准提供辩护。

卢克斯提出了对权力的三重描述。第一个维度的权力关心 A 强迫 B 去做某事的那些语境。这包括物质力量在内。第二个维度的权力关心 A 控制与 B 互动的日程的语境。日程控制可能会（但不必）基于强迫，即基于第一个维度的权力。它关心在一种互动中准备抓住什么、放弃什么，以及谁作决定。第二个维度的权力并不建立在强迫的基础上，理解这些情境是非常重要的。第三个维度的权力关心 A 控制 B 看到的世界的那些语境。这意味着 A 控制着对此情境的界定：什么是疾病，什么是正常的，等等。对权力的这种构想方式，在研究医疗保健中的权力时有一些优势。

一是在现代医疗保健中，除了一些例外情况（主要在精神病治疗中）之外，不存在太多的强迫（force）。在关于医护人员和患者之间的适当关系的文献中，有一种偏见是，这些讨论者往往使权力等同于强迫。这样，如果我们废除强迫，就剩下一个人人平等的共同体。围绕知情同意和患者的自由选择，也有许多的幻想。这些幻想是由使权力等同于强迫（即第一个维度的权力）的权力观促成的。因此，强迫的缺失能被解释成自由选择的机会。但是，这种假设是被误导的。权力要复杂得多。意图设置（agenda setting）和世界观的控制，对自由选择和知情同意来说，甚至和强迫一样重要。

二是通过强调意图设置和世界观的控制，卢克斯的进路能够解释医护人员和患者之间的大多数不对称关系。通常，这些不对称并非建立在强迫的基础上。一种不对称是认知的不对称，在这里，一方比另一方知道得多。因此，处于劣势地位的一方就缺少智力资源对互动的意图干扰（agenda interaction）和假定的世界观提出挑战。坎特（Canter）说，医疗权并不是能轻易地从医生转让给患者的货币。[①] 这方面的一个理由是这样的事实：医疗权不是建立在强迫的基础上。如果权力是建立在强迫的基础上，那么，消除强迫就会转让权力。但是，许多事实证明的主张是：即使不怎么使用强迫，通过协商来解决一切问题和做出一切决策的医护人员和患者也绝不会成为平等的。我不是建议说，协商模式在道德上是不正确的，我只是建议说，为了以理性的方式讨论协商模式，即使在医护人员和患者之间关系中没有强迫，我们也不能忽视不可避免的那些权力差别。

---

① Canter, R., "Patients and Medical Power", *British Medical Journal*, 2001, 323: 414.

卢克斯进路的第三个优势是，它能解释权力、信任和风险之间的联系。当我作为一名患者进入医疗保健体制时，有人能够为我检测、评估、承担、隐藏、造成或消除风险。不是他们运用强迫的能力创造了这些机会，相反，要理解这些现象，我们必须运用卢克斯进路中权力的另外两个维度。医护人员的这些机会主要是由我的无助和我相信他们比我有认知的优势地位创造的。

五

西方国家的医疗保健体制瞬息万变。我将挑出可以改变权力、信任和风险关系的几种趋势。这样一个视角说明了这种关系如何根深蒂固的问题。如果削弱家长作风和引入协商不可能使医疗互动转变成一种对称的伙伴关系，那么，什么能彻底改变这种互动关系呢？并且，它会向更好方向变化吗？对这个问题的回答是含糊不清的。或许，如果削弱家长作风，我们最好应该希望，医护人员变得对权力更有自我意识，更多地参与关于制度形式的道德和政治讨论。这能够转变这种关系，却不能消除。消除权力、信任和风险之间的关系，可能将医疗保健体制沿着一个或许我们不应该选择的方向改变得面目全非。

一种改变是医疗的女性化。在丹麦，在新的全科医生中，60％以上是女性。[①]这种现象并非丹麦特有。[②]包括少数医生在内的一些人认为，应该根据更"女性化"的价值观提出新的基于关怀的医学定义。但是，很难看到这会显著地改变权力、信任和风险的关系。如何定义医务工作这个职业的问题，主要是这个职业应该有哪些价值观的问题。价值观是重要的，但价值观本身改变不了像认知失衡之类的根深蒂固的结构现象。价值观能够让无助的人感觉更好，但是它们本身却不会让这些人不那么无助。

第二组变化涉及新型患者和知识的民主化。一种新型患者的文化正在发展中。患者接受过更好的教育。他们会更有机会（例如，通过网络）听取专家的意见。他们的身体状况更好，寿命更长。而且，他们向医护人员提出了更高的要求。技术的发展改变了医护人员和患者之间的关系。有人会说，和健康相关的知识的重要性提高了。但医护人员对这些知识的控制却降低了。尽管关于患者信任的文献非常难以解释（因为有这么多对信任的不同定义在使用），但在西方国家，患者对医疗保健体制和医护人员的信任似乎正在下降。这一切可能会导致医

---

① Hølge-Hazelton, B. , "En klassisk profession skifter køn", In *De professionelle—forskning i professioner og professionsdannelser*, K. Hjort, ed. , pp. 143-151. Roskilde: Roskilde Universitetsforlag, 2004, p143.

② Riska, E. , "Towards a Gender Balance: But Will Women Physicians Have an Impact on Medicine? " *Social Science and Medicine*, 52 (2), pp. 179-187, 2001, p. 181.

护人员失去对界定患者病情和根据这种界定分配医疗补助的垄断。[1] 这不仅会对道德专长的观念构成威胁，而且会对医学专长的观念构成威胁。这组变化可能向权力、信任和风险的关系提出一个严峻的挑战。它挑战着这一关系的核心，即职业自主的范围和合法性。它也对医护人员对他们工作的技术控制提出了挑战。与健康相关的知识日趋复杂的趋势，在现代社会的某些领域是很明显可见的，不过却有蔓延的可能性。

第三组变化涉及医生这个角色的结构性特征。这一角色沿着多个维度发生变化。每个医生都比过去治疗更多的患者，但花费在每个患者身上的时间却更少了。随着工作和闲暇时间之间的更明确的区分，医生改变了生活方式。做一名医生（或一名护士）越来越成为一项普通的工作，越来越不是一种职业了。一般的医疗实践也在发生变化。全科医生治疗无伤大雅的病。专家治疗更多的其他病。医学专业对医院里的劳动分工的影响正在消失。它也正在失去管理基础，一部分原因是由于改变了与其他职业（例如，护理，一个即将来临的职业）的关系，另一部分原因是医院里专业管理人员越来越多。在这些变化中，有些变化能够向权力、信任和风险的关系提出挑战，因为它们影响了职业自主权。劳动分工对工作的技术方面的控制或许是非常重要的。[2]

医护人员和患者之间的真正的协商预先假定了这些变化，因为首先他们之间的关系能够变成对称的。然后，就没有老师，也没有学生，只有平等。只有当权力、信任、风险的关系不复存在时，医护人员和患者之间那些关系，才可能成为对称关系。这一点，而不是家长作风，才是一个严肃的道德问题。但这种进展尚不明朗。或许我们不应该指望在这个方向上有更进一步的发展。

我将通过提及知情患者和知识的民主化来简要阐明我的观点。如果两者都出现了的话，它们都是好事。但是它们提出了一些严重的问题。

"知识民主化"的标签很可能是误称。它假定，我们正在见证知识的社会分配方面的一个进展，这类似于19世纪与20世纪西方国家的政治民主化。但是我们很可能不会见证这类事情。一个问题的民主化进程的目标，要么是各个相关方都有权在这个问题上平等投票（一人一票），要么是有权平等地参与讨论，就此问题做出决定。政治民主化是从权利分配或参与或两者的不均衡开始，最终达到均衡。但关于知识提出类似的主张是难以得到辩护的。西方国家的真实情况是，

---

① Hølge-Hazelton, B., "En klassisk profession skifter køn", In *De professionelle—forskning i professionerog professionsdannelser*, K. Hjort, ed., pp. 143-151. Roskilde: Roskilde Universitetsforlag, 2004, p. 150, n3; Hilden, P. K., *Risk and Late Modern Health: Socialities of a Crossed-Out Pancreas*, Oslo: Unipub, 2003.

② 有人会认为，循证医学也将改变权力、信任和风险的关系。循证医学不但是一个知识问题，而且是一个社会关系的问题。但是我认为，这不会沿着认知上更加对称的方向改变医生和患者的互动。

普通教育的水平提高了。因此，人们有机会接触到各种不同类型的信息。但是，我们有合理的理由认为，知识的分布——尤其是与健康相关的知识的分布——已经变得更加均衡了吗？我并不这么认为，理由主要有三。第一，人的能力各不相同，这与他们能够获得什么样的知识相关。单独这一点就对知识可以完全民主化的主张极为不利。不管我们是否喜欢，知识普及没有平衡自然能力。第二，不同类型的知识在社会中的传播方式是有差异的，而且，我们对为何会这样知之甚少。但是，知识的普及与权利的普及不一样。不同类型的知识以不同的方式传播。即便我们能够举一个例子来断言，理论医学的知识比以前的分布更加均衡了，但关于实践的医疗技能却无法举出一个相似的例子。某人对癌症的致病原因有所了解，这一事实并不意味着他能为自己或他人诊断和治疗。理论知识和实践技能是以不同的方式传播的。第三，前沿知识是一个不断前进的指标。当某些方面的知识分布变得更加均衡时，知识的前沿可能又遥遥领先了，旧的不均衡的分布维持在另一个层次。多数情况担保能断言，至少患者在某些时期变得更加了解情况。但是，主张这些患者已经变得或将要变得如此详细地了解情况，以致我们应该放弃医护人员的职业自主权，也就是说，向他们对自己工作的技术方面的控制权真正提出了挑战，是令人难以置信的。变得非常了解情况，与某一领域内的一位专家获得渊博的理论知识和精湛的实践技能，不是一回事。我们确实见证的不是知识的民主化，而是认知劳动的一种新的社会分工。在医疗保健体制中，我认为，我们看到了两个趋势，重要的是不要将这两种趋势相混淆：医学知识和外行（患者）的知识之间的知识鸿沟正在加深，而不是正在缩小。与此同时，医学专业正在失去对医学知识的控制权。这种控制权不是转让给外行，而是转让给其他专业人员。循证（evidence basing）就是这方面的一个有趣例子。循证把对知识的控制权转让给一类新型专家：那些元分析领域内的专门家。他们中的许多人根本不是医生。但是，正如拥护循证的那些人所断言的那样，循证是否可以使知识民主化，是十分可疑的。

对道德专长的观点提出挑战和对医学专长的观点提出挑战是不同的。有可靠的论据反对主张存在着真正的道德专长。很了解道德理论或积累了关于道德案例的大量知识的某人，未必就是一位更善良的人，而且，他作为道德问题的顾问，未必是最佳选择。道德理论和道德案例的知识没有在这方面转化为行动与品质。但是，有可靠的论据支持主张存在着真正的认知专长，例如，医学专长。已知有适当的医学技能，关于病因及其治疗的更多知识，的确可以转化为行动。有人会主张，一名有更多知识的医生，也是一名更好的医生，即便事实并非总是如此，因为要成为一名好医生，其他能力也很重要。因此，即便我们不会主张应该允许知晓很多道德理论的人控制我们的道德行为，但主张医护人员应该掌控他们工作的技术方面也是合理的，这包括以不同的方式为我们——病人——诊断和治疗在内。如果"知识的民主化"在很大程度上是一种幻想，那么，对权力、信任和风险关系的真正替代选择就是让"知情患者"占上风。但那会有副作用。让"知情患者"了解情况，但仅此而已。尽管知情患者占上风，可在医疗保健体制中仍然有一个根深蒂固的——尽管是正在发生变化的——认知的劳动分工。我认为，这并没有比以前更加民主化。

　　总而言之，我们应该祈求维持权力、信任和风险的关系，因而为医护人员的自主权祈祷。某些人比我更了解疾病，这并不是一件坏事。实际上，如果事实并非如此，我将会很担心。但是，我们应该认真地讨论这一关系的各种制度形式。它们真的至关重要。而且，医护人员应该参与这些讨论。事实上，他们应当是主要的参与者，因为在这些讨论中，他们的自我理解和世界观很可能比其他任何东西都重要。

（王巧贞 译，成素梅 校）